VOLUME EIGHTY EIGHT

CURRENT TOPICS IN MEMBRANES

New Methods and Sensors for Membrane and Cell Volume Research

CURRENT TOPICS IN MEMBRANES, VOLUME 88

Series Editor

IRENA LEVITAN
Division of Pulmonary and Critical Care Medicine
University of Illinois at Chicago
Chicago, IL, United States

VOLUME EIGHTY EIGHT

CURRENT TOPICS IN MEMBRANES

New Methods and Sensors for Membrane and Cell Volume Research

Edited by

MICHAEL A. MODEL

Kent State University, Kent, Ohio, United States

IRENA LEVITAN

Division of Pulmonary and Critical Care,
Department of Medicine, University of Illinois at Chicago,
Chicago, IL, United States

Academic Press is an imprint of Elsevier
50 Hampshire Street, 5th Floor, Cambridge, MA 02139, United States
525 B Street, Suite 1650, San Diego, CA 92101, United States
The Boulevard, Langford Lane, Kidlington, Oxford OX5 1GB, United Kingdom
125 London Wall, London, EC2Y 5AS, United Kingdom

First edition 2021

ISBN: 978-0-323-91114-6
ISSN: 1063-5823

For information on all Academic Press publications
visit our website at https://www.elsevier.com/books-and-journals

Publisher: Zoe Kruze
Acquisitions Editor: Ashlie Jackman
Developmental Editor: Jhon Michael Peñano
Production Project Manager: Abdulla Sait
Cover Designer: Greg Harris

Typeset by STRAIVE, India

Contents

Contributors

Francisco J. Barrantes
Biomedical Research Institute (BIOMED), Catholic University of Argentina (UCA)–National Scientific and Technical Research Council (CONICET), Buenos Aires, Argentina

Martin Brennan
University of Illinois at Chicago, Department of Cellular and Molecular Pharmacology and Regenerative Medicine, Chicago, IL, United States

Pei-Chuan Chao
Department of Civil, Structural and Environmental Engineering, University at Buffalo, The State University of New York, Buffalo, NY, United States

Stephanie M. Cologna
Department of Chemistry; Laboratory of Integrated Neuroscience, University of Illinois at Chicago, Chicago, IL, United States

Jordan Fauser
University of Illinois at Chicago, Department of Cellular and Molecular Pharmacology and Regenerative Medicine, Chicago, IL, United States

Lewis Frame
School of Natural Sciences, University of York, York, United Kingdom

Bauke F. Gaastra
Department of Biochemistry, University of Groningen, Groningen, the Netherlands

Sumit Garg
Physics Department, University of Illinois at Chicago, Chicago, IL, United States

Yuri Gerelli
Department of Life and Environmental Sciences, Universita' Politecnica delle Marche, Ancona, Italy

Imogen Hayton
Department of Biology, University of York, York, United Kingdom

Andrei V. Karginov
University of Illinois at Chicago, Department of Cellular and Molecular Pharmacology and Regenerative Medicine, Chicago, IL, United States

Nigina Khamidova
Department of Chemistry, University of Illinois at Chicago, Chicago, IL, United States

Yulia Kolobkova
Department of Human Medicine and Institute for Molecular Medicine, MSH Medical School Hamburg, Germany

Edgar E. Kooijman
Department of Biological Sciences, Kent State University, Kent, OH, United States

Mark C. Leake
Department of Physics; Department of Biology, University of York, York, United Kingdom

Sarah Lecinski
Department of Physics, University of York, York, United Kingdom

Irena Levitan
Division of Pulmonary and Critical Care, Department of Medicine, University of Illinois at Chicago, Chicago, IL, United States

Chris MacDonald
Department of Biology, University of York, York, United Kingdom

Luca Mantovanelli
Department of Biochemistry, University of Groningen, Groningen, the Netherlands

Michael A. Model
Kent State University, Kent, Ohio, United States

Thu T.A. Nguyen
Department of Chemistry, University of Illinois at Chicago, Chicago, IL, United States

Koralege C. Pathmasiri
Department of Chemistry, University of Illinois at Chicago, Chicago, IL, United States

Ursula Perez-Salas
Physics Department, University of Illinois at Chicago, Chicago, IL, United States

Sumaira Pervaiz
Institute of Chemistry and Biochemistry, Freie Universität Berlin, Germany

Bert Poolman
Department of Biochemistry, University of Groningen, Groningen, the Netherlands

Lionel Porcar
Institut Laue Langevin, Grenoble Cedex 9, France

Frederick Sachs
Department of Physiology and Biophysics, University at Buffalo, The State University of New York, Buffalo, NY, United States

Jack W. Shepherd
Department of Physics; Department of Biology, University of York, York, United Kingdom

Tobias Stauber
Department of Human Medicine and Institute for Molecular Medicine, MSH Medical School Hamburg; Institute of Chemistry and Biochemistry, Freie Universität Berlin, Germany

Amber R. Titus
Department of Biological Sciences, Kent State University, Kent, OH, United States

Denis Tsygankov
Georgia Institute of Technology, Wallace H. Coulter Department of Biomedical Engineering, Atlanta, GA, United States

On, in, and under membrane

Michael A. Model and Irena Levitan

We can only study what we know how to study, and methods reviews are an important niche of scientific literature. Volume 88 of *Current Topics in Membranes* comprises 11 chapters on experimental methods for studying the two most essential and indispensable cellular compartments: the outer plasma membrane and the cytosol; by the latter, we mean the organelle-free part of the cytoplasm, with its unique physical and physicochemical properties: high concentration of proteins, limited free space, and the presence of bound water. Although membrane biology has long become a standard textbook subject, the cytosol has for many years remained relatively in the shadows. Nevertheless, it has always had a strong appeal for some researchers, being both simple and mysterious and hinting at some basic principles of the organization of living matter.

The cytosol and membrane are not only equally indispensable, but they are also closely interconnected. Despite evidence that the cytosol can self-regulate to some extent (Orlov, Shiyan, Boudreault, Ponomarchuk, & Grygorczyk, 2018), the standard model envisions membrane channels and transporters as the critical valves that control the cytosolic composition and density. The reverse is less obvious, but there are reasons to believe that some membrane channels may in turn be regulated by macromolecular crowding, i.e., by the high density of intracellular proteins (Burg, 2000; Model, Hollembeak, & Kurokawa, 2021; Parker, 1993). Recently, liquid–liquid phase transitions began to draw the attention of cell biologists, adding another fascinating aspect to the cytosol science; a possible analogy to lipid–lipid phase separation within the membranes (Chapter 9) comes to mind.

However, "peaceful coexistence" (to borrow a political term) between the cytosol and membrane, or rather between the proponents of the cytosol and membrane theories of cellular control, has been lacking at times. This interesting topic deserves a separate study, but there are signs that the biological community is moving toward a more balanced view, in which the complexity of each compartment and their interrelationships become better appreciated. Hence, we offer this edition to the reader as a rough guide to some modern experimental trends in both research areas.

Chapter 1 by Mantovanelli, Gaastra, and Poolman provides a very broad overview of fluorescence methods in cell biology, including microscopy techniques, properties of fluorescent proteins, detection of ions and metabolites, approaches to labeling, and viscosity measurements, among others. Both membrane and cytosol researchers will find much useful information in the chapter.

The first half of the volume (Chapters 1–5) is mainly dedicated to methods of studying the cytosol. Two different approaches to the quantification of macromolecular crowding are presented by Lecinski et al. (Chapter 3) and Model (Chapter 5). Lecinski and coauthors use fluorescence sensors—specially designed FRET pairs of linked proteins that fold into a more compact shape in a crowded environment, resulting in an increase in the FRET signal. By combining this method with an original superresolution imaging technique, the authors investigated the behavior of macromolecular crowding in replicating yeast. Chapter 5 describes a method based on transmitted illumination in the presence of an external dye: an image taken through a red filter gives cell volume, and a couple of images taken through a blue filter give protein mass; taken together, they provide the necessary information to calculate protein density. Both chapters discuss some of the established or hypothetical roles of macromolecular crowding.

In some cases, high protein density results in a liquid–liquid phase transition, when denser condensates enriched in certain proteins become separated from the rest of the cytosol. Nucleoli are the best-known examples of such particles, but there are others, often without established names. Liquid condensates are now believed to play many important roles in biological processes. In Chapter 2, Titus and Kooijman focus on in vitro assays and, in particular, describe the use of pendant drop tensiometry for measuring interfacial tension, which is a critical parameter in condensate formation. They also review the existing databases and the computational approaches to predict phase transitions.

The volume-regulated anion channel (VRAC) is the subject of Chapter 4 by Kolobkova, Pervaiz, and Stauber. VRAC is activated by swelling and has a broad selectivity for inorganic anions and numerous organic substrates and osmolytes. Stauber was an author of one of the seminal papers that elucidated the molecular origin of VRAC (Voss et al., 2014), and their chapter provides an authoritative and in-depth review of VRAC properties, as well as experimental methods to study its activity: electrophysiological, radioactive, volume-based, and fluorescence, among others.

The second half of the volume (Chapters 6–11) is dedicated to methods in membrane research: the measurement of membrane tension by Chao and Sachs (Chapter 6), the characterization of lipid order by Levitan (Chapter 8), and the detection of membrane domains by Barrantes (Chapter 9). It also includes a chapter by Cologna (Chapter 10) introducing a novel method of membrane imaging by mass spectroscopy of lipids and a chapter by Perez-Salas (Chapter 11) describing a rarely used but powerful physical technique to study membrane organization via neutron scattering.

The measurement of membrane tension is described by Sachs (a pioneer in the discovery and biophysical characterization of mechanosensitive ion channels) and Chao. Membrane tension is determined by a cross talk between the lipid bilayer and the submembrane cytoskeleton and is critical in the regulation of cell shape, motility, and sensitivity to mechanical stimuli. The explanation of methods is complemented by a historical perspective describing how the field has been developing over the years, both in theoretical understanding and methodologically. A historical perspective is also given in Chapter 8 that describes the development of a method to assess lipid order—another fundamental property of lipid bilayers that dictates membrane biophysical properties. This method, known as Laurdan imaging, is named after a unique fluorescent probe whose spectrum is sensitive to the polarity of its environment, which enables the assessment of local penetration of water into the lipid bilayer. This method is increasingly used in cellular studies to assess membrane heterogeneity and distribution of local environments into ordered and disordered domains. It is important to note that the nature and even the existence of cholesterol-rich membrane domains in cellular membranes has been a matter of strong controversy for more than two decades. Laurdan imaging definitely shows a significant heterogeneity in the lipid order that constitutes a continuum; at the same time, this method does not allow for a more detailed exploration of the lipid composition and of cholesterol content of these domains. The state-of-the-art approaches to achieve the latter are described in Chapter 9 by Barrantes, a pioneer in the field of lipid regulation of ion channels. Barrantes describes the advances in the generation of fluorescent lipid sensors that have significantly moved the field forward. Mass spectrometry imaging (MSI) of membrane lipids is a very novel approach currently used only by a few groups; it is described in the chapter by Cologna. With MSI, one can assess not only the general lipid order of the local environment or identify specific lipids but also dive into the complexity of lipid composition of the membrane and provide a comprehensive unbiased lipid analysis. There is no doubt that this approach will be gaining more traction in future membrane research. Last but not least, Chapter 11 by Perez-Salas and coauthors presents a unique and highly sophisticated method to study membrane organization by time-resolved, small-angle neutron scattering. It enables a highly precise analysis of the relative amounts of isotopes and involves radioactive labeling of lipids. The past decades have seen a trend to replace radioactivity with chemical fluorescent tags; however, the advantage of radioactivity is that it leaves complex membrane architectures virtually undisturbed. Taken together, the chapters of this volume present the depth and breadth of the most up-to-date methods in membrane research.

References

Burg, M. B. (2000). Macromolecular crowding as a cell VolumeSensor. *Cellular Physiology and Biochemistry*, *10*(5–6), 251–256.

Model, M. A., Hollembeak, J. E., & Kurokawa, M. (2021). Macromolecular crowding: A hidden link between cell volume and everything else. *Cellular Physiology and Biochemistry*, *55*(S1), 25–40.

Orlov, S. N., Shiyan, A., Boudreault, F., Ponomarchuk, O., & Grygorczyk, R. (2018). Search for upstream cell volume sensors: The role of plasma membrane and cytoplasmic hydrogel. *Current Topics in Membranes*, *81*, 53–82.

Parker, J. C. (1993). In defense of cell volume? *American Journal of Physiology. Cell Physiology*, *265*(5), C1191–C1200.

Voss, F. K., Ullrich, F., Münch, J., Lazarow, K., Lutter, D., Mah, N., et al. (2014). Identification of LRRC8 heteromers as an essential component of the volume-regulated anion channel VRAC. *Science*, *344*(6184), 634–638.

CHAPTER ONE

Fluorescence-based sensing of the bioenergetic and physicochemical status of the cell

Luca Mantovanelli, Bauke F. Gaastra, and Bert Poolman*
Department of Biochemistry, University of Groningen, Groningen, the Netherlands
*Corresponding author: e-mail address: b.poolman@rug.nl

Contents

Current Topics in Membranes, Volume 88
ISSN 1063-5823
https://doi.org/10.1016/bs.ctm.2021.10.002

Abstract

Fluorescence-based sensors play a fundamental role in biological research. These sensors can be based on fluorescent proteins, fluorescent probes or they can be hybrid systems. The availability of a very large dataset of fluorescent molecules, both genetically encoded and synthetically produced, together with the structural insights on many sensing domains, allowed to rationally design a high variety of sensors, capable of monitoring both molecular and global changes in living cells or in *in vitro* systems. The advancements in the fluorescence-imaging field helped researchers to obtain a deeper understanding of how and where specific changes occur in a cell or *in vitro* by combining the readout of the fluorescent sensors with the spatial information provided by fluorescent microscopy techniques. In this review we give an overview of the state of the art in the field of fluorescent biosensors and fluorescence imaging techniques, and eventually guide the reader through the choice of the best combination of fluorescent tools and techniques to answer specific biological questions. We particularly focus on sensors for probing the bioenergetics and physicochemical status of the cell.

Abbreviations

cpFP circularly permuted fluorescent protein
CTPE chemogenetic tags with probe exchange
FAST fluorescence-activating and absorption shifting tag
FLIM fluorescence lifetime imaging
FP fluorescent protein
FRAP fast recovery after photobleaching
FRET Förster resonance energy transfer
HILO highly inclined and laminated optical sheet
PAFP photoactivable fluorescent protein
PALM photo-activated localization microscopy
POI protein of interest
PSFP photoswitchable fluorescent protein
SBP substrate-binding protein
SMDM single molecule displacement mapping
STORM stochastic optical reconstruction microscopy
TICT twisted intramolecular charge transfer
TCSPC time correlated single photon counting
TIRF total internal reflection fluorescence

1. Introduction

Fluorescence-based sensors can be divided into three groups: Fluorescent protein-based sensors, chemical probes and hybrid systems. Fluorescent proteins (FPs) have been known in the life sciences for several decades (Shimomura, Johnson, & Saiga, 1962). Since the cloning of the *Aequorea victoria* gene for Green Fluorescent Protein (GFP) in 1992 (Prasher, Eckenrode, Ward, Prendergast, & Cormier, 1992), and its subsequent use as an *in vivo* fluorescent tag in 1994 (Chalfie, Tu, Euskirchen, Ward, & Prasher, 1994), FPs have been extensively used to obtain a deeper understanding of the structures and biochemical processes of living organisms. Compared to, *e.g.*, electron microscopy (EM) techniques, it now became possible to track macromolecules and structures in living cells and tissues, albeit with lower resolution than in EM. The imaging potential of GFP sparked the interest of many scientists and led to the discovery of new FPs with diverse and improved photophysical properties (Chudakov, Matz, Lukyanov, & Lukyanov, 2010; Rodriguez et al., 2017; Shinoda et al., 2018; Zhang, Gurtu, & Kain, 1996), allowing to gain a better understanding of the mechanism of action (Tsien, 1998) of fluorescent proteins. A color palette of FPs ranging from near ultra-violet (Tomosugi et al., 2009) to far-red (Kamper, Ta, Jensen, Hell, & Jakobs, 2018) is now available (*vide infra*).

By specifically labeling proteins of interest, it became possible to localize and determine the dynamics of macromolecules (Chamberlain & Hahn, 2000; Day & Schaufele, 2008), to discover protein interaction partners, and to perform protein turnover analyses (Knop & Edgar, n.d.; Trauth et al., 2020). FPs have been engineered to obtain pH-insensitive (Roberts et al., 2016; Shinoda et al., 2018) and highly monomeric (Campbell et al., 2020; Shaner et al., 2013) variants. Subsequently, FPs have been used to develop a great variety of sensors (Berg, Hung, & Yellen, 2009; Miyawaki et al., 1997; Nadler, Morgan, Flamholz, Kortright, & Savage, 2016) to study both molecular and global changes in the tagged macromolecule or structure of the cell. The availability of FPs emitting at different wavelengths allowed the application of the Förster Resonance Energy Transfer (FRET) mechanism to create sensors (Calamera et al., 2019; Miyawaki et al., 1997; Otten et al., 2019; Sadoine, Reger, Wong, & Frommer, 2021) and to determine interaction between proteins or protein domains (Ivanusic, Eschricht, & Denner, 2014; Kaufmann et al., 2020). Circularly permuted fluorescent proteins (cpFPs) variants (Topell, Hennecke, & Glockshuber, 1999) have been developed to create sensors (Kostyuk, Demidovich, Kotova, Belousov, & Bilan, 2019) to detect variations in the concentration of specific molecules. Finally, the development of photoactivatable (Lippincott-Schwartz & Patterson, 2009; Lukyanov, Chudakov, Lukyanov, & Verkhusha, 2005; Wang, Moffitt, Dempsey, Xie, & Zhuang, 2014) and photoswitchable

(Brakemann et al., 2011; Wazawa et al., 2021; Zhou & Lin, 2013) FPs, coupled with technological advancement in the microscopy field, allowed performing experiments beyond the diffraction limit of light, and a resolution of 20–30 nm has been obtained with optical microscopy (Shcherbakova, Sengupta, Lippincott-Schwartz, & Verkhusha, 2014). In recent years, technological advancements allowed to push the boundaries of what can be observed via optical microscopy even further: the recently developed system MINFLUX allows to localize single molecules in living cells with a resolution of less than 2 nm (Balzarotti et al., 2017).

Organic fluorescent dyes can be used for imaging purposes (Strack, 2021) or as biosensors (Fu & Finney, 2018), but the tagging of specific macromolecules is more challenging than with genetically encoded probes. However, in recent years several approaches for specific labeling of proteins inside cells have been developed (Cole, 2013; Takaoka, Ojida, & Hamachi, 2013). Chemical probes are typically characterized by more narrow excitation and emission spectra than FPs. Moreover, chemical probes generally are much more photostable and emit at least an order of magnitude more photons than FPs. A wide range of chemical probes varying in their photochemical and physical properties is available (Alexa Fluor, 2021; BODIPY, 2021), allowing to construct different type of sensors (Barreto-Chang & Dolmetsch, 2009; Liu et al., 2020). Another use of fluorescent dyes is in combination with other macromolecules. These dyes can be directly conjugated with specific sensing domains to create hybrid biosensors (Hu et al., 2014), or to antibodies for imaging purposes (Mao & Mullins, 2010). An additional way of using fluorescent dyes in combination with macromolecules is by using probes capable of interacting with specific binding pockets. These dyes need to be virtually non-fluorescent when unbound, and increase their fluorescence and change their lifetime upon interaction with the macromolecular partner. This system has been applied to create both hybrid protein-dye systems (Iyer et al., 2021; Plamont et al., 2016), and RNA-dye systems (Autour et al., 2018; Pothoulakis, Ceroni, Reeve, & Ellis, 2014), which can be used for imaging purposes (Gautier et al., 2008; Ouellet, 2016) or to develop biosensors (Jepsen et al., 2018; Tebo et al., 2018; Wang, Wilson, & Hammond, 2016).

The goal of this review is to give an overview of the available systems and to highlight their strengths and their weaknesses. We will describe the major uses of fluorescent tools in biological research and we will describe examples of sensors to track molecular and global changes both in solution and in living cells. Finally, given the extreme variety of available sensors and techniques, we will provide the reader with a concept map to help choosing the best fluorescent tool and the best technique for analyzing the bioenergetics and physicochemical status of cells. We will focus on critical cellular parameters such as ATP (and other nucleotides), ionic strength, macromolecular crowding (excluded volume), membrane potential, NAD(P) and NAD(P)H levels, pH, temperature, viscosity and volume.

2. Currently available fluorescent molecules

The fluorescence molecules currently used in biological research can be divided into two categories: genetically encoded fluorescence proteins (Zhang et al., 1996) and fluorogenic molecules. The latter can be further distinguished into two subcategories: molecules that do not require a binding partner to emit detectable fluorescence upon excitation (Strack, 2021), and molecules that greatly increase their emission intensity after binding to a target molecule (Iyer et al., 2021; Plamont et al., 2016). Both fluorescent proteins and fluorogenic molecules have advantages and disadvantages, and they greatly differ in their mechanism of action.

2.1 Fluorescent proteins

2.1.1 Chromophore formation in fluorescent proteins

The structural studies performed on GFP (Yang, Moss, & Phillips, 1996), and subsequently on other FPs (Marshall, 2000; Park, Kang, & Yoon, 2016; Wachter, Elsliger, Kallio, Hanson, & Remington, 1998), unveiled the mechanisms through which these proteins gain fluorescence. All FPs have a β-barrel structure with an embedded α-helix (Fig. 1A) (Yang et al., 1996). The three amino acids located on this helix, which in wild type GFP are Ser65, Tyr66 and Gly67, undergo a post-translational self-modification that yields the chromophore. The proposed mechanism of formation of the GFP chromophore (Heim, Prasher, & Tsien, 1994) consists of a cyclization-dehydration-oxidation sequence (Fig. 1B). Briefly, the nucleophilic attack of the amino group of Gly67 onto the carbonyl group of Ser65 results in the formation of the imidazolidinone ring with elimination of water. Subsequently, the oxidation of the C^{α}-C^{β} bond of Tyr66 causes the formation of a large π system of which the electrons can be excited with photons of proper wavelength (Reid & Flynn, 1997). The radiative decay of the electrons to the ground energy level is responsible for the emission of light. Other FPs undergo different post-translational modification and there is variation in the three consecutive amino acids, but the first step

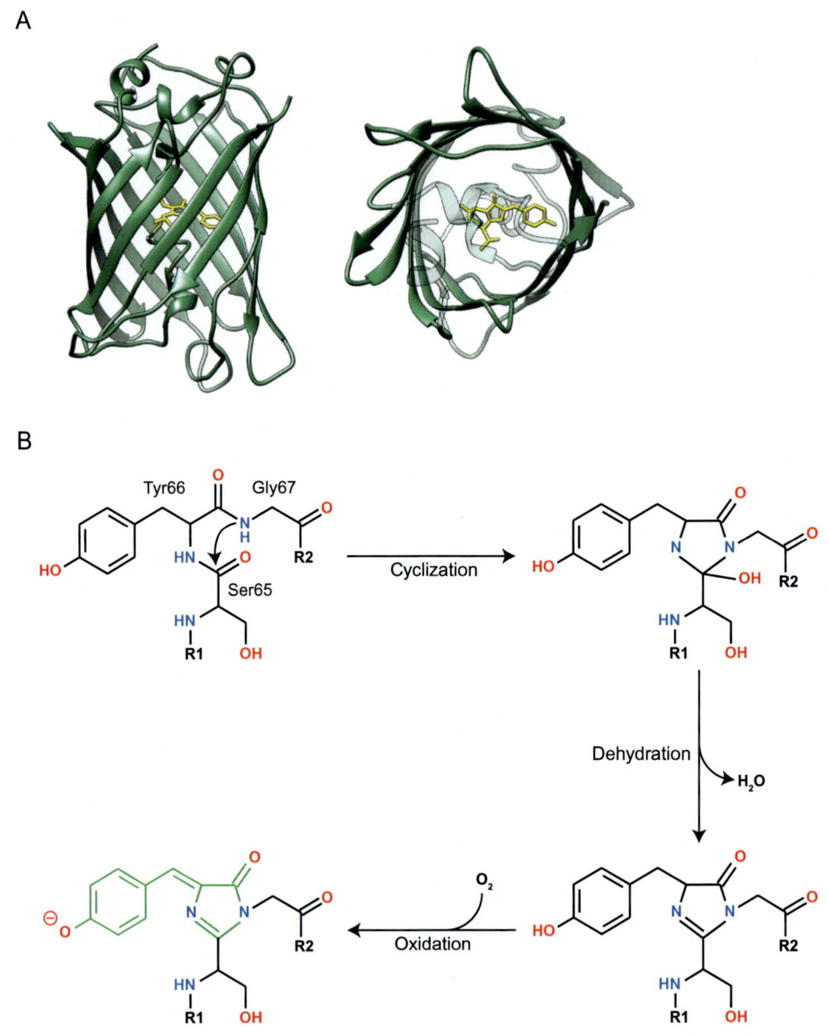

Fig. 1 (A) Structure of wild type GFP obtained via UCSF Chimera (Pettersen et al., 2004). Structure used for the representation: 1EMA. (B) Three step mechanism for the chromophore formation. Only the negatively charged phenolate is fluorescent. The uncharged protonated phenol group is not fluorescent. Scheme based on the work by Barondeau, Putnam, Kassmann, Tainer, and Getzoff (2003).

for all FPs is represented by the cyclization of the first and third residue that form the chromophore (residues 65 and 67 in wild type GFP). The chromophore of some red FPs develops directly from GFP-like chromophores, with a further oxidation step commonly referred to as oxidative redding pathway (Verkhusha, Chudakov, Gurskaya, Lukyanov, & Lukyanov, 2004; Wachter, Watkins, & Kim, 2010). In other cases, the red chromophore develops from GFP-like chromophores that are excited by a laser pulse in the ultraviolet range, generating the so-called photoconvertible FPs (Ando, Hama, Yamamoto-Hino, Mizuno, & Miyawaki, 2002; Mizuno et al., 2003). Finally, another proposed mechanism of formation of red chromophores is via the formation of blue emitting chromophore intermediates, which undergo a further oxidation step to generate the red emitting molecule (Pletnev, Subach, Dauter, Wlodawer, & Verkhusha, 2010; Strack, Strongin, Mets, Glick, & Keenan, 2010). An extensive overview of chromophore formation is provided by Stepanenko et al. (2011).

For many years it was believed that Gly67 in GFP and other FPs was essential for the formation of the chromophore (Barondeau, Kassmann, Tainer, & Getzoff, 2005; Sniegowski, Phail, & Wachter, 2005). However, a recent study described the first functional FP containing a G67A mutation (Roldán-Salgado, Sánchez-Barreto, & Gaytán, 2016). Nevertheless, both residues 66 and 67 are highly conserved within all FPs, and all naturally occurring GFP-like proteins are characterized by the presence of Tyr66 (Heim et al., 1994). Mutations of residue 66 are less rare, and in fact Tyr66 can be substituted with any aromatic amino acid, resulting in chromophores that have a blue-shifted emission compared to GFP (Heim et al., 1994). Amino acids at position 65 can be diverse among FPs, yielding chemically distinct chromophores and proteins that emit light at different wavelengths. Although residues 65–67 are crucial for chromophore formation, they are not the only players responsible for the photochemical characteristics of FPs. A Tyrosine in position 203, for example, has a key role in the formation of yellow emitting chromophores due to a π-stacking interaction between the aromatic ring of Tyr203 and the chromophore (Wachter et al., 1998). Residues Glu222 and His203 are crucial for the formation of photoactivatable FPs (Henderson et al., 2009), and in general amino acidic modifications both in the chromophore and in the β-barrel can lead to chromophores with different emission wavelength, quantum yield and brightness (Box 1). For a comprehensive overview of amino acidic substitutions in FPs that lead to different chromophores we refer to the review by Stepanenko et al (Stepanenko et al., 2011). Further modifications in the structure of FPs have been deployed to modify other physical-chemical parameters, such as pH sensitiveness (Shinoda et al., 2018) or tendency to dimerize (Shaner et al., 2013) (among others). More recently, modifications that rearrange the structure of FPs by

BOX 1 Fluorescence-related terms

Fluorescent quantum yield

The quantum yield of a fluorescent molecule represents the ratio between the number of photon emitted by a molecule and the number of photons absorbed.

Brightness

The brightness of a fluorescent molecule indicates the sensitivity and the signal-to-noise ratio for the detection. It is a value given by the Fluorescent Quantum Yield multiplied by the molar extinction coefficient. Since the molar extinction coefficient depends on the wavelength, the brightness will depend on the chosen excitation wavelength.

Quenching

Quenching is used to describe processes that decrease the fluorescence of a fluorescent molecule. It can depend on different factors: in a FRET pair the acceptor acts as a dynamic quencher for the donor, static quenching can occur when a fluorescent molecule aggregates, and a dark quencher can absorb the fluorescence of a fluorescent molecule and dissipate it as heat.

Photobleaching

Photobleaching occurs when a fluorescent molecule is photochemically altered. This can happen due to the cleavage of covalent bonds or due to reactions between the fluorophore and other molecules. The cleavage of covalent bonds can occur in the transition of the fluorophore from a singlet to the triplet state. Such transition is formally forbidden, and the probability of occurring differs for various fluorescence molecules: some molecules photobleach after absorbing a few photons, while other molecules can undergo several absorption and emission cycles before being destroyed.

generating circularly permuted FPs (cpFPs) (Kostyuk et al., 2019) and split FPs (Romei & Boxer, 2019) have been successfully deployed to develop sensors with specific (ligand-binding) properties.

2.1.2 Maturation of fluorescent proteins

Once a FP is expressed, it needs to go through several stages in a process called maturation to become functional as fluorophore (Remington, 2006). The first step involves the folding of the protein, and here the FP assumes its characteristic β-barrel conformation. Next, the three key amino acids that form the chromophore need to undergo the processes of cyclization, dehydration and oxidation (Fig. 1B). The protein folding is relatively fast and normally takes less than a minute (Naganathan & Muñoz, 2005).

On the other hand, the full formation of the chromophore can be very slow, with the final oxidation step being rate-limiting (Ma, Sun, & Smith, 2017). Maturation of FPs depends on different factors: the amino acidic sequence of the protein, which can be modified to obtain faster maturing FPs (Balleza, Kim, & Cluzel, 2018); the environment (host) in which the protein is expressed, which influences both the speed of protein folding and chromophore formation (Hebisch, Knebel, Landsberg, Frey, & Leisner, 2013); the temperature at which the organism expressing the protein is grown (Guo, Xu, & Gruebele, 2012); and the presence of oxygen in the environment (the β-barrel type fluorophore will not be able to mature in the absence of oxygen; Ma et al., 2017). Overall the maturation process can last from 5 minutes for optimized FPs, such as super-folder GFP expressed in *Escherichia coli* under optimal conditions (Pédelacq, Cabantous, Tran, Terwilliger, & Waldo, 2006), up to more than 1 hour (Balleza et al., 2018), depending on the aforementioned factors.

2.1.3 The effect of pH on fluorescent proteins

Most FPs are sensitive to changes in the environmental pH. Wild-type GFP fluorescence is stable in the pH range from 6 to 10, but the fluorescence decreases below pH 6 and increases at pH values higher than 10 (Campbell, 2001). Other GFP variants display greater pH sensitivity (Mahon, 2011). The effect of pH on FPs has been exploited to obtain FPs highly sensitive to pH changes, which have subsequently been employed as pH sensors (Kollenda et al., 2020; Liu et al., 2021; Mahon, 2011; Shen, Rosendale, Campbell, & Perrais, 2014). In parallel, FPs insensitive to pH in the physiological range (pH 6–8) have been developed (Roberts et al., 2016; Shinoda et al., 2018). A decrease in fluorescence stability at lower pH values has been observed in all β-barrel type fluorescent proteins, leading to the hypothesis that the protonated form of the chromophore has lost the ability to emit fluorescence (Haupts, Maiti, Schwille, & Webb, 1998; Ward, Prentice, Roth, Cody, & Reeves, 1982). Studies on denatured wild-type GFP showed that the chromophore has pH-dependent excitation and emission spectra due to the ionization of the phenolic group of Tyr66 (Campbell, 2001), with the phenolate form of the chromophore being responsible for the observed fluorescence. The pK_a for the transition from uncharged phenol to phenolate is 8.1 (Campbell, 2001). With such a high pK_a it is surprising that the stability of wild type GFP is maintained until pH 6. However, it is important to remember that the chromophore in non-denatured FPs is buried inside the β-barrel structure, which shields it from the external environment (Ward et al., 1982). Environmental pH changes need to first alter and partially denature the structure of the mature FP in order for protons to be able to interact with the chromophore and convert it from the anionic to the neutral form.

2.1.4 Photoactivatable and photoswitchable fluorescent proteins

In recent years the use of photoactivatable and photoswitchable FPs (PAFPs and PSFPs respectively) has grown, following the development of super-resolution imaging techniques (Nienhaus & Nienhaus, 2017). The use of PAFPs and PSFPs allow for the conversion of FPs from a dark form to a bright form or, *e.g.*, a green to red conversion by altering the hydrogen bonding and extending the conjugated double bond network of the fluorophore (Lukyanov et al., 2005). Briefly, PAFPs and PSFPs excited at a first specific wavelength fluoresce at a lower wavelength before being activated, or do not fluoresce at all. Activation occurs upon excitation with a laser pulse at a specific wavelength, typically 405 nm (Wang et al., 2014). By properly tuning the power and the duration of the laser pulse it is possible to activate only few molecules from the population of PAFPs and PSFPs, which can then be excited with the desired, specific wavelength and imaged. PSFPs can be reconverted to their dark state by exciting them with a third specific wavelength laser pulse (Zhou & Lin, 2013). The mechanism through which these proteins can gain fluorescence and switch from a dark to a bright state is via a *cis-trans* isomerization of the chromophore, which is triggered by the applied laser pulse. For a more detailed analysis of the mechanisms of photo-activation and photoswitching of PAFPs and PSFPs we redirect the reader to the review by Lukyanov et al. (2005) and by Zhou and Lin (2013).

2.1.5 Comparison of fluorescent proteins

In the last thirty years many different FPs have been developed, with different emission wavelength (Fig. 2), brightness, quantum yield, maturation time and pH sensitivity. We report in Table 1 an overview of the most commonly used FPs and related information (when available) about their chemical and photophysical properties.

2.2 Fluorogenic molecules

Fluorogenic molecules have been used extensively in biological research (Braut-Boucher et al., 1995; Gray, Mitchell, & Searles, 2015; Liu et al., 2020). Some of the main advantages of these molecules over FPs are that

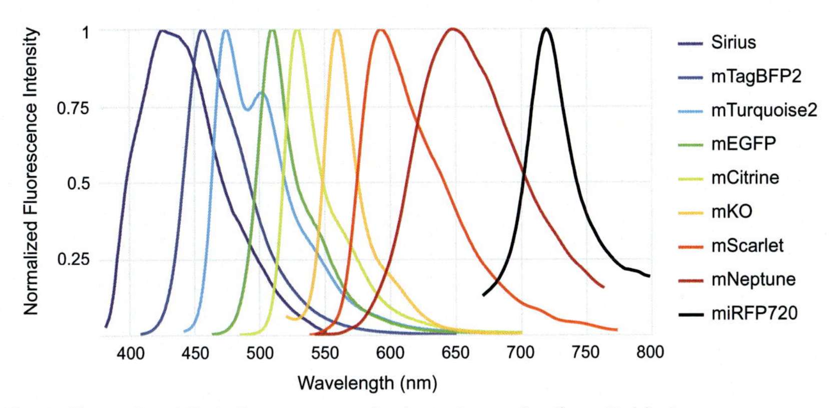

Fig. 2 Normalized Emission spectra of selected proteins from Table 1.

Table 1 List of some common fluorescent proteins with emission spectra ranging from dark blue to near infra-red.

Name	Ex max (nm)	Em max (nm)	Quantum Yield	Brightness	pKa	FPbase ID
Sirius*	355	424	0.24	3.6	3.0	BU5R3
EBFP2	383	448	0.56	17.9	5.3	DVMQ7
mTagBFP2*	399	454	0.64	32.3	2.7	ZO7NN
mTurquoise2*	434	474	0.93	27.9	3.1	7AV5G
mCerulean3	433	475	0.87	34.8	3.2	TWJXO
TagCFP	458	480	0.57	21.1	4.7	WRK8K
mEGFP*	488	507	0.60	33.6	6.0	QKFJN
mNeonGreen	506	517	0.80	92.8	5.7	ZRKRV
Gamillus	504	519	0.90	74.7	3.4	21PQ5
mVenus	515	527	0.64	66.6	5.5	WCSN6
mEYFP	515	528	0.62	49.0	6.9	SBLM5
mCitrine*	516	529	0.74	69.6	5.7	3Q37R
mKO*	548	559	0.60	31.0	5.0	RR1M4
mOrange2	548	562	0.69	49.0	6.5	5GR1V
mTangerine	568	585	0.30	11.4	5.7	N63O3
mScarlet*	569	594	0.70	70.0	5.3	FVS3D
mCherry	587	610	0.22	15.8	4.5	ZERB6
mNeptune*	600	650	0.20	13.4	5.4	1LT8G
miRFP	674	703	0.10	9.0	4.3	HO9GG
SNIFP	697	720	0.02	3.3	4.5	Q5F5J
miRFP720*	702	720	0.06	6.0	4.5	AJLWS
Photoswitchable Fluoresent Proteins						
mMaple3 (Green)	491	506	0.37	5.8	-	MNH1D
mEos3.2 (Green)	507	516	0.84	53.3	5.4	VUXRF
mEos3.2 (Red)	572	580	0.55	17.7	5.8	VUXRF
mMaple3 (Red)	568	583	0.52	12.5	-	MNH1D

The rows are color coded according to the emission wavelength of the fluorescent protein. Columns are, from left to right: Name of the fluorescent protein, Maximum Excitation wavelength, Maximum Emission wavelength, Fluorescent Quantum Yield, Brightness, pKa and FPbase ID. The normalized emission spectra of the proteins denoted with an asterisk (*) are visualized in Fig. 2.

they do not have to mature, have greater photostability, emit more photons and are much smaller. Some fluorogenic probes can be used as standalone molecules (Braut-Boucher et al., 1995; Liu et al., 2020) due to their natural high brightness and quantum yield, while others need to be paired with macromolecules (Iyer et al., 2021; Plamont et al., 2016; Pothoulakis et al., 2014), since only after interacting with a specific binding pocket they become (highly) fluorescent. Possible limitations of fluorogenic molecules, especially for *in vivo* research, are (i) their potential cytotoxicity and impact on the growth of cells, (ii) the non-trivial specific labeling of macromolecules or supramolecular structures and (iii) the targeting to specific compartments. Commonly employed fluorogenic molecules are the ATTO, Alexa Fluor and Cy-dye series with colors ranging from UV to infrared. The choice of the more appropriate dye depends on many factors, such as its photostability, brightness, quantum yield, lifetime, as well as its tendency to interact with different biological structures.

2.2.1 Organic dyes

Organic dyes as standalone molecules are available for different purposes. One of the main applications of these dyes is for cell imaging. For example, Calcein-AM and BCECF-AM (AM refers to acetoxymethyl ester form) are cell permeant dyes that are demethylated upon entering the cell and thereby trapped on the inside (Allen & Cleland, 1980). These dyes allow observing cell features that would not be as easy to distinguish with normal wide-field microscopy. BCECF and derivatives are commonly used to monitor the cytoplasmic pH of cells (Boens et al., 2006). Standalone dyes are also used to monitor physical chemical parameters of the environment in which they are present. For example, Calcein can be used to monitor volume changes (Gabba et al., 2020) due to its self-quenching (see Box 1) at high concentrations (Patel, Tscheka, & Heerklotz, 2009) or its quenching by proteins inside cells (Solenov, Watanabe, Manley, & Verkman, 2004), while molecular rotors such as BODIPY rotors can be used to monitor viscosity of solutions (bulk) or the local viscosity inside the cell (Liu et al., 2020). Molecular rotors can rotate one segment of their structure with respect to the rest of the molecule, causing a change in the ground-state and excited-state energy levels, which is reflected in a variation of the lifetime of the excited molecule. The amount of this energy change (and of lifetime variation) is dependent on the amount of intramolecular rotation, which depends on the environment (Liu et al., 2020).

Organic dyes also have a fundamental role in click chemistry. Here, a specific ligand conjugated with a dye can interact with its binding partner, allowing for detection of particular targets. These reactions can be performed both *in vitro* and *in vivo*. In particular, it is possible to perform *in vitro* bioconjugation reactions of natural amino acids on a protein surface, specifically targeting exposed cysteine (Kim et al., 2008), lysine (Larda, Pichugin, & Prosser, 2015), tyrosine (Dorta, Deniaud, Mével, & Gouin, 2020) and tryptophan (Ladner, Turner, & Edwards, 2007) residues. Several *in vivo* techniques are available for labeling molecules, such as the SNAP-tag method, in which a protein of interest (POI) is functionalized with an enzyme tag that allows the covalent labeling of the protein (Cole, 2013), the incorporation of unnatural amino acids in the sequence of a POI (Laxman, Ansari, Gaus, & Goyette, 2021), allowing their direct chemical modification, and the Staudinger-Bertozzi ligation reaction (Saxon, Armstrong, & Bertozzi, 2000), which consists in the ligation of a triarylphosphine conjugate reporter to an azide-functionalized biomolecular analogue that can be incorporated in cell structures, allowing for the detection of many different macromolecules, such as glycans, lipids, DNA and proteins (van Berkel, van Eldijk, & van Hest, 2011).

Another use of non-interacting organic dyes is via conjugation with antibodies (Mao & Mullins, 2010); these binding partner molecules can then be used to target specific structures of the cell, allowing multicolor imaging (Westerhof, Li, Bachman, & Nelson, 2016). A caveat of these techniques is that cells need to be permeabilized in order for the antibodies to reach their targets, making it impossible to be used *in vivo*. Fluorogenic dyes can also be conjugated with peptides (Cummings et al., 2002), allowing for example the detection of enzymatic reactions capable of digesting the polypeptide sequence. The list of fluorogenic dyes and their application is very extensive, and for a more detailed review on the topic we redirect the reader to the extensive works of Specht, Braselmann, and Palmer (2017), Takaoka et al. (2013) and Agouridas et al. (2019).

2.2.2 Hybrid systems

The use of hybrid systems as fluorescent tools can be very advantageous as they combine the best of FPs (specific targeting) and organic dyes (photophysical properties) (Fig. 3C). As long as a specific binding pocket is present in the protein or RNA, the dyes will interact with the macromolecule and emit fluorescence. Many dyes with different chemical structures and capable of binding to different cellular components are available, allowing for detection and imaging of membranes (Spötl, Sarti, Dierich, & Möst, 1995), DNA (Dirks & Tanke, 2006) and RNA (Dirks & Tanke, 2006) among other structures. Recently, proteins with more or less specific dye

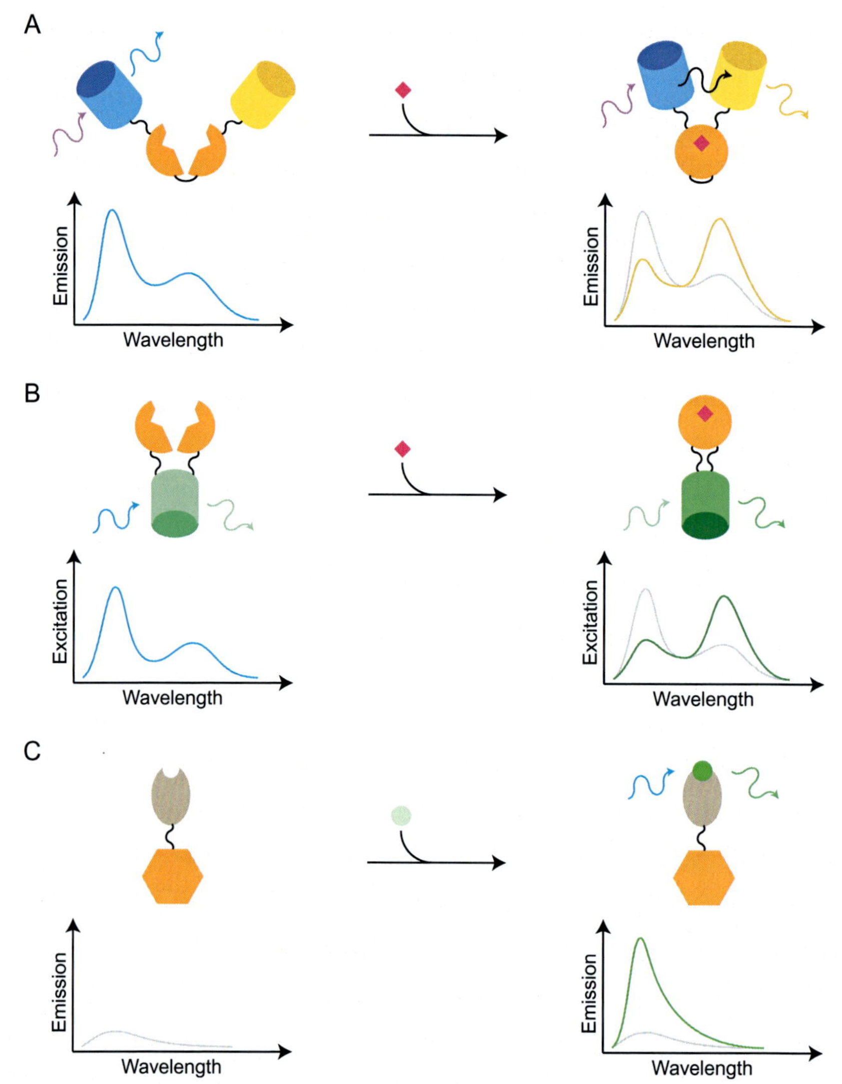

Fig. 3 Schematic of different types of sensors. (A) FRET sensor with a CFP as donor and a YFP as acceptor. Upon binding of the target molecule the two FPs are brought closer to each other with subsequent increased FRET, which can be detected in the emission spectrum as a decrease in intensity in the CFP peak or an increase in the YFP peak. (B) Ratiometric sensor based on a cpFP. After the binding of the POI with the substrate, the cpFP changes its conformation with a subsequent change in the excitation spectrum, where one of the peaks increases in intensity and the other decreases. (C) Hybrid system for labeling a POI. The POI is linked to a protein that can bind a fluorogenic dye. Before binding to its partner, the dye's brightness is very low. After the binding of the dye, the system increases its fluorescence emission.

binding motifs have been developed (Iyer et al., 2021; Plamont et al., 2016). These proteins can be used for genetic tagging of targets in the same way that FPs are used. The fluorescent dyes bind to these proteins, which leads to an instantaneous increase in fluorescence. These hybrid systems do not require protein maturation and labeling can be done in the absence of oxygen, allowing for studies in anaerobes (Iyer et al., 2021). Importantly, the reversible binding of organic fluorophores to these proteins and the superior photophysical properties of organic fluorophores enable long-term fluorescence microscopy of living cells. A caveat of this method compared to FPs is that the dyes need to be membrane permeable and they should not bind to other cell structures (Iyer et al., 2021). The hybrid systems, known as fluorescence-activating and absorption shifting tag (FAST) (Plamont et al., 2016) and Chemogenetic Tags with Probe Exchange (CTPEs) (Iyer et al., 2021), have been also used for multicolor imaging (Tebo et al., 2021) and subsequent development of FRET sensors.

Another recent innovation is the development of organic dyes and RNA aptamers capable of interacting with each other (Pothoulakis et al., 2014). Both green and red variants are available (Autour et al., 2018; Pothoulakis et al., 2014), which can be used concurrently, since different dyes interact with different aptamers. These systems provide a tool to image RNA in a way not achievable with other organic dyes or with the use of FPs (Zhang et al., 2015), allowing fast detection of the transcribed nucleic acid. These fluorescent systems have also been used to develop sensors, both in the form of single color sensors (Kellenberger, Chen, Whiteley, Portnoy, & Hammond, 2015) as well as FRET based (Jepsen et al., 2018), for detecting small molecules or other RNA sequences (Jepsen et al., 2018). An advantage of using RNA-type of sensors compared to protein-based sensors is that the detection time of the analytes is faster, since the RNA does not need to be translated and folded into a mature protein. RNA folding occurs during the transcription process, allowing for the fast formation of a binding pocket (Zhang et al., 2015). A major drawback in developing this kind of sensor, however, is that RNA prediction tools are much less advanced than the corresponding methods for proteins.

3. Labeling of (macro)molecules of interest

Tagging of molecules with fluorescent probes can be achieved in different ways, and below we briefly present the common strategies to label molecules and to selectively localize proteins, lipids and nucleic acids in the cell. This overview is not comprehensive and only outline possible scenarios. More extensive overviews are provided by Specht et al. (2017) and by Dirks and Tanke (2006).

3.1 Protein labeling

Tagging of proteins with fluorescent reporters can serve as a tool to localize proteins within the cell and visualize particular cell structures and cell compartments (Chalfie et al., 1994). The most common way of tagging proteins with fluorescent reporters is through the creation of a fusion construct of the POI with a FP (Chalfie et al., 1994). Once the construct is translated, the fully matured FP will emit fluorescence upon excitation. In case of GFP and derivatives the fluorescent proteins adds a 27kDa domain to the POI, and one needs to make sure that the localization of the POI is not affected by the tagging. If the POI needs to be localized immediately after translation, or if the conditions in which the POI needs to be expressed require absence of oxygen (such as in obligated anaerobes), the preferred imaging tool would be FAST (Plamont et al., 2016) or CTPE (Iyer et al., 2021), for which the fluorescence development is not dependent of oxygen and for which there is no delay between protein expression and fluorescence emission. Another way to visualize a POI is through the incorporation into proteins of fluorophores conjugated to unnatural amino acids (Laxman et al., 2021), but the number of dyes available for direct protein labeling is limited.

3.2 Membrane labeling

The use of membrane proteins coupled to fluorescent tags for imaging membranes can help understanding the functional dynamics and structure of this organelle (Costantini et al., 2015). This approach will not be discussed here as it essentially takes the route of protein labeling. Alternatively, one can visualize membranes by adding organic dyes capable of intercalating in the lipid bilayer (Collot, Boutant, Fam, Danglot, & Klymchenko, 2020). A drawback of using these dyes is in most cases the inability to specifically label a particular membrane. More advanced techniques such as *in vitro* conjugation of lipids with organic dyes and the application of a lipid binding domains have increased the specificity of membrane labeling (Kundu, Chandra, & Datta, 2021; Shi, Heegaard, Rasmussen, & Gilbert, 2004). For instance, the lactadherin C2 domain allows the detection of phosphatidylserine (PS) lipids, whereas other protein domains and antibodies (labeled with fluorescent probes) have been used for phosphoinositides (PI) (Várnai & Balla, 2006; Wakelam, 2014). Biochemical detection techniques now allow quantification of all seven phosphoinositides, and the use of fluorescently tagged PI-binding domains enables real-time visualization of most of them in intact cells.

3.3 Nucleic acids labeling

One of the first ways of labeling nucleic acids was via staining of the DNA through the use of organic molecules capable of entering the cell and intercalating into the DNA double helix. This method, however, prohibited long-term *in vivo* studies (Salic & Mitchison, 2008), such as monitoring the structure and dynamics of DNA during the cell cycle, due to the damage done to the DNA by the staining molecule, but recently less damaging dyes have been developed (Qu et al., 2011). Another method for labeling nucleic acids is via the use of probes capable of entering the cell and binding specific sequences of DNA or RNA (Dirks & Tanke, 2006). Linear Phosphodiester Oligodeoxynucleotides (Politz, Taneja, & Singer, 1995), Peptide Nucleic Acids (Molenaar et al., 2003), 2′-O-methyl RNAs (Molenaar, Abdulle, Gena, Tanke, & Dirks, 2004) and Locked Nucleic Acids (Darfeuille, Hansen, Orum, Primo, & Toulmé, 2004) are a few examples of probes that can target specific nucleic acid sequences. These probes are made of molecules that pair either with DNA or RNA, and that are conjugated with a fluorescence reporter. After entering the cells, these probes cannot be metabolized, allowing to visualize nucleic acids for an extended period of time. Another way to label nucleic acids is via the use of labeled DNA or RNA binding proteins. Labeling of proteins can be accomplished with any of the methods reported in Section 3.1. A drawback of the method is that some DNA regions can be elusive and not have a protein-binding partner. An elegant solution to this problem could be the use of CRISPR-Cas with a labeled inactive Cas protein (Deng, Shi, Tjian, Lionnet, & Singer, 2015). The guide RNA used in the system would then allow labeling virtually any location on the DNA and any RNA sequence. Finally, a series of tools for RNA imaging developed in the past few years is represented by aptamers capable of binding organic dyes, which in turn can emit fluorescence (Pothoulakis et al., 2014). Several versions of these aptamers have been developed (Ouellet, 2016), allowing for labeling of virtually all types of RNA (Okuda, Fourmy, & Yoshizawa, 2017) and for multicolor imaging.

4. Sensors and methods of detection

There are different types of fluorescence-based sensors: ratiometric sensors, intensiometric sensors and sensors based on Förster Resonance Energy Transfer (FRET). In ratiometric sensors (Fig. 3B), the readout is a ratio of two wavelengths (usually two spectral maxima or a maximum and isosbestic point) in either the excitation or the emission spectrum. Ratiometric sensors have the benefit that the concentration of the fluorescent protein or probe does not have a direct influence on the readout. In intensiometric sensors, the signal is defined by the change in emission intensity upon stable excitation. Knowing the concentration of these sensors is critical for the analysis, which is often not possible due to cell-to-cell variations and changes in sensor concentration in the course of an experiment. Intensiometric sensors are in most cases disfavored when ratiometric or Forster Resonance Energy Transfer (FRET)-based approaches are available.

FRET involves the transfer of resonance energy from a donor fluorophore to an acceptor fluorophore without emission of light (Förster, 1948) (Fig. 3A). This energy transfer occurs when (i) the donor's emission has a higher energy (shorter wavelength) than the acceptor's emission (Shrestha, Jenei, Nagy, Vereb, & Szöllősi, 2015); (ii) the donor emission spectrum overlaps with the acceptor excitation spectrum (Shrestha et al., 2015); and (iii) the donor and the acceptor are at a distance between 3 and 7 nm from each other, depending on the donor-acceptor pair. The choice of donor and acceptor can be complex, as it is necessary to take into account many factors, such as the quantum yield and brightness of both fluorophores as well as the overlap of their spectra. An excellent review on FRET couples is provided by Bajar, Wang, Zhang, Lin, and Chu (2016). The apparent FRET efficiency can be calculated according to Eq. (1):

$$E_{app} = \frac{I_A}{I_A + I_D} \tag{1}$$

where I_A and I_D are the emission intensity of the acceptor and of the donor, respectively. A good overlap of the donor's emission and acceptor's excitation spectrum is necessary for a high efficiency of energy transfer, which is never 100%. That means that after exciting the donor and detecting the emission of the acceptor, the emission of the donor will also be detected.

According to Eq. (1), when the brightness and the quantum yield of the donor are high compared to the values for the acceptor, the apparent FRET will be low.

A more accurate way of obtaining the FRET efficiency is by measurement of the lifetime of the donor (Periasamy, Mazumder, Sun, Christopher, & Day, 2015). The lifetime of fluorescent molecules is described by the reciprocal of the sum of radiative and non-radiative decays (Eq. 2), and it is directly related to the quantum yield (Eq. 3):

$$\tau = \frac{1}{k_r + k_{nr}} \tag{2}$$

$$\Phi = \frac{k_r}{k_r + k_{nr}} \tag{3}$$

where τ is the lifetime, k_r is the rate of the radiative decay, k_{nr} is the rate of the non-radiative decays and Φ is the quantum yield. In the case of FRET, the rate of energy transfer must also be taken into consideration (Eqs. 4 and 5):

$$\tau_{DA} = \frac{1}{k_r + k_{nr} + k_{ET}} \tag{4}$$

$$\Phi_{DA} = \frac{k_r}{k_r + k_{nr} + k_{ET}} \tag{5}$$

where τ_{DA} is the lifetime of the donor in the presence of the acceptor, Φ_{DA} is the quantum yield of the donor in the presence of the acceptor and k_{ET} is the rate of energy transfer. The real FRET efficiency is then obtained from Eq. (6):

$$E = 1 - \frac{\tau_{DA}}{\tau_D} \tag{6}$$

where τ_D is the lifetime of the donor alone.

FRET can be used both *in vitro* and *in vivo*, such as for co-localization studies (Augustinack, Sanders, Tsai, & Hyman, 2002), conformational changes in macromolecules (Gauer et al., 2016) and molecule detection by sensors (Otten et al., 2019). The FRET efficiency can also be calculated as a function of the distance between the donor and the acceptor (Eq. 7):

$$E = \frac{R_0^6}{R_0^6 + r^6} \tag{7}$$

where R_0, the Förster distance, is the distance between donor and acceptor at which 50% FRET efficiency, and r is the actual distance between donor and acceptor.

The Förster distance depends on the overlap integral of the donor emission spectrum and the acceptor absorption spectrum, as well as on their reciprocal molecular orientation and the refractive index of the medium (Eq. 8):

$$R_0^6 = \frac{2.07}{128\,\pi^5\,N_A}\frac{\kappa^2 Q_D}{n^4}\int F_D(\lambda)\varepsilon_A(\lambda)\lambda^4 d\lambda \tag{8}$$

where Q_D is the quantum yield of the donor in the absence of the acceptor, κ (Prasher et al., 1992) is the dipole orientation factor, n is the refractive index of the medium, N_A is the Avogadro constant, F_D is the donor emission spectrum normalized to an area of 1, ε_A is the acceptor molar extinction coefficient, and λ is the wavelength. The value κ can vary from 0 to 4, and normally it is assumed to be 2/3 for freely rotating fluorescent pairs, which reflects the average of all the possible orientations that the two molecules can take (van der Meer, 2002). Fluorescent anisotropy measurements can be performed to ascertain that the donor and the acceptor are freely rotating (van der Meer, 2002).

4.1 Organic dyes as sensors

Standalone organic dyes can be used to sense physical chemical parameters in solutions (Bittermann, Grzelka, Woutersen, Brouwer, & Bonn, 2021), vesicles (McNamara & Rosenzweig, 1998) and cells (Oliveira et al., 2018). Dyes have been developed that detect changes in the concentration of small molecules like oxygen (Mirabello, Cortezon-Tamarit, & Pascu, 2018), ATP (Wang et al., 2016), reactive oxygen species (Choi, Yang, & Weisshaar, 2015), protons (pH) (Ozkan & Mutharasan, 2002), iron (Ma, Abbate, & Hider, 2015) and calcium (Paredes, Etzler, Watts, & Lechleiter, 2008) among others, and dyes that can detect global changes in solutions, such as molecular rotors used for measuring the viscosity of a solution (Liu et al., 2020), or self-quenching probes that report volume changes (Gabba et al., 2020).

4.2 Fluorescent protein-based sensors

The physical chemical properties of FPs have been exploited extensively to develop different types of sensors, and for an extensive review on single fluorescent protein-based biosensors we redirect the reader to the review by Nasu et al (Nasu, Shen, Kramer, & Campbell, 2021). One of the best-documented properties of FPs is the pH dependent fluorescence. This property has been used to develop sensors that are highly sensitive to pH changes (Liu et al., 2021; Mahon, 2011), which have been used to determine the pH inside of living cells. Another use of FPs is their possible employment as viscosity sensors, as it has been observed that lifetime of some fluorescent proteins can change with the refractive index (Davidson et al., 2020) and that the time-resolved anisotropy is affected by the viscosity of the environment (Borst, Hink, van Hoek, & Visser, 2005; Suhling, Davis, & Phillips, 2002).

The development of circularly permuted FPs (cpFPs) provided researchers with tools for designing a whole new range of sensors (Kostyuk et al., 2019). In the structure of circularly permuted proteins the N- and the C-terminus are fused together with a linker. New N- and C-termini are created in another part of the protein, creating a gap in a portion that was originally continuous (Topell et al., 1999). A circularly permutated fluorescent protein is generated by rearranging the DNA. First, a 5′ region is shifted upstream, toward the start of the gene, creating a new N-terminus in the protein. Secondly, the original C- and N-termini are connected by a flexible linker. The new N- and C-termini have to be in a non-critical position and be flexible (Kostyuk et al., 2019). In the case of cpFPs, the new N- and C-termini are usually located on a region of the β-barrel that, if subjected to conformational rearrangement, causes the whole β-barrel to change its conformation, potentially exposing the chromophore to the external environment with subsequent changes in the photophysics of the chromophore (Kostyuk et al., 2019). cpFPs have been used as sensitive pH sensors (Deng et al., 2021), but the main application is by connecting the new N- and C-termini to a sensing domain for a specific metabolite (Berg et al., 2009; Honda & Kirimura, 2013; Nagai, Sawano, Park, & Miyawaki, 2001) (Fig. 3B). The conformational changes of these domains upon binding or unbinding of the metabolite cause a rearrangement of the β-barrel structure of the cpFPs, which in turn alters the fluorescence spectral properties. Below we discuss in more detail cpFPs specific for the detection of the calcium, potassium, ATP/ADP ratio, NAD(P)H and pH variations.

Another structural modification of FPs is based on the creation of split molecules (Romei & Boxer, 2019), which represent a technological advancement of cpFPs. The structure of split FPs is represented by a circularly permuted β-barrel in which one of the β-sheets fundamental for the maturation of the chromophore is lacking. Genetically, this can be achieved in the same way as the cpFPs are obtained, with subsequent deletion of the DNA region corresponding to the β-sheet. The β-sheet is then expressed from another plasmid. Once the β-sheet encounters the β-barrel, it complements the structure, allowing for the maturation of the chromophore with subsequent emission of fluorescence. As for normal FPs, the oxidation of the chromophore is the rate limiting step for split-FPs, hence there will be a lag time between the moment in which the β-barrel has been complemented and the emission of fluorescence. Split FPs can be used for different purposes, such as verifying the co-localization of target proteins (Bader et al., 2020) or the co-transcription of target genes (Romei & Boxer, 2019).

4.3 RNA-based sensors

RNA-based sensors have also been developed for detecting small molecules inside cells. These aptamer-based sensors contain a binding domain for the metabolite or signaling molecule, such as cyclic di-AMP (Kellenberger et al., 2015) or cyclic di-GMP (Wang, Wilson, & Hammond, 2016), and an aptamer capable of binding a fluorogenic molecule (Pothoulakis et al., 2014). The aptamer is in an unfolded state prior to the binding of the small molecule. Once the interaction with the analyte occurs, the RNA sensor changes its conformation, allowing the formation of the aptamer and binding of the fluorophore. DFHBI is the main fluorophore used in the so-called Spinach-type structures (Pothoulakis et al., 2014), whereas TO1-biotin is used in Mango-like (Autour et al., 2018). These fluorescent molecules can permeate the cell membrane without affecting the growth of the cells. An advantage of these sensors compared to protein based ones is the rapid detection of metabolites, since the formation of the binding pocket occurs during transcription and does not depend on protein translation and folding.

4.4 FRET-based sensors

FRET sensors can be obtained with any combination of fluorescent molecules, as long as a donor and an acceptor fluorophore are present (Fig. 3A). Co-localization studies can be performed by assessing the presence or absence of FRET between two molecules tagged with a donor and an acceptor (Augustinack et al., 2002). Conformational dynamics studies can be performed by labeling two amino acids on the same protein with a donor and an acceptor fluorophore and measuring variations in FRET efficiency upon binding of, *e.g.*, a ligand (Götz et al., 2021). Finally, a donor and an acceptor can be connected through a linker containing a sensing domain for specific metabolite, ion or physical factor such as temperature or excluded volume. A change in conformation in this linker will generally result in a change in the positions of the two fluorescent probes relative to one another, bringing them closer together or further apart, depending on the sensor's design (Imamura et al., 2009; Sadoine et al., 2021).

Most genetically encoded FRET sensors use a pair of FPs as donor and acceptors. However, there are three challenges for application of FRET-based sensors: (i) Different fluorescent proteins are affected differently by variations in pH (Campbell, 2001). Hence, one needs to make sure that the pH of cell does not change; (ii) The maturation time of the fluorescent proteins may differ, which affects FRET efficiency (Liu et al., 2018). A possible way to overcome this limitation is by using homo-FRET (Tramier & Coppey-Moisan, 2008). In homo-FRET, the donor and the acceptor are the same molecule, which has the disadvantage that the donor and acceptor cannot be discriminated. However, it is possible to determine variations in fluorescence polarization. When polarized light is used to excite the fluorophore, the emission energy can be transferred via FRET to an identical fluorophore in close proximity. In this energy transfer the emission light gets depolarized, causing a decrease in fluorescence polarization as a function of the distance between two identical fluorophores (Tramier & Coppey-Moisan, 2008). Alternatively, one can use RNA-based FRET sensors, in which two aptamers binding two different dyes change distance upon a conformational change in the sensing domain (Jepsen et al., 2018). These sensors have been used to sense *in vivo* variations in concentrations of small molecules and the detection of specific RNA sequences (Jepsen et al., 2018). (iii) The donor and acceptor can differ in their sensitivity to photobleaching. If the acceptor bleaches more rapidly, then the apparent FRET ratio will be skewed toward lower values. On the contrary, if the donor is more sensitive to bleaching, then the apparent FRET ratio will be higher.

A final note on FRET measurements relates to the environment in which the measurements are performed. It is known that the rate of radiative decay is influenced by the refractive index of the environment (Davidson et al., 2020; Suhling et al., 2002; Tregidgo, Levitt, & Suhling, 2008), and different fluorophores are affected differently by this effect (Borst et al., 2005). Other studies point out a dependency between the time-resolved anisotropy of fluorophores and the viscosity of solutions (Borst et al.,

2005; Suhling, Davis, & Phillips, 2002). If FRET measurements need to be conducted in environments in which the viscosity is expected to change, it is important to consider this effect on the measured FRET efficiency and, if possible, correct the values accordingly.

5. Tracking of molecular and global changes

Fluorescence-based sensors can be used to track a great variety of molecules and physicochemical conditions in the cell (or cell membranes). Such changes can be divided into three categories: (i) detection of molecules, including changes in concentrations of small molecules such as metabolites, ions and signaling molecules; (ii) detection of global changes in the physicochemical properties of the cell such as pH, ionic strength, macromolecular crowding (excluded volume effects), membrane potential, viscosity and volume changes: and (iii) detection of macromolecular interactions and conformational dynamics. The list of available fluorescent sensors is so vast that it is impossible to report them all in a single document. Here, we focus on the tools to track, arguably, the most important molecules and global factors that report the bioenergetics and physicochemistry of the cell and synthetic cell like systems (Table 2).

5.1 Detection of small molecules

Measuring the concentration of specific molecules requires the development of sensors that bind the compound of interest within the expected concentration range and buffer system. In other words, the concentration of the compound must be relatively close to the dissociation constant of the sensor for the molecule. Binding affinities of some sensors, such as ion probes, are often different when measured *in vitro* or *in vivo*. Substrate-binding proteins associated with ATP-binding cassette importers and other types of transporters and ion channels (Scheepers, Nijeholt, & Poolman, 2016) have been a popular source of proteins to which a fluorescence donor and acceptor can be engineered, either by gene fusion with FPs (Isoda et al., 2021; Okumoto, Jones, & Frommer, 2012; Sadoine et al., 2021) or via chemical modification of strategically engineered Cys pairs (de Boer et al., 2019; Gouridis et al., 2015). These proteins are readily engineered to obtain sensors in the appropriate affinity range(s), but other ligand-specific proteins have also been used to design ratiometric or FRET-based sensors (*vide infra*). In addition, we present here a series of chemical probe-based sensors that are taken by

Table 2 List of selected sensors from the main text.

Parameter	Name	Fluorophore	Read-out	Spectral maxima (nm)		Comments
				Excitation	Emission	
Ca^{2+}	Cameleons	BFP, GFP or CFP, YFP	Ratiometric FRET	370 or 440	440, 510 or 480, 535	Multiple versions with varying affinities
Ca^{2+}	Pericam	GFP	Ratiometric	418, 494	511	Intensiometric version available
Ca^{2+}	Fura-2	Stilbene	Ratiometric	340, 380	510	Commercially available
K^+	KIRIN1	mCerulean3 cpVenus	Ratiometric FRET	410	475, 530	Selective for K^+ over Na^+
K^+	KIRIN1-GR	Clover mRuby2	Ratiometric FRET	470	520, 600	Small FRET change
K^+	GINKO1	EGFP	Ratiometric	400, 500	520	Sensitive to high concentrations of Na^+
ATP	Ateam	CFP mVenus	Ratiometric FRET	435	475, 527	Moderately pH sensitive
ATP	GO-Ateam	GFP OFP	Ratiometric FRET	470	510, 560	Red shifted ATeam sensor
ATP	yAT1.03	mTurquoise2 tdTomato	Ratiometric FRET	430	483, 570	pH stable
ATP	Queen	cpEGFP	Ratiometric	400, 494	513	Moderately pH sensitive
ATP/ADP	PercevalHR	cpmVenus	Ratiometric	420, 500	515	pH sensitive
ATP	iATPSnFR	cpSFGFP	Intensiometric	490	512	Ratiometric when fused to mRuby, moderately pH sensitive
ATP	ATPOS	Cy3	Intensiometric	556	566	Hybrid sensor
NADH	Frex	cpYFP	Ratiometric	410, 500	518	pH sensitive

Continued

Table 2 List of selected sensors from the main text.—cont'd

Parameter	Name	Fluorophore	Read-out	Spectral maxima (nm)		Comments
				Excitation	Emission	
$NADH/NAD^+$	Peredox	T-Sapphire mCherry	Ratiometric	400, 575	528, 635	pH stable
$NADH/NAD^+$	SoNar	cpYFP	Ratiometric	420, 485	530	pH insensitive
NADPH	iNAP	cpYFP	Ratiometric	420, 485	530	pH insensitive
Sucrose	FLIPsuc	eCFP eYFP	Ratiometric FRET	435	475, 530	Multiple versions with varying affinities
Cyclic di-AMP	YuaA-Spinac2	Spinach	intensiometric	455	505	RNA-based biosensor
Cyclic di-GMP	Vc2-Spinach	Spinach	intensiometric	455	505	RNA-based biosensor
pH	pHluorin	GFP	Ratiometric	410, 470	535	Intensiometric variant available
pH	pHred	mKeima	Ratiometric	440, 585	610	Compatible with PercevalHR
pH	pyranine	Arylsulfonate	Ratiometric	400, 450	510	Commercially available
pH	BCECF	fluorescein	Ratiometric	439, 490	530	Commercially available
Membrane potential	$DiSC_3$-5	Carbocyanine	Intensiometric	653	676	Commercially available, negative inside potentials
Membrane potential	Oxonol VI	Polymethine	Intensiometric	599	634	Commercially available, positive inside potentials
Viscosity	Various					Various classes available, including ratiometric variants
Vesicle volume/ leakage	Calcein	Fluorescein	Intensiometric	495	520	Self-quenches at high concentrations

Parameter	Sensor	Fluorophore(s)	Type	Excitation	Emission	Comments
Excluded volume	Crowding sensor	Cerulean Citrine	Ratiometric FRET	420	475, 525	Sensors differing in crowding sensitivity are available; different designs available
Excluded volume	Synthetic crowding sensor	Atto488 Atto565	Ratiometric FRET	470, 555	512, 630	Not commercially available
Ionic strength	I-sensor	Cerulean Citrine	Ratiometric FRET	420	475, 525	Different designs available
Temperature	ER thermoyellow		Intensiometric	560	584	Monitor temperature in Enodplasmic Reticulum of eukaryotic cells

Columns are, from left to right: parameter that is sensed by the sensor, name of the sensor, fluorophore(s) used in the sensor, type of sensor (intensiometric, ratiometric or FRET-based), maximum excitation wavelength, maximum emission wavelength, comments on the sensor.

cells via endocytosis or as acetoxymethyl ester derivative. In case of synthetic vesicles or cell-like systems the protein or chemical probe-based sensors are encapsulated in the lumen during the reconstitution procedure.

5.1.1 Calcium sensors

The very first fluorescent genetically encoded sensors were developed to monitor Ca^{2+} ions, which is a key signaling molecule in many cell types (Miyawaki et al., 1997). These FRET-based sensors consist of blue, and green or yellow emitting GFP analogues. Calmodulin fused to the calmodulin-binding peptide of myosin light-chain kinase (M13) are used as Ca^{2+} binding domain. Ca^{2+} binding switches the conformation from an extended to a compact and globular shape, which draws the two fluorophores closer toward each other, increasing the FRET efficiency. Pericam Ca^{2+} sensors have one fluorescent protein (Nagai et al., 2001) and use calmodulin fused to M13 to bind calcium ions. In addition to an intensiometric and an inverted intensiometric sensor (fluorescence intensity decreases upon binding of Ca^{2+}), a ratiometric version is available with an affinity constant for Ca^{2+} binding of 1.7 μM. The Pericam sensors are sensitive toward pH, so care must be taken to keep the pH constant during experiments or to correct for the pH bias.

A popular chemical probe for calcium is Fura-2 (Grynkiewicz, Poenie, & Tsien, 1985). It has high affinity for Ca^{2+} ions ($K_D \sim 0.1$ μM). The excitation spectrum of Fura-2 is ratiometric. Similar to many other chemical dyes, an acetoxymethyl ester form is available, which is membrane-permeable. When used *in vivo*, the ester bond is cleaved intracellularly, which traps the dye inside of the cell.

5.1.2 Potassium and sodium ion sensors

Potassium ions can be measured with protein-based sensors (Shen et al., 2019). KIRIN1, KIRIN1-GR and GINKO1 are sensors based on K^+ binding protein Kbp from *Escherichia coli*. KIRIN1 and KIRIN1-GR are similar in design as a FRET sensor, but differ in fluorophores. KIRIN1 uses CFP and mVenus, while KIRIN1-GR has GFP and mRuby2 as fluorescent proteins. GINKO1 only has a single fluorophore and uses the excitation ratio as readout. As the binding protein is the same for all three sensors, the resulting dissociation constants for K^+ are very similar, *i.e.*, between 0.4 and 2.5 mM, which is well below the physiological levels of potassium in most cells. Here, it would be highly desirable to engineer the sensors toward a lower affinity for K^+ ions (K_D in the 100 mM range)

The measurement of sodium ions is more problematic as no protein-based sensors are available. Some chemical probes have been developed, but they suffer from poor selectivity toward either K^+ or Na^+, poor affinity and an intensiometric readout (Meier, Kovalchuk, & Rose, 2006; Minta & Tsien, 1989; Szmacinski & Lakowicz, 1997). The ions are chelated via crown ethers, ring structures that consist of several ether groups. The size of the crown determines which alkali ion is preferred, but usually multiple ions are accepted. These probes can be useful in a scenario where only one of the two ions is present in the reaction mixture.

5.1.3 ATP sensors

One of the first protein sensors for ATP is ATeam (Imamura et al., 2009), which is a FRET-based sensor. The ε-subunit of the F_0F_1-ATP synthase of *Bacillus subtilis* is flanked by two fluorophores, CFP and mVenus. The latter GFP analogue was circularly permutated to improve the dynamic range of ATP concentrations. Without ATP, the sensor adopts an extended and flexible conformation with low FRET efficiency between the two fluorescent proteins. When ATP is bound, the two fluorophores are drawn closer to each other, causing an increase in acceptor emission. The sensor has a millimolar affinity for ATP and is thus suitable for use with ATP concentrations in the physiological range. By using the ε-subunit of a thermophilic *Bacillus* sp. PS3, the affinity for ATP was increased to the micromolar range. A more pH stable ATP sensor was developed by Botman, van Heerden, and Teusink (2020). The donor fluorophore was replaced by mTurquoise2 and the acceptor with tdTomato, which are both pH stable fluorescent proteins. A newer version of the ATeam sensor, Queen, contains only a single circularly permuted fluorophore (Yaginuma et al., 2014). The Queen sensors have the same affinities for ATP as the ATeam sensors, but are less sensitive to molecular crowding and bleaching.

PercevalHR is an ATP to ADP ratio sensor; it binds both ATP and ADP with similar micromolar affinity (Tantama, Martínez-François, Mongeon, & Yellen, 2013). It consist of a circularly permutated mVenus as the fluorophore, and GlnK as the nucleotide-binding domain. GlnK has a role in ammonia transport in prokaryotes and binds both nucleotides, but only ATP binding stabilizes the conformation of the loop structure near the binding site. This results in increased fluorescence with excitation at 500 nm, whereas ADP binding causes a slight increase in fluorescence with excitation at 420 nm. The ratio between these wavelengths in the excitation spectrum can be used as readout for the ATP to ADP ratio. Knowing the actual

BOX 2 Energy currencies of the cell

All known forms of life use mostly two forms of energy currency: ATP and electrochemical ion gradients. In order describe the energy status of a cell, the concentrations of ATP, ADP and inorganic phosphate and the electrochemical gradients of protons and sodium ions across the membrane need to be measured. The amount of free energy released upon hydrolysis of ATP to ADP plus inorganic phosphate is given by the phosphorylation potential (ΔG_p or $\Delta G_p/F$):

$$\Delta G_p = \Delta G^{0\prime} + 2.3RTlog\frac{[ADP][Pi]}{[ATP]}(\text{kJ/mol})$$

$$or\ \frac{\Delta G_p}{F} = \frac{\Delta G^{0\prime}}{F} + \frac{2.3RT}{F}\ log\frac{[ADP][Pi]}{[ATP]}(\text{mV})$$

Electrochemical proton or sodium ion gradients are most often used to drive membrane-bound processes, even though other types of ion and solute gradients exist. The F_0F_1-ATP synthase/hydrolase interconverts the free energy of the phosphorylation potential into an electrochemical proton gradient, hereafter referred to as proton motive force (Δp):

$$\Delta p = \Delta\Psi + \frac{2.3RT}{F}\ log\frac{[H^+]_{in}}{[H^+]_{out}} = \Delta\Psi - Z\Delta pH(\text{mV})$$

where $2.3RT/F$ equals 58 mV (at T = 298 K) and is abbreviated as Z; F is the Faraday constant, R the gas constant and T is the absolute temperature. ΔΨ is the membrane potential, and ΔpH refers to the pH gradient across the membrane. $\Delta G^{0\prime} = -30.5$ kJ/mol, and typically ΔG_p ranges from −50 to −65 kJ/mol (or $\Delta G_p/F$ varies from −520 to −670 mV). A sodium motive force (Δs) can be formed in a similar manner:

$$\Delta s = \Delta\Psi + \frac{2.3RT}{F}\ log\frac{[Na^+]_{in}}{[Na^+]_{out}} = \Delta\Psi - Z\Delta pNa(\text{mV})$$

concentrations of ATP and ADP (and inorganic phosphate) is required to calculate the phosphorylation potential (Box 2), which is possible if the sum of ATP plus ADP is known and the ATP/ADP ratio is determined (Pols et al., 2019).

Lastly, ATPOS is a hybrid sensor (Kitajima et al., 2020). Similar to the ATeam and Queen sensor variants, ATPOS uses the ε-subunit of the F_0F_1-ATP synthase of thermophilic *Bacillus* PS3 for ATP binding. The readout, however, is done by Cy3, a small molecule fluorophore instead of a fluorescent protein. The resulting sensor has very high affinity for ATP (a K_D of 150 nM) and is insensitive to changes in pH between pH values of 6 and 8.5.

5.1.4 NAD(P)H sensors

The first sensors for NADH and NAD^+/NADH were Frex (Zhao et al., 2011) and Peredox (Hung, Albeck, Tantama, & Yellen, 2011), respectively. Both sensors utilize the bacterial protein Rex, which binds to NADH and regulates metabolism based on the NAD^+/NADH levels (Somerville & Proctor, 2009). Although Rex itself is a homodimeric protein, the sensors contain an in tandem dimeric version. In the Frex sensor, part of one of the monomers is replaced by cpYFP, whereas in Peredox the cpFP is inserted between two complete monomers. Frex reports the NADH concentration with a K_D of 3.7 μM, while Peredox reports the NAD^+ to NADH ratio despite having poor affinity for NAD^+.

Zhao et al. (2015) introduced the NAD^+/NADH ratio sensor SoNar, which is also based on the Rex protein and includes two cpFPs instead of just one. The affinity of SoNar for NADH and NAD^+ differs 20-fold (higher affinity for NADH), which compares to a ±8000 fold difference in Peredox. Therefore, the sensor is sensitive to changes in concentrations of both NADH and NAD^+. Additionally, SoNar is not sensitive to pH changes between pH 7 and 8. The sensor was successfully used in finding compounds that cause oxidative stress in cancer cells.

To be able to measure NADPH levels in cells, the iNAP sensors were developed (Tao et al., 2017). The binding site of SoNar was mutated to selectively bind NADPH instead of NADH by introducing positive charges to accommodate the negative charge of the phosphate group of NADPH and reducing the steric hindrance caused by this group. This resulted in four iNAP sensors with affinities for NADPH varying from 2 to 120 μM.

5.1.5 Metabolite sensors based on substrate-binding proteins

Substrate binding protein (SBP)-based sensors have been developed for many different molecules, including sugars (Otten et al., 2019), amino acids (Ko, Kim, & Lee, 2017) and vitamins (Edwards, 2021). Most are FRET-based sensors, in which a donor and an acceptor are fused to flexible domains of the SBP. These sensors exploit the conformational change of SBPs upon binding of their substrate. The conformational changes can either drive closer or further apart the two fluorescent proteins, with subsequent increases or decreases of FRET.

An example of an SBP-based sensor is the periplasmic leucine-binding protein (LBP) fused to an FP and chemical modified with a fluorescence donor (Ko et al., 2017). The protein was genetically engineered to introduce a fluorescent unnatural amino acid, L-(7-hydroxycoumarin-4-yl) ethylglycine (CouA), which acts as FRET donor, and a YFP as fluorescence acceptor fused to the N-terminus of the protein. The donor and acceptor moieties are brought closer together upon binding of Leu to the LBP, with subsequent increase of FRET.

A recently developed series of sucrose-specific sensors allowed to expand the detection range of sucrose from micro- to millimolar concentrations (Sadoine et al., 2021). In these sensors a sucrose binding protein, ThuE, is genetically engineered by fusing to it an eCFP, which acts as FRET donor, and an eYFP, which acts as FRET acceptor. Binding of sucrose by ThuE drives the two FPs further apart, with subsequent decrease of FRET. The currently available SBP-based sensors are too many to be listed in this work, but we redirect the reader to the review by Specht et al. (Specht et al., 2017) for a more extensive overview.

5.1.6 RNA-based sensors

RNA-based sensors are a new development, and at the writing of this manuscript only a few examples are available. Some recently developed RNA-based fluorescent sensors are capable of sensing variations in concentration of cyclic di-AMP (Kellenberger et al., 2015), cyclic di-GMP (Wang, Wilson, & Hammond, 2016) and cyclic AMP-GMP (Kellenberger, Wilson, Sales-Lee, & Hammond, 2013). The design of RNA-based sensors is based on the presence of an aptamer and a sensing domain. In the unbound state, the aptamer is unfolded. Upon binding of the ligand, the sensor undergoes a conformational change that causes the folding of the aptamer, which can then bind an organic dye and emit fluorescence. The aptamer of currently available sensors are often Spinach- (Pothoulakis et al., 2014) or Broccoli-derived (Filonov, Moon, Svensen, & Jaffrey, 2014). Spinach and Broccoli are two aptamers characterized by a GFP-like emission spectrum upon binding of the organic dye DFHBI.

5.2 Detection of general physicochemical factors

The general physicochemical state of the cell is characterized by the internal pH (the difference in pH between two compartments yields a pH gradient or ΔpH; Box 2), macromolecular crowding (or excluded volume effects), ionic strength, membrane potential (see also Box 2), temperature, volume and viscosity. These general or global factors impact the growth of any cell and influence the efficiency of biochemical reactions.

5.2.1 pH sensors

pHluorin is one of the first GFP analogues specifically developed to measure pH inside cells (Miesenböck, De Angelis, & Rothman, 1998). The excitation spectrum of wild type GFP is virtually unaffected by changes in pH. By introducing nine mutations, the excitation spectrum displays a change in the ratio of two distinct maxima between pH 5.5 and 7.5. In addition to this ratiometric version, an intensiometric version of pHluorin has been created which loses fluorescence at low pH values.

Another pH sensitive fluorescent protein is pHred (Tantama, Hung, & Yellen, 2011), and this sensor was developed for the simultaneous use with other GFP-based sensors. pHred is based on the RFP mKeima, which has a long Stokes shift. As with pHluorin, the ratio of the intensity of the excitation peaks changes from pH 5.5 to 9. pHred was successfully used simultaneously with PercevalHR (Tantama et al., 2013), which is pH sensitive. The pH data from pHred were used to correct for the influence of pH on the ATP/ADP ratio data reported by PercevalHR (Tantama et al., 2013).

Commercial chemical probes are also available to monitor pH values. These probes allow imaging for longer periods of time compared to the protein-based sensors, because they are less sensitive to photobleaching. Both Pyranine (Kano & Fendler, 1978) and BCECF (James-Kracke, 1992) are ratiometric and have pK_a values of 7.2 and 7.0, respectively. The BCECF-AM derivative can be used to introduce and trap BCECF in living cells.

5.2.2 Membrane potential sensors

The membrane potential is one of the components of the proton and sodium motive force (Box 2) and an important energy currency of all cells. It is used as driving force (often in combination with a pH or sodium gradient) for numerous membrane-bound processes such as the synthesis of ATP, solute transport, reverse electron transport, protein translocation and others. The membrane potential can be measured with chemical probes such as $DiSC_3(5)$ (Sims, Waggoner, Wang, & Hoffman, 1974). $DiSC_3(5)$ or 3,3′-Dipropylthiadicarbocyanine Iodide belongs to the group of so-called Nernstian probes that have a delocalized positive charge. The monomeric form of the probe is fluorescent, and the molecule distributes between the extracellular medium and the lipid membrane. In vesicle suspensions without applied membrane potential the dye distributes between the extracellular medium and the lipid membrane. Upon hyperpolarization of the membrane the dye will accumulate on the side of membrane where the potential is negative and dimers and higher order aggregates will form, resulting in a decrease in fluorescence. When the vesicles are depolarized the dye redistributes over both membrane leaflets and is partially released back into the extracellular medium.

Oxonol VI is a similar probe as $DiSC_3(5)$, but has a negative charge instead (Apell & Bersch, 1987). Therefore this dye is suited for measuring inside positive membrane potentials. Upon polarization (inside positive), the fluorescence increases, in contrast to the decrease in fluorescence of $DiSC_3(5)$ upon the formation of inside negative potentials.

5.2.3 Viscosity

Viscosity can be measured using fluorescent molecular rotors (Kuimova, 2012; Lee et al., 2018; Liu et al., 2020). When excited, these sensors can either relax to a lower energy state by emitting a photon, or adopting a twisted intramolecular charge transfer (TICT) state. Part of the molecule rotates to adopt this TICT state and the rotation process is affected by viscosity. At high viscosity the TICT state is less favorable, thus fluorescence is increased. Depending on the probe used, fluorescent lifetime, the emission intensity or the ratio of maxima in the emission spectrum can be used as readout.

5.2.4 Volume sensing

When subjected to an osmotic shock or upon the transport of large amounts of solutes over the membrane, the size of a cell or vesicle will change. While the size of cells can be measured by microscopy, this is not possible for very small cells or, *e.g.*, large-unilamellar vesicles with diameters in the range from 100 to 200 nm, which is below the diffraction limit. Therefore the size of these vesicles can only be measured by indirect methods. A way to measure vesicle size is by taking advantage of the self-quenching characteristics of fluorophores. When dyes like Calcein are encapsulated at high (approximately 10 mM) self-quenching concentrations, the fluorescence readout signal becomes dependent of the volume of the compartment. When the internal volume decreases, the calcein fluorescence decreases; similarly, the signal increases when the vesicles swell (Gabba et al., 2020; van der Heide, Stuart, & Poolman, 2001).

5.2.5 Excluded volume sensors

Excluded volume or macromolecular crowding can affect the conformation of proteins and reactions' efficiency (van den Berg, Boersma, & Poolman, 2017). The excluded volume is the volume taken by all the macromolecules of the cell, which is therefore not available for a given molecule added to the system. Compaction of a macromolecule by, *e.g.*, the coming closer of two or more protein domains is favoured due to an entropic gain. This principle was used to develop FRET sensors capable of probing the excluded volume of the cell (Boersma, Zuhorn, & Poolman, 2015; Liu et al., 2017, 2018). The genetically encoded sensors consist of mCitrine (YFP, yellow fluorescent protein) and mCerulean (CFP, cyan fluorescent protein), which are connected via a flexible linker, including two α-helices. At high excluded volume levels the sensor adopt a more condensed conformation, bringing the two fluorophores closer to each other, which is observed as an increase in apparent FRET efficiency. In a follow up study, a set of nine systematically varied sensors have been developed and the crowding-induced compression of the proteins has been investigated (Liu et al., 2017).

Using the same principle, a sensor has been created from a polymer linker coupled to synthetic fluorophores (Gnutt, Gao, Brylski, Heyden, & Ebbinghaus, 2015). Here, the linker consists of a 10 kDa PEG molecule, labelled with Atto488 and Atto565 at either end of the polymer. Both the genetically encoded and polymer-linked probe sensors are especially sensitive to crowding by macromolecules or synthetic polymers.

5.2.6 Ionic strength sensors

Ionic strength can be measured via a FRET-based sensor that acts similarly to the macromolecular crowding sensor (Liu, Poolman, & Boersma, 2017). This sensor also consists of two fluorescent proteins joined by a flexible linker. Here, the linker consists of two α-helices with opposite charges. At low ionic strength levels, the opposite charges of the helices attract each other, increasing the FRET efficiency. At higher ionic strength levels, the charges of the linker are shielded by ions, which allows the FPs to stay further apart, which lowers the apparent FRET ratio.

5.2.7 Temperature sensors

Temperature in solutions or in living cells can be measured using chemical probes. Different probes have recently been developed (Arai, Lee, Zhai, Suzuki, & Chang, 2014; Maksimov et al., 2019; Okabe et al., 2012), and of particular interest for physiological studies is the use of polymer-encapsulated quantum dots (Fan et al., 2015), which show a high resistance to pH and ionic strength changes in the physiological range. They can enter mammalian cells by endocytosis but they have not been applied in lower eukaryotes or prokaryotes.

5.3 Detection of macromolecular interactions and conformational dynamics

Fluorescent tools are typically used to track changes in molecular interactions, localization, conformation and concentration. Localization of macromolecules can be achieved by adding a fluorescent tag to the molecule of interest (Chalfie et al., 1994; van Berkel et al., 2011). This strategy has also been employed to study proteins and RNA turnover (Trauth et al., 2020). Interactions between macromolecules can be observed by tagging different putative interacting partners with different fluorescent reporters and subsequently measuring the FRET efficiency, which will increase as a function of the proximity of the two fluorescent molecules (Kaufmann et al., 2020). In a similar way, changes in conformation of macromolecules can be studied via FRET measurements by tagging with different fluorescent reporters different parts of the analyzed macromolecules (Götz et al., 2021). We refer to a set of papers (Ploetz et al., n.d.; Asher et al., 2021; de Boer et al., 2019; Lerner et al., 2021) for determining interactions between macromolecules and conformational dynamics within proteins.

6. Microscopy techniques

In this section we present some of the most common techniques to measure the fluorescence of the sensors reported heretofore, highlighting the differences and the advantages and disadvantages of the various methods. Reporting all possible fluorescence detection methods would be beyond the scope of this review, and for a more thorough characterization of the available techniques we redirect the reader to different works (Combs, 2010; Datta, Heaster, Sharick, Gillette, & Skala, 2020; Huang, Bates, & Zhuang, 2009; Lichtman & Conchello, 2005; Renz, 2013).

For most *in vitro* (in solution or in vesicles) measurements, a spectrophotometer is typically used. A spectrophotometer allows exciting a sample at a specific wavelength or range of wavelengths and acquires the emission at the desired wavelength or range of wavelengths. It can be used to analyze both the emission and the excitation spectra, and it is fundamental to study how the spectra of fluorescent probes are affected by changes in the environment.

Time Correlated Single Photon Counting (TCSPC) allows obtaining information about the lifetime of the fluorescent species in a solution (Phillips, Drake, O'Connor, & Christensen, 1985), to perform different *in vitro* studies, such as accurately determining the FRET efficiency of a FRET pair, or measuring the viscosity of a solution with molecular rotors (Liu et al., 2020).

Fluorescence microscopy, on the other hand, is the most utilized technique for *in vivo* measurements, as it allows obtaining spatial information on the localization of the analyzed probes.

6.1 Confocal microscopy

Confocal microscopy is an imaging technique that allows increased contrast and resolution of a micrograph compared to classical fluorescence microscopy (Jonkman, Brown, Wright, Anderson, & North, 2020). Confocal microscopes use point illumination in combination with a pinhole to filter out the out-of-focus signal. This allows obtaining images at higher resolution (yet still diffraction limited), at the cost of reduced emission intensity. Such limitation can be overcome by increasing the pinhole size (hence lowering the resolution), increase the exposure time (hence encountering the possibility of blurring effects due to particles diffusion) or using probes with high brightness and quantum yield.

Since only a single point is illuminated in the field of view, confocal microscopy requires 2D scanning of the confocal plane to obtain an image. The confocal plane can then be moved along the z-axis, allowing to obtain multiple 2D images across the same sample, which can then be stacked together to obtain a 3D reconstruction of the sample (Jonkman et al., 2020). Other than for imaging purposes, confocal microscopes can be used to perform different types of measurements as summarized in the next subsections.

6.1.1 Fluorescence recovery after photobleaching

Fluorescence recovery after photobleaching (FRAP) (Carnell, Macmillan, & Whan, 2015) is a technique used to study diffusion and interactions of macromolecules, hence it can be used to determine whether a sensor is freely diffusing or if it is confined within specific regions of the cell. Briefly, a high intensity laser pulse at the excitation wavelength of the imaged probe is used to bleach a region of the imaged sample. Once bleached (see Box 1), the probes localized in that region undergo a structural change and lose the ability to emit photons. The bleached region appears dark upon bleaching of the probes, and gradually the fluorescence increases due to the diffusion of undamaged probes into the bleached are. The kinetics of the recovery of fluorescence can be used to calculate the ensemble diffusion coefficient and the fraction of freely diffusing macromolecules. Due to their lower photostability and tendency to rapidly photobleach, FPs can be more suitable for FRAP experiments in living cells than photostable dyes. FRAP can be used to track diffusion of proteins in the cytoplasm (Mika, Krasnikov, van den Bogaart, de Haan, & Poolman, 2011; Schavemaker, Śmigiel, & Poolman, 2017) and in the cell membrane (Goehring, Chowdhury, Hyman, & Grill, 2010) of both eukaryotic and prokaryotic cells, and in small compartments of eukaryotes such as mitochondria (Sukhorukov et al., 2010). At the same time, it has also been proven useful to track diffusion in various membrane environments such as lipid bilayers (Pincet et al., 2016), and giant-unilamellar vesicles (Göpfrich et al., 2019).

6.1.2 Fluorescence lifetime imaging microscopy

Fluorescence lifetime imaging microscopy (FLIM) (Datta et al., 2020) allows measuring the lifetime of the excited fluorescent probes, which is important for establishing that the probes in FRET-based sensors are freely rotating (see Section 4.4). FLIM employs pulsed illumination, using ultrashort pulses of light. TCSPC equipment is required for obtaining pulses at a sub-picosecond time resolution. The time between the laser pulse and the emission of the photon by the fluorescent probe can then be calculated. Thereby information is obtained on the permanence of the fluorophore in the excited state. Billions of data points are accumulated over a short period of time and then used to generate a histogram that follows a Poisson distribution (Datta et al., 2020). The data points are then fitted with an exponential model, which allows determining the fluorescence lifetime of the sample and the eventual presence of multiple lifetimes (Poudel, Mela, & Kaminski, 2020). Multiple lifetimes can be observed when the probes are present in different conformational states (Borst et al., 2005). FLIM can be used to study protein dynamics (Sun, Hays, Periasamy, Davidson, & Day, 2012) and the environmental conditions of solutions and cells, using lifetime-based probes. In these probes it is not the shape of the excitation or emission spectra that change in response to variations in the measured parameter, rather it is the lifetime of the fluorescent molecule. FLIM is also commonly used to measure FRET efficiencies *in vivo* via FLIM-FRET (see Section 6.3). Lifetime of fluorescent molecules can be affected by changes in the environment (Kashirina et al., 2020) or by changes in the structure of the fluorescent molecule itself (Hirata, Hirakawa, Shimada, Watanabe, & Ohtsuki, 2021), allowing for the development of probes that can sense environmental changes or molecular changes.

6.2 Super resolution microscopy

Super resolution microscopy encompass a set of techniques that allow obtaining wide-field images of the analyzed probes at a resolution level beyond the diffraction limit via detection of single molecules (Khater, Nabi, & Hamarneh, 2020). Detection of single molecules can be achieved by using Total Internal Reflection Fluorescence (TIRF) microscopes (Fish, 2009), which use a source of light to illuminate the sample at a sufficiently oblique angle such that the light wave is totally reflected without refraction into the sample, allowing to image a very thin region of the cell, usually less than 200 nm. TIRF is an extremely powerful technique to image fluorescently labeled molecules that are in the vicinity of the glass slide onto which the cell or vesicle sample is loaded (Fish, 2009). To measure fluorescence deeper inside the vesicles or cells, it is necessary for the light to pass through the sample and excite the fluorescent probes in a confined area. A technique called Highly Inclined and Laminated Optical sheet (HILO) microscopy (Tokunaga, Imamoto, & Sakata-Sogawa, 2008) can be used to achieve single-molecule images in these areas. Here the light beam encounters the sample at an angle slightly below the critical angle for total internal reflection, allowing for some light to be refracted into the sample, increasing the image intensity and decreasing background fluorescence (Tokunaga et al., 2008).

There are several super-resolution optical microscopy techniques and here we describe the ones that we are frequently using in conjunction with the probing of the physicochemical state of the cells with the heretofore-reported sensors.

6.2.1 Photo-activated localization microscopy and stochastic optical reconstruction microscopy

Photo-activated localization microscopy (PALM) (Gould, Verkhusha, & Hess, 2009) and Stochastic Optical Reconstruction Microscopy (STORM) (Rust, Bates, & Zhuang, 2006) are both based on the use of photoblinking of fluorescent molecules, such as photoactivatable or photoswitchable FPs or fluorescent dyes. Photoblinking allows obtaining spatially separated spots of fluorescence, thereby overcoming the diffraction limit (Khater et al., 2020). Briefly, a low intensity laser pulse of the proper wavelength is used to stochastically photoactivate a few fluorescent molecules, converting their fluorophore from its inactive-OFF to its active-ON state. A second laser pulse is then used to excite the active fluorescent molecule and its emission is measured as a single distinguishable spot. Laser pulses are spaced at milliseconds intervals and repeated for thousands of frames, allowing detecting multiple single molecules (Khater et al., 2020). After measuring its emission, the fluorescent molecule can then either be brought back to its OFF-inactive state by a third laser pulse at the proper wavelength (Wazawa et al., 2021) or photobleached, to avoid recording the same molecule more than once. In the case of photobleaching, it is necessary to ensure a high concentration of fluorescent molecule prior of the experiment, such as with a high expression system in the case of FPs, as the number of activated molecules will decrease over time. The single spots are then analyzed and a spatial map at resolution well beyond the diffraction limit of light is obtained. Stacking together all the acquired frames allows obtaining a super-resolution image. Recent advancements allowed to reconstruct super resolution 3D images, for example by assigning a different z coordinate to a spot as a function of its intensity (Huang, Wang, Bates, & Zhuang, 2008).

6.2.2 Single molecule displacement mapping

Single molecule displacement mapping (SMDM) (Xiang, Chen, Yan, Li, & Xu, 2020) is a recently developed technique that allows obtaining maps of diffusion coefficients at a nanometer scale resolution, providing an insight on how static structures or interactions affect the motion of particles *in vitro* and in living cells (Xiang et al., 2020). Briefly, photoactivable FPs are stochastically switched to their active-ON state by a short laser pulse of low intensity and the proper wavelength. Subsequently, two short pulses that excite the active FPs at a short time distance from each other allow monitoring the position of the same FP at two distinct moments. Knowing the time step and the displacement allows to reconstruct the diffusion coefficient by fitting a probability distribution function for a two dimensional random walk (Eq. 9):

$$p(x,t) = \frac{2r}{4Dt} e^{-\frac{r^2}{4Dt}} + kr \tag{9}$$

where t is the time step, r is the displacement, k is a factor used to correct for the background fluorescence, and D is the lateral diffusion coefficient. This technique allows observing heterogeneities in diffusion at single molecule resolution, which may reveal static structures or confined regions in the cell (Xiang et al., 2020).

6.3 FRET imaging

Confocal FRET imaging allows measuring the fluorescence intensity of donor and acceptor separately, which are then used to calculate the apparent FRET efficiency as in Eq. (1). This technique is used to perform colocalization (Augustinack et al., 2002) and interaction (Margineanu et al., 2016) studies. The FRET signal is sensitive to the concentration of sensor molecule analyzed. Moreover, fluorescent molecules with partially overlapping emission spectra can lead to a lower apparent FRET efficiency. Therefore, if FRET efficiency is used for quantitative measurements, such as with FRET-based biosensors, we believe that FLIM-FRET is a more powerful tool as it is not dependent on the concentration of the fluorescent species and only requires measurement of the donor lifetime (Periasamy et al., 2015). With this technique it is possible to calculate the exact FRET efficiency of a FRET pair, as per Eq. (6). A drawback of this method, however, is the necessity of having to measure the lifetime of the donor alone, isolated from the FRET pair. Since the lifetime and the anisotropy decay of fluorescent molecules, in particular of fluorescent proteins, are dependent on the environment (Borst et al., 2005; Suhling, Davis, & Phillips, 2002), it is not possible to use *in vitro*-obtained values of the donor's lifetime, but it is necessary to measure the donor's lifetime in the same system in which the FRET pair is analyzed.

7. A map to navigate the fluorescent sea

The amount of fluorescent tools and available techniques to detect them is extremely vast. Depending on the requirements and the condition of a specific study, one should accurately choose the proper probe and the proper method. However, finding the proper tool with the right characteristic and pairing it with the proper technique can be overwhelming, as many factors need to be taken into account: is the study going to be performed *in vitro* or *in vivo*? Is the environmental pH going to change? Are other environmental parameters such as the viscosity expected to change? Is the experiment going to be based on photobleaching? Will the study assess quantitative FRET changes as a function of variations in the concentration of metabolites? Recently, an algorithm for the selection of fluorescent reporters depending on the instrument settings has been published (Vaidyanathan et al., 2021), helping in the choice of the fluorescent molecules as a function of their spectral properties. Below (Fig. 4) we consider the

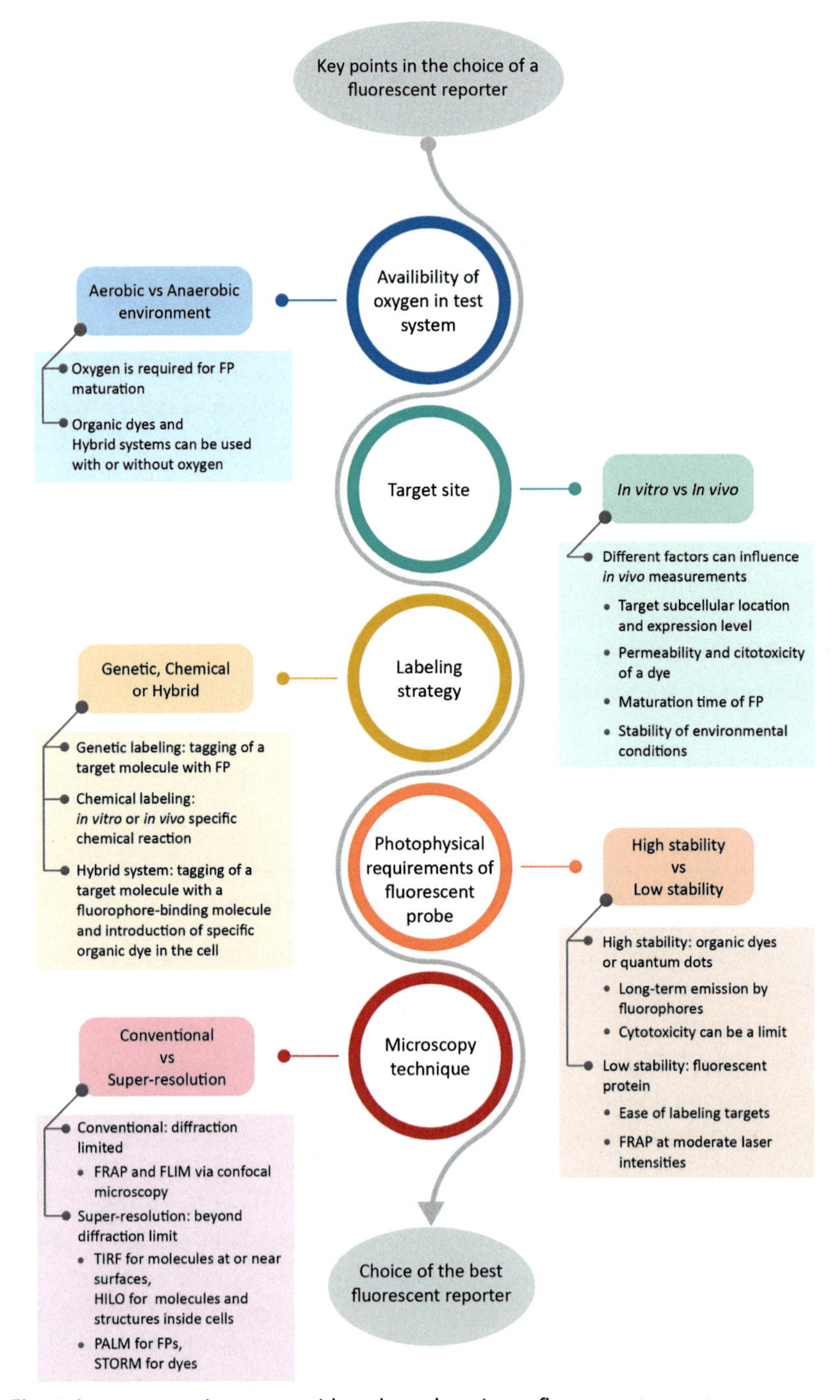

Fig. 4 Important points to consider when choosing a fluorescent reporter.

problem from a broader perspective, and we provide a flowchart of the important steps to consider for fluorescence-based sensing of living cells or artificial systems.

8. Conclusions

The field of biological fluorescence has evolved rapidly. New tools are constantly being developed, and new techniques allow obtaining more accurate data. In this review we have tried to summarize the state of the art on fluorescent probes used for studying the physicochemical state of the cell, be it a living cell, vesicle or cell-like system. We have tried to give a complete, yet very brief overview of the most useful methods for biological imaging. Finally, we provided a series of guidelines to help in the choice of the proper fluorescent tool and imaging technique depending on the purpose of the study.

Although the field of fluorescence probes and fluorescence microscopy has made huge steps forward in the last several years, we are approaching the limits of the possibilities provided by these instruments. While more and more molecules will be detected by the development of new molecule-specific sensors, answering questions on the general chemical physical status of a cell at high spatiotemporal resolution has proven to be much more complicated. Recently developed techniques such as SMDM allow obtaining detailed spatial information in living cells, while single molecule FRET allows observing heterogeneities in the FRET efficiency within the same system. These methods, however, pose technical hurdles that need to be overcome before application on systems scale is possible. Yet, the future lies in the high-throughput biochemical analysis of the cell at high spatiotemporal resolution, for which the further development of fluorescence-based sensors and methods will remain crucial.

Acknowledgments

The research was funded by an ERC Advanced grant (ABCVolume; #670578) and the EU Marie-Curie ITN project SynCrop (project number 764591).

References

Agouridas, V., et al. (2019). Native chemical ligation and extended methods: Mechanisms, catalysis, scope, and limitations. *Chemical Reviews*, *119*, 7328–7443.

Alexa Fluor. (2021). *Alexa fluor dyes spanning the visible and infrared spectrum—Section 1.3—NL.* www.thermofisher.com/uk/en/home/references/molecular-probes-the-handbook/fluorophores-and-their-amine-reactive-derivatives/alexa-fluor-dyes-spanning-the-visible-and-infrared-spectrum.html.

Allen, T. M., & Cleland, L. G. (1980). Serum-induced leakage of liposome contents. *Biochimica et Biophysica Acta (BBA) - Biomembranes*, *597*, 418–426.

Ando, R., Hama, H., Yamamoto-Hino, M., Mizuno, H., & Miyawaki, A. (2002). An optical marker based on the UV-induced green-to-red photoconversion of a fluorescent protein. *Proceedings of the National Academy of Sciences*, *99*, 12651–12656.

Apell, H.-J., & Bersch, B. (1987). Oxonol VI as an optical indicator for membrane potentials in lipid vesicles. *Biochimica et Biophysica Acta (BBA) - Biomembranes*, *903*, 480–494.

Arai, S., Lee, S.-C., Zhai, D., Suzuki, M., & Chang, Y. T. (2014). A molecular fluorescent probe for targeted visualization of temperature at the endoplasmic reticulum. *Scientific Reports*, *4*, 6701.

Asher, W. B., et al. (2021). Single-molecule FRET imaging of GPCR dimers in living cells. *Nature Methods*, *18*, 397–405.

Augustinack, J. C., Sanders, J. L., Tsai, L.-H., & Hyman, B. T. (2002). Colocalization and fluorescence resonance energy transfer between cdk5 and AT8 suggests a close association in pre-neurofibrillary tangles and neurofibrillary tangles. *Journal of Neuropathology and Experimental Neurology*, *61*, 557–564.

Autour, A., et al. (2018). Fluorogenic RNA mango aptamers for imaging small non-coding RNAs in mammalian cells. *Nature Communications*, *9*, 656.

Bader, G., et al. (2020). Assigning mitochondrial localization of dual localized proteins using a yeast Bi-Genomic Mitochondrial-Split-GFP. *eLife*, *9*, e56649.

Bajar, B. T., Wang, E. S., Zhang, S., Lin, M. Z., & Chu, J. (2016). A guide to fluorescent protein FRET pairs. *Sensors*, *16*, 1488.

Balleza, E., Kim, J. M., & Cluzel, P. (2018). Systematic characterization of maturation time of fluorescent proteins in living cells. *Nature Methods*, *15*, 47–51.

Balzarotti, F., et al. (2017). Nanometer resolution imaging and tracking of fluorescent molecules with minimal photon fluxes. *Science*, *355*(6325), 606–612.

Barondeau, D. P., Kassmann, C. J., Tainer, J. A., & Getzoff, E. D. (2005). Understanding GFP chromophore biosynthesis: Controlling backbone cyclization and modifying post-translational chemistry. *Biochemistry*, *44*, 1960–1970.

Barondeau, D. P., Putnam, C. D., Kassmann, C. J., Tainer, J. A., & Getzoff, E. D. (2003). Mechanism and energetics of green fluorescent protein chromophore synthesis revealed by trapped intermediate structures. *Proceedings of the National Academy of Sciences*, *100*, 12111–12116.

Barreto-Chang, O. L., & Dolmetsch, R. E. (2009). Calcium imaging of cortical neurons using fura-2 AM. *Journal of Visualized Experiments*, *23*, e1067. https://doi.org/10.3791/1067.

Berg, J., Hung, Y. P., & Yellen, G. (2009). A genetically encoded fluorescent reporter of ATP/ADP ratio. *Nature Methods*, *6*, 161–166.

Bittermann, M. R., Grzelka, M., Woutersen, S., Brouwer, A. M., & Bonn, D. (2021). Disentangling nano- and macroscopic viscosities of aqueous polymer solutions using a fluorescent molecular rotor. *Journal of Physical Chemistry Letters*, *12*, 3182–3186.

BODIPY. (2021). *Dye series—Section 1.4—NL*. www.thermofisher.com/uk/en/home/references/molecular-probes-the-handbook/fluorophores-and-their-amine-reactive-derivatives/bodipy-dye-series.html.

Boens, N., et al. (2006). Photophysics of the fluorescent pH indicator BCECF. *The Journal of Physical Chemistry. A*, *110*, 9334–9343.

Boersma, A. J., Zuhorn, I. S., & Poolman, B. (2015). A sensor for quantification of macromolecular crowding in living cells. *Nature Methods*, *12*, 227–229.

Borst, J. W., Hink, M. A., van Hoek, A., & Visser, A. J. W. G. (2005). Effects of refractive index and viscosity on fluorescence and anisotropy decays of enhanced cyan and yellow fluorescent proteins. *Journal of Fluorescence*, *15*, 153–160.

Botman, D., van Heerden, J. H., & Teusink, B. (2020). An improved ATP FRET sensor for yeast shows heterogeneity during nutrient transitions. *ACS Sensors*, *5*, 814–822.
Brakemann, T., et al. (2011). A reversibly photoswitchable GFP-like protein with fluorescence excitation decoupled from switching. *Nature Biotechnology*, *29*, 942–947.
Braut-Boucher, F., et al. (1995). A non-isotopic, highly sensitive, fluorimetric, cell-cell adhesion microplate assay using calcein AM-labeled lymphocytes. *Journal of Immunological Methods*, *178*, 41–51.
Calamera, G., et al. (2019). FRET-based cyclic GMP biosensors measure low cGMP concentrations in cardiomyocytes and neurons. *Communications Biology*, *2*, 1–12.
Campbell, T. (2001). *The Effect of pH on Green Fluorescent Protein: A Brief Review*. undefined.
Campbell, B. C., et al. (2020). mGreenLantern: A bright monomeric fluorescent protein with rapid expression and cell filling properties for neuronal imaging. *Proceedings of the National Academy of Sciences*, *117*, 30710–30721.
Carnell, M., Macmillan, A. & Whan, R. Fluorescence recovery after photobleaching (frap): Acquisition, analysis, and applications. in Methods in Membrane Lipids (ed. Owen, D. M.) 255–271 (Springer, 2015). doi:10.1007/978-1-4939-1752-5_18.
Chalfie, M., Tu, Y., Euskirchen, G., Ward, W. W., & Prasher, D. C. (1994). Green fluorescent protein as a marker for gene expression. *Science*, *263*, 802–805.
Chamberlain, C., & Hahn, K. M. (2000). Watching proteins in the wild: Fluorescence methods to study protein dynamics in living cells. *Traffic*, *1*, 755–762.
Choi, H., Yang, Z., & Weisshaar, J. C. (2015). Single-cell, real-time detection of oxidative stress induced in Escherichia coli by the antimicrobial peptide CM15. *Proceedings of the National Academy of Sciences*, *112*, E303–E310.
Chudakov, D. M., Matz, M. V., Lukyanov, S., & Lukyanov, K. A. (2010). Fluorescent proteins and their applications in imaging living cells and tissues. *Physiological Reviews*, *90*, 1103–1163.
Cole, N. B. (2013). Site-specific protein labeling with SNAP-tags. *Current Protocols in Protein Science*, *73*, 30.1.1–30.1.16.
Collot, M., Boutant, E., Fam, K. T., Danglot, L., & Klymchenko, A. S. (2020). Molecular tuning of styryl dyes leads to versatile and efficient plasma membrane probes for cell and tissue imaging. *Bioconjugate Chemistry*, *31*, 875–883.
Combs, C. A. (2010). Fluorescence microscopy: A concise guide to current imaging methods. *Current Protocols in Neuroscience*, *50*, 2.1.1–2.1.14.
Costantini, L. M., et al. (2015). A palette of fluorescent proteins optimized for diverse cellular environments. *Nature Communications*, *6*, 7670.
Cummings, R. T., et al. (2002). A peptide-based fluorescence resonance energy transfer assay for Bacillus anthracis lethal factor protease. *Proceedings of the National Academy of Sciences*, *99*, 6603–6606.
Darfeuille, F., Hansen, J. B., Orum, H., Primo, C. D., & Toulmé, J. (2004). LNA/DNA chimeric oligomers mimic RNA aptamers targeted to the TAR RNA element of HIV-1. *Nucleic Acids Research*, *32*, 3101–3107.
Datta, R., Heaster, T. M., Sharick, J. T., Gillette, A. A., & Skala, M. C. (2020). Fluorescence lifetime imaging microscopy: Fundamentals and advances in instrumentation, analysis, and applications. *Journal of Biomedical Optics*, *25*, 071203.
Davidson, M. N., et al. (2020). Measurement of the fluorescence lifetime of GFP in high refractive index levitated droplets using FLIM. *Physical Chemistry Chemical Physics*, *22*, 14704–14711.
Day, R. N., & Schaufele, F. (2008). Fluorescent protein tools for studying protein dynamics in living cells: A review. *Journal of Biomedical Optics*, *13*, 031202.
de Boer, M., et al. (2019). Conformational and dynamic plasticity in substrate-binding proteins underlies selective transport in ABC importers. *eLife*, *8*, e44652.

Deng, W., Shi, X., Tjian, R., Lionnet, T., & Singer, R. H. (2015). CASFISH: CRISPR/Cas9-mediated in situ labeling of genomic loci in fixed cells. *Proceedings of the National Academy of Sciences*, *112*, 11870–11875.

Deng, H., et al. (2021). Genetic engineering of circularly permuted yellow fluorescent protein reveals intracellular acidification in response to nitric oxide stimuli. *Redox Biology*, *41*, 101943.

Dirks, R. W., & Tanke, H. J. (2006). Advances in fluorescent tracking of nucleic acids in living cell. *BioTechniques*, *40*, 489–496.

Dorta, D. A., Deniaud, D., Mével, M., & Gouin, S. G. (2020). Tyrosine conjugation methods for protein labelling. *Chemistry - A European Journal*, *26*, 14257–14269.

Edwards, K. A. (2021). Periplasmic-binding protein-based biosensors and bioanalytical assay platforms: Advances, considerations, and strategies for optimal utility. *Talanta Open*, *3*, 100038.

Fan, Y., et al. (2015). Extremely high brightness from polymer-encapsulated quantum dots for two-photon cellular and deep-tissue imaging. *Scientific Reports*, *5*, 9908.

Filonov, G. S., Moon, J. D., Svensen, N., & Jaffrey, S. R. (2014). Broccoli: Rapid selection of an RNA mimic of green fluorescent protein by fluorescence-based selection and directed evolution. *Journal of the American Chemical Society*, *136*, 16299–16308.

Fish, K. N. (2009). Total internal reflection fluorescence (TIRF) microscopy. *Current Protocols in Cytometry/Editorial Board, J Paul Robinson, Managing Editor. [et al]*, *12*. Unit 12.18.

Förster, T. (1948). Zwischenmolekulare Energiewanderung und Fluoreszenz. *Annals of Physics*, *437*, 55–75.

Fu, Y., & Finney, N. S. (2018). Small-molecule fluorescent probes and their design. *RSC Advances*, *8*, 29051–29061.

Gabba, M., et al. (2020). Weak acid permeation in synthetic lipid vesicles and across the yeast plasma membrane. *Biophysical Journal*, *118*, 422–434.

Gauer, J. W., et al. (2016). Chapter ten—Single-molecule FRET to measure conformational dynamics of DNA mismatch repair proteins. In M. Spies, & Y. R. Chemla (Eds.), *581*. *Methods in Enzymology* (pp. 285–315). Academic Press.

Gautier, A., et al. (2008). An engineered protein tag for multiprotein labeling in living cells. *Chemistry & Biology*, *15*, 128–136.

Gnutt, D., Gao, M., Brylski, O., Heyden, M., & Ebbinghaus, S. (2015). Excluded-volume effects in living cells. *Angewandte Chemie, International Edition*, *54*, 2548–2551.

Goehring, N. W., Chowdhury, D., Hyman, A. A., & Grill, S. W. (2010). FRAP analysis of membrane-associated proteins: Lateral diffusion and membrane-cytoplasmic exchange. *Biophysical Journal*, *99*, 2443–2452.

Göpfrich, K., et al. (2019). One-pot assembly of complex giant unilamellar vesicle-based synthetic cells. *ACS Synthetic Biology*, *8*, 937–947.

Götz, C., et al. (2021). Conformational dynamics of the dengue virus protease revealed by fluorescence correlation and single-molecule FRET studies. *The Journal of Physical Chemistry B*, *125*, 6837–6846.

Gould, T. J., Verkhusha, V. V., & Hess, S. T. (2009). Imaging biological structures with fluorescence photoactivation localization microscopy. *Nature Protocols*, *4*, 291–308.

Gouridis, G., et al. (2015). Conformational dynamics in substrate-binding domains influences transport in the ABC importer GlnPQ. *Nature Structural & Molecular Biology*, *22*, 57–64.

Gray, W. D., Mitchell, A. J., & Searles, C. D. (2015). An accurate, precise method for general labeling of extracellular vesicles. *MethodsX*, *2*, 360–367.

Grynkiewicz, G., Poenie, M., & Tsien, R. Y. (1985). A new generation of Ca2+ indicators with greatly improved fluorescence properties. *The Journal of Biological Chemistry*, *260*, 3440–3450.

Guo, M., Xu, Y., & Gruebele, M. (2012). Temperature dependence of protein folding kinetics in living cells. *Proceedings of the National Academy of Sciences*, *109*, 17863–17867.

Haupts, U., Maiti, S., Schwille, P., & Webb, W. W. (1998). Dynamics of fluorescence fluctuations in green fluorescent protein observed by fluorescence correlation spectroscopy. *Proceedings of the National Academy of Sciences of the United States of America*, *95*, 13573–13578.

Hebisch, E., Knebel, J., Landsberg, J., Frey, E., & Leisner, M. (2013). High variation of fluorescence protein maturation times in closely related escherichia coli strains. *PLoS One*, *8*, e75991.

Heim, R., Prasher, D. C., & Tsien, R. Y. (1994). Wavelength mutations and posttranslational autoxidation of green fluorescent protein. *Proceedings of the National Academy of Sciences*, *91*, 12501–12504.

Henderson, J. N., et al. (2009). Structure and mechanism of the photoactivatable green fluorescent protein. *Journal of the American Chemical Society*, *131*, 4176–4177.

Hirata, R., Hirakawa, K., Shimada, N., Watanabe, K., & Ohtsuki, T. (2021). Fluorescence lifetime probes for detection of RNA degradation. *Analyst*, *146*, 277–282.

Honda, Y., & Kirimura, K. (2013). Generation of circularly permuted fluorescent-protein-based indicators for in vitro and in vivo detection of citrate. *PLoS One*, *8*, e64597.

Hu, R., et al. (2014). Multicolor fluorescent biosensor for multiplexed detection of DNA. *Analytical Chemistry*, *86*, 5009–5016.

Huang, B., Bates, M., & Zhuang, X. (2009). Super-resolution fluorescence microscopy. *Annual Review of Biochemistry*, *78*, 993–1016.

Huang, B., Wang, W., Bates, M., & Zhuang, X. (2008). Three-dimensional super-resolution imaging by stochastic optical reconstruction microscopy. *Science*, *319*, 810–813.

Hung, Y. P., Albeck, J. G., Tantama, M., & Yellen, G. (2011). Imaging cytosolic NADH-NAD+ redox state with a genetically encoded fluorescent biosensor. *Cell Metabolism*, *14*, 545–554.

Imamura, H., et al. (2009). Visualization of ATP levels inside single living cells with fluorescence resonance energy transfer-based genetically encoded indicators. *Proceedings of the National Academy of Sciences*, *106*, 15651–15656.

Isoda, R., et al. (2021). Sensors for the quantification, localization and analysis of the dynamics of plant hormones. *The Plant Journal*, *105*, 542–557.

Ivanusic, D., Eschricht, M., & Denner, J. (2014). Investigation of membrane protein–protein interactions using correlative FRET-PLA. *BioTechniques*, *57*, 188–198.

Iyer, A., et al. (2021). Chemogenetic tags with probe exchange for live-cell fluorescence microscopy. *ACS Chemical Biology*. https://doi.org/10.1021/acschembio.1c00100.

James-Kracke, M. R. (1992). Quick and accurate method to convert BCECF fluorescence to pHi: Calibration in three different types of cell preparations. *Journal of Cellular Physiology*, *151*, 596–603.

Jepsen, M. D. E., et al. (2018). Development of a genetically encodable FRET system using fluorescent RNA aptamers. *Nature Communications*, *9*.

Jonkman, J., Brown, C. M., Wright, G. D., Anderson, K. I., & North, A. J. (2020). Tutorial: Guidance for quantitative confocal microscopy. *Nature Protocols*, *15*, 1585–1611.

Kamper, M., Ta, H., Jensen, N. A., Hell, S. W., & Jakobs, S. (2018). Near-infrared STED nanoscopy with an engineered bacterial phytochrome. *Nature Communications*, *9*, 4762.

Kano, K., & Fendler, J. H. (1978). Pyranine as a sensitive pH probe for liposome interiors and surfaces. pH gradients across phospholipid vesicles. *Biochimica et Biophysica Acta (BBA) - Biomembranes*, *509*, 289–299.

Kashirina, A. S., et al. (2020). Monitoring membrane viscosity in differentiating stem cells using BODIPY-based molecular rotors and FLIM. *Scientific Reports*, *10*, 14063.

Kaufmann, T., et al. (2020). Direct measurement of protein–protein interactions by FLIM-FRET at UV laser-induced DNA damage sites in living cells. *Nucleic Acids Research*, *48*(21), e122.

Kellenberger, C. A., Chen, C., Whiteley, A. T., Portnoy, D. A., & Hammond, M. C. (2015). RNA-based fluorescent biosensors for live cell imaging of second messenger cyclic di-AMP. *Journal of the American Chemical Society*, *137*, 6432–6435.

Kellenberger, C. A., Wilson, S. C., Sales-Lee, J., & Hammond, M. C. (2013). RNA-based fluorescent biosensors for live cell imaging of second messengers cyclic di-GMP and cyclic AMP-GMP. *Journal of the American Chemical Society*, *135*, 4906–4909.

Khater, I. M., Nabi, I. R., & Hamarneh, G. (2020). A review of super-resolution single-molecule localization microscopy cluster analysis and quantification methods. *Patterns*, *1*, 100038.

Kim, Y., et al. (2008). Efficient site-specific labeling of proteins via cysteines. *Bioconjugate Chemistry*, *19*, 786–791.

Kitajima, N., et al. (2020). Real-time in vivo imaging of extracellular ATP in the brain with a hybrid-type fluorescent sensor. *eLife*, *9*, e57544.

Knop, M. & Edgar, B. A. Tracking protein turnover and degradation by microscopy: Photo-switchable versus time-encoded fluorescent proteins. Open Biology 4, 140002.

Ko, W., Kim, S., & Lee, H. S. (2017). Engineering a periplasmic binding protein for amino acid sensors with improved binding properties. *Organic & Biomolecular Chemistry*, *15*, 8761–8769.

Kollenda, S., et al. (2020). A pH-sensitive fluorescent protein sensor to follow the pathway of calcium phosphate nanoparticles into cells. *Acta Biomaterialia*, *111*, 406–417.

Kostyuk, A. I., Demidovich, A. D., Kotova, D. A., Belousov, V. V., & Bilan, D. S. (2019). Circularly permuted fluorescent protein-based indicators: History, principles, and classification. *International Journal of Molecular Sciences*, *20*, 4200.

Kuimova, M. K. (2012). Mapping viscosity in cells using molecular rotors. *Physical Chemistry Chemical Physics*, *14*, 12671–12686.

Kundu, R., Chandra, A., & Datta, A. (2021). Fluorescent chemical tools for tracking anionic phospholipids. *Israel Journal of Chemistry*, *61*, 199–216.

Ladner, C. L., Turner, R. J., & Edwards, R. A. (2007). Development of indole chemistry to label tryptophan residues in protein for determination of tryptophan surface accessibility. *Protein Science: A Publication of the Protein Society*, *16*, 1204–1213.

Larda, S. T., Pichugin, D., & Prosser, R. S. (2015). Site-specific labeling of protein lysine residues and N-terminal amino groups with indoles and indole-derivatives. *Bioconjugate Chemistry*, *26*, 2376–2383.

Laxman, P., Ansari, S., Gaus, K., & Goyette, J. (2021). The benefits of unnatural amino acid incorporation as protein labels for single molecule localization microscopy. *Frontiers in Chemistry*, *9*, 161.

Lee, S.-C., et al. (2018). Fluorescent molecular rotors for viscosity sensors. *Chemistry - A European Journal*, *24*, 13706–13718.

Lerner, E., et al. (2021). FRET-based dynamic structural biology: Challenges, perspectives and an appeal for open-science practices. *eLife*, *10*, e60416.

Lichtman, J. W., & Conchello, J.-A. (2005). Fluorescence microscopy. *Nature Methods*, *2*, 910–919.

Lippincott-Schwartz, J., & Patterson, G. H. (2009). Photoactivatable fluorescent proteins for diffraction-limited and super-resolution imaging. *Trends in Cell Biology*, *19*, 555–565.

Liu, B., Poolman, B., & Boersma, A. J. (2017). Ionic strength sensing in living cells. *ACS Chemical Biology*, *12*, 2510–2514.

Liu, B., et al. (2017). Design and properties of genetically encoded probes for sensing macromolecular crowding. *Biophysical Journal*, *112*, 1929–1939.

Liu, B., et al. (2018). Influence of fluorescent protein maturation on FRET measurements in living cells. *ACS Sensors*, *3*, 1735–1742.

Liu, X., et al. (2020). Molecular mechanism of viscosity sensitivity in BODIPY rotors and application to motion-based fluorescent sensors. *ACS Sensors*, *5*, 731–739.

Liu, A., et al. (2021). pHmScarlet is a pH-sensitive red fluorescent protein to monitor exocytosis docking and fusion steps. *Nature Communications*, *12*, 1413.

Lukyanov, K. A., Chudakov, D. M., Lukyanov, S., & Verkhusha, V. V. (2005). Photoactivatable fluorescent proteins. *Nature Reviews. Molecular Cell Biology*, *6*, 885–890.

Ma, Y., Abbate, V., & Hider, R. C. (2015). Iron-sensitive fluorescent probes: Monitoring intracellular iron pools. *Metallomics*, *7*, 212–222.

Ma, Y., Sun, Q., & Smith, S. C. (2017). The mechanism of oxidation in chromophore maturation of wild-type green fluorescent protein: A theoretical study. *Physical Chemistry Chemical Physics*, *19*, 12942–12952.

Mahon, M. J. (2011). pHluorin2: An enhanced, ratiometric, pH-sensitive green florescent protein. *Advances in Bioscience and Biotechnology*, *2*, 132–137.

Maksimov, E. G., et al. (2019). A genetically encoded fluorescent temperature sensor derived from the photoactive Orange Carotenoid Protein. *Scientific Reports*, *9*, 8937.

Mao, S.-Y., & Mullins, J. M. (2010). Conjugation of fluorochromes to antibodies. In C. Oliver, & M. C. Jamur (Eds.), *Immunocytochemical Methods and Protocols* (pp. 43–48). Humana Press. https://doi.org/10.1007/978-1-59745-324-0_6.

Margineanu, A., et al. (2016). Screening for protein-protein interactions using Förster resonance energy transfer (FRET) and fluorescence lifetime imaging microscopy (FLIM). *Scientific Reports*, *6*, 28186.

Marshall, A. (2000). Red fluorescent protein structure. *Nature Biotechnology*, *18*, 1231.

McNamara, K. P., & Rosenzweig, Z. (1998). Dye-encapsulating liposomes as fluorescence-based oxygen nanosensors. *Analytical Chemistry*, *70*, 4853–4859.

Meier, S. D., Kovalchuk, Y., & Rose, C. R. (2006). Properties of the new fluorescent Na + indicator CoroNa Green: Comparison with SBFI and confocal Na + imaging. *Journal of Neuroscience Methods*, *155*, 251–259.

Miesenböck, G., De Angelis, D. A., & Rothman, J. E. (1998). Visualizing secretion and synaptic transmission with pH-sensitive green fluorescent proteins. *Nature*, *394*, 192–195.

Mika, J. T., Krasnikov, V., van den Bogaart, G., de Haan, F., & Poolman, B. (2011). Evaluation of pulsed-FRAP and conventional-FRAP for determination of protein mobility in prokaryotic cells. *PLoS One*, *6*, e25664.

Minta, A., & Tsien, R. Y. (1989). Fluorescent indicators for cytosolic sodium*. *The Journal of Biological Chemistry*, *264*, 19449–19457.

Mirabello, V., Cortezon-Tamarit, F., & Pascu, S. I. (2018). Oxygen sensing, hypoxia tracing and in vivo imaging with functional metalloprobes for the early detection of non-communicable diseases. *Frontiers in Chemistry*, *6*, 27.

Miyawaki, A., et al. (1997). Fluorescent indicators for Ca2 + based on green fluorescent proteins and calmodulin. *Nature*, *388*, 882–887.

Mizuno, H., et al. (2003). Photo-induced peptide cleavage in the green-to-red conversion of a fluorescent protein. *Molecular Cell*, *12*, 1051–1058.

Molenaar, C., Abdulle, A., Gena, A., Tanke, H. J., & Dirks, R. W. (2004). Poly(A)+ RNAs roam the cell nucleus and pass through speckle domains in transcriptionally active and inactive cells. *The Journal of Cell Biology*, *165*, 191–202.

Molenaar, C., Wiesmeijer, K., Verwoerd, N. P., Khazen, S., Eils, R., Tanke, H. J., et al. (2003). Visualizing telomere dynamics in living mammalian cells using PNA probes. *The EMBO Journal*, *22*, 6631–6641.

Nadler, D. C., Morgan, S.-A., Flamholz, A., Kortright, K. E., & Savage, D. F. (2016). Rapid construction of metabolite biosensors using domain-insertion profiling. *Nature Communications*, *7*, 12266.

Nagai, T., Sawano, A., Park, E. S., & Miyawaki, A. (2001). Circularly permuted green fluorescent proteins engineered to sense Ca2+. *Proceedings of the National Academy of Sciences*, *98*, 3197–3202.
Naganathan, A. N., & Muñoz, V. (2005). Scaling of folding times with protein size. *Journal of the American Chemical Society*, *127*, 480–481.
Nasu, Y., Shen, Y., Kramer, L., & Campbell, R. E. (2021). Structure- and mechanism-guided design of single fluorescent protein-based biosensors. *Nature Chemical Biology*, *17*, 509–518.
Nienhaus, K., & Nienhaus, G. U. (2017). Fluorescent proteins for super-resolution microscopy. *Biophysical Journal*, *112*(453a).
Okabe, K., et al. (2012). Intracellular temperature mapping with a fluorescent polymeric thermometer and fluorescence lifetime imaging microscopy. *Nature Communications*, *3*, 705.
Okuda, M., Fourmy, D., & Yoshizawa, S. (2017). Use of baby spinach and broccoli for imaging of structured cellular RNAs. *Nucleic Acids Research*, *45*, 1404–1415.
Okumoto, S., Jones, A., & Frommer, W. B. (2012). Quantitative imaging with fluorescent biosensors. *Annual Review of Plant Biology*, *63*, 663–706.
Oliveira, E., et al. (2018). Green and red fluorescent dyes for translational applications in imaging and sensing analytes: A dual-color flag. *ChemistryOpen*, *7*, 9–52.
Otten, J., et al. (2019). A FRET-based biosensor for the quantification of glucose in culture supernatants of mL scale microbial cultivations. *Microbial Cell Factories*, *18*, 143.
Ouellet, J. (2016). RNA fluorescence with light-up aptamers. *Frontiers in Chemistry*, *4*, 29.
Ozkan, P., & Mutharasan, R. (2002). A rapid method for measuring intracellular pH using BCECF-AM. *Biochimica et Biophysica Acta*, *1572*, 143–148.
Paredes, R. M., Etzler, J. C., Watts, L. T., & Lechleiter, J. D. (2008). Chemical calcium indicators. *Methods San Diego Calif*, *46*, 143–151.
Park, S., Kang, S., & Yoon, T.-S. (2016). Crystal structure of the cyan fluorescent protein Cerulean-S175G. *Acta Crystallographica Section F: Structural Biology Communications*, *72*, 516–522.
Patel, H., Tscheka, C., & Heerklotz, H. (2009). Characterizing vesicle leakage by fluorescence lifetime measurements. *Soft Matter*, *5*, 2849–2851.
Pédelacq, J.-D., Cabantous, S., Tran, T., Terwilliger, T. C., & Waldo, G. S. (2006). Engineering and characterization of a superfolder green fluorescent protein. *Nature Biotechnology*, *24*, 79–88.
Periasamy, A., Mazumder, N., Sun, Y., Christopher, K. G., & Day, R. N. (2015). FRET microscopy: Basics, issues and advantages of FLIM-FRET imaging. In W. Becker (Ed.), *Advanced time-correlated single photon counting applications* (pp. 249–276). Springer International Publishing. https://doi.org/10.1007/978-3-319-14929-5_7.
Pettersen, E. F., et al. (2004). UCSF Chimera—A visualization system for exploratory research and analysis. *Journal of Computational Chemistry*, *25*, 1605–1612.
Phillips, D., Drake, R. C., O'Connor, D. V., & Christensen, R. L. (1985). Time correlated single-photon counting (Tcspc) using laser excitation. *Instrumentation Science and Technology*, *14*, 267–292.
Pincet, F., et al. (2016). FRAP to characterize molecular diffusion and interaction in various membrane environments. *PLoS One*, *11*, e0158457.
Plamont, M.-A., et al. (2016). Small fluorescence-activating and absorption-shifting tag for tunable protein imaging in vivo. *Proceedings of the National Academy of Sciences*, *113*, 497–502.
Pletnev, S., Subach, F. V., Dauter, Z., Wlodawer, A., & Verkhusha, V. V. (2010). Understanding blue-to-red conversion in monomeric fluorescent timers and hydrolytic degradation of their chromophores. *Journal of the American Chemical Society*, *132*, 2243–2253.

Ploetz, E. *et al.* Structural and biophysical characterization of the tandem substrate-binding domains of the ABC importer GlnPQ. Open Biology 11, 200406.

Politz, J. C., Taneja, K. L., & Singer, R. H. (1995). Characterization of hybridization between synthetic oligodeoxynuclotides and RNA in living cells. *Nucleic Acids Research*, *23*, 4946–4953.

Pols, T., et al. (2019). A synthetic metabolic network for physicochemical homeostasis. *Nature Communications*, *10*, 4239.

Pothoulakis, G., Ceroni, F., Reeve, B., & Ellis, T. (2014). The spinach RNA aptamer as a characterization tool for synthetic biology. *ACS Synthetic Biology*, *3*, 182–187.

Poudel, C., Mela, I., & Kaminski, C. F. (2020). High-throughput, multi-parametric, and correlative fluorescence lifetime imaging. *Methods and Applications in Fluorescence*, *8*, 024005.

Prasher, D. C., Eckenrode, V. K., Ward, W. W., Prendergast, F. G., & Cormier, M. J. (1992). Primary structure of the Aequorea victoria green-fluorescent protein. *Gene*, *111*, 229–233.

Qu, D., et al. (2011). 5-Ethynyl-2′-deoxycytidine as a new agent for DNA labeling: Detection of proliferating cells. *Analytical Biochemistry*, *417*, 112–121.

Reid, B. G., & Flynn, G. C. (1997). Chromophore formation in green fluorescent protein. *Biochemistry*, *36*, 6786–6791.

Remington, S. J. (2006). Fluorescent proteins: Maturation, photochemistry and photophysics. *Current Opinion in Structural Biology*, *16*, 714–721.

Renz, M. (2013). Fluorescence microscopy—A historical and technical perspective. *Cytometry. Part A*, *83*, 767–779.

Roberts, T. M., et al. (2016). Identification and Characterisation of a pH-stable GFP. *Scientific Reports*, *6*, 28166.

Rodriguez, E. A., et al. (2017). The growing and glowing toolbox of fluorescent and photoactive proteins. *Trends in Biochemical Sciences*, *42*, 111–129.

Roldán-Salgado, A., Sánchez-Barreto, C., & Gaytán, P. (2016). LanFP10-A, first functional fluorescent protein whose chromophore contains the elusive mutation G67A. *Gene*, *592*, 281–290.

Romei, M. G., & Boxer, S. G. (2019). Split green fluorescent proteins: Scope, limitations, and outlook. *Annual Review of Biophysics*, *48*, 19–44.

Rust, M. J., Bates, M., & Zhuang, X. (2006). Sub-diffraction-limit imaging by stochastic optical reconstruction microscopy (STORM). *Nature Methods*, *3*, 793–796.

Sadoine, M., Reger, M., Wong, K. M., & Frommer, W. B. (2021). Affinity series of genetically encoded Förster resonance energy-transfer sensors for sucrose. *ACS Sensors*, *6*, 1779–1784.

Salic, A., & Mitchison, T. J. (2008). A chemical method for fast and sensitive detection of DNA synthesis in vivo. *Proceedings of the National Academy of Sciences*, *105*, 2415–2420.

Saxon, E., Armstrong, J. I., & Bertozzi, C. R. (2000). A "traceless" Staudinger ligation for the chemoselective synthesis of amide bonds. *Organic Letters*, *2*, 2141–2143.

Schavemaker, P. E., Śmigiel, W. M., & Poolman, B. (2017). Ribosome surface properties may impose limits on the nature of the cytoplasmic proteome. *eLife*, *6*, e30084.

Scheepers, G. H., Nijeholt, J. A. L. A., & Poolman, B. (2016). An updated structural classification of substrate-binding proteins. *FEBS Letters*, *590*, 4393–4401.

Shaner, N. C., et al. (2013). A bright monomeric green fluorescent protein derived from Branchiostoma lanceolatum. *Nature Methods*, *10*, 407–409.

Shcherbakova, D. M., Sengupta, P., Lippincott-Schwartz, J., & Verkhusha, V. V. (2014). Photocontrollable fluorescent proteins for superresolution imaging. *Annual Review of Biophysics*, *43*, 303–329.

Shen, Y., Rosendale, M., Campbell, R. E., & Perrais, D. (2014). pHuji, a pH-sensitive red fluorescent protein for imaging of exo- and endocytosis. *The Journal of Cell Biology*, *207*, 419–432.

Shen, Y., et al. (2019). Genetically encoded fluorescent indicators for imaging intracellular potassium ion concentration. *Communications Biology*, *2*, 1–10.

Shi, J., Heegaard, C. W., Rasmussen, J. T., & Gilbert, G. E. (2004). Lactadherin binds selectively to membranes containing phosphatidyl-l-serine and increased curvature. *Biochimica et Biophysica Acta (BBA) - Biomembranes*, *1667*, 82–90.

Shimomura, O., Johnson, F. H., & Saiga, Y. (1962). Extraction, purification and properties of aequorin, a bioluminescent protein from the luminous hydromedusan, aequorea. *Journal of Cellular and Comparative Physiology*, *59*, 223–239.

Shinoda, H., et al. (2018). Acid-Tolerant Monomeric GFP from Olindias formosa. *Cell Chemical Biology*, *25*, 330–338. e7.

Shrestha, D., Jenei, A., Nagy, P., Vereb, G., & Szöllősi, J. (2015). Understanding FRET as a research tool for cellular studies. *International Journal of Molecular Sciences*, *16*, 6718–6756.

Sims, P. J., Waggoner, A. S., Wang, C.-H., & Hoffman, J. F. (1974). Mechanism by which cyanine dyes measure membrane potential in red blood cells and phosphatidylcholine vesicles. *Biochemistry*, *13*, 3315–3330.

Sniegowski, J. A., Phail, M. E., & Wachter, R. M. (2005). Maturation efficiency, trypsin sensitivity, and optical properties of Arg96, Glu222, and Gly67 variants of green fluorescent protein. *Biochemical and Biophysical Research Communications*, *332*, 657–663.

Solenov, E., Watanabe, H., Manley, G. T., & Verkman, A. S. (2004). Sevenfold-reduced osmotic water permeability in primary astrocyte cultures from AQP-4-deficient mice, measured by a fluorescence quenching method. *American Journal of Physiology. Cell Physiology*, *286*, C426–C432.

Somerville, G. A., & Proctor, R. A. (2009). At the crossroads of bacterial metabolism and virulence factor synthesis in Staphylococci. *Microbiology and Molecular Biology Reviews*, *73*, 233–248.

Specht, E. A., Braselmann, E., & Palmer, A. E. (2017). A critical and comparative review of fluorescent tools for live-cell imaging. *Annual Review of Physiology*, *79*, 93–117.

Spötl, L., Sarti, A., Dierich, M. P., & Möst, J. (1995). Cell membrane labeling with fluorescent dyes for the demonstration of cytokine-induced fusion between monocytes and tumor cells. *Cytometry*, *21*, 160–169.

Stepanenko, O. V., et al. (2011). Modern fluorescent proteins: From chromophore formation to novel intracellular applications. *BioTechniques*, *51*, 313–327.

Strack, R. (2021). Organic dyes for live imaging. *Nature Methods*, *18*, 30.

Strack, R. L., Strongin, D. E., Mets, L., Glick, B. S., & Keenan, R. J. (2010). Chromophore formation in DsRed occurs by a branched pathway. *Journal of the American Chemical Society*, *132*, 8496–8505.

Suhling, K., Davis, D. M., & Phillips, D. (2002). The influence of solvent viscosity on the fluorescence decay and time-resolved anisotropy of green fluorescent protein. *Journal of Fluorescence*, *12*, 91–95.

Suhling, K., et al. (2002). Imaging the environment of green fluorescent protein. *Biophysical Journal*, *83*, 3589–3595.

Sukhorukov, V. M., et al. (2010). Determination of protein mobility in mitochondrial membranes of living cells. *Biochimica et Biophysica Acta (BBA) - Biomembranes*, *1798*, 2022–2032.

Sun, Y., Hays, N. M., Periasamy, A., Davidson, M. W., & Day, R. N. (2012). Chapter nineteen—Monitoring protein interactions in living cells with fluorescence lifetime imaging microscopy. In P. M. Conn (Ed.), *504*. *Methods in Enzymology* (pp. 371–391). Academic Press.

Szmacinski, H., & Lakowicz, J. R. (1997). Sodium green as a potential probe for intracellular sodium imaging based on fluorescence lifetime. *Analytical Biochemistry*, *250*, 131–138.

Takaoka, Y., Ojida, A., & Hamachi, I. (2013). Protein organic chemistry and applications for labeling and engineering in live-cell systems. *Angewandte Chemie, International Edition*, *52*, 4088–4106.

Tantama, M., Hung, Y. P., & Yellen, G. (2011). Imaging intracellular pH in live cells with a genetically encoded red fluorescent protein sensor. *Journal of the American Chemical Society, 133*, 10034–10037.

Tantama, M., Martínez-François, J. R., Mongeon, R., & Yellen, G. (2013). Imaging energy status in live cells with a fluorescent biosensor of the intracellular ATP-to-ADP ratio. *Nature Communications, 4*, 2550.

Tao, R., et al. (2017). Genetically encoded fluorescent sensors reveal dynamic regulation of NADPH metabolism. *Nature Methods, 14*, 720–728.

Tebo, A. G., et al. (2018). Circularly permuted fluorogenic proteins for the design of modular biosensors. *ACS Chemical Biology, 13*, 2392–2397.

Tebo, A. G., et al. (2021). Orthogonal fluorescent chemogenetic reporters for multicolor imaging. *Nature Chemical Biology, 17*, 30–38.

Tokunaga, M., Imamoto, N., & Sakata-Sogawa, K. (2008). Highly inclined thin illumination enables clear single-molecule imaging in cells. *Nature Methods, 5*, 159–161.

Tomosugi, W., et al. (2009). An ultramarine fluorescent protein with increased photostability and pH insensitivity. *Nature Methods, 6*, 351–353.

Topell, S., Hennecke, J., & Glockshuber, R. (1999). Circularly permuted variants of the green fluorescent protein. *FEBS Letters, 457*, 283–289.

Tramier, M., & Coppey-Moisan, M. (2008). Fluorescence anisotropy imaging microscopy for homo-FRET in living cells. In *85. Methods in Cell Biology* (pp. 395–414). Academic Press.

Trauth, J., et al. (2020). Strategies to investigate protein turnover with fluorescent protein reporters in eukaryotic organisms. *AIMS Biophysics, 7*, 90–118.

Tregidgo, C. L., Levitt, J. A., & Suhling, K. (2008). Effect of refractive index on the fluorescence lifetime of green fluorescent protein. *Journal of Biomedical Optics, 13*, 031218.

Tsien, R. Y. (1998). The green fluorescent protein. *Annual Review of Biochemistry, 67*, 509–544.

Vaidyanathan, P., et al. (2021). Algorithms for the selection of fluorescent reporters. *Communications Biology, 4*, 1–8.

van Berkel, S. S., van Eldijk, M. B., & van Hest, J. C. M. (2011). Staudinger ligation as a method for bioconjugation. *Angewandte Chemie, International Edition, 50*, 8806–8827.

van den Berg, J., Boersma, A. J., & Poolman, B. (2017). Microorganisms maintain crowding homeostasis. *Nature Reviews. Microbiology, 15*, 309–318.

van der Heide, T., Stuart, M. C., & Poolman, B. (2001). On the osmotic signal and osmosensing mechanism of an ABC transport system for glycine betaine. *The EMBO Journal, 20*, 7022–7032.

van der Meer, B. W. (2002). Kappa-squared: From nuisance to new sense. *Reviews in Molecular Biotechnology, 82*, 181–196.

Várnai, P., & Balla, T. (2006). Live cell imaging of phosphoinositide dynamics with fluorescent protein domains. *Biochimica et Biophysica Acta (BBA) - Molecular and Cell Biology of Lipids, 1761*, 957–967.

Verkhusha, V. V., Chudakov, D. M., Gurskaya, N. G., Lukyanov, S., & Lukyanov, K. A. (2004). Common pathway for the red chromophore formation in fluorescent proteins and chromoproteins. *Chemistry & Biology, 11*, 845–854.

Wachter, R. M., Elsliger, M.-A., Kallio, K., Hanson, G. T., & Remington, S. J. (1998). Structural basis of spectral shifts in the yellow-emission variants of green fluorescent protein. *Structure, 6*, 1267–1277.

Wachter, R. M., Watkins, J. L., & Kim, H. (2010). Mechanistic diversity of red fluorescence acquisition by GFP-like proteins. *Biochemistry, 49*, 7417–7427.

Wakelam, M. J. O. (2014). The uses and limitations of the analysis of cellular phosphoinositides by lipidomic and imaging methodologies. *Biochimica et Biophysica Acta (BBA) - Molecular and Cell Biology of Lipids, 1841*, 1102–1107.

Wang, S., Moffitt, J. R., Dempsey, G. T., Xie, X. S., & Zhuang, X. (2014). Characterization and development of photoactivatable fluorescent proteins for single-molecule–based superresolution imaging. *Proceedings of the National Academy of Sciences*, *111*, 8452–8457.

Wang, L., et al. (2016). A multisite-binding switchable fluorescent probe for monitoring mitochondrial ATP level fluctuation in live cells. *Angewandte Chemie, International Edition*, *55*, 1773–1776.

Wang, X. C., Wilson, S. C., & Hammond, M. C. (2016). Next-generation RNA-based fluorescent biosensors enable anaerobic detection of cyclic di-GMP. *Nucleic Acids Research*, *44*(17), e139.

Ward, W. W., Prentice, H. J., Roth, A. F., Cody, C. W., & Reeves, S. C. (1982). Spectral perturbations of the aequorea green-fluorescent protein. *Photochemistry and Photobiology*, *35*, 803–808.

Wazawa, T., et al. (2021). A photoswitchable fluorescent protein for hours-time-lapse and sub-second-resolved super-resolution imaging. *Microscopy*, *70*, 340–352.

Westerhof, T. M., Li, G.-P., Bachman, M., & Nelson, E. L. (2016). Multicolor immunofluorescent imaging of complex cellular mixtures on micropallet arrays enables the identification of single cells of defined phenotype. *Advanced Healthcare Materials*, *5*, 767–771.

Xiang, L., Chen, K., Yan, R., Li, W., & Xu, K. (2020). Single-molecule displacement mapping unveils nanoscale heterogeneities in intracellular diffusivity. *Nature Methods*, *17*, 524–530.

Yaginuma, H., et al. (2014). Diversity in ATP concentrations in a single bacterial cell population revealed by quantitative single-cell imaging. *Scientific Reports*, *4*, 6522.

Yang, F., Moss, L. G., & Phillips, G. N. (1996). The molecular structure of green fluorescent protein. *Nature Biotechnology*, *14*, 1246–1251.

Zhang, G., Gurtu, V., & Kain, S. R. (1996). An enhanced green fluorescent protein allows sensitive detection of gene transfer in mammalian cells. *Biochemical and Biophysical Research Communications*, *227*, 707–711.

Zhang, J., et al. (2015). Tandem spinach array for mRNA imaging in living bacterial cells. *Scientific Reports*, *5*, 17295.

Zhao, Y., et al. (2011). Genetically encoded fluorescent sensors for intracellular NADH detection. *Cell Metabolism*, *14*, 555–566.

Zhao, Y., et al. (2015). SoNar, a highly responsive NAD+/NADH sensor, allows high-throughput metabolic screening of anti-tumor agents. *Cell Metabolism*, *21*, 777–789.

Zhou, X. X., & Lin, M. Z. (2013). Photoswitchable fluorescent proteins: Ten years of colorful chemistry and exciting applications. *Current Opinion in Chemical Biology*, *17*, 682–690.

CHAPTER TWO

Current methods for studying intracellular liquid-liquid phase separation

Amber R. Titus* **and Edgar E. Kooijman**
Department of Biological Sciences, Kent State University, Kent, OH, United States
*Corresponding author: e-mail address: atitus3@kent.edu

Contents

Abstract

Liquid-liquid phase separation (LLPS) is a ubiquitous process that drives the formation of membrane-less intracellular compartments. This compartmentalization contains vastly different protein/RNA/macromolecule concentrations compared to the surrounding cytosol despite the absence of a lipid boundary. Because of this, LLPS is important for many cellular signaling processes and may play a role in their dysregulation. This chapter highlights recent advances in the understanding of intracellular phase transitions along with current methods used to identify LLPS *in vitro* and model LLPS *in situ*.

1. Characteristics of liquid-liquid phase separation

Intracellular components have liquid-like properties, which are driven by interfacial tension, the mechanical tension at the boundary of two phases, and viscoelasticity (Asherie, Lomakin, & Benedek, 1996; Hyman & Simons, 2012; Hyman, Weber, & Julicher, 2014; Shin & Brangwynne, 2017). Interfaces, or phase boundaries, occur if there are two different compositions of molecular species in both phases. Commonly, intracellular cytosol has

Current Topics in Membranes, Volume 88
ISSN 1063-5823
https://doi.org/10.1016/bs.ctm.2021.09.003

been treated as a well-mixed homogenous environment, but this is not necessarily true. Recently, aqueous-aqueous separation has been observed in cell cytoplasm and nucleoplasm resulting in the compartmentalization of biological macromolecules (Banani et al., 2016; Brangwynne et al., 2009; Ditlev, Case, & Rosen, 2018; Hyman et al., 2014; Shin & Brangwynne, 2017; Weber & Brangwynne, 2012). The process of intracellular aqueous phase separation is referred to as liquid-liquid phase separation (LLPS). LLPS is a ubiquitous process, driven by macromolecular interactions, which results in the formation of biomolecular condensates, referred to as membrane-less organelles (Banani et al., 2017; Boeynaems et al., 2018; Shin & Brangwynne, 2017; Wheeler & Hyman, 2018). Because membrane-less organelles lack lipid membranes, their formation and stability depend on specific physicochemical properties of their components (Stroberg & Schnell, 2017).

When intracellular macromolecules, like proteins or nucleic acids, undergo LLPS, they condense into liquid droplets where the concentration of the macromolecules is higher inside the liquid droplets compared to those in the surrounding cytosol. LLPS is unique because typically entropy will drive a system to be well-mixed and disordered (Hyman et al., 2014; Shin & Brangwynne, 2017). However, we also know of liquids that can demix. For example, when oil and water are mixed and allowed to sit for an extended period of time, they will demix into two different phases: an oil phase and an aqueous phase. The physical interactions between the oil and water molecules are what drives the separation of two phases. When oil molecules neighbor other hydrophobic molecules, this system has a lower total energy compared to a system where oil molecules neighbor water molecules. This energy reduction by demixing is what opposes entropy-driven mixing and favors phase separation (Hyman et al., 2014).

This specific type of demixing can also be seen in biological systems containing lipid droplets. Lipid droplets contain free fatty acids (oil) surrounded by a lipid monolayer in the aqueous cell cytosol. The lipid monolayer, along with binding proteins, that surround lipid droplets is what helps stabilize the oil drop in the aqueous cell environment due to the amphipathic nature of these molecules. Because of the absence of a lipid phase boundary, LLPS relies solely on the phase separation between two aqueous environments within the cell. LLPS is also unique as its phase separation does not primarily rely on the difference in hydrophobicity of the two resulting phases, as in the oil-water example. LLPS is one of the fundamental mechanisms for organizing intracellular space and although the full mechanism behind LLPS remains unknown, some of the driving forces have been identified. One major driving force is the interactions between the biological macromolecules and water (Alberti, Gladfelter, & Mittag, 2019; Titus et al., 2020). Another driving force for phase separation is weak, transient interactions between molecules with multivalent domains (Shin & Brangwynne, 2017). Phase transitions are familiar in systems made up of water and oil but have also been seen in water itself with and without additives (Shin & Brangwynne, 2017; Titus et al., 2020). Some parameters of LLPS phase transitions have also been suggested to fall into two categories: Non-covalent interactions, like those referred to above, and global free energy which is governed by environmental factors like temperature and pH (Bentley, Frey, & Deniz, 2019; Berry, Brangwynne, & Haataja, 2018; Falahati & Haji-Akbari, 2019). It has been established that pH controls intracellular signaling in yeast, and more recently on the metabolic regulation of the trans-Golgi network, and may thus be a driving force for intracellular LLPS formation (Shin et al., 2020; Young et al., 2010). Macromolecular crowding also appears to favor the initiation of LLPS in cells (André & Spruijt, 2020). By nature, the intracellular environment is highly crowded with the cytosol containing a complex mixture of macromolecules such as proteins, nucleic acids, and sugars. This dense crowding has been shown to determine the stability of biomolecular mixtures and promote the formation of LLPS by enhancing intermolecular interactions (André & Spruijt, 2020). By gaining a better understanding of the physical chemistry and biophysics behind LLPS formation, we can better mimic the biological processes that rely on LLPS.

2. Liquid-liquid phase separation in biology

LLPS occurs when a complex set of proteins and RNA segregate from the cytoplasm due to a higher affinity with each other compared to other cytoplasmic molecules and aqueous surroundings. LLPS is regulated by active cell signaling processes since the system needs attractive interactions between molecules, like proteins, DNA, and RNA, to drive phase separation (e.g., transcription, and posttranslational modifications) (Hnisz et al., 2017; Shin & Brangwynne, 2017). LLPS drives the formation of multiple membrane-less organelles, such as the nucleolus (Pederson, 2011) and ribonucleoprotein bodies/granules (e.g., nuclear bodies, Cajal bodies, and P granules) (Brangwynne et al., 2009; Gall, 2003; Handwerger & Gall, 2006; Kedersha & Anderson, 2007; Mao, Zhang, & Spector, 2011). Recent studies have shown liquid-like properties in the nucleolus, where nucleoli are associated with nucleolar organizer regions of chromosomal DNA and show fusion, flowing, dissociation, and regrowth during various stages of the cell cycle (Bentley et al., 2019; Brangwynne, Mitchison, & Hyman, 2011; Mitrea et al., 2018). Weber et al. discussed in detail evidence for the five nuclear components: nucleolus, heterochromatin, paraspeckles, transcriptional condensates, and the replication compartment (Weber, 2019). In order to distinguish LLPS from other organizational processes in the nucleus, for these components to participate in LLPS they must: maintain a spherical shape, fuse after touching, and contain mobile molecules that undergo internal rearrangement and external exchange (Weber, 2019). LLPS has been suggested as a mechanism for chromatin organization, however, studies using purified human heterochromatin protein 1α (HP1α) show mixed results (Larson et al., 2017; Shakya et al., 2020; Strom et al., 2017). Strom et al. showed, using fluorescence correlation spectroscopy, that HP1α diffuses slowly across the heterochromatin-euchromatin border, suggesting the presence of a phase boundary (Strom et al., 2017). The H1 (linker histone) may also play an important role in the higher-order structuring of nucleosome core particles as it binds directly to HP1α. Shakya et al. showed using GFP-tagged H1 in Hela cells that H1 condensations colocalized with HP1α and heterochromatin domains, showing liquid-like behavior (Shakya et al., 2020). Skin barrier function, enzyme regulation, and autophagy have also been suggested to rely on LLPS for their formation and function (Alberti, 2017; Nakashima, Vibhute, & Spruijt, 2019; Noda, Wang, & Zhang, 2020; O'Flynn & Mittag, 2021; Prouteau & Loewith, 2018; Quiroz et al., 2020; Shin & Brangwynne, 2017). Because of the ubiquitous nature of LLPS, it is critical to gain a better understanding of how it originates and the macromolecules that affect its formation and function.

An active component of intracellular LLPS systems is an intrinsically disordered protein/region (IDP/IDR). Typically, structured proteins have several folded protein–protein interaction domains (Alberti & Dormann, 2019). IDPs/IDRs are prion-like and do not have a fixed globular structure due to the absence of aliphatic and aromatic residues (Alberti & Dormann, 2019; Wang et al., 2018). Because of this, IDPs/IDRs exhibit conformational heterogeneity and are able to rearrange their conformation more easily (Berry et al., 2018). IDPs/IDRs are commonly described to follow laws of polymer chemistry where IDPs/IDRs contain attractive interactions (i.e., stickers) or areas that promote conformational flexibility (i.e., linkers/spacers) (Alberti & Dormann, 2019). The primary sequence of IDPs/IDRs is composed of low complexity regions and is likely to determine phase behavior. These sequences are typically enriched in polar amino acid residues, such as serine, tyrosine, glutamine, or asparagine, but the interactions between tyrosine and arginine residues are usually what result in phase-separation behavior (Alberti et al., 2009). These tyrosine and arginine residues are the stickers that determine the saturation concentration of a phase-separating protein. These stickers are connected by spacer residues that determine the material properties of condensates, presumably by introducing conformational constraints on the polypeptide chain (Alberti et al., 2019, 2009).

Many neurodegenerative diseases are caused/characterized by protein aggregations (see Fig. 1A). Progressive loss of neurons and synapses in distinct brain regions have been associated with aggregation of either cytosolic or nuclear proteins (Jucker & Walker, 2018; Taylor, Hardy, & Fischbeck, 2002). These protein associations may form *via* LLPS but in a diseased state,

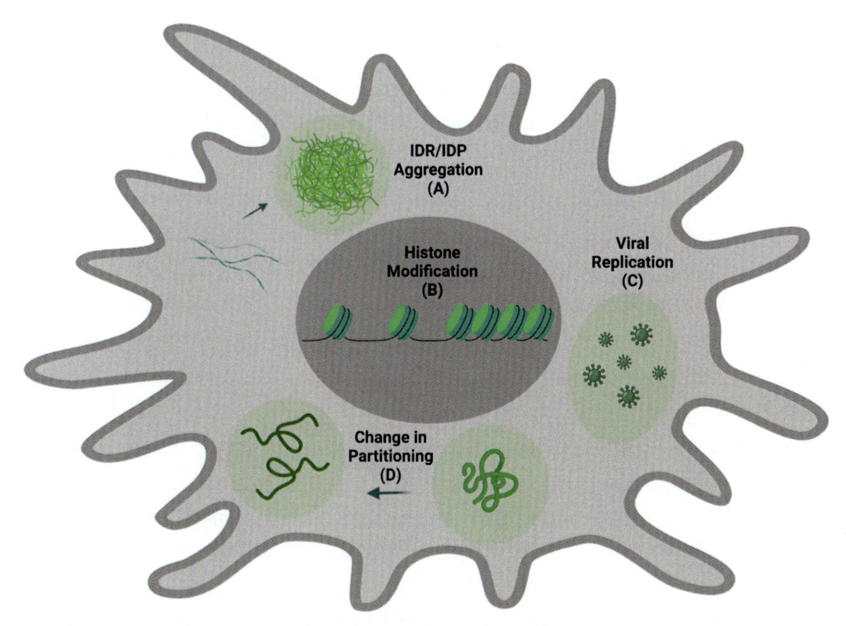

Fig. 1 Diagram inspired by Alberti et al. describing the evidence for LLPS in neurodegenerative diseases, cancer, infectious disease, and histone modification (Alberti & Dormann, 2019). In neurodegenerative disease states (A), abnormal posttranslational modifications and quality control in IDRs/IDPs can promote their condensates to become solid-like rather than liquid-like leading to protein aggregates. LLPS appears to play a role in the reorganization of euchromatin to heterochromatin (B) and in the formation of viral replication centers in host cells (C). In cancer diseased states, mutated proteins phase separate differently in LLPS (D). Figure created using Biorender.com.

these aggregations can take on more solid-like properties which change their function in neurons (Hyman & Brangwynne, 2011; Hyman et al., 2014; Li et al., 2013; Malinovska, Kroschwald, & Alberti, 2013; Shin & Brangwynne, 2017; Shulman, De Jager, & Feany, 2011; Weber & Brangwynne, 2012). For example, Alzheimer's disease is characterized by toxic aggregates of amyloid and tau. Tau is an IDP that lacks a stable, well-defined structure. In healthy neurons, tau plays important physiological functions, where it promotes microtubule assembly and stability (Cohen et al., 2011). In a diseased state, tau forms condensates which subsequently harden and aggregate, which leads to impaired tau-microtubule interactions (Alberti & Dormann, 2019; Ávila et al., 2002; Cohen et al., 2011; Dong et al., 2021). Lin et al. found that tau undergoes LLPS at high salt concentrations using fluorescence lifetime imaging measurements, suggesting that LLPS may play a role in the aggregation of tau in the onset of Alzheimer's disease (Lin et al., 2021). Amyotrophic lateral sclerosis (ALS) is another neurodegenerative disease characterized by the aggregation of proteins which affect motor neuron function (Blokhuis et al., 2013). These protein aggregates have been shown to accumulate in stress granules, whose formation is driven by LLPS, and which are precursors to pathological RNA-binding protein aggregates (Alberti & Dormann, 2019; Ambadipudi et al., 2017; Bosco et al., 2010; Dewey et al., 2011; Dormann et al., 2010; Kim et al., 2013). The cytoplasmic aggregation of α-synuclein in Parkinson's disease has also been suggested to be driven by LLPS (Ray et al., 2020).

Not only are neurodegenerative diseases caused/associated with LLPS, but one study investigated the link between protein phase separation and cancer, Fig. 1D (Bouchard et al., 2018). Cancer mutations were recently suggested to have "prion-like" behavior with mutated/misfolded proteins containing IDRs or fully behaving as IDPs (Costa et al., 2016; Darling, Zaslavsky, & Uversky, 2019; Iakoucheva et al., 2002). Bouchard et al. found that cancer mutations in the tumor suppressor, speckle-type BTB/POZ protein (SPOP), were linked to specific phase separation defects (Bouchard et al., 2018). Under healthy conditions, SPOP is associated with membrane-less nuclear bodies but when mutated, SPOP is unable to self-assemble (*via* LLPS) into higher-order oligomers which leads to the development of breast, prostate, and other cancerous tumors (Bouchard et al., 2018; Marzahn et al., 2016). Infectious diseases have also been suggested to operate under a LLPS framework, Fig. 1C (Alberti & Dormann, 2019; McSwiggen et al., 2019). Viruses like rabies, herpes simplex virus 1, and vesicular stomatitis virus, induce the formation of compartments (viral replication centers) when they infect host cells (McSwiggen et al., 2019; Netherton & Wileman, 2011; Novoa et al., 2005; Schmid et al., 2014). These viral replication centers are subcellular compartments that direct the replication of the viral genome in host cells (Knipe & Cliffe, 2008). Recently, SARS-CoV-2 has also been suggested to operate under a LLPS framework when invading and hijacking host cells (Chen et al., 2020; Noda, Wang, & Zhang, 2020; Perdikari et al., 2020). Fig. 1 summarizes the current evidence for LLPS in diseased cells, like neurodegenerative diseases, cancer, infectious disease, and histone modification.

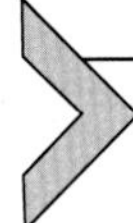

3. *In vivo* and *in vitro* methods of liquid-liquid phase separation detection

In vitro LLPS measurements are often carried out using non-natural systems and conditions, and while these measurements do not directly describe cellular processes, they can provide insights into the biophysical aspects of LLPS. Aqueous salt, protein, and/or RNA solutions provide a less complex environment that can help pinpoint demixing behavior and the environmental and molecular factors behind them. Because LLPS in cells contain complex mixtures of proteins, nucleic acids, and metabolites, aqueous two-phase systems (ATPS) have been proposed as a simpler *in vitro* model solution (Zaslavsky, 1994). ATPS are formed in water when two additives, typically polymers, exceed a certain threshold and two phases emerge. ATPS formed by two non-ionic polymers and salts have been used to separate and analyze multiple biological macromolecules (da Silva et al., 2015, 2019; Ferreira et al., 2015, 2017; Titus et al., 2020). The emergence of interfacial tension is a necessary condition for LLPS to occur in cells, and values of interfacial tension have been reported to be similar between ATPS and isolated membrane-less organelles (Feric et al., 2016; Jawerth et al., 2018). Titus et al. established that the interfacial tension values of various ATPS can be described by the differences between the solvent features and coexisting phases of ATPS, with the most important interactions being ion-ion, dipole-dipole, dipole-induced-dipole, ion-dipole, and hydrogen bonding (Titus et al., 2020).

Due to the importance of interfacial tension in LLPS formation and function, interfacial tension measurements are critical for both model systems, like ATPS, and isolated membrane-less organelles. One option for *in vitro* interfacial tension measurements is pendant drop tensiometry, see Fig. 2 (Berry et al., 2015). Pendant drop tensiometry uses dimensional drop shape

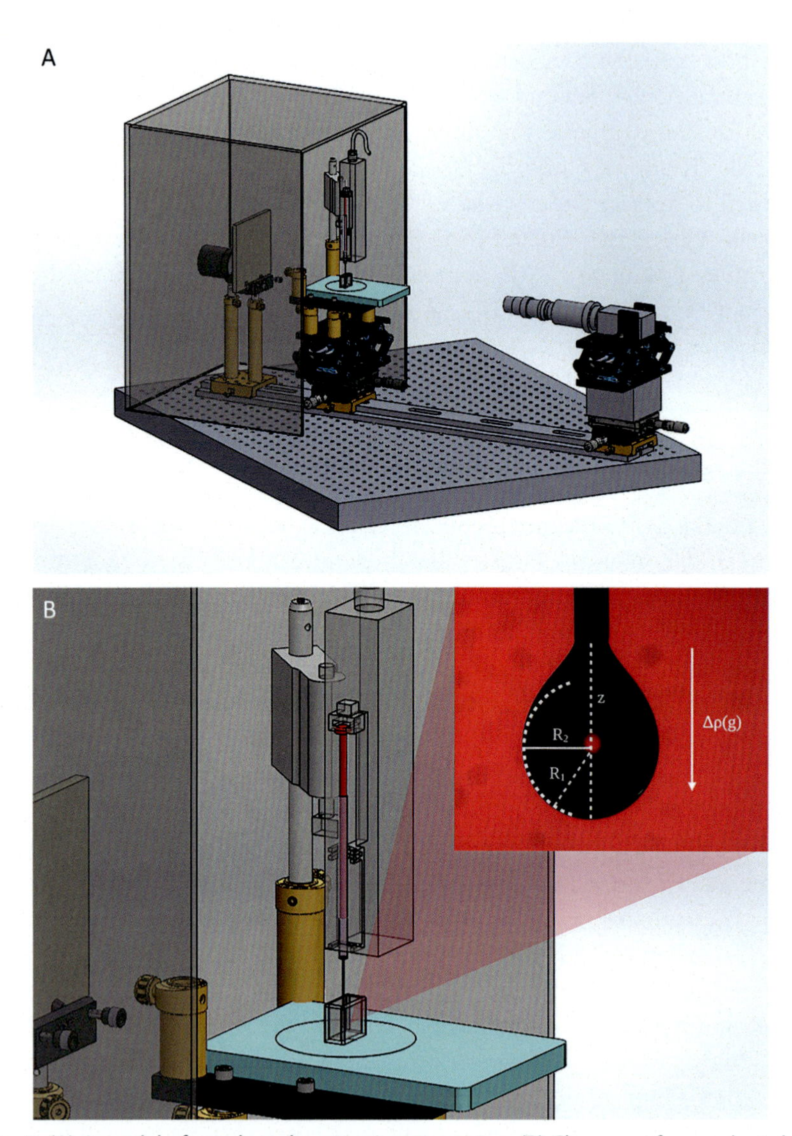

Fig. 2 (A) A model of pendant drop tensiometry setup. (B) Close-up of a pendant drop tensiometry setup which contains a syringe attached to a motorized pump along with a droplet image (inset) indicating the two principle radii at one point on the droplet. R_1 is the radius of the best-fit circle in the image plane and R_2 is the radius of the circular horizontal droplet cross-section along the axis of symmetry, z. *Pane (A) reprinted from Titus, A.R., et al., Interfacial tension and mechanism of liquid-liquid phase separation in aqueous media. Phys Chem Chem Phys, 2020. 22(8): p. 4574–4580.*

analysis to determine surface and interfacial tensions along with contact angle (when the droplet sits on a flat surface). The shape of the pendant drop is reliant on the balance between gravity and the interfacial tension of the system. The interfacial tension will promote a more spherical drop while gravity favors elongation. By analyzing the silhouette of the drop using Axisymmetric Drop Shape Analysis (ADSA) we can gain accurate measurements of interfacial tension by applying an optimized fit to the silhouette *via* the Young-Laplace equation of capillarity (Saad & Neumann, 2016; Titus et al., 2020, 2021):

$$\Delta P = \gamma\left(\frac{1}{R_1} + \frac{1}{R_2}\right) = (\Delta\rho)gz + \Delta P_{0.}$$

In this equation, ΔP refers to the Laplace, or capillary, pressure across the surface of the drop at any point; γ represents the droplet interfacial tension; $\frac{1}{R_1}$ and $\frac{1}{R_2}$ are the principal radii of curvature at a specific point on the drop, x; $\Delta\rho$ is the density difference between the drop solution and surrounding medium; g is the gravitational acceleration; z is the distance along the axis of symmetry between the point, x, and a reference point where the pressure difference is ΔP_0. Fig. 2B provides a representative description of this equation. Rather than measuring interfacial tension directly, pendant drop tensiometry measures the capillary length, λ_c:

$$\lambda_c = \left(\frac{\gamma}{g\Delta\rho}\right)^{1/2}$$

Because of this, the uncertainty of pendant drop tensiometry relies on the density measurements of the solutions/systems. This uncertainty can be minimized with a tightly controlled temperature regulation system thereby minimizing any fluctuations in solution density. Pendant drop tensiometry has multiple advantages over other interfacial tension measurement systems as it only requires small amounts of sample and can measure both static and dynamic interfacial tensions (Titus et al., 2020, 2021; Yang, Yu, & Zuo, 2017). In ATPS, the values of the interfacial tension are very low (<1 mN/m) which limits the size of the drop formed and increases the frequency of snap off (where the drop separates from the needle) (Cohen et al., 1999). This makes dynamic measurements difficult, so we recommend taking static measurements of multiple drops (volumes all within 0.5% of each other) as conducted in Titus et al. (2020). Taking measurements of a fully formed static drop over a short time period (~5–10 min should be sufficient) can also allow drops to equilibrate to ensure there are no artifacts that may alter interfacial tension values. In order to obtain accurate interfacial tension measurements using pendant drop tensiometry, we recommend calculating a dimensionless number to quantify the deformation of drops as ADSA can only be applied to well-deformed droplets. There are multiple dimensionless numbers that can be used for this, such as Bond numbers, Worthington numbers, and Neumann numbers (Berg, 2010; Berry et al., 2015; Hua & Lou, 2007; Yang et al., 2017). Titus et al. has used Neumann numbers to quantify droplets formed using ATPS and found that even with small volumes, the Neumann number of these droplets is of the order of 1, which is expected to give good accuracy when using ADSA (Titus et al., 2020; Yang et al., 2017).

Various groups have measured droplet formation and turbidity in systems containing salt, polymers and water (Berry et al., 2018; Wang, Zhang, & Zhang, 2019; Wang et al., 2018). Polymers like Ficoll, dextran, and PEG are most commonly used in ATPS formation due to their well-established biophysical and biochemical properties, however, it has been suggested that polymers like these may not mimic cell cytosol very well (André & Spruijt, 2020). Along with ATPS, multiple groups have devised their own phase forming assays to measure the effects of particular macromolecules suggested to play a part in intracellular LLPS (Alberti et al., 2018). Once these systems are formed, multiple well-established techniques can be used to measure the emergence and stability of phase separation in *in vitro* assays. Most commonly used are microscopy techniques, such as fluorescence recovery after photobleaching (FRAP), electron microscopy (EM), and atomic force microscopy (AFM) (Babinchak & Surewicz, 2020a; Le Ferrand et al., 2019; O'Brien et al., 2015; Zhang et al., 2005). Improved fluorescence imaging, single-molecule, cryo-EM, NMR, and mass-spectrometry tools in conjunction with biochemical and biophysical techniques can provide information from small to large scales (Bentley et al., 2019). Microscopy techniques tend to be paired with theoretical procedures, for example, with liquid-phase EM and self-consistent mean field theory (Ianiro et al., 2019). It is also common that microscopy techniques are paired with assays measuring turbidity or droplet coalescence/ripening (Babinchak & Surewicz, 2020a). These *in vitro* assays tend to use more complex (i.e., taking into account stereochemistry) biopolymers rather than the more well-studied but simpler polymers used in ATPS studies (Perry et al., 2015). FRAP has been used to show that several nucleolar subcompartment proteins are able to phase separate *in vitro*, but its most common use is for measuring droplet maturation (Babinchak & Surewicz, 2020b;

Feric et al., 2016; Shin & Brangwynne, 2017; Wang et al., 2018). Droplet maturation is typically considered the final stage of LLPS and droplets formed *via* aggregation-prone IDRs/IDPs tend to have a quicker maturation rate compared to globular proteins (Babinchak & Surewicz, 2020b; Lin et al., 2015; Lu & Weitz, 2013; Peskett et al., 2018). Fluorescence microscopy is also used to measure the formation of LLPS through time in the presence of a fluorescently tagged IDP/IDR. Kanaan et al. observed that fluorescently-labeled tau forms phase-separated liquid droplets *in vitro* when exposed to prolonged, crowded conditions (Kanaan et al., 2020). Förster resonant energy transfer (FRET) is an advanced technique that utilizes the energy transfer of one fluorescent donor to an acceptor species and can be carried out intracellularly. This technique works when two proteins, labeled with fluorophores, are in close proximity allowing for distance-mediated fluorescence emission. Single-molecule FRET can be useful for shedding light on protein conformational changes during phase separation (Bentley et al., 2019; Mitrea et al., 2016). Beyond fluorescence microscopy, X-ray chromatography and cryo-EM have also been used to measure LLPS (Girelli et al., 2021). Because these are static experiments, they cannot provide information about dynamic IDPs/IDRs. Murthy et al. showed that solution-state nuclear magnetic resonance (NMR) can be used to provide information on the structure and motion inside liquid-like assemblies, thus potentially making this an interesting technique to study IDRs/IDPs (Murthy & Fawzi, 2020).

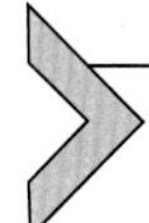

4. Computational methods for liquid-liquid phase separation prediction and modeling

In general, for all computational simulation techniques (see Fig. 3) there is a tradeoff between model detail and simulation efficiency, where there is better accuracy in smaller (i.e., subatomic-scale) systems vs larger (i.e., molecular scale) systems. Quantum mechanics (QM) is the most accurate computational technique but it is limited to short peptide sequences or approximately several hundred atoms (Dignon, Zheng, & Mittal, 2019). Because of this, QM is not commonly used for LLPS modeling. However, if the level of complexity in a system is reduced to just classical mechanics, all-atom simulations can still be a very accurate technique due to its ability to explicitly represent solvent molecules. All-atom representations are very commonly used in biomolecular systems because they can provide a high level of detail—like characteristics of protein sequences and inter-residue interactions in agreement with experimental measures. This technique

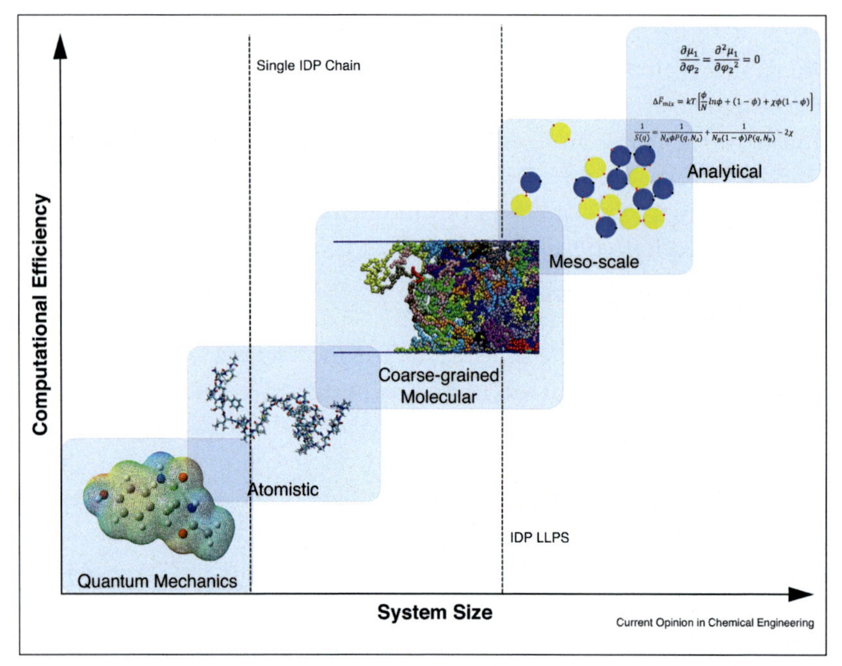

Fig. 3 Reprinted from Dignon et al. describing computational models from high to low resolution (Dignon et al., 2019). There is a tradeoff between detail and model size where LLPS systems fit best using atomistic and CG models.

can provide the most accurate representation applicable to IDRs/IDPs and has been used to study the conformational properties of tau, but it is still limited to smaller systems (Dong et al., 2021). Coarse grained (CG) models implicitly modify protein–protein interactions to account for protein–solvent interactions and can be system specific. Residue-level CG models are a promising technique for LLPS as they are able to represent specific amino acid sequences and allow users to observe phase coexistence (Dignon et al., 2019). Because of this CG models have been applied to IDP phase separation and assembly (Ando et al., 2014; Dignon et al., 2018; Ghavami, van der Giessen, & Onck, 2013; Roberts et al., 2018; Samanta, Chakraborty, & Thirumalai, 2018).

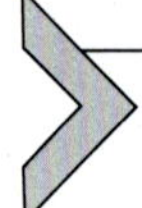

5. Databases on liquid-liquid phase separation and intrinsically disordered proteins

Due to increased interest in LLPS formation multiple databases of IDRs/IDPs, and other LLPS-related proteins have been formed. The most pertinent databases containing LLPS-related proteins are PhaSepDB, LLPSDB, PhaSePro, and DrLLPS (Farahi et al., 2021; Li et al., 2020a,

2020b; Mészáros et al., 2020; Ning et al., 2020; You et al., 2020). Forty six proteins appear across all databases, which are commonly referred to as the core LLPS dataset. There are benefits/drawbacks to any database, a few will be outlined here. PhaSepDB (http://db.phasep.pro/) is a LLPS-dedicated database that sorts proteins according to the specific intracellular membrane-less compartment that they are associated with (Farahi et al., 2021; Li et al., 2020b; Pancsa, Vranken, & Mészáros, 2021; You et al., 2020). This database is relatively large, storing 6981 membrane-less organelle-related proteins as of August 2021 and is useful to anyone who wants to know if a particular protein has been associated with membrane-less organelles, however PhaSepDB does not distinguish between driver and regulators of LLPS and therefore the formation and regulation of intracellular membrane-less organelles (Farahi et al., 2021; Pancsa et al., 2021). LLPSDB (http://bio-comp.org.cn/llpsdb/) is another LLPS-dedicated database but unlike PhaSepDB, it contains collections of *in vitro* experiments rather than proteins (Li et al., 2020b; Pancsa et al., 2021). LLPSDB stores 1175 *in vitro* experiments as of August 2021. Although the listings are experiments, LLPSDB does contain information on the conformation of protein constructs along with what nucleic acids were used in experiments. LLPSDB contains systems where the condensates have been observed to show liquid-like properties (i.e., flow, fuse, drop, wet and reverse), or in which liquid morphology was identified by FRAP, EM and other techniques (Li et al., 2020b). LLPSDB is a good source for anyone who is interested in the biophysics surrounding LLPS and LLPS-related proteins; however, there tends to be little information regarding the biological context of each experiment or membrane-less organelles. DrLLPS (http://llps.biocuckoo.cn/) is the largest database listed here, with 437,887 known and predicted LLPS-related proteins as of August 2021. This database also distinguishes LLPS-related proteins into scaffolds, regulators, and clients and contains both high-throughput and low-throughput experiments which address: The physical/functional association of proteins with membrane-less organelles, phenotypic effects of their knockout, silencing, or overexpression on membrane-less organelles, and those used in dedicated LLPS experiments. (Farahi et al., 2021; Ning et al., 2020). The smallest of the databases listed here is PhaSePro (https://phasepro.elte.hu/) which contains 121 LLPS-related proteins verified to drive phase separation as of August 2021 (Mészáros et al., 2020). Although this database is significantly smaller than the other three, the curation criteria is more stringent as it needs to consider the physiological relevance and the conditions reported for experimental data in order to categorize a protein as a LLPS driver.

6. Summary

LLPS is a ubiquitous cell-organization mechanism driven by the liquid properties of intracellular components and macromolecular interactions. Although LLPS regulates multiple cellular functions, the specific mechanism and driving factors behind its formation are still poorly understood. Simplified LLPS model systems, like ATPS provide an opportunity to identify the physicochemical and biophysical factors behind its formation. ATPS paired with theoretical models are useful for the identification of the physicochemical properties behind phase separation and phase partitioning of nucleic acids and proteins, but these model systems may not mimic biological (highly complex) systems very well. Other *in vitro* assays that use more complex biopolymers and additives paired with microscopy can help in putting together a story about specific cell signaling pathways but also do not tell the full story due to the limitations of each technique. Computational techniques can help fill in some gaps between theory, *in vitro*, and *in vivo* data but even these techniques have their own limitations. Currently, probing LLPS *in vivo* is limited to fluorescence microscopy in live cells. The continued refinement of present-day techniques and the development of new methods will continue to be important for both *in vitro* and *in vivo* experiments probing LLPS.

Acknowledgments

We gratefully acknowledge Dr. Elizabeth Mann for critical discussion on pendant drop tensiometry and Dr. Boris Zaslavsky for providing us with ATPS samples and giving insight on these systems as a model for LLPS. This work was supported by the National Science Foundation under Grant No. CHE-1808281.

References

Alberti, S. (2017). The wisdom of crowds: Regulating cell function through condensed states of living matter. *Journal of Cell Science*, *130*(17), 2789–2796.

Alberti, S., & Dormann, D. (2019). Liquid–liquid phase separation in disease. *Annual Review of Genetics*, *53*, 171–194.

Alberti, S., Gladfelter, A., & Mittag, T. (2019). Considerations and challenges in studying liquid-liquid phase separation and biomolecular condensates. *Cell*, *176*(3), 419–434.

Alberti, S., et al. (2009). A systematic survey identifies prions and illuminates sequence features of prionogenic proteins. *Cell*, *137*(1), 146–158.

Alberti, S., et al. (2018). A User's guide for phase separation assays with purified proteins. *Journal of Molecular Biology*, *430*(23), 4806–4820.

Ambadipudi, S., et al. (2017). Liquid-liquid phase separation of the microtubule-binding repeats of the Alzheimer-related protein tau. *Nature Communications*, *8*(1), 275.

Ando, D., et al. (2014). Nuclear pore complex protein sequences determine overall copolymer brush structure and function. *Biophysical Journal, 106*(9), 1997–2007.
André, A. A., & Spruijt, E. (2020). Liquid–liquid phase separation in crowded environments. *International Journal of Molecular Sciences, 21*(16), 5908.
Asherie, N., Lomakin, A., & Benedek, G. B. (1996). Phase diagram of colloidal solutions. *Physical Review Letters*, 77(23), 4832–4835.
Ávila, J., et al. (2002). Tau function and dysfunction in neurons. *Molecular Neurobiology, 25*(3), 213–231.
Babinchak, W. M., & Surewicz, W. K. (2020a). Studying protein aggregation in the context of liquid-liquid phase separation using fluorescence and atomic force microscopy, fluorescence and turbidity assays, and FRAP. *Bio-Protocol, 10*(2), e3489. https://doi.org/10.21769/BioProtoc.3489.
Babinchak, W. M., & Surewicz, W. K. (2020b). Liquid–liquid phase separation and its mechanistic role in pathological protein aggregation. *Journal of Molecular Biology, 432*(7), 1910–1925.
Banani, S. F., et al. (2016). Compositional control of phase-separated cellular bodies. *Cell, 166*(3), 651–663.
Banani, S. F., et al. (2017). Biomolecular condensates: Organizers of cellular biochemistry. *Nature Reviews. Molecular Cell Biology, 18*(5), 285–298.
Bentley, E. P., Frey, B. B., & Deniz, A. A. (2019). Physical chemistry of cellular liquid-phase separation. *Chemistry–A European Journal, 25*(22), 5600–5610.
Berg, J. C. (2010). *An introduction to interfaces & colloids: The bridge to nanoscience*. World Scientific.
Berry, J., Brangwynne, C. P., & Haataja, M. (2018). Physical principles of intracellular organization via active and passive phase transitions. *Reports on Progress in Physics, 81*(4), 046601.
Berry, J. D., et al. (2015). Measurement of surface and interfacial tension using pendant drop tensiometry. *Journal of Colloid and Interface Science, 454*, 226–237.
Blokhuis, A. M., et al. (2013). Protein aggregation in amyotrophic lateral sclerosis. *Acta Neuropathologica, 125*(6), 777–794.
Boeynaems, S., et al. (2018). Protein phase separation: A new phase in cell biology. *Trends in Cell Biology, 28*(6), 420–435.
Bosco, D. A., et al. (2010). Mutant FUS proteins that cause amyotrophic lateral sclerosis incorporate into stress granules. *Human Molecular Genetics, 19*(21), 4160–4175.
Bouchard, J. J., et al. (2018). Cancer mutations of the tumor suppressor SPOP disrupt the formation of active, phase-separated compartments. *Molecular Cell, 72*(1), 19–36 (e8).
Brangwynne, C. P., Mitchison, T. J., & Hyman, A. A. (2011). Active liquid-like behavior of nucleoli determines their size and shape in Xenopus laevis oocytes. *Proceedings of the National Academy of Sciences of the United States of America, 108*(11), 4334–4339.
Brangwynne, C. P., et al. (2009). Germline P granules are liquid droplets that localize by controlled dissolution/condensation. *Science, 324*(5935), 1729–1732.
Chen, H., et al. (2020). Liquid–liquid phase separation by SARS-CoV-2 nucleocapsid protein and RNA. *Cell Research, 30*(12), 1143–1145.
Cohen, I., et al. (1999). Two fluid drop snap-off problem: Experiments and theory. *Physical Review Letters, 83*(6), 1147–1150.
Cohen, T. J., et al. (2011). The acetylation of tau inhibits its function and promotes pathological tau aggregation. *Nature Communications, 2*(1), 1–9.
Costa, D. C. F., et al. (2016). Aggregation and prion-like properties of misfolded tumor suppressors: Is cancer a prion disease? *Cold Spring Harbor Perspectives in Biology, 8*(10), a023614.
da Silva, N. R., et al. (2015). Analysis of partitioning of organic compounds and proteins in aqueous polyethylene glycol-sodium sulfate aqueous two-phase systems in terms of solute–solvent interactions. *Journal of Chromatography A, 1415*, 1–10.

da Silva, N. R., et al. (2019). Effects of sodium chloride and sodium perchlorate on properties and partition behavior of solutes in aqueous dextran-polyethylene glycol and polyethylene glycol-sodium sulfate two-phase systems. *Journal of Chromatography A*, *1583*, 28–38.

Darling, A. L., Zaslavsky, B. Y., & Uversky, V. N. (2019). Intrinsic disorder-based emergence in cellular biology: Physiological and pathological liquid-liquid phase transitions in cells. *Polymers*, *11*(6), 990.

Dewey, C. M., et al. (2011). TDP-43 is directed to stress granules by sorbitol, a novel physiological osmotic and oxidative stressor. *Molecular and Cellular Biology*, *31*(5), 1098–1108.

Dignon, G. L., Zheng, W., & Mittal, J. (2019). Simulation methods for liquid–liquid phase separation of disordered proteins. *Current Opinion in Chemical Engineering*, *23*, 92–98.

Dignon, G. L., et al. (2018). Sequence determinants of protein phase behavior from a coarse-grained model. *PLoS Computational Biology*, *14*(1), e1005941.

Ditlev, J. A., Case, L. B., & Rosen, M. K. (2018). Who's in and who's out—Compositional control of biomolecular condensates. *Journal of Molecular Biology*, *430*(23), 4666–4684.

Dong, X., et al. (2021). Liquid–liquid phase separation of tau protein is encoded at the monomeric level. *The Journal of Physical Chemistry Letters*, *12*, 2576–2586.

Dormann, D., et al. (2010). ALS-associated fused in sarcoma (FUS) mutations disrupt Transportin-mediated nuclear import. *The EMBO Journal*, *29*(16), 2841–2857.

Falahati, H., & Haji-Akbari, A. (2019). Thermodynamically driven assemblies and liquid–liquid phase separations in biology. *Soft Matter*, *15*(6), 1135–1154.

Farahi, N., et al. (2021). Integration of data from liquid-liquid phase separation databases highlights concentration and dosage sensitivity of LLPS drivers. *International Journal of Molecular Sciences*, *22*(6), 3017.

Feric, M., et al. (2016). Coexisting liquid phases underlie nucleolar subcompartments. *Cell*, *165*(7), 1686–1697.

Ferreira, L. A., et al. (2015). Analyzing the effects of protecting osmolytes on solute–water interactions by solvatochromic comparison method: I. small organic compounds. *RSC Advances*, *5*(74), 59812–59822.

Ferreira, L. A., et al. (2017). Effects of osmolytes on solvent features of water in aqueous solutions. *Journal of Biomolecular Structure and Dynamics*, *35*(5), 1055–1068.

Gall, J. G. (2003). A role for Cajal bodies in assembly of the nuclear transcription machinery. *Tsitologiia*, *45*(10), 971–975.

Ghavami, A., van der Giessen, E., & Onck, P. R. (2013). Coarse-grained potentials for local interactions in unfolded proteins. *Journal of Chemical Theory and Computation*, *9*(1), 432–440.

Girelli, A., et al. (2021). Microscopic dynamics of liquid-liquid phase separation and domain coarsening in a protein solution revealed by X-Ray photon correlation spectroscopy. *Physical Review Letters*, *126*(13), 138004.

Handwerger, K. E., & Gall, J. G. (2006). Subnuclear organelles: New insights into form and function. *Trends in Cell Biology*, *16*(1), 19–26.

Hnisz, D., et al. (2017). A phase separation model for transcriptional control. *Cell*, *169*(1), 13–23.

Hua, J., & Lou, J. (2007). Numerical simulation of bubble rising in viscous liquid. *Journal of Computational Physics*, *222*(2), 769–795.

Hyman, A. A., & Brangwynne, C. P. (2011). Beyond stereospecificity: Liquids and mesoscale organization of cytoplasm. *Developmental Cell*, *21*(1), 14–16.

Hyman, A. A., & Simons, K. (2012). Cell biology. Beyond oil and water--phase transitions in cells. *Science*, *337*(6098), 1047–1049.

Hyman, A. A., Weber, C. A., & Julicher, F. (2014). Liquid-liquid phase separation in biology. *Annual Review of Cell and Developmental Biology*, *30*, 39–58.

Iakoucheva, L. M., et al. (2002). Intrinsic disorder in cell-signaling and cancer-associated proteins. *Journal of Molecular Biology*, *323*(3), 573–584.

Ianiro, A., et al. (2019). Liquid–liquid phase separation during amphiphilic self-assembly. *Nature Chemistry, 11*(4), 320–328.
Jawerth, L. M., et al. (2018). Salt-dependent rheology and surface tension of protein condensates using optical traps. *Physical Review Letters, 121*(25), 258101.
Jucker, M., & Walker, L. C. (2018). Propagation and spread of pathogenic protein assemblies in neurodegenerative diseases. *Nature Neuroscience, 21*(10), 1341–1349.
Kanaan, N. M., et al. (2020). Liquid-liquid phase separation induces pathogenic tau conformations in vitro. *Nature Communications, 11*(1), 1–16.
Kedersha, N., & Anderson, P. (2007). Mammalian stress granules and processing bodies. In *Methods in enzymology* (pp. 61–81). Academic Press.
Kim, H. J., et al. (2013). Mutations in prion-like domains in hnRNPA2B1 and hnRNPA1 cause multisystem proteinopathy and ALS. *Nature, 495*(7442), 467–473.
Knipe, D. M., & Cliffe, A. (2008). Chromatin control of herpes simplex virus lytic and latent infection. *Nature Reviews Microbiology, 6*(3), 211–221.
Larson, A. G., et al. (2017). Liquid droplet formation by HP1α suggests a role for phase separation in heterochromatin. *Nature, 547*(7662), 236–240.
Le Ferrand, H., et al. (2019). Time-resolved observations of liquid–liquid phase separation at the nanoscale using in situ liquid transmission electron microscopy. *Journal of the American Chemical Society, 141*(17), 7202–7210.
Li, Y. R., et al. (2013). Stress granules as crucibles of ALS pathogenesis. *Journal of Cell Biology, 201*(3), 361–372.
Li, Q., et al. (2020a). LLPSDB: A database of proteins undergoing liquid–liquid phase separation in vitro. *Nucleic Acids Research, 48*(D1), D320–D327.
Li, Q., et al. (2020b). Protein databases related to liquid-liquid phase separation. *International Journal of Molecular Sciences, 21*(18), 6796.
Lin, Y., et al. (2015). Formation and maturation of phase-separated liquid droplets by RNA-binding proteins. *Molecular Cell, 60*(2), 208–219.
Lin, Y., et al. (2021). Liquid-liquid phase separation of tau driven by hydrophobic interaction facilitates fibrillization of tau. *Journal of Molecular Biology, 433*(2), 166731.
Lu, P. J., & Weitz, D. A. (2013). Colloidal particles: Crystals, glasses, and gels. *Annual Review of Condensed Matter Physics, 4*(1), 217–233.
Malinovska, L., Kroschwald, S., & Alberti, S. (2013). Protein disorder, prion propensities, and self-organizing macromolecular collectives. *Biochimica et Biophysica Acta (BBA) - Proteins and Proteomics, 1834*(5), 918–931.
Mao, Y. S., Zhang, B., & Spector, D. L. (2011). Biogenesis and function of nuclear bodies. *Trends in Genetics, 27*(8), 295–306.
Marzahn, M. R., et al. (2016). Higher-order oligomerization promotes localization of SPOP to liquid nuclear speckles. *The EMBO Journal, 35*(12), 1254–1275.
McSwiggen, D. T., et al. (2019). Evidence for DNA-mediated nuclear compartmentalization distinct from phase separation. *eLife, 8*, e47098.
Mészáros, B., et al. (2020). PhaSePro: The database of proteins driving liquid–liquid phase separation. *Nucleic Acids Research, 48*(D1), D360–D367.
Mitrea, D. M., et al. (2016). Nucleophosmin integrates within the nucleolus via multi-modal interactions with proteins displaying R-rich linear motifs and rRNA. *eLife, 5*, e13571.
Mitrea, D. M., et al. (2018). Self-interaction of NPM1 modulates multiple mechanisms of liquid–liquid phase separation. *Nature Communications, 9*(1), 842.
Murthy, A. C., & Fawzi, N. L. (2020). The (un) structural biology of biomolecular liquid-liquid phase separation using NMR spectroscopy. *Journal of Biological Chemistry, 295*(8), 2375–2384.
Nakashima, K. K., Vibhute, M. A., & Spruijt, E. (2019). Biomolecular chemistry in liquid phase separated compartments. *Frontiers in Molecular Biosciences, 6*(21), 21.

Netherton, C. L., & Wileman, T. (2011). Virus factories, double membrane vesicles and viroplasm generated in animal cells. *Current Opinion in Virology*, *1*(5), 381–387.
Ning, W., et al. (2020). DrLLPS: A data resource of liquid-liquid phase separation in eukaryotes. *Nucleic Acids Research*, *48*(D1), D288–d295.
Noda, N. N., Wang, Z., & Zhang, H. (2020). Liquid–liquid phase separation in autophagy. *Journal of Cell Biology*, *219*(8), e202004062. https://doi.org/10.1083/jcb.202004062.
Novoa, R. R., et al. (2005). Virus factories: Associations of cell organelles for viral replication and morphogenesis. *Biology of the Cell*, *97*(2), 147–172.
O'Brien, R. E., et al. (2015). Liquid–liquid phase separation in aerosol particles: Imaging at the nanometer scale. *Environmental Science & Technology*, *49*(8), 4995–5002.
O'Flynn, B. G., & Mittag, T. (2021). The role of liquid–liquid phase separation in regulating enzyme activity. *Current Opinion in Cell Biology*, *69*, 70–79.
Pancsa, R., Vranken, W., & Mészáros, B. (2021). Computational resources for identifying and describing proteins driving liquid–liquid phase separation. *Briefings in Bioinformatics*.
Pederson, T. (2011). The nucleolus. *Cold Spring Harbor Perspectives in Biology*, *3*(3), a000638.
Perdikari, T. M., et al. (2020). SARS-CoV-2 nucleocapsid protein undergoes liquid-liquid phase separation stimulated by RNA and partitions into phases of human ribonucleoproteins. *BioRxiv*. https://doi.org/10.1101/2020.06.09.141101.
Perry, S. L., et al. (2015). Chirality-selected phase behaviour in ionic polypeptide complexes. *Nature Communications*, *6*(1), 6052.
Peskett, T. R., et al. (2018). A liquid to solid phase transition underlying pathological huntingtin Exon1 aggregation. *Molecular Cell*, *70*(4), 588–601 (e6).
Prouteau, M., & Loewith, R. (2018). Regulation of cellular metabolism through phase separation of enzymes. *Biomolecules*, *8*(4), 160.
Quiroz, F. G., et al. (2020). Liquid-liquid phase separation drives skin barrier formation. *Science*, *367*(6483). https://doi.org/10.1126/science.aax9554.
Ray, S., et al. (2020). α-Synuclein aggregation nucleates through liquid–liquid phase separation. *Nature Chemistry*, *12*(8), 705–716.
Roberts, S., et al. (2018). Injectable tissue integrating networks from recombinant polypeptides with tunable order. *Nature Materials*, *17*(12), 1154–1163.
Saad, S. M., & Neumann, A. W. (2016). Axisymmetric drop shape analysis (ADSA): An outline. *Advances in Colloid and Interface Science*, *238*, 62–87.
Samanta, H. S., Chakraborty, D., & Thirumalai, D. (2018). Charge fluctuation effects on the shape of flexible polyampholytes with applications to intrinsically disordered proteins. *The Journal of Chemical Physics*, *149*(16), 163323.
Schmid, M., et al. (2014). DNA virus replication compartments. *Journal of Virology*, *88*(3), 1404–1420.
Shakya, A., et al. (2020). Liquid-liquid phase separation of histone proteins in cells: Role in chromatin organization. *Biophysical Journal*, *118*(3), 753–764.
Shin, J. J., et al. (2020). pH biosensing by PI4P regulates cargo sorting at the TGN. *Developmental Cell*, *52*(4), 461–476 (e4).
Shin, Y., & Brangwynne, C. P. (2017). Liquid phase condensation in cell physiology and disease. *Science*, *357*(6357). https://doi.org/10.1126/science.aaf4382.
Shulman, J. M., De Jager, P. L., & Feany, M. B. (2011). Parkinson's disease: Genetics and pathogenesis. *Annual Review of Pathology*, *6*, 193–222.
Stroberg, W., & Schnell, S. (2017). On the origin of non-membrane-bound organelles, and their physiological function. *Journal of Theoretical Biology*, *434*, 42–49.
Strom, A. R., et al. (2017). Phase separation drives heterochromatin domain formation. *Nature*, *547*(7662), 241–245.
Taylor, J. P., Hardy, J., & Fischbeck, K. H. (2002). Toxic proteins in neurodegenerative disease. *Science*, *296*(5575), 1991–1995.

Titus, A. R., et al. (2020). Interfacial tension and mechanism of liquid-liquid phase separation in aqueous media. *Physical Chemistry Chemical Physics*, *22*(8), 4574–4580.

Wang, Z., Zhang, G., & Zhang, H. (2019). Protocol for analyzing protein liquid–liquid phase separation. *Biophysics Reports*, *5*(1), 1–9.

Titus, A. R., et al. (2021). The C-terminus of Perilipin 3 shows distinct lipid binding at phospholipid-oil-aqueous interfaces. *Membranes (Basel)*, *11*(4), 265.

Wang, J., et al. (2018). A molecular grammar governing the driving forces for phase separation of prion-like RNA binding proteins. *Cell*, *174*(3), 688–699 (e16).

Weber, S. C. (2019). Evidence for and against liquid-liquid phase separation in the nucleus. *Non-coding RNA*, *5*(4), 50.

Weber, S. C., & Brangwynne, C. P. (2012). Getting RNA and protein in phase. *Cell*, *149*(6), 1188–1191.

Wheeler, R. J., & Hyman, A. A. (2018). Controlling compartmentalization by non-membrane-bound organelles. *Philosophical Transactions of the Royal Society of London. Series B, Biological Sciences*, *373*(1747), 20170193.

Yang, J., Yu, K., & Zuo, Y. Y. (2017). Accuracy of axisymmetric drop shape analysis in determining surface and interfacial tensions. *Langmuir*, *33*(36), 8914–8923.

You, K., et al. (2020). PhaSepDB: A database of liquid-liquid phase separation related proteins. *Nucleic Acids Research*, *48*(D1), D354–d359.

Young, B. P., et al. (2010). Phosphatidic acid is a pH biosensor that links membrane biogenesis to metabolism. *Science*, *329*(5995), 1085–1088.

Zaslavsky, B. Y. (1994). *Aqueous two-phase partitioning: Physical chemistry and bioanalytical applications*. CRC press.

Zhang, X., et al. (2005). Fluctuation-assisted crystallization: In a simultaneous phase separation and crystallization polyolefin blend system. *Macromolecular Rapid Communications*, *26*(16), 1285–1288.

CHAPTER THREE

Investigating molecular crowding during cell division and hyperosmotic stress in budding yeast with FRET

Sarah Lecinski[a,†], Jack W. Shepherd[a,b,†], Lewis Frame[c], Imogen Hayton[b], Chris MacDonald[b], and Mark C. Leake[a,b,*]
[a]Department of Physics, University of York, York, United Kingdom
[b]Department of Biology, University of York, York, United Kingdom
[c]School of Natural Sciences, University of York, York, United Kingdom
*Corresponding author: e-mail address: mark.leake@york.ac.uk

Contents

† These authors contributed equally.

Current Topics in Membranes, Volume 88
ISSN 1063-5823
https://doi.org/10.1016/bs.ctm.2021.09.001

Abstract

Cell division, aging, and stress recovery triggers spatial reorganization of cellular components in the cytoplasm, including membrane bound organelles, with molecular changes in their compositions and structures. However, it is not clear how these events are coordinated and how they integrate with regulation of molecular crowding. We use the budding yeast *Saccharomyces cerevisiae* as a model system to study these questions using recent progress in optical fluorescence microscopy and crowding sensing probe technology. We used a Förster Resonance Energy Transfer (FRET) based sensor, illuminated by confocal microscopy for high throughput analyses and Slimfield microscopy for single-molecule resolution, to quantify molecular crowding. We determine crowding in response to cellular growth of both mother and daughter cells, in addition to osmotic stress, and reveal hot spots of crowding across the bud neck in the burgeoning daughter cell. This crowding might be rationalized by the packing of inherited material, like the vacuole, from mother cells. We discuss recent advances in understanding the role of crowding in cellular regulation and key current challenges and conclude by presenting our recent advances in optimizing FRET-based measurements of crowding while simultaneously imaging a third color, which can be used as a marker that labels organelle membranes. Our approaches can be combined with synchronized cell populations to increase experimental throughput and correlate molecular crowding information with different stages in the cell cycle.

Abbreviations

FRET Förster resonance energy transfer
PM plasma membrane
ConA Concanavalin A
OD optical density
YPD yeast extract-peptone-dextrose medium
SD synthetic defined medium
DIC differential interference contrast

1. Introduction

The term "molecular crowding" describes the range of molecular confinement-induced effects (e.g., mobility, soft attraction, and repulsion forces) observed in a closed system of concentrated molecules. Cells are highly crowded membrane-bound environments containing a range of biomolecular species including proteins, polysaccharides and nucleic acids. Typically, these molecules occupy a huge volume of the cell (up to 40%), equivalent to a concentration of up to 400 mg/mL (Fulton, 1982; Zimmerman & Trach, 1991). Two terms can be encountered in the literature: "*macromolecular crowding*" referring to dynamic effects of volume exclusion encountered between two molecules and "*macromolecular confinement*" referring to the same effect caused by the static shape and size of the system (Huan-Xiang, Rivas, & Minton, 2008; Sanfelice et al., 2013; Zhang et al., 2019). Both describe a typical free-space limitation occurring in a highly concentrated environment of molecules, which leads to non-specific interactions between macromolecules in close proximity (Sarkar, Li, & Pielak, 2013). Excluded volume theory is a key concept to understanding macromolecular crowding (Garner & Burg, 1994; Kuznetsova, Turoverov, & Uversky, 2014). By their presence, molecules exclude access to the solvent/surface of other molecules. This imposed volume restriction where exclusion is dependent on the molecule's size and shape (Fig. 1). As a result, if each molecule excludes a certain volume from every other, and the mobility of each is also reduced—a molecule can only diffuse into an available volume, which effectively slows the time scale of the overall diffusive process. Note also that the overall effect

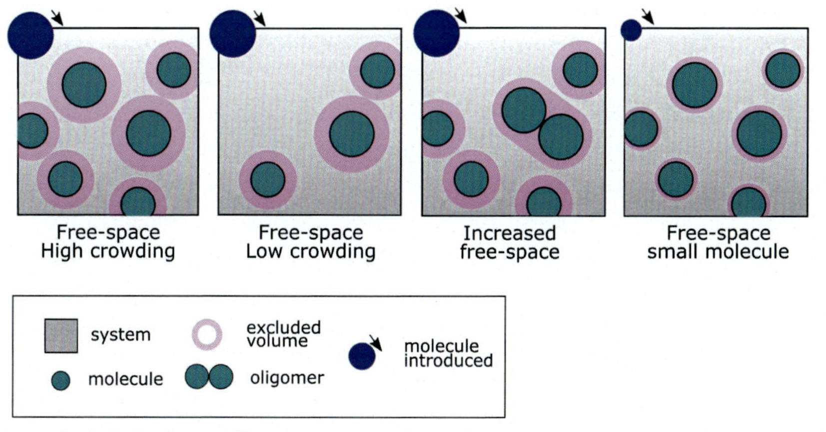

Fig. 1 Excluded volume effect, molecules in a confined space. Illustration of the concept of excluded volume, solvent availability and molecular crowding for molecules in a confined environment. In pink the excluded volume for each molecule in the system therefore limiting mobility and solvent accessibility for the molecules introduced (dark blue molecules on top). The third panel shows how protein oligomerization may modify solvent availability for another molecule and last panel shows the different excluded volume observed in the system for a molecule introduced of a smaller size, excluded volume is dependant of its shape and size.

of the multiple biomolecular species in vivo is referred to exclusively as crowding with the term "concentration" used generally for individual protein species, which may be at a low concentration in an overall high crowding environment or vice versa (Ellis, 2013; Minton, 2006).

An extensive number of studies on molecular crowding have been conducted in vitro mimicking the cytoplasmic crowded environment (Benny & Raghunath, 2017; Davis, Deutsch, & Gruebele, 2020) using inert and highly water-soluble crowding agents such as glycerol (Zhou, Liao, Chen, & Liang, 2006), Polyethylene glycol (PEG) (Benny & Raghunath, 2017), and Ficoll (Christiansen, Wang, Cheung, & Wittung-Stafshede, 2013; Van Den Berg, Ellis, & Dobson, 1999). Several of these studies have described the role of crowding in protein stability (König, Soranno, Nettels, & Schuler, 2021) and its impact on the free-energy of folding (Minton, 2006; Nguemaha, Qin, & Zhou, 2019) and kinetics of interaction (Phillip & Schreiber, 2013; Stagg, Zhang, Cheung, & Wittung-Stafshede, 2007). In vivo, dynamics of repulsion and attraction forces between macromolecules has been described as one of the effects of crowding (Garner & Burg, 1994), with a role on molecular rearrangement (Rivas, Ferrone, & Herzfeld, 2004), promoting oligomerization and aggregation (Christiansen et al., 2013). Interestingly, in theory the excluded volume of two monomers does not overlap, but if molecules can biochemically interact to form a stable new entity, this new molecule will now have its own excluded volume changing the solvent availability for other proteins (Allen, 2001; André & Spruijt, 2020; Poland, 1998)—see Fig. 1). These dynamic effects have raised new interest in the context of protein aggregation (Jing, Qin, & Tong, 2020), intracellular organization (Löwe, Kalacheva, Boersma, & Kedrov, 2020), and membrane-less compartments in the cytoplasm, for example the formation of liquid-liquid phase separation in the cytoplasm (Franzmann et al., 2018; Jin et al., 2021; Park et al., 2020). More generally, macromolecular crowding has been linked to a number of biological process such as protein-protein interaction (Bhattacharya, Kim, & Mittal, 2013), protein conformational changes (Dong, Qin, & Zhou, 2010), stability and folding (Despa, Orgill, & Lee, 2006; Stagg et al., 2007; Tokuriki, 2004), promoting signaling cascades (Rohwer, Postma, Kholodenko, & Westerhoff, 1998), regulating biomolecular diffusion (Tabaka, Kalwarczyk, Szymanski, Hou, & Holyst, 2014; Tan, Saurabh, Bruchez, Schwartz, & LeDuc, 2013), and intracellular transport (Nettesheim et al., 2020).

Increased crowding in the cytoplasm has also been associated with promoting binding interaction (Bhattacharya, Kim, & Mittal, 2013) and complexation with enzymatic reactions (Akabayov, Akabayov, Lee, Wagner, & Richardson, 2013), and enhancing gene expression (Hancock, 2014; Mager & Siderius, 2002; Shim et al., 2020) and protein phosphorylation (Ouldridge & Rein ten Wolde, 2014). However, a strong shift in cellular crowding may also have the opposite effect to lower cellular metabolic activities with gene expression and entry to cell division compromised (André & Spruijt, 2020; Ma & Nussinov, 2013). Such metabolic reactions are typically encountered in cells exposed to acute stress. External stress perturbs the enclosed cell and triggers physical, morphological, and metabolic changes (Campisi & D'Adda Di Fagagna, 2007; Canetta, Walker, & Adya, 2006; Marini, Nüske, Leng, Alberti, & Pigino, 2020). In general, cellular stress responses are mediated by complex signaling pathways that involve a coordinated set of biomolecular reactions and interactions that modify gene expression, metabolism, protein localization, and in some cases cell morphology. Taken together these determine the cell's ultimate fate, either recovery over time or programmed cell death (Fulda, Gorman, Hori, & Samali, 2010; Galluzzi, Yamazaki, & Kroemer, 2018) (see Fig. 2A).

In general, cellular stresses may be sorted into two categories, either environmental such as for heat shock or nutrient starvation; or of an intracellular nature due to for example toxic by-products of metabolism produced due to aging (Lumpkin et al., 1985; Sinclair & Guarente, 1997), the presence of reactive oxygen species (ROS) in the cytosol (Liochev, 2013; Santos, Sinha, & Lindner, 2018), or the formation of protein aggregates (Groh et al., 2017; López-Otín, Blasco, Partridge, Serrano, & Kroemer, 2013). Stresses of these types are encountered by the majority of cells populations, which could explain the highly conserved nature of response mechanisms (Kültz, 2020; Lockshin & Zakeri, 2007). Indeed, response to thermal stress mediated by heat shock proteins and associated chaperones, conserved between prokaryotes and even complex eukaryotic cells, is a perfect example (Li & Srivastava, 2003; Verghese, Abrams, Wang, & Morano, 2012). Other highly conserved mechanisms include DNA damage repair systems (essential to preserve cell integrity), e.g., the case of the highly conserved Rad52 protein which operates via the Mre11-Rad50-Nbs1 (MRN) complex to effect double strand break repair (Syed & Tainer, 2018) (arising for example from UV stress (Cadet, Sage, & Douki, 2005)) and homologous recombination which is essential for cell integrity (Hakem, 2008; Lamarche, Orazio, & Weitzman, 2010).

Upon osmotic stress, when cells are exposed to a high ionic strength environment, the cell volume suddenly reduces because of the osmotic

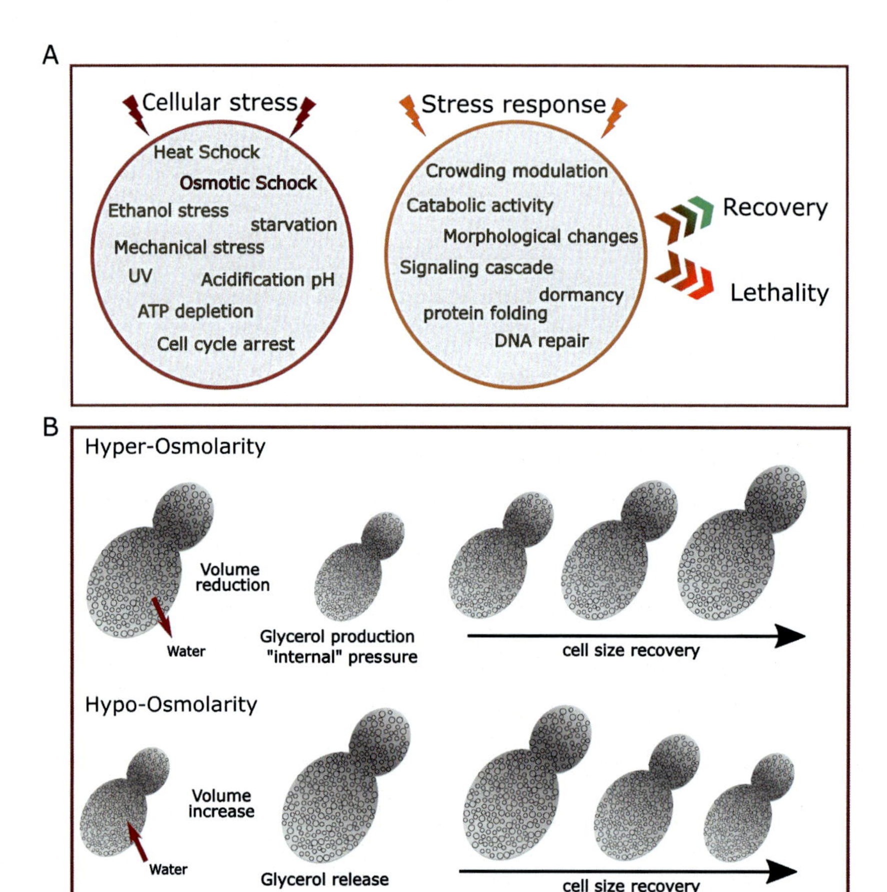

Fig. 2 Cellular stress and stress responses. (A) General listing of main cellular stress (left) and stress responses (right) that can either trigger a set of metabolic adjustment leading to cell survival and recovery or, if the stress cannot be resolved, leads to cellular death mainly via apoptosis. (B) Schematic osmotic shock response in yeast. Under hyperosmotic stress, cell size drastically reduces as water flows out the cells, resulting in an increased crowding effect in the cytoplasm and the activating osmoregulation Hog1 pathways, the subsequent production of glycerol in the cytoplasm pressure back the cell to recover its initial size. On the contrary during hypo-tonic osmotic stress water flows in and the cell volume increase, less glycerol produced and transporters such as aquaporin are upregulated to allow water to flows in and help the cell recover its original size.

pressure generated, which triggers a diffusion-driven water exchange from the cell cytosol to the external environment. This drastic volume reduction spatially confines the pool of macromolecules inside the cells and directly increases molecular crowding (see Fig. 2B). In response, cells activate biochemical pathways such as that involving the protein Hog1 to produce a gain in volume via internal pressure generation (Tamás, Rep, Thevelein, & Hohmann, 2000)—this occurs notably via the production of glycerol (Hohmann, 2002) and the regulation of transporters at the plasma membrane such as aquaporins to control the glycerol/water ratio (Hohmann, 2015; Saito & Posas, 2012). In eukaryotic cells, osmostress and osmoregulation have been extensively studied using yeast as a model (Babazadeh et al., 2017; Gibson, Lawrence, Leclaire, Powell, & Smart, 2007; Hsu et al., 2015; Özcan & Johnston, 1999). As a model eukaryote (Tissenbaum & Guarente, 2002), yeast has many advantages: it is easy to manipulate, rapid to culture (Petranovic, Tyo, Vemuri, & Nielsen, 2010), and as a unicellular organism is particularly susceptible to osmotic shock—the entire yeast cell is exposed to the media as opposed to only a fraction of cells in a multicellular model. Moreover, yeast genetics and essential metabolic pathways are conserved within the eukaryotic kingdom (Duina, Miller, & Keeney, 2014; Dujon, 1996) and have been extensively studied (Foury, 1997; He, Zhou, & Kennedy, 2018).

In *Saccharomyces cerevisiae* (budding yeast) cells, mitosis and cytokinesis take place via a budding process, where a bud emerges and grows from the mother cell (Chen, Howell, Robesand, & Lew, 2011; Juanes & Piatti, 2016). This process requires the establishment of a cell polarity between the mother cell and daughter cell (Bi & Park, 2012) and therefore close regulation of the cytoplasmic content and spatial organization (Nasmyth, 1996), including changes in membrane morphology. Cell division is therefore a highly asymmetrical process (Higuchi-Sanabria et al., 2014) associated with a range of biochemical processes involving protein-protein interaction (Chen et al., 2011) and significant material and compartment transport (Champion, Linder, & Kutay, 2017; Yeong, 2005). All these processes rely on biomolecular rearrangement and volume changes, and as such may impact local macromolecular crowding conditions. Cell growth is often inhibited during stress (Li & Cai, 1999; Taïeb et al., 2021), correlated to diversion of metabolic resources diverted to mitigate the resultant detrimental physiological effects (Bonny, Kochanowski, Diether, & El-Samad, 2020; Chasman et al., 2014). Growth capability is therefore a key parameter in measuring cellular recovery, with those that fail to reverse the arrest and re-enter a replicative cycle generally dying (Petranovic & Ganley, 2014). Yeast, however, can enter a semi-dormant "quiescent" state, in which cells don't divide and maintain only minimal metabolic functions (Nurse & Broek, 1993). In essence, this is a survival state to economize resources when exposed to prolonged periods of starvation (Valcourt et al., 2012). The cytoplasm of this new state has been shown to transition from a liquid-like to a solid-like state, with an associated acidification of the environment which also results in lower diffusion rates (Munder et al., 2016).

The budding process in *S. cerevisiae* is specifically characterized by the formation of a septin ring on the cell membrane early in the replication cycle (Byers & Goetsch, 1976; Chen et al., 2011). This septin ring acts as a junction between the mother cell and the daughter cell (Vrabioiu & Mitchison, 2006) and defines the mother cell/daughter cell polarity (Juanes & Piatti, 2016) (Fig. 3). During this process, the role of the plasma-associated GTPase protein Cdc42 is key to initiate the polarization process (Okada et al., 2013) and trigger recruitment of proteins to form the septin ring at the interface between mother and daughter cell in a region generally known as the bud neck (Faty, Fink, & Barral, 2002; McMurray et al., 2011). Several septins are recruited to make this hetero-oligomeric structure, with key proteins Shs1, Cdc3, Cdc10, Cdc11, Cdc12 forming a characteristic double ring shape (McMurray & Thorner, 2009; Vrabioiu & Mitchison, 2006). Meanwhile, the cytoskeleton network is restructured such that it is polarized along the axis of the bud neck between mother and daughter cell (Moseley & Goode, 2006) to facilitate active transport of complexes and organelles (Juanes & Piatti, 2016; Warren & Wickner, 1996) to the daughter cell, mainly accomplished by myosin transport along actin cables (Knoblach & Rachubinski, 2015; Macara & Mili, 2008).

During mitosis essential cellular complexes and molecules travels from the mother cell to the daughter cell and are inherited progressively (Cramer, 2008; Yaffe, 1991). This involves the selection of biomolecules and molecular complexes to travel from mother to daughter, especially in the case of key structures the daughter cell cannot synthesize de novo

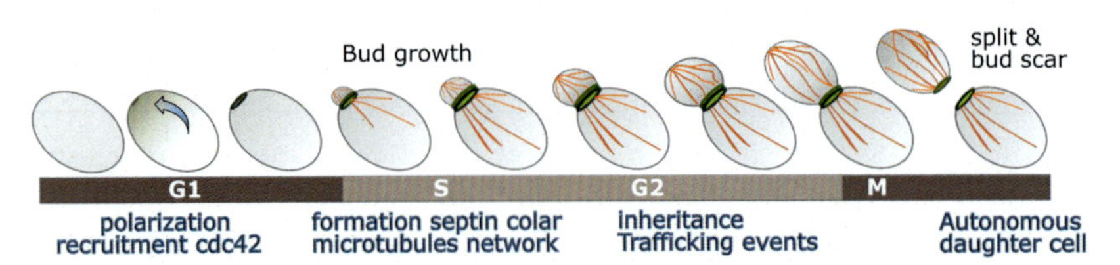

Fig. 3 Schematic budding yeast polarized cell division. Simple representation of a single yeast cell dividing, polarization to form the bud neck and the septin rings (green) and cytoskeleton polarized through the bud neck in (orange).

including metabolic compartments such as the vacuole, mitochondria, nucleus, and endoplasmic reticulum (Menendez-Benito et al., 2013; Warren & Wickner, 1996). A recent study (Li, Lu, Iwamoto, Drubin, & Pedersen, 2021) reported organelles inherited in a specific and ordered sequence during cell division. Endoplasmic reticulum (ER) and peroxisomes being transported first in the newly formed bud, followed by the vacuole and mitochondria while the bud grow, and at last, in large buds near cytokinesis, the nucleus is transported to the daughter cell. Decoupling cell cycle progression and bud growth, they showed an overall maintained order and timing of migration, supporting a multi-factorial model for organelle inheritance, thus only partially controlled by metabolic pathways associated to cell division. Also, age- or stress-damaged molecules such as extrachromosomal rDNA circles (Lumpkin et al., 1985; Sinclair & Guarente, 1997) and damaged proteins (Chiti & Dobson, 2017) are excluded from trafficking and remain in the mother cell (Higuchi-Sanabria et al., 2014; López-Otín et al., 2013; Zhou et al., 2014). The bud-neck region has been described as a "diffusion barrier" influencing trafficking events and the relative cytosolic volume between mother cell and daughter cell (Clay et al., 2014; Gladfelter, Pringle, & Lew, 2001; Shcheprova, Baldi, Frei, Gonnet, & Barral, 2008; Sugiyama & Tanaka, 2019; Valdez-Taubas & Pelham, 2003) although given the active super-diffusive transport which dominates trafficking, the existence of a meaningful physical diffusion barrier is still an open debate (Nyström & Liu, 2014; Zhou et al., 2011).

The influence of macromolecular crowding at the highly localized region between the mother and daughter cell interface also remains unknown. Quantifying crowding dynamics—or other physicochemical properties—in living cells is challenging, but recent improvements in optical microscope technology and the development of synthetic fluorescent protein sensors have enabled whole cell measurements of crowding during aging and under osmotic stress (Mouton et al., 2019). In essence, these sensors consist of a FRET-pair of fluorescent proteins on an alpha-helical "spring" (Boersma, Zuhorn, & Poolman, 2015). In crowded conditions, the proteins are pushed closer together, and the FRET efficiency increases, while in less crowded conditions the FRET efficiency is reduced (Fig. 4A). By imaging with dual-color fluorescence microscopy, the FRET efficiency can be quantified and used as a signature for molecular crowding (Fig. 4B). In general, the DNA coding for these sensors is transformed into the cell by homologous recombination (Liu et al., 2017; Mouton et al., 2019), are expressed endogenously, and localize largely to the cytoplasm

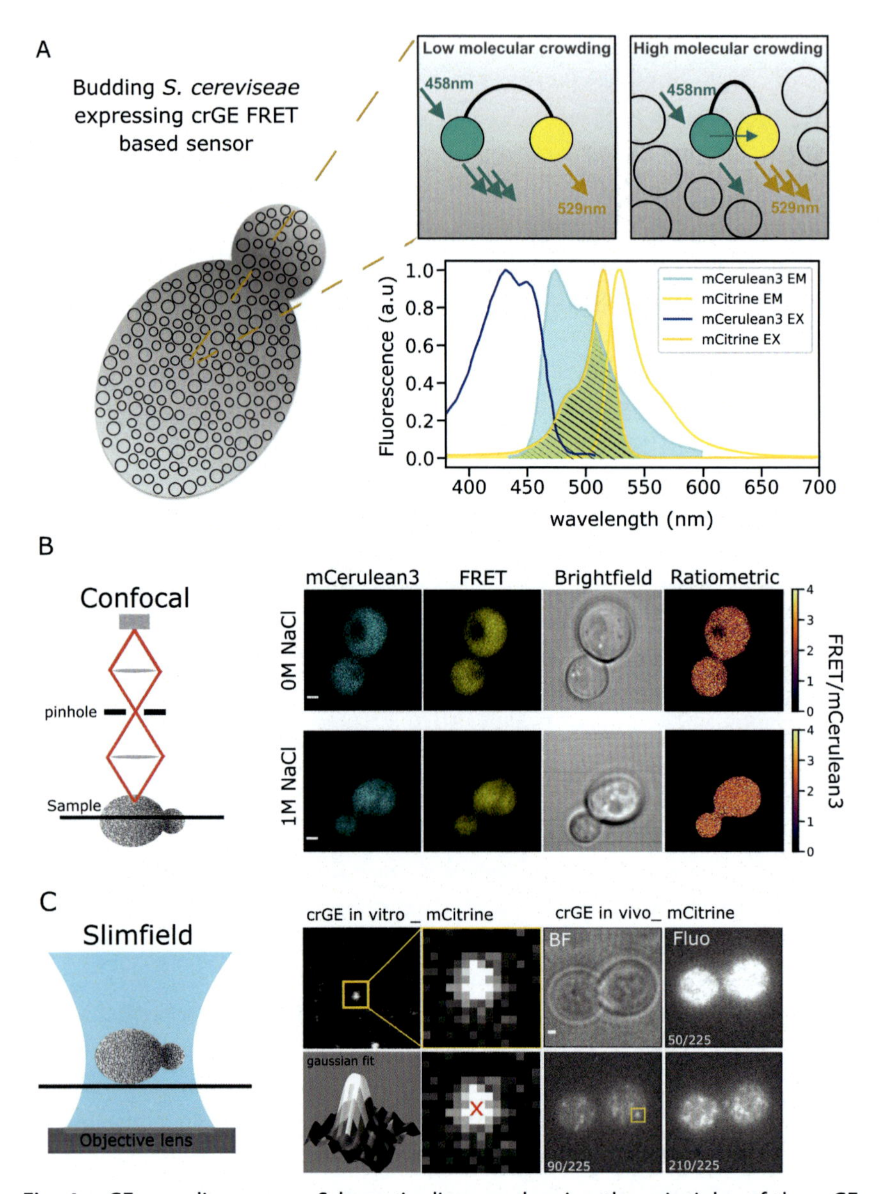

Fig. 4 crGE crowding sensor. Schematic diagram showing the principles of the crGE FRET sensor. In a crowded environment, the two probes are brought closer to each other which favors the non-radiative energy transfer from mCerulean3 to mCitrine, increasing the FRET efficiency. Inset: the mCerulean3 emission spectrum overlaid with the mCitrine excitation spectrum. (A) Left: schematic representation of confocal microscopy. Right: fluorescence micrographs of yeast cells expressing crGE in both 0 and 1 M NaCl. Last micrographs on the right show the respective ratiometric maps from the fluorescence images. Scale bar: 1 μm. (B) Left: schematic representation of Slimfield illumination. Right: Visualization of mCitrine single molecules from crGE sensor in vitro and in vivo expressed in yeast. Iterative Gaussian fits for each spot detected allows us to reach a localization precision around 30 nm.

with both the vacuole in yeast and plasma membrane being excluded volumes (Mouton et al., 2020; Shepherd et al., 2020). While some sensors, notably the crGE mCerulean3-mCitrine crowding sensor are not ideal for super-resolution localization microscopy (SMLM) due to photophysical limitations, new generations of crowding sensors such as crGE2.3 based on mEGFP and mScarlet-I have fluorescent proteins which undergo FRET and are separately super-resolvable through techniques such as Slimfield microscopy (Lenn & Leake, 2016). Slimfield uses a bespoke microscope similar to a standard epifluorescence setup, but with a small lateral excitation field (Chiu & Leake, 2011; Lenn & Leake, 2012; Plank, Wadhams, & Leake, 2009; Wollman et al., 2016). The laser beam generates a Gaussian intensity profile over the sample with the field of excitation reduced to a small area of effective width $< 10\,\mu m$ (full width at half maximum). The excitation intensity is strongly increased as a result compared to conventional epifluorescence excitation, up to ~100 times comparing to widefield image intensities, since the area targeted is smaller by approximatively a factor 10. This feature enables single-molecule detection at very high speeds, helped by precision localization through iterative Gaussian masking of detected bright spots (Shepherd, Higgins, Wollman, & Leake, 2021b). Typically, a stack of images is acquired, first the intense illumination excites typically a large number of fluorophores. These fluorophores are then stochastically photobleached in a stepwise fashion, and it is this step size that is characteristic of the brightness of a single fluorophore in that specific microenvironment that ultimately enables us to determine the stoichiometry of fluorescently-labeled molecular complexes that may comprise multiple copies of the same type of fluorophore. To overcome diffusion in the cell cytoplasm the fluorescence signal needs to be captures within a short sampling window, and we typically use a camera exposure times of ~5 ms. This position of single molecules emitting fluorescence can be resolved via Gaussian fitting to a precision that is better than the optical resolution limit, typically a few tens of nm. Dual-color imaging can also be performed with this technology, opening possibilities for performing colocalization experiments as well as single-molecule FRET quantification for individual sensors localized to a super-resolved spatial precision to map molecular crowding throughout the cell.

Here, we implemented Slimfield using a narrowfield illumination mode (Wollman & Leake, 2016), based around a standard epifluorescence microscope with high excitation powers at the sample plane in excess of $1800\,W/cm^2$ which we have demonstrated previously may be used for single molecule localization (Miller, Zhou, Wollman, & Leake, 2015). This high intensity excitation promotes excitation of individual fluorophores even at very rapid millisecond sampling time scales, and thus enables very rapid single-molecule tracking in live cells relevant to diffusion not only in relatively viscous membranes but also in the cell cytoplasm for which traditional widefield methods are too slow (Fig. 4C). Typically, the field of view is confined to the area encompassing just a few individual cells to allow rapid readout from the camera pixel array, allowing exploration of millisecond time scale cellular dynamics and accurate diffusion coefficient calculation (Plank, Wadhams, & Leake, 2009; Wollman et al., 2016). Alongside this, tracking single molecules with Slimfield microscopy has been used to resolve oligomerization states of molecular assemblies in vivo (Badrinarayanan, Reyes-Lamothe, Uphoff, Leake, & Sherratt, 2012; Laidlaw et al., 2021; Shashkova, Nyström, Leake, & Wollman, 2020; Shashkova, Wollman, Leake, & Hohmann, 2017; Sun, Wollman, Huang, Leake, & Liu, 2019; Syeda et al., 2019; Wollman et al., 2017). The crGE sensor copy numbers (i.e., number of sensor molecules present per cell) were also measured in previous work (Shepherd et al., 2020), demonstrating that crowding readout in the cell was independent of local sensor concentration. Single crGE molecules were also tracked through the low peak emission intensity of mCerulean3 this had precluded single-molecule FRET measurement.

In our present work here, we present novel methods for quantifying subcellular crowding differences between mother and daughter cells during the *S. cerevisiae* budding process, that have particular relevance to the cell dependent changes exhibited in membrane morphologies. We begin by comparing the average molecular crowding between mother and daughter cells using confocal microscopy of the crGE crowding FRET sensor, and correlate these with cell area. We then use deep learning segmentation and a bespoke semi-automatic Python 3 image analysis framework to quantify subcellular regions along and parallel to the cell division axis. We also quantify crowding during cell recovery from osmotic stress and during cell growth, as well as quantifying crowding FRET readout at the bud-neck region around which the plasma membranes of mother and daughter cells are very tightly pinched. We additionally present a new imaging method with combining crowding sensing with mCerulean3-mCitrine FRET sensor and FM4–64, a fluorescent dye that labels endolysosomal lipids (Vida & Emr, 1995), allowing simultaneous visualization of the vacuole membrane. Finally, we present our most recent data in which we have successfully quantified single molecule FRET in vivo using the crGE2.3 crowding sensor and our super-resolution image analysis software suite PySTACHIO (Shepherd, Higgins, Wollman, & Leake, 2021a).

2. Results and discussion

2.1 Macromolecular crowding stability between mother cell and daughter cells in budding yeast

To investigate if polarization and asymmetry between mother cells and budding daughter cells in dividing *S. cerevisiae* is reflected in the crowding environment we expressed the crGE crowding sensor and grew cells to mid-log phase ($OD_{600} = 0.4–0.6$) in synthetic complete media containing glucose as previously described (Mouton et al., 2020; Shepherd et al., 2020) and performed confocal microscopy. Fig. 5 shows the average ratiometric FRET characterized between mother cells and daughter cells/ growing buds. This analysis showed a similar FRET readout for mother cells and daughter cells (Fig. 5A and B and Statistic Table 1A in Supplementary Material in the online version at https://doi.org/10.1016/bs.ctm.2021.09.001). We found that for exponentially dividing cells, the individual cellular area was not correlated with FRET (Fig. 5C and D), consistent with the essential role of crowding stability in cellular integrity and survival (Mouton et al., 2020; Van Den Berg, Boersma, & Poolman, 2017). However, there was a greater range of ratiometric FRET values found for smaller cells, which includes the growing daughter cells and buds (Fig. 5C). We hypothesize that this is due to a lower cell volume leading to smaller cells having a greater sensitivity to their immediate environment during initial growth. Moreover, in growing/budding cells the exact stage in the replication cycle may be correlated with crowding, as transport of large material such as inherited organelles may lead to relatively high short-term crowding variability.

Fig. 2D shows two sub-populations of budding yeast plotted against bud area. We have split the analysis into two groups, the first displaying higher ratiometric FRET in the mother cell than in the daughter cell and the second where lower ratiometric FRET was measured in the mother cell than in the daughter cell. For those two categories, we see that cell area between the two conditions is statistically equivalent with no statistical difference as measured by both Student's *t*-test and the non-parametric Brunner-Munzel test between the two populations (Statistic Table 1B in Supplementary Material in the online version at https://doi.org/10.1016/bs.ctm.2021.09.001).

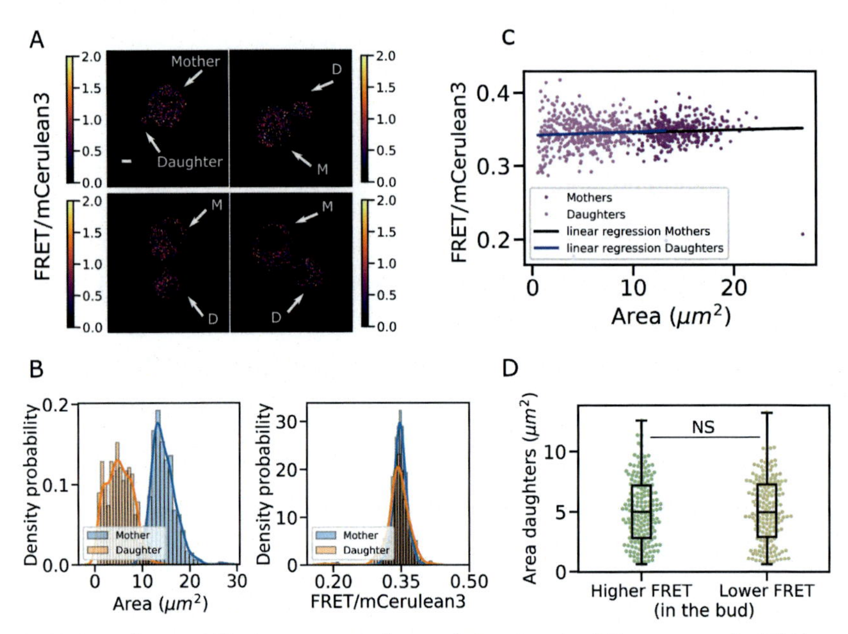

Fig. 5 Crowding stability between mother and daughter budding yeast cells. (A) Yeast cells expressing the crGE sensor. Only budding yeasts were analyzed, defined by each mother cell still being attached to its bud or daughter cell as shown in the ratiometric FRET (FRET/mCerulean3) images sample, with white arrows to show the mother cell and the daughter cell of each budding yeast. (B) Scatter plot of mother and daughter cell ratiometric FRET against cell area. (C) From left to right, histograms comparing the cell area between mother cells and daughter cells and comparison of the FRET efficiency between mother cells and daughter cells. (D) Crowding and cell size dependency in the buds. Two population of cells were measured, one with buds displaying a higher FRET efficiency than their mother cells, the other with buds display a lower FRET efficiency than their mother cells.

We therefore conclude that the cell size during normal growth is not a predictive factor for subcellular crowding.

2.2 Local crowding readout and cell-by-cell tracking for timeline response to cellular growth and osmotic stress

To investigate how crowding progresses through the whole replicative cycle we performed confocal imaging with cells immobilized with Concanavalin A (Con A) on a bespoke microfluidics system (Laidlaw et al., 2021) and performed time lapse experiment on growing cells. Analysis was performed using a bespoke Python 3 utility that generated FRET/mCerulean3 ratiometric heat maps showing local regions of high FRET intensity. Fig. 6A shows

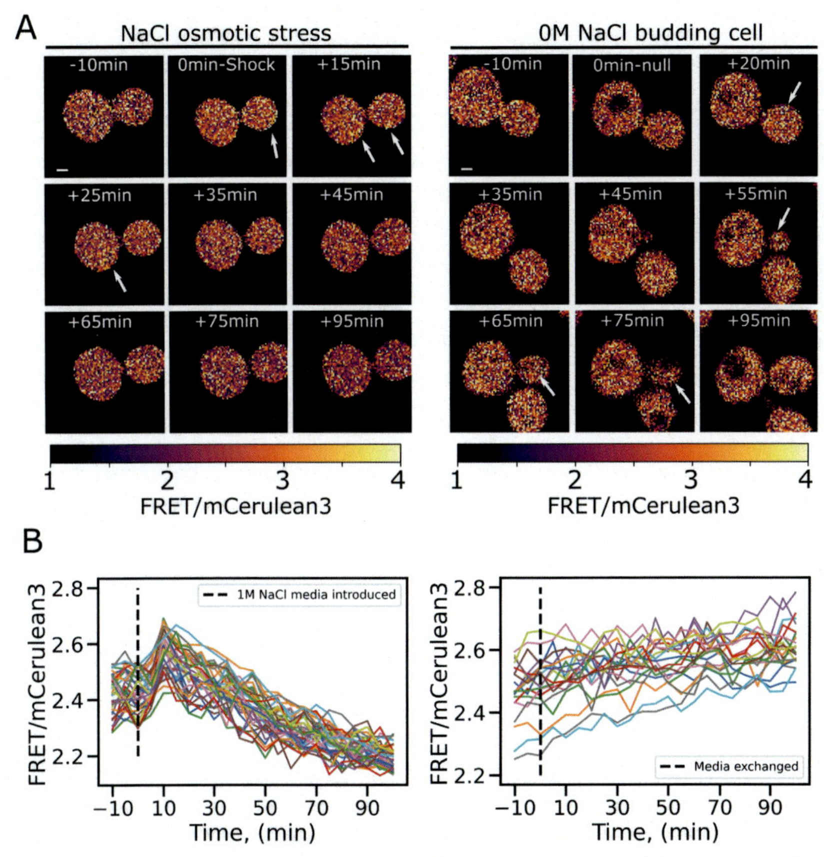

Fig. 6 Mapping molecular crowding in single cells. (A) Heatmap of the ratio FRET from confocal images acquired during time lapse experiment over 90 min. The panel on the left shows cells experiencing osmotic shock with 1 M NaCl after 20 min and imaged during recovery. On the right panel micrographs show cells left to grow for 90 min in non-stress media where a budding event can be observed on the right panel (white arrow). (B) Ratiometric plot through time for each cell revealing the crowding homogeneous behavior across the cell population.

ratiometric heat maps of cells that have undergone 1 M NaCl osmotic shock after 20 min in media lacking salt (Fig. 6A, left panel) and ratiometric maps of a cell budding in standard growth conditions (Fig. 6A, right panel). In both conditions, we see a heterogeneous distribution of values across the cytoplasmic volume, and an overall increase of crowding values when 1 M NaCl shock is introduced. We also qualitatively observe low and high localized regions of crowding values in budding cells in the absence of osmotic stress (Fig. 6, white arrows highlighting hot-spot regions).

After image registration and heat map analysis, cell segmentation was performed using the YeastSpotter deep learning model (Lu, Zarin, Hsu, & Moses, 2019), and whole cell ratiometric FRET values were individually tracked using a simple centroid tracking method to plot cell crowding through time. Fig. 6B shows cell-by-cell tracking analysis, where the mean FRET efficiency for each cell is plotted against time. We see that in the shocked population crowding rises sharply shortly after stress media is introduced, but over time the cells recover and the ratiometric FRET reduces to and beyond its initial value (Fig. 6B, on the left). Meanwhile, ratiometric FRET increases slowly in non-stress media which we hypothesize is associated with the replicative cycle of *S. cerevisiae* (Fig. 6B, on the right).

2.3 Macromolecular crowding in the region of the bud neck

To further investigate local crowding dynamics during cell division, we focussed on the bud neck, as the narrow region connecting mother cell and daughter cell during mitosis, key to establishment polarity and molecular traffic between the two cells (Faty et al., 2002; Perez & Thorner, 2019). We selected several markers of the bud neck (Myo1, Cdc1, Cdc12 and Hof1) tagged with super-folder GFP (sfGFP) and confirmed their localization to this region (Weill et al., 2018) by confocal microscopy, and further demonstrated that the tag did not perturb growth (Figs. S1A and S1B in Supplementary Material in the online version at https://doi.org/10.1016/bs.ctm.2021.09.001). We observed ring formation along bud emergence and the splitting event at the end of division for cells expressing sfGFP-Hof1 (Fig. 7A; Video 1a and 1b in Supplementary Material in the online version at https://doi.org/10.1016/bs.ctm.2021.09.001).

We then resolved the 3D bud neck structure using time-lapse AiryScan microscopy (Huff, 2015) to image the fluorescent reporter sfGFP-Hof1. The 3D structure showed the characteristic bud neck structure which consists of two parallel doughnut-like septin rings (Fig. 7B; Video 2 in Supplementary Material in the online version at https://doi.org/10.1016/bs.ctm.2021.09.001). The measured average dimension of bud necks were 0.57 μm in thickness along the mother daughter axis and 0.89 μm in apparent diameter (Fig. 7C) consistent with other measurement (Li et al., 2021) and gave an indication of bud neck dimensions that we used to set spatial parameters in subsequent analysis. Measurement of local crowding from either side of the bud neck was performed using a bespoke semi-automatic analysis workflow which automatically generated regions of set width around a

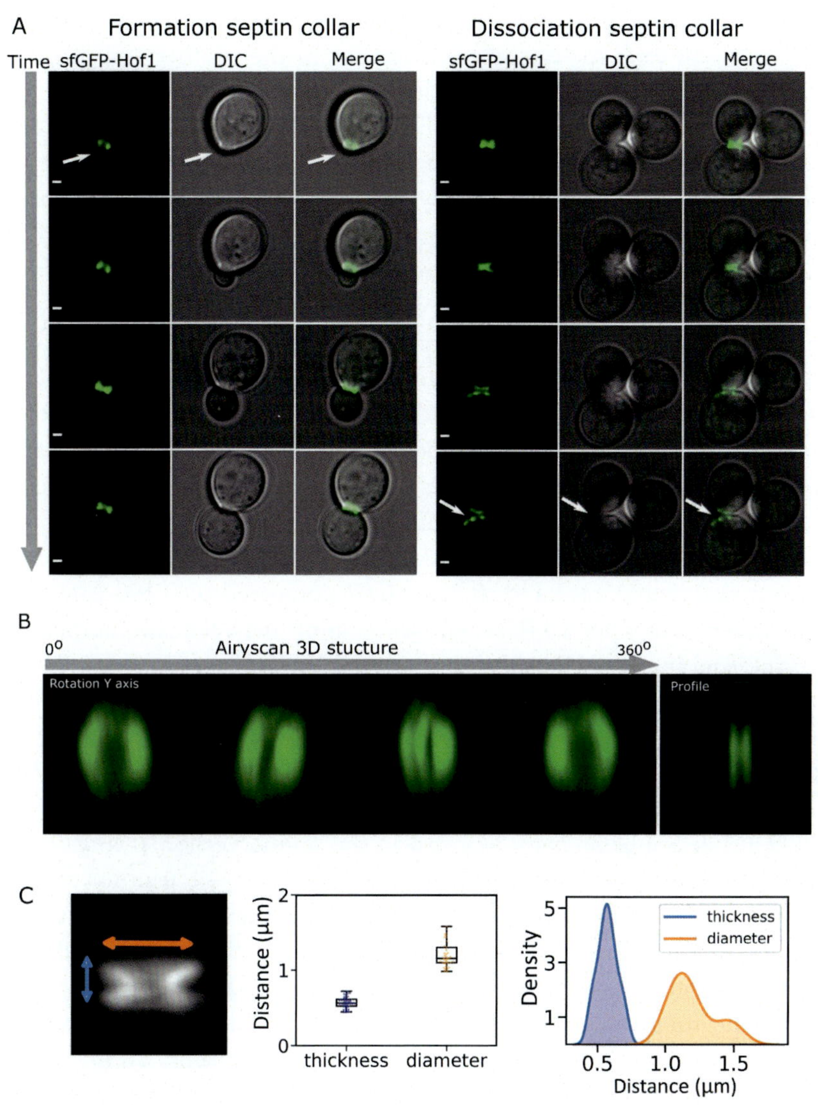

Fig. 7 Bud neck imaging. (A) Confocal image of sfGFP-Hof1 expressed in budding yeast. Showing the fluorescence channel with sfGFP-Hof1, the DIC gray channel and the merge between the two channels. Scale bar: 1 μm. Right: visualization formation of the bud neck. Left: Visualization cytokinesis event with dissociation of the septin rings- Both indicated with white arrows. (B) 3D structure of the bud neck resolved using Airyscan confocal microscopy, 24 slices of 0.18 μm spacing allowing to fully capturing the bud neck volume. The micrograph shows the 3D volume at different angle of rotation along the y-axis, revealing the doughnut like structure and the profile picture showing the width of the two visible septin contractile rings. (C) Bud neck dimensions. Showing visual of z-projected image of 3D bud neck airy scan stack image (left) with Jitter (center) and KDE plot distribution (right) of the bud neck measured thickness and diameter.

user-specified bud neck. Three regions were defined—one for the bud neck itself and one each for the adjacent regions in the mother and daughter cells. A width of 0.5 μm was specified for the bud neck area as indicated by our length and width quantification. In the mother and daughter cells we analyzed only 200 nm region immediately adjacent to the bud neck in each cell (Fig. 8A). We additionally separated cells into three categories, one grouping small buds at the beginning of the division process which we defined as buds with an area below 3 μm^2. Large buds, with cell volume comparable to the mother cell and which are close to scission event were defined as buds with area above 7 μm^2. All remaining bud sizes were defined as medium size (Fig. 8B). For all three categories the daughter cell maintains a higher ratiometric FRET readout through the cell cycle at the immediate region next to the bud neck while there is a significantly lower FRET at the equivalent region in the mother cell. We measured a mean FRET efficiency of 0.263 ± 0.08 ($\pm$ SD) for the daughter cell and 0.188 ± 0.05 for the mother cell for the small bud, a 28% difference. For medium bud sizes, the buds have a mean ratiometric FRET of 0.285 ± 0.07 compared to 0.190 ± 0.04 for the mother cells a 68% FRET efficiency jump between the mother cell and the daughter cell. For large buds, we find a mean ratiometric FRET of 0.294 ± 0.07 for the daughter and 0.195 ± 0.05 for the daughter equivalent to a 66% jump. Therefore, using our automatic workflow for analysis, all conditions show significant differences either side of the bud neck, highlighting a polarized crowding trend during replication (Statistic Table 1C in Supplementary Material in the online version at https://doi.org/10.1016/bs.ctm.2021.09.001). We found this to be true for all cells imaged which indicates that the asymmetry between mother and daughter cells is a stable and maintained condition throughout division (Fig. 8B).

We hypothesized that the cytoplasmic region adjacent to the plasma membrane might exhibit higher molecular crowding as surface proteins with cytoplasmic regions are organized in specific domains. To investigate if there were any local crowding difference close to the plasma membrane we analyzed regions expressing the crGE sensor immediately next to the cellular periphery. For each segmented cell we also found the average ratiometric values inside a virtual ring of 200 μm thickness at the periphery of the membrane (outer ring) vs the rest of the cells signal (without the outer ring). This analysis showed non-significant differences between the outer ring, the cell area excluding the outer ring and the whole cell measurement (Fig. 8C; Statistic Table 1D in Supplementary Material in the online version at https://doi.org/10.1016/bs.ctm.2021.09.001). The overall region near

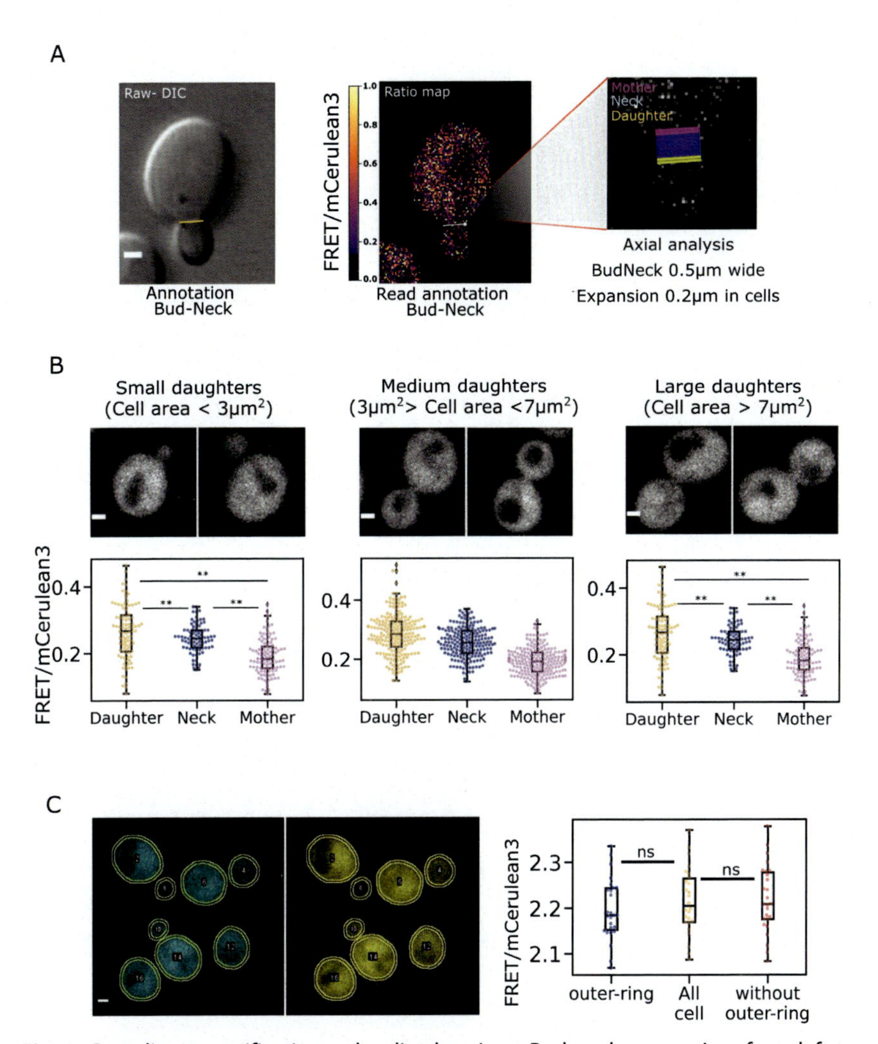

Fig. 8 Crowding quantification at localized regions. Bud neck annotation, from left to right showing raw image (DIC channel) with manually drawn line between mother cell and daughter cell to define the region of interest (yellow line), Scale bar: 1 μm. The raw annotated image is read by our bespoke python utility, visual in the center showing ratiometric map with bud neck line in white, white dot as the indicator of the line orientation, to determine mother and daughter cell position. In the left output visual of the area measured using our python-based analysis code, in pink and yellow the area of 200 nm, respectively, entering the mother cell and the daughter cell, from the defined region of the bud neck in blue. (A) Crowding readout at the bud neck. The figure show an example of each cell category, the small bud with bud size inferior at 3 μm^2, large buds with area above 7 μm^2 and medium category with all the intermediated daughter cell measured. Bellow each category, the respective Jitter plots representative of the FRET efficiency ensured at the bud neck extremity of the mother cells, the defined bud region and the daughter cell. Double asterisk indicates non-parametric Brunner-Munzel test less than 0.005. Scale bar: 1 μm. (B) Average ratiometric FRET in the area underneath the plasma membrane. 200 nm ring at the cytoplasmic periphery of the cell drawn and measured by ImageJ/Fiji macro as presented in the images with segmentation outlines in yellow. (C) Intensity values measured in both channels to calculate FRET/mCerulean3 ratio plotted on the right against the outer ring, the whole cell and the cell without this outer ring region. NS indicate a non-significant correlation between the data with a non-parametric Brunner-Munzel test *P* value greater than 0.05.

the membrane therefore appears to experience crowding equivalent to the rest of the cytoplasmic space in the cell. Although the cell membrane is known to be a crowded and highly dynamic region, processes happen on a millisecond timescale and from these data it appears that only highly constrained environment such as the bud neck can sustain a measurable difference in crowding.

2.4 Vacuole inheritance during replication imaged with FM4–64

To assess if these bud neck crowding events correlate with organelle inheritance, we used time lapse microscopy to follow the inheritance of the vacuole, which is conveniently both a large organelle that is inherited early in the budding process (Li et al., 2021). Using a mNeonGreen tagged version of the uracil permease Fur4, which localizes to the plasma membrane and also the vacuolar lumen (Paine, Ecclestone, & MacDonald, 2021), we could track inheritance over time (Video 3 in Supplementary Material in the online version at https://doi.org/10.1016/bs.ctm.2021.09.001). We also labeled the vacuole by performing a pulse-chase with media containing the red-fluorescent dye FM4–64 (Vida & Emr, 1995). This dye does not diffuse freely through the plasma membrane but instead gets internalized by endocytosis and stains the yeast vacuolar membranes (Fischer-Parton et al., 2000; Petranovic et al., 2010). By tracking the inheritance of FM4–64 labeled vacuoles over time in cells co-expressing either the polarized v-SNARE protein Snc1, or the bud neck marked sfGFP-Hof1 strain we observe vacuole inheritance events. The process observed can be decoupled in three steps. First, the initial vacuole deformation forming an apparent protrusion migrating toward the bud neck region. Then, following by a crossing event, where the vacuole starts occupying the bud neck region and cross the bud neck cell with progressive transport and relaxation toward the daughter cell cytoplasm. Finally, the scission event occurs freeing almost instantly at the bud neck site (Fig. 9 and Video 4a and 4b in Supplementary Material in the online version at https://doi.org/10.1016/bs.ctm.2021.09.001). We also noticed some rare events of maternal vacuole retraction at the earlier stage of the crossing phase, and thus failure of efficient vacuolar inheritance during these experiments (Video 5a and 5b in Supplementary Material in the online version at https://doi.org/10.1016/bs.ctm.2021.09.001). Stable and timely occupancy of the near bud neck region appears therefore critical to undergo vacuolar migration, this suggests the existence of an adaptive transition before organelles engage crossing. This hypothesis is consistent with our crowding observation at the bud neck. The crowding

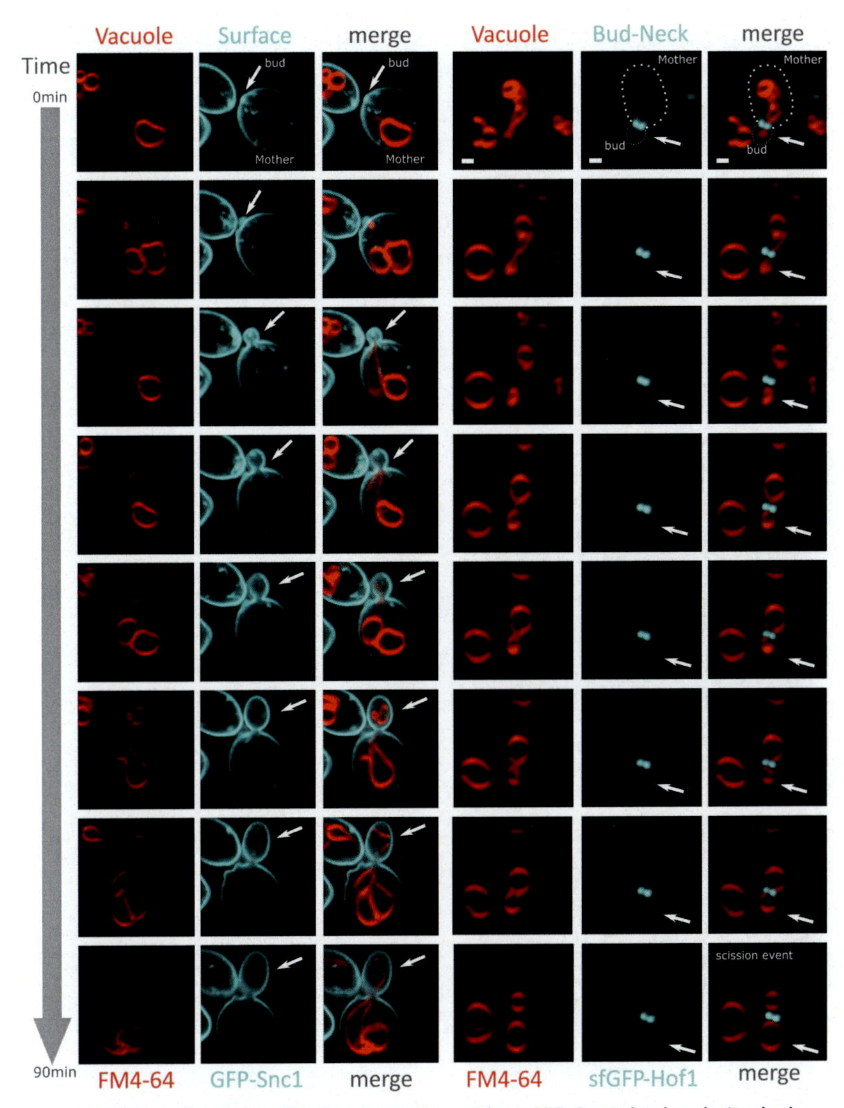

Fig. 9 Vacuole inheritance. Strain expressing either GFP-Snc1 bud polarized plasma membrane protein and sfGFP-Hof1 bud neck specific protein, combined with pulse and chase staining with FM4–64 to image the vacuole. Micrographs show each channel for the vacuole in red (FM4-64) and local marker in cyan (GFP & sfGFP), as well as the merge. White arrows to indicate bud/daughter cell position. Scale bar: 1 μm.

differential between the two regions boarding the bud neck, in the mother and in the daughter cell, could influence and disturb organelles diffusivity (Video 6a and 6b in Supplementary Material in the online version at https://doi.org/10.1016/bs.ctm.2021.09.001).

2.5 Simultaneous crowding sensing and vacuole visualization during cell division: Combined crGE and FM4–64 labelling

We developed a three-color experiment where the crGE sensor is expressed in cells labeled with FM4–64. We optimized imaging settings so that FRET signal is captured first using the FRET emission filter and excitation condition (see Section 4). Following this we performed an immediate acquisition of FM4–46 exited by a 561 nm argon laser. Fluorescence micrographs and excitation/emission spectrum for these experiments are shown Fig. 10A and B. We tested this set up for cells exposed to strong osmotic stress to confirm we have not impaired crowding sensing quality of our

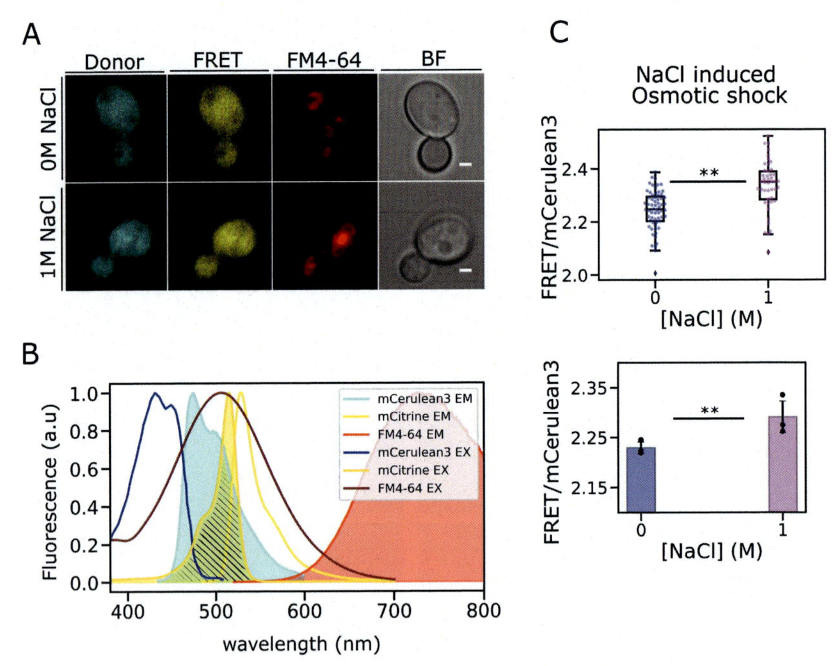

Fig. 10 Crowding quantification with simultaneous vacuole imaging. (A) Micrographs of three-color imaging for 0 M NaCl and 1 M NaCl. Scale bar: 1 μm. (B) Excitation and emission spectra for mCerulean3, mCitrine and FM4–64 M. The excitation laser for FM4–64 is 561 nm which is above the excitation spectra of mCerulean3 (dark blue) and mCitrine (golden yellow). However, acquisition is set so that the FRET signal is acquired before the vacuole marker to minimize impact on the other fluorophore. (C) Quantified crowding for *S. cerevisiae* grown in 2% glucose expressing crGE and labeled with FM4–64, imaged with 0 or 1 M NaCl. Inset jitter plot shows a representative crowding dataset for these conditions. Below: box plot of three biological replicates with standard deviation error bar. Double asterisk represents non-parametric Brunner-Munzel test with $P < 0.05$.

crGE sensor, analysis showed we maintained a statistically significant increase of FRET efficiency between the two conditions, reflecting the typical crowding response occurring upon osmotic stress (Fig. 10C and Statistic Table 1E in Supplementary Material in the online version at https://doi.org/10.1016/bs.ctm.2021.09.001).

Secondly, we optimized cell growth synchronization of wild-type MATa cells to focus on cell division events during crowding analyses. Cell cycle arrest at the G1 phase of the cell cycle was performed by incubating cells in 10 μM α-factor for 2 h (Marina Robles, Millán-Pacheco, Pastor, & Del Río, 2017; Udden & Finkelstein, 1978), leading to the forming of polarized protrusions termed shmoos (Merlini, Dudin, & Martin, 2013) (Fig. 11). After exchanging buffer to one without α-factor, we observed a return to cell division (Fig. 11B and Video 7 in Supplementary Material in the online version at https://doi.org/10.1016/bs.ctm.2021.09.001). After 8 h growth time we confirmed that all cells were back to a budding elliptical shape but conserving cell cycle synchronicity across the population (Fig. 11C). We noticed that 4 h post α-factor release some cells were still dividing from a shmoo-like phenotype, forming an unwanted subpopulation of cells which were dividing but without forming a regular bud neck. We believe this method is valuable in the field to study yeast dynamics in a synchronized population as it can be applied to any blue-yellow FRET sensor or could be used with green and red fluorescent reporter tags for protein localization studies.

2.6 Tracking single FRET reporters in vivo

Finally, we used our Slimfield microscope to track individual crGE2.3 which were endogenously expressed in the same way as the crGE. The two sensors are identical except for the fluorescent proteins used, with crGE2.3 making use of mEGFP and mScarletI as donor and acceptor respectively to make use of the greater single-protein intensities compared to mCerulean3 and mCitrine (Mouton et al., 2020), with the CrGE2.3 thus being suitable for single-molecule tracking. Here we used our redeveloped single molecule tracking code PySTACHIO (Shepherd, Higgins, et al., 2021b) which performs two-channel tracking as well as colocalization analysis using overlap integrals alongside straightforward distance cutoffs as described in Section 4. The localization was used to calculate the normalized fret NFRET which is possible for this FRET pair due to low spectral overlap and cross-excitation. Detected foci and FRET localizations are shown in Fig. 12A.

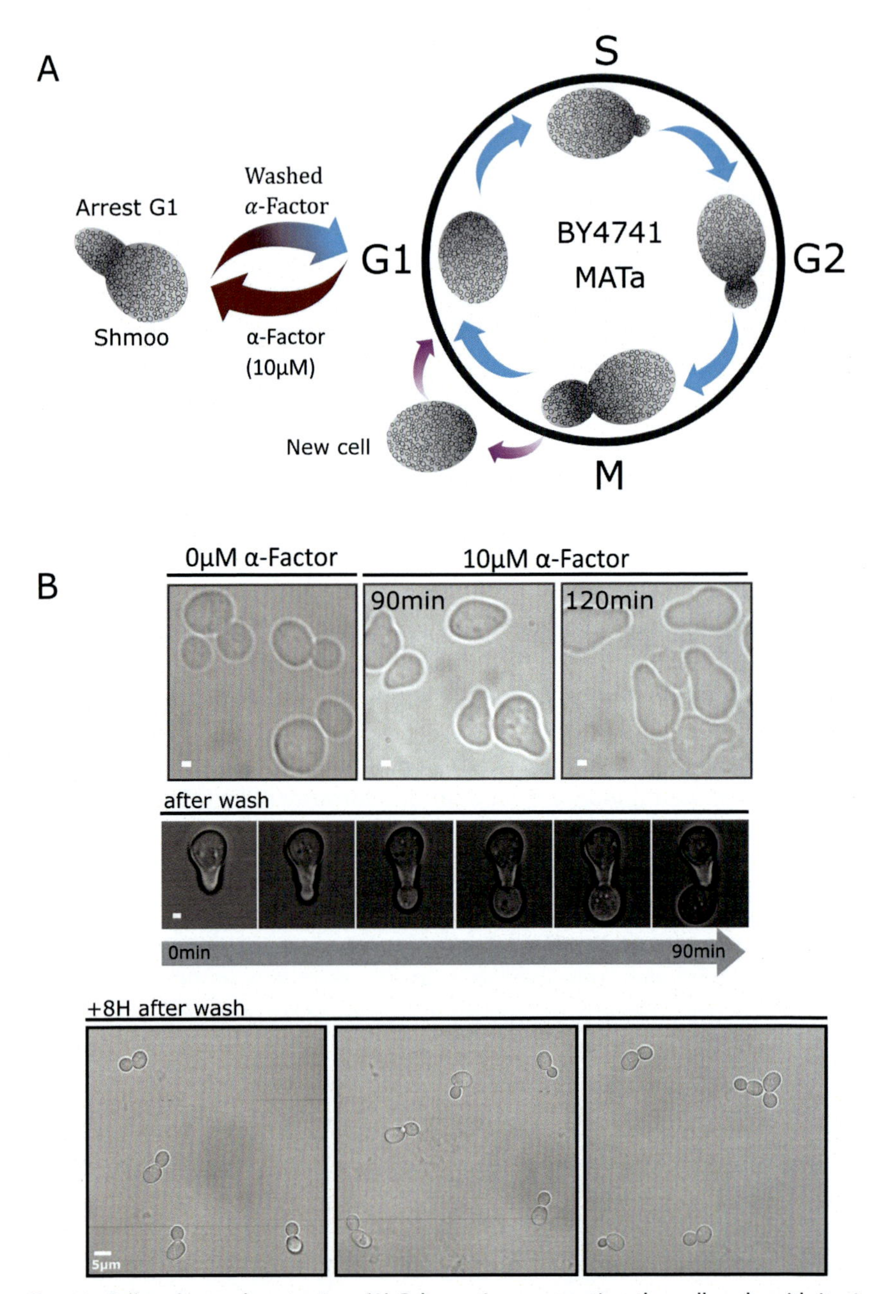

Fig. 11 Cell cycle synchronization. (A) Schematic representing the cell cycle with its 4 phases (S,G1,M,G2) highlighting arrest of the cycle at G1 phase when MATa yeast cells are exposed to 10 μM α-factor, cells stop dividing and adopted the shmoo phenotype. (B) Brightfield images of shmoos formed after 90 and 120 min incubations with 10 μM α-facto (scale bar: 1 μm). Micrograph below shows a shmoo entering cell division immediately after removal of α-factor from the media (scale bar: 1 μm), and a micrograph 8 h after removal showing synchronized budding of yeast cells (scale bar: 5 μm).

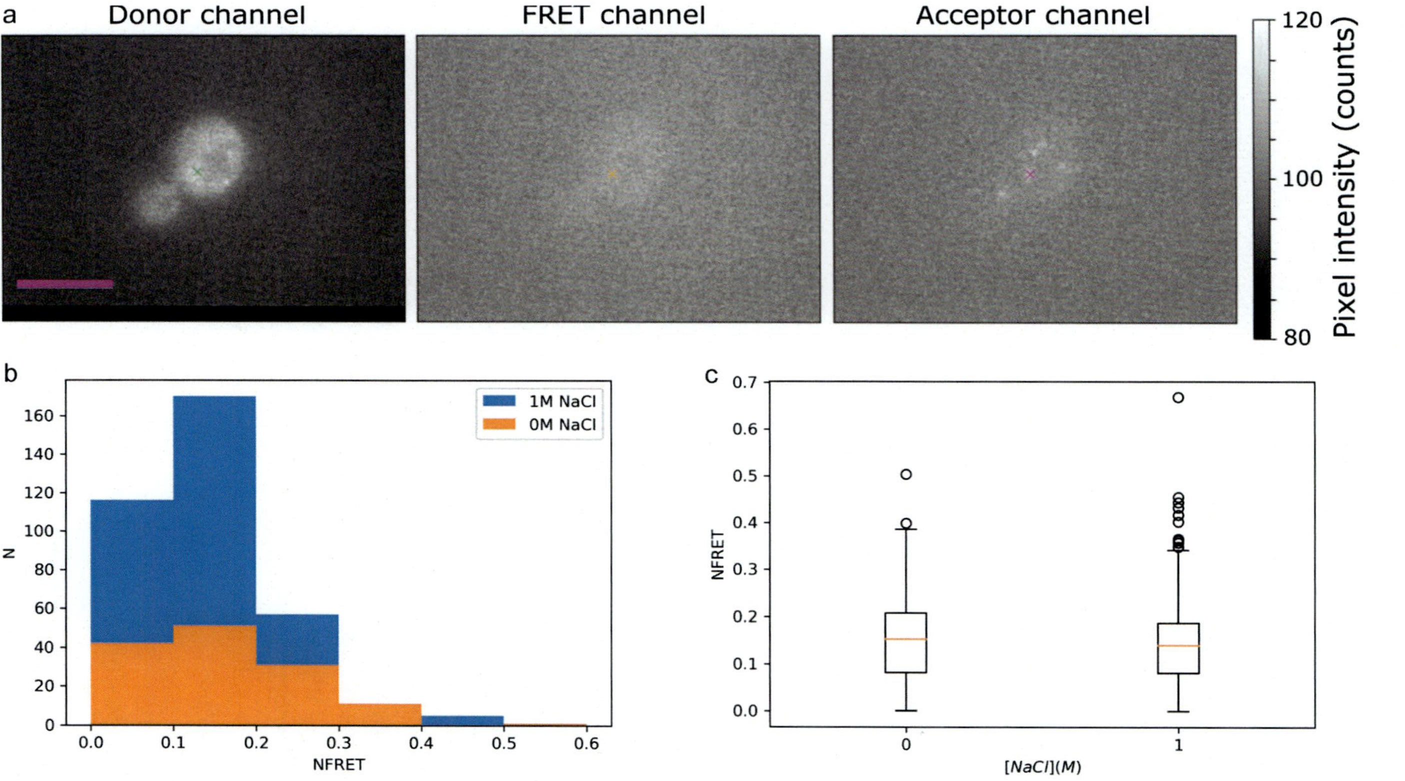

Fig. 12 Slimfield microscopy for single molecule detection inVivo of crGE2.3. (A) Representative colocalized data from our Slimfield experiments in the three channels post-registration. In the donor channel the localized focus is shown with a green gross, an orange cross is used in the FRET channel to show the average position of the donor and acceptor which is used to measure FRET intensity, and in the acceptor channel we represent the magenta cross shows the localized position of the acceptor. (B) Histogram of the NFRET data for *S. cerevisiae* in 0 and 1 M NaCl. (C) Boxplot of the same data. Bar: 5 μm.

Fig. 12B and C show the histogram and box plot of single molecule FRET values taken from yeast cells in both 0 and 1 M NaCl conditions. We see that the distributions of the two conditions' smFRET values is highly similar with means 0.17 and 0.16, respectively, in contrast to our whole-cell measurements which clearly show a high shift in FRET under osmotic stress. This can be explained by the photophysics of the system however. Given that FRET pairs bleach asymmetrically we ensure that we are tracking a functional FRET pair by colocalizing the donor and acceptor channels. However, as FRET increases, more energy is transferred to the acceptor from the donor and the donor intensity decreases in that image channel. In general, we are able to localize foci with intensity above ca. 0.7 GFP molecules (Miller et al., 2015). With FRET increasing, this limit will quickly be reached, and the donor intensity will drop, reducing the SNR and making localization of the donor fluorophore impossible. In effect we are therefore only sampling the low FRET pairs, not the full distribution. We believe that this is the cause of our semi-anomalous result, though we also note that in general Slimfield image acquisitions take several minutes and therefore some of the yeast cells may have begun to recover. To counteract this in the future it may be necessary to perform single-molecule imaging only on sensors which are tagged to known (semi-)static structures or organelles so that the exposure time can be increased, and to make use of microfluidics so that cells are imaged only immediately after the stress condition is introduced. Here we used an exposure of 5 ms and found FRET colocalizations for only ca. 1% of the localized foci. We could not therefore increase exposure significantly in this highly diffusive regime of cytosolic crGE2.3 sensors. However, an increase to, e.g., 40 ms would result in better sampling of the system as well as brighter foci for improved tracking and signal to noise ratios. Similarly, using the even brighter CRONOS sensor (Miyagi et al., 2021) which uses mNeonGreen in place of mEGFP would improve sampling efficiency. We note that this explanation of the NFRET similarity is supported by the higher number of high FRET outliers in the 1 M NaCl data in Fig. 12c—with greater sampling we may observe a greater difference between conditions.

3. Conclusion and discussion

Understanding intracellular crowding dynamics is highly challenging with crowding playing a role with a range of biomolecular processes and therefore difficult to isolate. Recent progress in optical microscopy and the development of FRET crowding sensor now allows us to access real-time quantification of molecular crowding under stress conditions or during replication.

In our present study, we performed a range of complementary characterization using a FRET based crowding sensor to quantify intracellular crowding between daughter cells and mother cells in budding yeast. We show a stable level of crowding between mother and daughter cells, irrespective of cell size. No particular trends were identified either between mother and daughter cells as a function of cell area (Fig. 5). This supports the idea that crowding is a critical homeostasis parameter which is closely linked to cell viability and integrity. Previous studies with the FRET sensor (Mouton et al., 2019) tracking macromolecular crowding through replicative aging also showed crowding stability between mother and daughter cells was maintained. Aging and the budding process are both physiological activities with morphological and cytoplasmic transformation generating two distinct groups with many differences such as organelle size and the accumulation of aging and metabolic by-products. Even with these differences however, the overall physical crowding remains stable.

However, in the face of sudden cellular stress changes in cell morphology or cellular content affects crowding and cell integrity. To test crowding recovery under osmotic stress we salt shocked yeast and imaged their crowding readout through time, which we analyzed on whole-cell levels with deep learning segmentation but also in a semi-quantitative subcellular way through heatmap generation. Here we see that crowding readout is not uniformly distributed throughout the cytosol but has a range distribution of values with local hotspots and dynamic evolution through time and along cell division (Fig. 6). Complementary whole cell analysis verified however that the population-level dynamics were largely homogeneous.

We finally targeted the local region at the bud neck and found a difference in ratiometric FRET and hence molecular crowding between the mother and daughter cells at the immediate region bordering the bud neck, with the mother having lower crowding readout than the daughter cell. This interesting result suggests a local crowding polarity between the mother cell and the daughter cell at the bud neck. We hypothesize that this is a consequence of content packaging in the daughter combined with its small volume in expansion. One can however hypothesize a greater role for this observed crowding difference, such as being an effective diffusion barrier during cell division ensuring that only actively transported cargoes reach the growing bud. We also noticed at the bud neck FRET efficiency values are intermediate between the immediate region in mother cells and the one in the daughter cell, which can be interpreted as a local crowding gradient. We therefore speculate crowding in this region to be a stable active marker of the mother/daughter polarity maybe contributing to actively limit the diffusion of freely diffusing material from the mother cell to the daughter cell for example limiting the free diffusion of aging harmful component such as protein aggregates, and other aging by product (Fig. S2 in Supplementary Material in the online version at https://doi.org/10.1016/bs.ctm.2021.09.001). As opposed to metabolically transported complexes such as the vacuole inherited from the mother cell to the daughter with acute organization and control, mainly via polarized cytoskeleton dependant transport processes involving molecular motors (Fig. 9). Noticeably, essentials cellular compartment are reported inherited via metabolic active transport such as the nucleus (Spichal & Fabre, 2017), the endoplasmic reticulum (Du, Ferro-Novick, & Novick, 2004) and even mitochondria (Boldogh, Yang, & Pon, 2001) shown to interact with the cytoskeleton. These observations supporting the presence of a diffusion barrier at the bud neck sorting elements viable to enter in the daughter cells or to retain in the mother cells, where we effectively report locally a stable crowding profile between the two cells.

Globally, the presence of various cellular components and organelles progressively trafficked and accumulating in the daughter cell volume are to consider a possible driven force for the crowding gradient observed. Compartment such as mitochondria, and liposomes shows distinct densities if compared to the cytoplasm (Wang, Lilley, & Oliver, 2014), with distinct shape, size and mobility. If the connection between local crowding dynamic and physical properties within organelles (e.g., density, composition) was not clearly identified and elucidated in the field. Our analysis and reflexions however converge to consider occupancy rate and physical presence close to the bud neck as the main influence on local crowding in the cytoplasm and the diffusion of other molecules and elements in the area.

Organelles are shown inherited in a timely manner, vacuole and mitochondria are inherited at a similar stage of cell division (Li et al., 2021) with other compartment being inherited earlier such as peroxisomes when the bud start growing or inherited later such as the nucleus at the end of division. The crowding profile we report between the mother cell and daughter cell is also constant throughout cell division, within the three bud size categories measured and representing different state of the cell cycle (Fig. 8). This observation supports an effect via element occupancy between the two cells.

As part of our overarching aim to correlate molecular crowding and organelle trafficking events during cell division we have also presented here our latest methodological progress to read out macromolecular crowding while simultaneously visualizing the vacuole during cell division (Fig. 7), demonstrating compatibility for a three-color imaging experiment where crGE (cyan mCerulean3 donor and yellow mCitrine acceptor) conserves its sensing properties when the vacuole was labeled with red dye FM4–64 (Fig. 8). Additionally, we have synchronized yeast cells using α-factor to arrest cell division at the G1 stage, and optimized conditions to ensure that no cells remain in a shmoo state after α-factor release (Fig. 10). Synchronizing the cell populations will be a crucial step for future experimentation to correlate organelle inheritance events with molecular crowding, and more generally to follow cycle timed events and investigate crowding dependant diffusion barrier between mother cell and daughter cell.

Lateral diffusion barriers in budding yeast have mainly been described as membrane dependant, at the bud neck it was shown that stressed ER confinement in the mother cell was Bud1-GTPase and sphingolipid dependent (Clay et al., 2014). Meanwhile other studies have shown that a diffusion barrier does not exist for the vesicle membrane receptor Snc1 which is locally retained via diffusion at the daughter cell membrane independent from the bud neck structure (Valdez-Taubas & Pelham, 2003; Fig. 9). Very little is known about these diffusion barrier dynamics in cells and how they are regulated. It however raises lots of questions and hypotheses regarding crowding modulation at the plasma membrane and cooperation between plasma membrane and cytoplasm dynamics. These powerful and complex processes tied to cellular organization and localization dynamic are closely linked to cellular fate. With new technology and sophisticated fluorescent physicochemical sensors we anticipate that these methods will enable us in the near future to unravel details of the physics of life at the single-molecule level (Leake, 2013), especially when used for emerging correlative microscopy approaches that allow multiple orthogonal data types to be acquired (Leake, 2021) including potential molecular orientational information from super-resolved fluorescence polarization microscopy (Shepherd, Payne-Dwyer, Lee, Syeda, & Leake, 2021). This may ultimately enable the crucial interplay of trafficking, diffusion, and crowding to be finally elucidated.

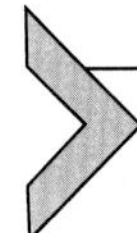

4. Materials and methods

4.1 Cell strains used

Yeast cells harboring a stable integration of the mCerulean3/mCitrine FRET crGE sensor at the *HIS3* locus, expressed from the constitutive *TEF1* promoter were used for molecular crowding measurements as described previously (Shepherd et al., 2020). The BY4741 strain from the genome SWAp-tag yeast library were used to visualize the bud neck, specifically sfGFP-Hof1, sfGFP-Cdc1, sfGFP-Cdc2; sfGFP-Myo1 expressed from the *NOP1*-promoter (Weill et al., 2018).

4.2 Cell culturing

Yeast expressing the FRET sensor were grown in synthetic drop-out media lacking Histidine (2% glucose, 1 × yeast nitrogen base; 1 × amino acid and base drop-out compositions (SD-His, Formedium Ltd., UK). Cultures were grown to mid-log phase (OD_{600} =0.4–0.6) prior to harvesting to preparation for optical microscopy.

4.3 FM4–63 vacuole labelling

Prior to imaging live cells were labeled with 0.8 μM FM4–64, incubated in YPD rich media for 1 h, and then washed twice with drop-out media prior to a 1-h chase period in synthetic drop-out media, before sample preparation and imaging.

4.4 Yeast cell synchronization

Mid log phase BY4741 cells of mating type "a" were arrested in G1 stage (Chen & Davis, 2000), after incubation for 120 min with 10 μM α-factor pheromone (Zymo-Research Corp, distributed by Cambridge Bioscience LTD, UK). Cells were spun down and washed twice in synthetic drop-out media and left to grow for up to 8 h before imaging to ensure several replications occurred and no cells exhibited a shmoo phenotype.

4.5 CrGE in vitro

CrGE purified as originally previously describe in the literature (Boersma et al., 2015). Phenylmethylsulfonyl fluoride (PMSF) in the lysis buffer was replaced by the same concentration of 4-(2-aminoethyl)benzenesulfonyl fluoride hydrochloride (AEBSF), a less toxic alternative for protease inhibition.

4.6 Sample preparation

Cells were imaged either in flow channel tunnel slides using 22 × 22 mm glass coverslip (No. 1.5 BK7 Menzel-Glazer glass coverslips, Germany) coated with 20 μL of 1 mg/mL Concanavalin A (ConA). Slide preparation was perform as previously described (Shepherd et al., 2020). After washing the ConA with 200 μL of imaging media, 20 μL of cells were flowed in, the slide was incubated, inverted for 5 min in a humidified chamber for adhesion to the ConA coated coverslip and washed again with 200 μL of imaging media, sealed with nail varnish ready for imaging.

Or using 35 mm glass-bottom dishes (Ibidi GmbH, Germany) were first treated with 300 μL of 1 mg/mL Concanavalin A (ConA) for 5 min, washed x3 with sterile distilled water and dried in a sterile laminar flow hood. Mid-log phase culture was diluted to $OD_{600} = \sim 0.2$ before being adhered to treated coverslips for 5 min at ambient temperature and then washed three times with imaging media to remove any unattached cells. Media exchanges were performed using microfluidics as previously described (Laidlaw et al., 2021).

4.7 Confocal microscopy

Budding yeasts for mother/daughter cells analysis were acquired using a commercial laser scanning confocal microscope (LMS 710 & Axio Imager2, Zeiss) with 1.4 NA Nikon Plan-Apochromat 63× oil-immersion objective lens. mCerulean3 was excited using a 458 nm argon laser at 2.1% of maximum laser power with mCerulean3 emission filter set to 454–515 nm and FRET emission filter set to 524–601 nm. The pinhole size was set to 0.83 Zeiss standard Airy-Units According as in previous work (Shepherd et al., 2020).

Time lapse and three-color microscopy experiments were performed on a commercial laser scanning microscope (LSM880, Zeiss) equipped with an AiryScan module with a 1.4 NA Nikon Plan-Apochromat 63× oil-immersion objective lens. The pinhole size was set to 4.61 Zeiss standard Airy-Units for optimal cytoplasmic signal (equivalent of 3 μm slice/section). Images were taken using the following excitation lasers and imaging wavelength ranges: mCerulean3 458 nm (argon laser) at 1.5% of maximum power and emission filter set to 463/500 nm, FRET 458 nm and emission filter set to 525/606 nm.

For time lapse experiment cells were imaged at 5-min intervals for time-lapse experiments, including 10–15 min prior to exchange in standard SD media and for 90 min after introduction of stress media (SD supplemented by 1 M NaCl).

For crGE cells dyed with FM4–64, three-color experiments were performed by imaging sequentially the FRET signal as described above followed by a frame mode scanning acquisition exciting FM4–64 with a 561 nm argon laser at 5% of maximum power and emission filter set to 578/731 nm.

The 3D bud neck structure was resolved by AiryScan confocal microscopy, 24 slices of 0.18 μm spacing were acquired. Images were processed for 3D reconstruction with the commercial Zeiss ZEN software and further visualized with the Volume Viewer ImageJ plugin (Barthel, 2005) to generate micrographs.

4.8 Image analysis

Analysis of confocal data was performed with a bespoke Python 3 utility based around the YeastSpotter deep learning segmentation tool (Lu et al., 2019). For each detected cell, mean pixel intensities in the FRET and donor channels were extracted to calculate ratiometric FRET values. All plots were generated using the matplotlib (Hunter, 2007) and seaborn (Waskom, 2021) Python libraries.

For tracking individual cells through time, all DIC images were registered to the first frame using scikit-image (van der Walt et al., 2014). After the first frame was segmented by YeastSpotter, those labels were used for tracking each cell thereafter. To determine if cells were the same between subsequent frames n and $n+1$, we found the center of mass of a given cell in frame n and found that pixel in frame $n+1$. If that pixel was within a segmented region in frame $n+1$ we assumed the cells to be the same. Though in general this is an unreliable assumption, the surface immobilization by ConA and image registration meant that in this case we found it to be reliable. If cells were pairs between frames, they were given the same label and at the end of the analysis, each labeled region could then be tracked for FRET ratio through time and plotted individually. Here, any cells, which appeared part-way through the acquisition or were lost part-way through the acquisition to drift or surface detachment were neglected—that is, we only analyze through time those cells which were successfully segmented and tracked throughout the full 105 min experiment.

For axial analysis of the bud neck, images were first manually annotated in ImageJ/Fiji to define the bud neck area. The line tool was used to draw bud neck lines between mother cells and daughter cell. To identify bud position, lines are drawn in orientated manner: left to right for buds situated in the top part of the image, right to left if bud below the mother cell in the lower part of the image, top to bottom if bud in the right side and bottom to top if on the left. Specifying start and end points here effectively defines a vector, the vector product of which with the z axis (0,0,1) gives the vector pointing axially into the mother cell. Annotated images were analyzed with a bespoke Python 3 script. Here, a standard pre-defined area is applied around the bud neckline. From either side of the bud area, in the mother cells and in the daughter cell for a pre-set distance into each cell ratiometric FRET values were extracted. Here to ensure that no noise pixels were included in analysis, we segmented the cells through a Gaussian blur of 3 pixels width followed by Otsu thresholding. This generates an approximate mask which excludes most of the background, but which does not exclude noise pixels in the cells themselves. To do this we did not analyze any pixel with intensity less than the mean background plus three background noise standard deviations as we have reported previously (Shepherd et al., 2020). Again, plots were generated using the matplotlib and seaborn python libraries.

4.9 Single-molecule microscopy and analysis

Samples with surface immobilized log-phase *S. cerevisiae* were constructed as described above, using our crGE2.3-incorporating strain described previously (Shepherd et al., 2020). We used our Slimfield single-molecule microscope (Wollman & Leake, 2015) equipped with 488 nm and 561 nm emitting lasers (Obis LS and LX series, respectively) with power set to be around 20 mW at the sample plane for both lasers. We performed all experiments in epifluorescence mode to quickly photobleach most emitters and enter the single molecule regime. We took 5000 frames at 10 ms exposure, and analyzed frames 1000–2000 which we found to have a relatively high number of identifiable foci in both donor and acceptor channels.

Data was analyzed using PySTACHIO (Shepherd, Higgins, et al., 2021b) with snr_min_threshold = 0.5 and struct_disk_radius = 9 in Alternating Laser Excitation (ALEX) mode. We relaxed our usual constraints on trajectory length because ALEX mode with 10 ms exposure allows considerable diffusion between successive captures in one channel and thus trajectory linking is compromised in this single-molecule regime. Following PySTACHIO analysis, we began by finding the translation-only registration transformation between channels using brightfield images and the pystackreg library (Lichtner & Thévenaz, 2021). We then applied the registration transformation to the trajectories identified by PySTACHIO as well as to the fluorescence stacks. We performed colocalization using PySTACHIO with the distance cut-off set to 2 pixels and the overlap integral set to 0.75—this is the "true" colocalization metric, with the distance cut-off used to decrease computational cost by not calculating overlap integrals for foci which will fall beneath the overlap integral threshold. If spots in the donor and acceptor channel were accepted as being one FRET sensor we then estimated the FRET position using the mean position of the donor and acceptor, and found the summed intensity with local background correction as in PySTACHIO. Finally, we measured the normalized FRET parameter NFRET:

$$\mathrm{NFRET} = \frac{I_{FRET}}{\sqrt{I_D I_A}}$$

which has been reported as a crowding proxy previously (Mouton et al., 2020). These values were plotted using matplotlib (Hunter, 2007). Data analysis here was performed using a bespoke Python routine which made use of scikit-image (van der Walt et al., 2014), Pillow (Clark, 2015), numpy (Oliphant, 2010), and openCV (Bradski, 2000) for image data handling.

Acknowledgments

Arnold J Boersma (DWI-Leibniz Institute for Interactive Materials, Aachen, Germany) and Bert Poolman (European Research Institute for the Biology of Ageing, University of Groningen, The Netherlands) for crGE plasmid donation. Maya Schuldiner (Weizmann Institute of Science, Rehovot, Israel) for the shared SWAp-tag yeast library. Dr. Payne-Dwyer for assistance with slimfield microscopy, and Karen Hogg, Grant Calder, Graeme Park, and Karen Hodgkinson (Bioscience Technology Facility, University of York) for support with confocal microscopy.

Funding sources

This project has received funding from the European Union's Horizon 2020 research and innovation programme under the Marie Skłodowska Curie grant agreement no. 764591 (SynCrop), the Leverhulme Trust (reference RPG-2019-156), and BBSRC (reference BB/R001235/1) and Wellcome Trust and the Royal Society grant no. 204636/Z/16/Z.

References

Akabayov, B., Akabayov, S. R., Lee, S.-J., Wagner, G., & Richardson, C. C. (2013). Impact of macromolecular crowding on DNA replication. *Nature Communications*, *4*(1), 1–10. https://doi.org/10.1038/ncomms2620.

Allen, P. M. (2001). The influence of macromolecular crowding and macromolecular confinement on biochemical reactions in physiological media. *The Journal of Biological Chemistry*, *276*(14), 10577–10580. https://doi.org/10.1074/JBC.R100005200.

André, A. A. M., & Spruijt, E. (2020). Liquid–liquid phase separation in crowded environments. *International Journal of Molecular Sciences*, *21*(16), 1–20. https://doi.org/10.3390/ijms21165908.

Babazadeh, R., Lahtvee, P.-J., Adiels, C. B., Goksör, M., Nielsen, J. B., & Hohmann, S. (2017). The yeast osmostress response is carbon source dependent. *Scientific Reports*, 7(1), 990. https://doi.org/10.1038/s41598-017-01141-4.

Badrinarayanan, A., Reyes-Lamothe, R., Uphoff, S., Leake, M. C., & Sherratt, D. J. (2012). In vivo architecture and action of bacterial structural maintenance of chromosome proteins. *Science*, *338*(6106), 528–531. https://doi.org/10.1126/science.1227126.

Zimmerman, S. B., & Trach, S. O. (1991). Estimation of macromolecule concentrations and excluded volume effects for the cytoplasm of Escherichia coli. *Journal of Molecular Biology*, *222*(3), 599–620. https://doi.org/10.1016/0022-2836(91)90499-V.

Barthel, K. U. (2005). Volume viewer—Plugin. https://imagej.nih.gov/ij/plugins/volume-viewer.html.

Benny, P., & Raghunath, M. (2017). Making microenvironments: A look into incorporating macromolecular crowding into in vitro experiments, to generate biomimetic microenvironments which are capable of directing cell function for tissue engineering applications. *Journal of Tissue Engineering*, *8*, 1–8. https://doi.org/10.1177/2041731417730467.

Bhattacharya, A., Kim, Y. C., & Mittal, J. (2013). Protein-protein interactions in a crowded environment. *Biophysical Reviews*, *5*(2), 99–108. Springer https://doi.org/10.1007/s12551-013-0111-5.

Bi, E., & Park, H. O. (2012). Cell polarization and cytokinesis in budding yeast. *Genetics*, *191*(2), 347–387. https://doi.org/10.1534/genetics.111.132886.

Boersma, A. J., Zuhorn, I. S., & Poolman, B. (2015). A sensor for quantification of macromolecular crowding in living cells. *Nature Methods*, *12*(3), 227–229. https://doi.org/10.1038/nmeth.3257.

Boldogh, I. R., Yang, H.-C., & Pon, L. A. (2001). Mitochondrial inheritance in budding yeast. *Traffic*, *2*(6), 368–374. https://doi.org/10.1034/J.1600-0854.2001.002006368.X.

Bonny, A. R., Kochanowski, K., Diether, M., & El-Samad, H. (2020). Stress-induced transient cell cycle arrest coordinates metabolic resource allocation to balance adaptive tradeoffs. *BioRxiv*, 1–33. version. 1. https://doi.org/10.1101/2020.04.08.033035.

Bradski, G. (2000). The openCV library. *Dr. Dobb's Journal: Software Tools for the Professional Programmer*, *11*, 120–123.

Byers, B., & Goetsch, L. E. (1976). A highly ordered ring of membrane-associated filaments in budding yeast. *Journal of Cell Biology*, *69*(3), 717–721. https://doi.org/10.1083/jcb.69.3.717.

Cadet, J., Sage, E., & Douki, T. (2005). Ultraviolet radiation-mediated damage to cellular DNA. In *Vol. 571, issues 1–2. Mutation research—Fundamental and molecular mechanisms of mutagenesis* (pp. 3–17). Elsevier. SPEC. ISS. https://doi.org/10.1016/j.mrfmmm.2004.09.012.

Campisi, J., & D'Adda Di Fagagna, F. (2007). Cellular senescence: When bad things happen to good cells. *Nature Reviews. Molecular Cell Biology*, *8*(9), 729–740. https://doi.org/10.1038/nrm2233.

Canetta, E., Walker, G. M., & Adya, A. K. (2006). Correlating yeast cell stress physiology to changes in the cell surface morphology: Atomic force microscopic studies. *TheScientificWorldJOURNAL*, *6*, 777–780. https://doi.org/10.1100/tsw.2006.166.

Champion, L., Linder, M. I., & Kutay, U. (2017). Cellular reorganization during mitotic entry. *Trends in Cell Biology*, *27*(1), 26–41. Elsevier Ltd. https://doi.org/10.1016/j.tcb.2016.07.004.

Chasman, D., Ho, Y., Berry, D. B., Nemec, C. M., MacGilvray, M. E., Hose, J., et al. (2014). Pathway connectivity and signaling coordination in the yeast stress-activated signaling network. *Molecular Systems Biology*, *10*(11), 759. https://doi.org/10.15252/msb.20145120.

Chen, L., & Davis, N. G. (2000). Recycling of the yeast a-factor receptor. *The Journal of Cell Biology*, *151*(3), 731. https://doi.org/10.1083/JCB.151.3.731.

Chen, H., Howell, A. S., Robesand, A., & Lew, D. J. (2011). Dynamics of septin ring and collar formation in Saccharomyces cerevisiae. *Biological Chemistry*, *392*, 689–697. https://doi.org/10.1038/jid.2014.371.
Chiti, F., & Dobson, C. M. (2017). Protein misfolding, amyloid formation, and human disease: A summary of progress over the last decade. *Annual Review of Biochemistry*, *86*(1), 27–68. https://doi.org/10.1146/annurev-biochem-061516-045115.
Chiu, S. W., & Leake, M. C. (2011). Functioning nanomachines seen in real-time in living bacteria using single-molecule and super-resolution fluorescence imaging. *International Journal of Molecular Sciences*, *12*(4), 2518–2542. Molecular Diversity Preservation International. https://doi.org/10.3390/ijms12042518.
Christiansen, A., Wang, Q., Cheung, M. S., & Wittung-Stafshede, P. (2013). Effects of macromolecular crowding agents on protein folding in vitro and in silico. *Biophysical Reviews*, *5*(2), 137–145. Springer. https://doi.org/10.1007/s12551-013-0108-0.
Clark, A. (2015). Pillow *(pil fork) documentation*. Readthedocs. Https://Buildmedia. Readthedocs. Org/media/pdf/pillow/latest/pillow. Pdf.
Clay, L., Caudron, F., Denoth-Lippuner, A., Boettcher, B., Frei, S. B., Snapp, E. L., et al. (2014). A sphingolipid-dependent diffusion barrier confines ER stress to the yeast mother cell. *eLife*, *2014*(3), 1–23. https://doi.org/10.7554/eLife.01883.
Cramer, L. (2008). Organelle transport: Dynamic actin tracks for myosin motors. *Current Biology*, *18*(22), R1066–R1068. Cell Press https://doi.org/10.1016/j.cub.2008.09.048.
Davis, C. M., Deutsch, J., & Gruebele, M. (2020). An in vitro mimic of in-cell solvation for protein folding studies. *Protein Science*, *29*(4), 1060–1068. https://doi.org/10.1002/pro.3833.
Despa, F., Orgill, D. P., & Lee, R. C. (2006). Molecular crowding effects on protein stability. *Annals of the New York Academy of Sciences*, *1066*, 54–66. https://doi.org/10.1196/annals.1363.005.
Dong, H., Qin, S., & Zhou, H.-X. (2010). Effects of macromolecular crowding on protein conformational changes. *PLoS Computational Biology*, *6*(7), e1000833. https://doi.org/10.1371/JOURNAL.PCBI.1000833.
Du, Y., Ferro-Novick, S., & Novick, P. (2004). Dynamics and inheritance of the endoplasmic reticulum. *Journal of Cell Science*, *117*(14), 2871–2878. https://doi.org/10.1242/JCS.01286.
Duina, A. A., Miller, M. E., & Keeney, J. B. (2014). Budding yeast for budding geneticists: A primer on the Saccharomyces cerevisiae model system. *Genetics*, *197*(1), 33–48. https://doi.org/10.1534/genetics.114.163188.
Dujon, B. (1996). The yeast genome project: What did we learn? *Trends in Genetics*, *12*(7), 263–270. https://doi.org/10.1016/0168-9525(96)10027-5.
Ellis, J. R. (2013). *Protein misassembly: Macromolecular crowding and molecular chaperones*. https://www.ncbi.nlm.nih.gov/books/NBK6375/.
Faty, M., Fink, M., & Barral, Y. (2002). Septins: A ring to part mother and daughter. *Current Genetics*, *41*(3), 123–131. Springer https://doi.org/10.1007/s00294-002-0304-0.
Fischer-Parton, S., Parton, R. M., Hickey, P. C., Dijksterhuis, J., Atkinson, H. A., & Read, N. D. (2000). Confocal microscopy of FM4-64 as a tool for analysing endocytosis and vesicle trafficking in living fungal hyphae. *Journal of Microscopy*, *198*(3), 246–259. https://doi.org/10.1046/j.1365-2818.2000.00708.x.
Foury, F. (1997). Human genetic diseases a cross-talk between man and yeast. *Gene*, *195*(1), 1–10. https://doi.org/10.1016/S0378-1119(97)00140-6.
Franzmann, T. M., Jahnel, M., Pozniakovsky, A., Mahamid, J., Holehouse, A. S., Nüske, E., et al. (2018). Phase separation of a yeast prion protein promotes cellular fitness. *Science*, *359*(6371), eaao5654. https://doi.org/10.1126/science.aao5654.
Fulda, S., Gorman, A. M., Hori, O., & Samali, A. (2010). Cellular stress responses: Cell survival and cell death. *International Journal of Cell Biology*, 1–23. https://doi.org/10.1155/2010/214074.

Fulton, A. B. (1982). How crowded is the cytoplasm? *Cell, 30*(2), 345–347. Cell Press https://doi.org/10.1016/0092-8674(82)90231-8.

Galluzzi, L., Yamazaki, T., & Kroemer, G. (2018). Linking cellular stress responses to systemic homeostasis. *Nature Reviews. Molecular Cell Biology*, *19*(11), 731–745. https://doi.org/10.1038/s41580-018-0068-0.

Garner, M. M., & Burg, M. B. (1994). Macromolecular crowding and confinement in cells exposed to hypertonicity. *American Journal of Physiology—Cell Physiology*, *266*(4), C877–C892. https://doi.org/10.1152/ajpcell.1994.266.4.c877.

Gibson, B. R., Lawrence, S. J., Leclaire, J. P. R., Powell, C. D., & Smart, K. A. (2007). Yeast responses to stresses associated with industrial brewery handling. *FEMS Microbiology Reviews*, *31*(5). https://doi.org/10.1111/j.1574-6976.2007.00076.x.

Gladfelter, A. S., Pringle, J. R., & Lew, D. J. (2001). The septin cortex at the yeast mother-bud neck. *Current Opinion in Microbiology*, *4*(6), 681–689. https://doi.org/10.1016/S1369-5274(01)00269-7.

Groh, N., Bühler, A., Huang, C., Li, K. W., van Nierop, P., Smit, A. B., et al. (2017). Age-dependent protein aggregation initiates amyloid-β aggregation. *Frontiers in Aging Neuroscience*, *9*, 138. https://doi.org/10.3389/fnagi.2017.00138.

Hakem, R. (2008). DNA-damage repair; the good, the bad, and the ugly. *The EMBO Journal*, *27*(4), 589. https://doi.org/10.1038/EMBOJ.2008.15.

Hancock, R. (2014). Structures and functions in the crowded nucleus: New biophysical insights. *Frontiers in Physics*, *2*, 53. https://doi.org/10.3389/FPHY.2014.00053.

He, C., Zhou, C., & Kennedy, B. K. (2018). The yeast replicative aging model. *Biochimica et Biophysica Acta—Molecular Basis of Disease*, *1864*(9). https://doi.org/10.1016/j.bbadis.2018.02.023.

Higuchi-Sanabria, R., Pernice, W. M. A., Vevea, J. D., Alessi Wolken, D. M., Boldogh, I. R., & Pon, L. A. (2014). Role of asymmetric cell division in lifespan control in Saccharomyces cerevisiae. *FEMS Yeast Research*, *14*(8), 1133–1146. https://doi.org/10.1111/1567-1364.12216.

Hohmann, S. (2002). Osmotic stress signaling and osmoadaptation in yeasts. *Microbiology and Molecular Biology Reviews*, *66*(2), 300–372. https://doi.org/10.1128/mmbr.66.2.300-372.2002.

Hohmann, S. (2015). An integrated view on a eukaryotic osmoregulation system. *Current Genetics*, *61*(3), 373–382. https://doi.org/10.1007/s00294-015-0475-0.

Hsu, H. E., Liu, T. N., Yeh, C. S., Chang, T. H., Lo, Y. C., & Kao, C. F. (2015). Feedback control of Snf1 protein and its phosphorylation is necessary for adaptation to environmental stress. *Journal of Biological Chemistry*, *290*(27), 16786–16796. https://doi.org/10.1074/jbc.M115.639443.

Huan-Xiang, Z., Rivas, G., & Minton, A. P. (2008). Macromolecular crowding and confinement: Biochemical, biophysical, and potential physiological consequences. *Annual Review of Biophysics*, *37*, 375–397. https://doi.org/10.1146/ANNUREV.BIOPHYS.37.032807.125817.

Huff, J. (2015). The Airyscan detector from ZEISS: Confocal imaging with improved signal-to-noise ratio and super-resolution. *Nature Methods*, *12(12)*, i–ii. https://doi.org/10.1038/nmeth.f.388.

Hunter, J. D. (2007). Matplotlib: A 2D graphics environment. *Computing in Science and Engineering*, *9*(3), 90–95. https://doi.org/10.1109/MCSE.2007.55.

Jin, X., Lee, J. E., Schaefer, C., Luo, X., Wollman, A., Payne-Dwyer, A. L., et al. (2021). Membraneless organelles formed by liquid-liquid phase separation increase bacterial fitness. *Science Advances*, 7(43), eabh2929. https://doi.org/10.1126/sciadv.abh2929.

Jing, W., Qin, Y., & Tong, J. (2020). Effects of macromolecular crowding on the folding and aggregation of glycosylated MUC5AC. *Biochemical and Biophysical Research Communications*, *529*(4), 984–990. https://doi.org/10.1016/j.bbrc.2020.06.156.

Juanes, M. A., & Piatti, S. (2016). The final cut: Cell polarity meets cytokinesis at the bud neck in S. cerevisiae. *Cellular and Molecular Life Sciences*, *73*(16), 3115–3136. Springer. https://doi.org/10.1007/s00018-016-2220-3.

Knoblach, B., & Rachubinski, R. A. (2015). Sharing the cell's bounty—Organelle inheritance in yeast. *Journal of Cell Science*, *128*(4), 621–630. The Company of Biologists https://doi.org/10.1242/jcs.151423.

König, I., Soranno, A., Nettels, D., & Schuler, B. (2021). Impact of in-cell and in-vitro crowding on the conformations and dynamics of an intrinsically disordered protein. *Angewandte Chemie International Edition*, *60*(19), 10724–10729. https://doi.org/10.1002/ANIE.202016804.

Kültz, D. (2020). Evolution of cellular stress response mechanisms. *Journal of Experimental Zoology Part A: Ecological and Integrative Physiology*, *333*(6), 359–378. https://doi.org/10.1002/JEZ.2347.

Kuznetsova, I. M., Turoverov, K. K., & Uversky, V. N. (2014). What macromolecular crowding can do to a protein. *International Journal of Molecular Sciences*, *15*(12), 23090–23140. https://doi.org/10.3390/ijms151223090.

Laidlaw, K. M. E., Bisinski, D. D., Shashkova, S., Paine, K. M., Veillon, M. A., Leake, M. C., et al. (2021). A glucose-starvation response governs endocytic trafficking and eisosomal retention of surface cargoes in budding yeast. *Journal of Cell Science*, *134*(2), 1–16. https://doi.org/10.1242/jcs.257733.

Lamarche, B. J., Orazio, N. I., & Weitzman, M. D. (2010). The MRN complex in double-strand break repair and telomere maintenance. *FEBS Letters*, *584*(17), 3682–3695. NIH Public Access. https://doi.org/10.1016/j.febslet.2010.07.029.

Leake, M. C. (2013). The physics of life: One molecule at a time. *Philosophical Transactions of the Royal Society B: Biological Sciences.*, *368*(1611), 20120248. https://doi.org/10.1098/rstb.2012.0248.

Leake, M. C. (2021). Correlative approaches in single-molecule biophysics: A review of the progress in methods and applications. *Methods*, *193*, 1–4. https://doi.org/10.1016/j.ymeth.2021.06.012.

Lenn, T., & Leake, M. C. (2012). Experimental approaches for addressing fundamental biological questions in living, functioning cells with single molecule precision. *Open Biology*, *2*(6), 120090. https://doi.org/10.1098/rsob.120090.

Lenn, T., & Leake, M. C. (2016). Single-molecule studies of the dynamics and interactions of bacterial OXPHOS complexes. *Biochimica et Biophysica Acta: Bioenergetics*, *1857*(3), 224–231. https://doi.org/10.1016/j.bbabio.2015.10.008.

Li, X., & Cai, M. (1999). Recovery of the yeast cell cycle from heat shock-induced G1 arrest involves a positive regulation of G1 cyclin expression by the S phase cyclin Clb5. *Journal of Biological Chemistry*, *274*(34), 24220–24231. https://doi.org/10.1074/jbc.274.34.24220.

Li, K. W., Lu, M. S., Iwamoto, Y., Drubin, D. G., & Pedersen, R. (2021). A preferred sequence for organelle inheritance during polarized cell growth. *Journal of Cell Science*, jcs.258856. https://doi.org/10.1242/jcs.258856.

Gregor Lichtner; Philippe Thévenaz. (2021). *pyStackReg*.

Li, Z., & Srivastava, P. (2003). Heat-shock proteins. *Current Protocols in Immunology*, *58*(1), 1–6. https://doi.org/10.1002/0471142735.ima01ts58.

Liochev, S. I. (2013). Reactive oxygen species and the free radical theory of aging. *Free Radical Biology and Medicine*, *60*, 1–4. https://doi.org/10.1016/J.FREERADBIOMED.2013.02.011.

Liu, B., Åberg, C., van Eerden, F. J., Marrink, S. J., Poolman, B., & Boersma, A. J. (2017). Design and properties of genetically encoded probes for sensing macromolecular crowding. *Biophysical Journal*, *112*(9), 1929–1939. https://doi.org/10.1016/j.bpj.2017.04.004.

Lockshin, R. A., & Zakeri, Z. (2007). Cell death in health and disease. *Journal of Cellular and Molecular Medicine*, *11*(6), 1214–1224. https://doi.org/10.1111/J.1582-4934.2007.00150.X.

López-Otín, C., Blasco, M. A., Partridge, L., Serrano, M., & Kroemer, G. (2013). The hallmarks of aging. *Cell*, *153*(6), 1194. Europe PMC Funders. https://doi.org/10.1016/j.cell.2013.05.039.

Löwe, M., Kalacheva, M., Boersma, A. J., & Kedrov, A. (2020). The more the merrier: Effects of macromolecular crowding on the structure and dynamics of biological membranes. *The FEBS Journal*, *287*(23), 5039–5067. https://doi.org/10.1111/febs.15429.

Lu, A. X., Zarin, T., Hsu, I. S., & Moses, A. M. (2019). YeastSpotter: Accurate and parameter-free web segmentation for microscopy images of yeast cells. *Bioinformatics*, *35*(21), 4525–4527. https://doi.org/10.1093/bioinformatics/btz402.

Lumpkin, C. K., McGill, J. R., Riabowol, K. T., Moerman, E. J., Reis, R. J., & Goldstein, S. (1985). Extrachromosomal circular DNA and aging cells. *Advances in Experimental Medicine and Biology*, *190*, 479–493. https://doi.org/10.1007/978-1-4684-7853-2_24.

Ma, B., & Nussinov, R. (2013). Structured crowding and its effects on enzyme catalysis. *Topics in Current Chemistry*, *337*, 123–138. https://doi.org/10.1007/128_2012_316.

Macara, I. G., & Mili, S. (2008). Polarity and differential inheritance- -universal attributes of life? *Cell*, *135*(5), 801–812. https://doi.org/10.1016/J.CELL.2008.11.006.

Mager, W. H., & Siderius, M. (2002). Novel insights into the osmotic stress response of yeast. *FEMS Yeast Research*, *2*(3), 251–257. https://doi.org/10.1016/S1567-1356(02)00116-2.

Marina Robles, L., Millán-Pacheco, C., Pastor, N., & Del Río, G. (2017). Structure-function studies of the ALPHA pheromone receptor from yeast. *TIP*, *20*(1), 16–26. https://doi.org/10.1016/j.recqb.2016.11.002.

Marini, G., Nüske, E., Leng, W., Alberti, S., & Pigino, G. (2020). Reorganization of budding yeast cytoplasm upon energy depletion. *BioRxiv*, *31*(12), 1232–1245. https://doi.org/10.1091/mbc.E20-02-0125.

McMurray, M., Bertin, A., Garcia, G., Lam, L., Nogales, E., & Thorner, J. (2011). Septin filament formation is essential in budding yeast. *Developmental Cell*, *20*(4), 540–549. https://doi.org/10.1016/j.devcel.2011.02.004.

McMurray, M., & Thorner, J. (2009). Septins: Molecular partitioning and the generation of cellular asymmetry. *Cell Division*, *4*, 18. BioMed Central. https://doi.org/10.1186/1747-1028-4-18.

Menendez-Benito, V., van Deventer, S. J., Jimenez-Garcia, V., Roy-Luzarraga, M., van Leeuwen, F., & Neefjes, J. (2013). Spatiotemporal analysis of organelle and macromolecular complex inheritance. *Proceedings of the National Academy of Sciences of the United States of America*, *110*(1), 175–180. https://doi.org/10.1073/pnas.1207424110.

Merlini, L., Dudin, O., & Martin, S. G. (2013). Mate and fuse: How yeast cells do it. *Open Biology*, *3*(MAR). The Royal Society https://doi.org/10.1098/rsob.130008.

Miller, H., Zhou, Z., Wollman, A. J. M., & Leake, M. C. (2015). Superresolution imaging of single DNA molecules using stochastic photoblinking of minor groove and intercalating dyes. *Methods*, *88*, 81–88. https://doi.org/10.1016/j.ymeth.2015.01.010.

Minton, A. P. (2006). How can biochemical reactions within cells differ from those in test tubes? *Journal of Cell Science*, *119*(14), 2863–2869. https://doi.org/10.1242/jcs.03063.

Miyagi, T., Yamanaka, Y., Harada, Y., Narumi, S., Hayamizu, Y., Kuroda, M., et al. (2021). An improved molecular crowding sensor CRONOS for detection of crowding changes in membrane-less organelles under pathological conditions. *BioRxiv*, 1–26. version. 1. https://doi.org/10.1101/2021.03.31.437991.

Moseley, J. B., & Goode, B. L. (2006). The yeast actin cytoskeleton: From cellular function to biochemical mechanism. *Microbiology and Molecular Biology Reviews*, *70*(3), 605–645. https://doi.org/10.1128/mmbr.00013-06.

Mouton, S. N., Thaller, D. J., Crane, M. M., Rempel, I. L., Kaeberlein, M., Lusk, C. P., et al. (2019). A physicochemical roadmap of yeast replicative aging. *BioRxiv*, 1–33. Version 1.

Mouton, S. N., Thaller, D. J., Crane, M. M., Rempel, I. L., Terpstra, O., Steen, A., et al. (2020). A physicochemical perspective of aging from single-cell analysis of ph, macromolecular and organellar crowding in yeast. *eLife*, *9*, 1–42. https://doi.org/10.7554/ELIFE.54707.

Munder, M. C., Midtvedt, D., Franzmann, T., Nüske, E., Otto, O., Herbig, M., et al. (2016). A pH-driven transition of the cytoplasm from a fluid- to a solid-like state promotes entry into dormancy. *ELife*, *5*, e09347. https://doi.org/10.7554/eLife.09347.

Nasmyth, K. (1996). At the heart of the budding yeast cell cycle. *Trends in Genetics*, *12*(10), 405–412. Trends Genet https://doi.org/10.1016/0168-9525(96)10041-X.

Nettesheim, G., Nabti, I., Murade, C. U., Jaffe, G. R., King, S. J., & Shubeita, G. T. (2020). Macromolecular crowding acts as a physical regulator of intracellular transport. *Nature Physics*, *16*(11), 1–8. https://doi.org/10.1038/s41567-020-0957-y.

Nguemaha, V., Qin, S., & Zhou, H.-X. (2019). Transfer free energies of test proteins into crowded protein solutions have simple dependence on crowder concentration. *Frontiers in Molecular Biosciences*, *6*, 39–48. https://doi.org/10.3389/FMOLB.2019.00039.

Nurse, P., & Broek, D. (1993). Yeast cells can enter a quiescent state through G,, S, G2, or M phase of the cell cycle. *Cancer Research*, *53*(8), 1867–1870.

Nyström, T., & Liu, B. (2014). Protein quality control in time and space—Links to cellular aging. *FEMS Yeast Research*, *14*(1), 40–48. https://doi.org/10.1111/1567-1364.12095.

Okada, S., Leda, M., Hanna, J., Savage, N. S., Bi, E., & Goryachev, A. B. (2013). Daughter cell identity emerges from the interplay of Cdc42, septins, and exocytosis. *Developmental Cell*, *26*(2), 148–161. https://doi.org/10.1016/j.devcel.2013.06.015.

Oliphant, T. E. (2010). Guide to NumPy. In *Vol. 1. Methods*. USA: Trelgol Publishing.

Ouldridge, T. E., & Rein ten Wolde, P. (2014). The robustness of proofreading to crowding-induced pseudo-processivity in the MAPK pathway. *Biophysical Journal*, *107*(10), 2425. https://doi.org/10.1016/J.BPJ.2014.10.020.

Özcan, S., & Johnston, M. (1999). Function and regulation of yeast hexose transporters. *Microbiology and Molecular Biology Reviews*, *63*(3), 554–569. https://doi.org/10.1128/mmbr.63.3.554-569.1999.

Paine, K. M., Ecclestone, G. B., & MacDonald, C. (2021). Fur4 mediated uracil-scavenging to screen for surface protein regulators. *BioRxiv*, *2021*(05), 27.445995. https://doi.org/10.1101/2021.05.27.445995.

Park, S., Barnes, R., Lin, Y., Jeon, B.-j., Najafi, S., Delaney, K. T., et al. (2020). Dehydration entropy drives liquid-liquid phase separation by molecular crowding. *Communications Chemistry*, *3*(1), 1–12. https://doi.org/10.1038/s42004-020-0328-8.

Perez, A. M., & Thorner, J. (2019). Septin-associated proteins Aim44 and Nis1 traffic between the bud neck and the nucleus in the yeast Saccharomyces cerevisiae. *Cytoskeleton*, *76*(1), 15–32. https://doi.org/10.1002/cm.21500.

Petranovic, D., & Ganley, A. (2014). Yeast cell aging and death. *FEMS Yeast Research*, *14*(1), 1. https://doi.org/10.1111/1567-1364.12130.

Petranovic, D., Tyo, K., Vemuri, G. N., & Nielsen, J. (2010). Prospects of yeast systems biology for human health: Integrating lipid, protein and energy metabolism. *FEMS Yeast Research*, *10*(8), 1046–1059. https://doi.org/10.1111/j.1567-1364.2010.00689.x.

Phillip, Y., & Schreiber, G. (2013). Formation of protein complexes in crowded environments-from in vitro to in vivo. *FEBS Letters*, *587*(8), 1046–1052. No longer published by Elsevier. https://doi.org/10.1016/j.febslet.2013.01.007.

Plank, M., Wadhams, G. H., & Leake, M. C. (2009). Millisecond timescale slimfield imaging and automated quantification of single fluorescent protein molecules for use in probing complex biological processes. *Integrative Biology*, *1*(10), 602–612. https://doi.org/10.1039/b907837a.

Poland, D. (1998). The effect of excluded volume on aggregation kinetics. *The Journal of Chemical Physics*, *97*(1), 470. https://doi.org/10.1063/1.463593.
Rivas, G., Ferrone, F., & Herzfeld, J. (2004). Life in a crowded world. *EMBO Reports*, *5*(1), 23–27. https://doi.org/10.1038/sj.embor.7400056.
Rohwer, J. M., Postma, P. W., Kholodenko, B. N., & Westerhoff, H. V. (1998). Implications of macromolecular crowding for signal transduction and metabolite channeling. *Proceedings of the National Academy of Sciences of the United States of America*, *95*(18), 10547–10552. https://doi.org/10.1073/pnas.95.18.10547.
Saito, H., & Posas, F. (2012). Response to hyperosmotic stress. *Genetics*, *192*(2), 289–318. https://doi.org/10.1534/genetics.112.140863.
Sanfelice, D., Politou, A., Martin, S. R., De Los Rios, P., Temussi, P., & Pastore, A. (2013). The effect of crowding and confinement: A comparison of Yfh1 stability in different environments. *Physical Biology*, *10*(4), 045002. https://doi.org/10.1088/1478-3975/10/4/045002.
Santos, A. L., Sinha, S., & Lindner, A. B. (2018). The good, the bad, and the ugly of ROS: New insights on aging and aging-related diseases from eukaryotic and prokaryotic model organisms. *Oxidative Medicine and Cellular Longevity*, *2018*, 1941285. Hindawi Limited. https://doi.org/10.1155/2018/1941285.
Sarkar, M., Li, C., & Pielak, G. J. (2013). Soft interactions and crowding. *Biophysical Reviews*, *5*(2), 187–194. Springer. https://doi.org/10.1007/s12551-013-0104-4.
Shashkova, S., Nyström, T., Leake, M. C., & Wollman, A. J. M. (2020). Correlative single-molecule fluorescence barcoding of gene regulation in Saccharomyces cerevisiae. *Methods*, *193*(62–67). https://doi.org/10.1016/j.ymeth.2020.10.009.
Shashkova, S., Wollman, A. J. M., Leake, M. C., & Hohmann, S. (2017). The yeast Mig1 transcriptional repressor is dephosphorylated by glucose-dependent and -independent mechanisms. *FEMS Microbiology Letters*, *364*(14). https://doi.org/10.1093/femsle/fnx133.
Shcheprova, Z., Baldi, S., Frei, S. B., Gonnet, G., & Barral, Y. (2008). A mechanism for asymmetric segregation of age during yeast budding. *Nature*, *454*(7205), 728–734. https://doi.org/10.1038/nature07212.
Shepherd, J. W., Higgins, E. J., Wollman, A. J. M., & Leake, M. C. (2021a). PySTACHIO: Python single-molecule TrAcking stoiCHiometry intensity and simulatiOn, a flexible, extensible, beginner-friendly and optimized program for analysis of single-molecule microscopy. *BioRxiv*, *19*. 2021.03.18.435952. https://doi.org/10.1016/j.csbj.2021.07.004.
Shepherd, J. W., Higgins, E. J., Wollman, A. J. M., & Leake, M. C. (2021b). PySTACHIO: Python single-molecule TrAcking stoiCHiometry intensity and simulatiOn, a flexible, extensible, beginner-friendly and optimized program for analysis of single-molecule microscopy data. *Computational and Structural Biotechnology Journal*, *19*, 4049–4058. https://doi.org/10.1016/j.csbj.2021.07.004.
Shepherd, J. W., Lecinski, S., Wragg, J., Shashkova, S., MacDonald, C., & Leake, M. C. (2020). Molecular crowding in single eukaryotic cells: Using cell environment biosensing and single-molecule optical microscopy to probe dependence on extracellular ionic strength, local glucose conditions, and sensor copy number. *Methods*, *193*, 54–61. https://doi.org/10.1016/j.ymeth.2020.10.015.
Shepherd, J. W., Payne-Dwyer, A. L., Lee, J.-E., Syeda, A., & Leake, M. C. (2021). Combining single-molecule super-resolved localization microscopy with fluorescence polarization imaging to study cellular processes. *Journal of Physics: Photonics*, *3*, 34010. https://doi.org/10.1088/2515-7647/ac015d.
Shim, A. R., Nap, R. J., Huang, K., Almassalha, L. M., Matusda, H., Backman, V., et al. (2020). Dynamic crowding regulates transcription. *Biophysical Journal*, *118*(9), 2117–2129. https://doi.org/10.1016/j.bpj.2019.11.007.
Sinclair, D. A., & Guarente, L. (1997). Extrachromosomal rDNA circles—A cause of aging in yeast. *Cell*, *91*(7), 1033–1042. https://doi.org/10.1016/S0092-8674(00)80493-6.

Spichal, M., & Fabre, E. (2017). The emerging role of the cytoskeleton in chromosome dynamics. *Frontiers in Genetics*, *8*, 60. https://doi.org/10.3389/FGENE.2017.00060.

Stagg, L., Zhang, S. Q., Cheung, M. S., & Wittung-Stafshede, P. (2007). Molecular crowding enhances native structure and stability of α/β protein flavodoxin. *Proceedings of the National Academy of Sciences of the United States of America*, *104*(48), 18976–18981. https://doi.org/10.1073/pnas.0705127104.

Sugiyama, S., & Tanaka, M. (2019). Distinct segregation patterns of yeast cell-peripheral proteins uncovered by a method for protein segregatome analysis. *Proceedings of the National Academy of Sciences of the United States of America*, *116*(18), 8909–8918. https://doi.org/10.1073/pnas.1819715116.

Sun, Y., Wollman, A. J. M., Huang, F., Leake, M. C., & Liu, L. N. (2019). Single-organelle quantification reveals stoichiometric and structural variability of carboxysomes dependent on the environment. *Plant Cell*, *31*(7), 1648–1664. https://doi.org/10.1105/tpc.18.00787.

Syed, A., & Tainer, J. A. (2018). The MRE11-RAD50-NBS1 complex conducts the orchestration of damage signaling and outcomes to stress in DNA replication and repair. *Annual Review of Biochemistry*, *87*, 263–294. https://doi.org/10.1146/annurev-biochem-062917-012415.

Syeda, A. H., Wollman, A. J. M., Hargreaves, A. L., Howard, J. A. L., Brüning, J. G., McGlynn, P., et al. (2019). Single-molecule live cell imaging of Rep reveals the dynamic interplay between an accessory replicative helicase and the replisome. *Nucleic Acids Research*, *47*(12), 6287–6298. https://doi.org/10.1093/nar/gkz298.

Tabaka, M., Kalwarczyk, T., Szymanski, J., Hou, S., & Holyst, R. (2014). The effect of macromolecular crowding on mobility of biomolecules, association kinetics, and gene expression in living cells. *Frontiers in Physics*, *2*, 54. https://doi.org/10.3389/FPHY.2014.00054.

Taïeb, H. M., Garske, D. S., Contzen, J., Gossen, M., Bertinetti, L., Robinson, T., et al. (2021). Osmotic pressure modulates single cell cycle dynamics inducing reversible growth arrest and reactivation of human metastatic cells. *Scientific Reports*, *11*(1), 13455. https://doi.org/10.1038/s41598-021-92054-w.

Tamás, M. J., Rep, M., Thevelein, J. M., & Hohmann, S. (2000). Stimulation of the yeast high osmolarity glycerol (HOG) pathway: Evidence for a signal generated by a change in turgor rather than by water stress. *FEBS Letters*, *472*(1), 159–165. https://doi.org/10.1016/S0014-5793(00)01445-9.

Tan, C., Saurabh, S., Bruchez, M. P., Schwartz, R., & LeDuc, P. (2013). Molecular crowding shapes gene expression in synthetic cellular nanosystems. *Nature Nanotechnology*, *8*(8), 602–608. https://doi.org/10.1038/nnano.2013.132.

Tissenbaum, H. A., & Guarente, L. (2002). Model organisms as a guide to mammalian aging. *Developmental Cell*, *2*(1), 9–19. https://doi.org/10.1016/S1534-5807(01)00098-3.

Tokuriki, N. (2004). Protein folding by the effects of macromolecular crowding. *Protein Science*, *13*(1), 125–133. https://doi.org/10.1110/ps.03288104.

Udden, M. M., & Finkelstein, D. B. (1978). Reaction order of Saccharomyces cerevisiae alpha factor mediated cell cycle arrest and mating inhibition. *Journal of Bacteriology*, *133*(3), 1501–1507. https://doi.org/10.1128/jb.133.3.1501-1507.1978.

Valcourt, J. R., Lemons, J. M. S., Haley, E. M., Kojima, M., Demuren, O. O., & Coller, H. A. (2012). Staying alive: Metabolic adaptations to quiescence. *Cell Cycle*, *11*(9), 1680–1696. Taylor & Francis https://doi.org/10.4161/cc.19879.

Valdez-Taubas, J., & Pelham, H. R. B. (2003). Slow diffusion of proteins in the yeast plasma membrane allows polarity to be maintained by endocytic cycling. *Current Biology*, *13*(18), 1636–1640. https://doi.org/10.1016/j.cub.2003.09.001.

Van Den Berg, J., Boersma, A. J., & Poolman, B. (2017). Microorganisms maintain crowding homeostasis. *Nature Reviews. Microbiology*, *15*(5), 309–318. https://doi.org/10.1038/nrmicro.2017.17.

Van Den Berg, B., Ellis, R. J., & Dobson, C. M. (1999). Effects of macromolecular crowding on protein folding and aggregation. *EMBO Journal*, *18*(24), 6927–6933. https://doi.org/10.1093/emboj/18.24.6927.

van der Walt, S., Schönberger, J. L., Nunez-Iglesias, J., Boulogne, F., Warner, J. D., Yager, N., et al. (2014). Scikit-image: Image processing in Python. *PeerJ*, *2*, e453. https://doi.org/10.7717/peerj.453.

Verghese, J., Abrams, J., Wang, Y., & Morano, K. A. (2012). Biology of the heat shock response and protein chaperones: Budding yeast (Saccharomyces cerevisiae) as a model system. *Microbiology and Molecular Biology Reviews*, *76*(2), 115–158. https://doi.org/10.1128/mmbr.05018-11.

Vida, T. A., & Emr, S. D. (1995). A new vital stain for visualizing vacuolar membrane dynamics and endocytosis in yeast. *Journal of Cell Biology*, *128*(5), 779–792. https://doi.org/10.1083/jcb.128.5.779.

Vrabioiu, A. M., & Mitchison, T. J. (2006). Structural insights into yeast septin organization from polarized fluorescence microscopy. *Nature*, *443*(7110), 466–469. https://doi.org/10.1038/nature05109.

Wang, Y., Lilley, K. S., & Oliver, S. G. (2014). A protocol for the subcellular fractionation of Saccharomyces cerevisiae using nitrogen cavitation and density gradient centrifugation. *Yeast (Chichester, England)*, *31*(4), 127. https://doi.org/10.1002/YEA.3002.

Warren, G., & Wickner, W. (1996). Organelle inheritance. *Cell*, *84*(3), 395–400. https://doi.org/10.1016/S0092-8674(00)81284-2.

Waskom, M. L. (2021). Seaborn: Statistical data visualization. https://doi.org/10.21105/joss.03021.

Weill, U., Yofe, I., Sass, E., Stynen, B., Davidi, D., Natarajan, J., et al. (2018). Genome-wide SWAp-Tag yeast libraries for proteome exploration. *Nature Methods*, *15*(8), 617–622. https://doi.org/10.1038/s41592-018-0044-9.

Wollman, A. J. M., et al. (2016). An automated image analysis framework for segmentation and division plane detection of single live Staphylococcus aureus cells which can operate at millisecond sampling time scales using bespoke Slimfield microscopy. *Physical Biology*, *13*(5). https://doi.org/10.1088/1478-3975/13/5/055002.

Wollman, A. J. M., & Leake, M. C. (2015). Millisecond single-molecule localization microscopy combined with convolution analysis and automated image segmentation to determine protein concentrations in complexly structured, functional cells, one cell at a time. *Faraday Discussions*, *184*, 401–424. https://doi.org/10.1039/c5fd00077g.

Wollman, A. J. M., & Leake, M. C. (2016). Single-molecule narrow-field microscopy of protein-DNA binding dynamics in glucose signal transduction of live yeast cells. *Methods in Molecular Biology*, *1431*, 5–15. https://doi.org/10.1007/978-1-4939-3631-1_2.

Wollman, A. J. M., Shashkova, S., Hedlund, E. G., Friemann, R., Hohmann, S., & Leake, M. C. (2017). Transcription factor clusters regulate genes in eukaryotic cells. *eLife*, *6*, 1–36. https://doi.org/10.7554/eLife.27451.

Yaffe, M. P. (1991). Organelle inheritance in the yeast cell cycle. *Trends in Cell Biology*, *1*(6), 160–164. https://doi.org/10.1016/0962-8924(91)90017-4.

Yeong, F. M. (2005). Severing all ties between mother and daughter: Cell separation in budding yeast. *Molecular Microbiology*, *55*(5), 1325–1331. https://doi.org/10.1111/j.1365-2958.2005.04507.x.

Zhang, Q., Bai, Q., Zhu, L., Hou, T., Zhao, J., & Liang, D. (2019). Macromolecular crowding and confinement effect on the growth of DNA nanotubes in dextran and hyaluronic acid media. *ACS Applied Bio Materials*, *3*(1), 412–420. https://doi.org/10.1021/ACSABM.9B00892.

Zhou, Y. L., Liao, J. M., Chen, J., & Liang, Y. (2006). Macromolecular crowding enhances the binding of superoxide dismutase to xanthine oxidase: Implications for protein-protein interactions in intracellular environments. *International Journal of Biochemistry and Cell Biology*, *38*(11), 1986–1994. https://doi.org/10.1016/j.biocel.2006.05.012.

Zhou, C., Slaughter, B. D., Unruh, J. R., Eldakak, A., Rubinstein, B., & Li, R. (2011). Motility and segregation of Hsp104-associated protein aggregates in budding yeast. *Cell*, *147*(5), 1186–1196. https://doi.org/10.1016/j.cell.2011.11.002.

Zhou, C., Slaughter, B. D., Unruh, J. R., Guo, F., Yu, Z., Mickey, K., et al. (2014). Organelle-based aggregation and retention of damaged proteins in asymmetrically dividing cells. *Cell*, *159*(3), 530–542. https://doi.org/10.1016/j.cell.2014.09.026.

The expanding toolbox to study the LRRC8-formed volume-regulated anion channel VRAC

Yulia Kolobkova[a], Sumaira Pervaiz[b], and Tobias Stauber[a,b,*]
[a]Department of Human Medicine and Institute for Molecular Medicine, MSH Medical School Hamburg, Germany
[b]Institute of Chemistry and Biochemistry, Freie Universität Berlin, Germany
*Corresponding author: e-mail address: tobias.stauber@medicalschool-hamburg.de

Contents

Abstract

The volume-regulated anion channel (VRAC) is activated upon cell swelling and facilitates the passive movement of anions across the plasma membrane in cells. VRAC function underlies many critical homeostatic processes in vertebrate cells. Among them are the regulation of cell volume and membrane potential, glutamate release and apoptosis. VRAC is also permeable for organic osmolytes and metabolites including some anti-cancer drugs and antibiotics. Therefore, a fundamental understanding of VRAC's structure-function relationships, its physiological roles, its utility for therapy of diseases, and the development of compounds modulating its activity are important research

Current Topics in Membranes, Volume 88
ISSN 1063-5823
https://doi.org/10.1016/bs.ctm.2021.10.001

frontiers. Here, we describe approaches that have been applied to study VRAC since it was first described more than 30 years ago, providing an overview of the recent methodological progress. The diverse applications reflecting a compromise between the physiological situation, biochemical definition, and biophysical resolution range from the study of VRAC activity using a classic electrophysiology approach, to the measurement of osmolytes transport by various means and the investigation of its activation using a novel biophysical approach based on fluorescence resonance energy transfer.

1. Introduction

The plasma membrane of the majority of cells is highly permeable to water, therefore intracellular and extracellular osmotic perturbations lead to either cell swelling or shrinkage. To counteract these processes, cells engage several compensatory mechanisms to adjust their volume via activation of plasma membrane solute flux mechanisms. In most vertebrate cells, the volume-regulated anion channel (VRAC) mediates regulatory volume decrease (RVD) through the swelling-induced release of Cl^- and organic solutes (Jentsch, Lutter, Planells-Cases, Ullrich, & Voss, 2016; Pedersen, Okada, & Nilius, 2016; Stauber, 2015; Strange, Yamada, & Denton, 2019). Although various K^+ channels are known to be volume-sensitive (Wehner, 2006), the basal conductance for K^+ is usually much higher than that for Cl^- in most cells, rendering VRAC activity an important determinant for RVD. VRAC fulfills its physiological functions by distinct basic mechanisms (Fig. 1). First, the release of Cl^- and organic osmolytes through VRAC, which is followed by osmotic water efflux, leads to a shrinking of the cell. This is not only required for RVD in response to osmotic cell swelling, but has also been implicated in isosmotic processes such as cell proliferation, migration, apoptotic volume decrease (AVD), and steroidogenic process in Leydig cells (Akita & Okada, 2014; Jentsch, 2016; Lang & Hoffmann, 2012; Okada, Sato, & Numata, 2009; Pedersen, Hoffmann, & Novak, 2013; Pedersen, Klausen, & Nilius, 2015; Poletto Chaves & Varanda, 2008; Stauber, 2015). Second, the activation of VRAC anion currents shifts the membrane potential towards the equilibrium potential of chloride, thereby changing the driving force for other ions and affecting their channels. VRAC-mediated regulation of membrane potential was reported to play an important role for endothelial cells (Nilius & Droogmans, 2001), epithelial transport (Hoffmann, Schettino, & Marshall, 2007), and to contribute to glucose-stimulated insulin secretion in pancreatic β-cells (Best, Brown, Sener, & Malaisse, 2010; Kang et al.,

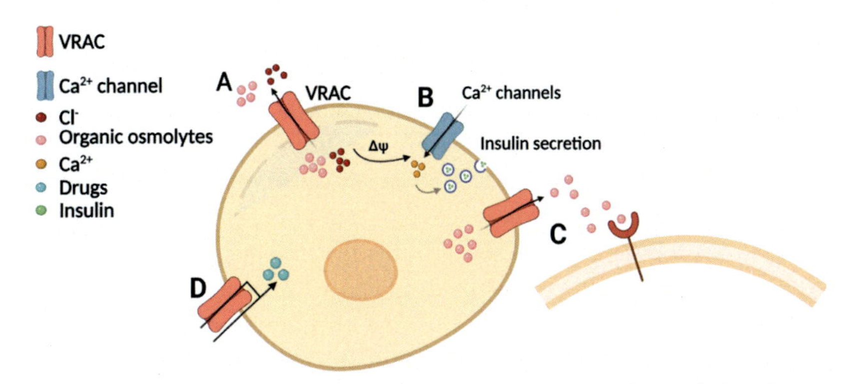

Fig. 1 VRACs exert their function by various mechanisms. (A) Release of Cl^- and organic osmolytes leads to osmotic efflux of water and cell volume decrease. (B) The opening of VRACs shifts the plasma membrane potential towards the equilibrium potential of chloride, affecting the transport of other ions and the activity of ion channels and transporters. (C) VRAC-conducted osmolytes, such as ATP, cGAMP or excitatory amino acids, act as signaling molecules. (D) VRAC contributes to the uptake of drugs (e.g., of blasticidin S or cisplatin) and signaling molecules.

2018; Stuhlmann, Planells-Cases, & Jentsch, 2018). Third, VRAC facilitates communication between cells by mediating the transport of signaling molecules. It was shown to release ATP (Burow, Klapperstück, & Markwardt, 2015; Gaitán-Peñas et al., 2016; Hisadome et al., 2002) and to mediate release and import of the signaling cyclic dinucleotide cGAMP (Chen et al., 2021; Lahey et al., 2020; Zhou et al., 2020). By conducting the excitatory amino acids glutamate and aspartate, VRAC is involved in neuron-glia crosstalk as well as in excitotoxicity under ischemic conditions (Feustel, Jin, & Kimelberg, 2004; Lutter, Ullrich, Lueck, Kempa, & Jentsch, 2017; Mongin, 2016; Schober, Wilson, & Mongin, 2017; Yang et al., 2019). Additionally, VRAC contributes to the cellular uptake of antibiotics such as blasticidin S, anticancer drugs such as cisplatin and carboplatin (Lee, Freinkman, Sabatini, & Ploegh, 2014; Planells-Cases et al., 2015).

After a long quest for the molecular identity of VRAC (Pedersen et al., 2015; Stauber, 2015), it was shown to be formed by heteromeric complexes of LRRC8 proteins, with LRRC8A being an essential subunit that needs to heteromerize with at least one other paralog, LRRC8B-E (Qiu et al., 2014; Voss et al., 2014). The solved structures of LRRC8A or LRRC8D homomers and of LRRC8A/C heteromers confirmed the predicted hexameric architecture and revealed important structural insight into the nature of the pore and selectivity (Deneka, Sawicka, Lam, Paulino, & Dutzler, 2018; Kasuya et al., 2018; Kefauver et al., 2018; Kern, Oh,

Hite, & Brohawn, 2019; König & Stauber, 2019; Nakamura et al., 2020). However, LRRC8A homomers do not recapitulate important biophysical VRAC properties (Yamada, Figueroa, Denton, & Strange, 2021). Normally, LRRC8A heteromerizes with at least one paralog to form functional VRACs (Syeda et al., 2016; Voss et al., 2014), but the combination and stoichiometry remain unknown. Single-step photobleaching of fluorescently tagged LRRC8 proteins showed that the stoichiometry of heterologously expressed LRRC8 proteins correlated with their expression levels (Gaitán-Peñas et al., 2016), and sequential co-immunoprecipitation revealed that LRRC8A can combine with more than one paralog isoform within one complex (Lutter et al., 2017). Quantitative immunoblotting of cell and tissue lysates and of LRRC8A co-precipitates suggested that a given LRRC8 complex may contain only one or two LRRC8A subunits (Pervaiz, Kopp, von Kleist, & Stauber, 2019). The variability in subunit combinations allows for the formation of a potentially large number of differently composed VRACs. The combination of LRRC8A with the other subunits confers variability in biophysical properties such as depolarization-dependent inactivation and open-probability (Syeda et al., 2016; Ullrich, Reincke, Voss, Stauber, & Jentsch, 2016; Voss et al., 2014), in response to activity-modulating signals (Gradogna, Gavazzo, Boccaccio, & Pusch, 2017) and substrate conductance (Gaitán-Peñas et al., 2016; Lutter et al., 2017; Planells-Cases et al., 2015; Schober et al., 2017).

The molecular identification of the channel obviously paved the ground for multiple approaches to study physiology, biophysics, and structure of VRAC. Here, we will focus on the methods that are applied to study VRAC activity. After a short overview of animal models that were used to investigate the physiological roles of VRAC, we will primarily discuss the various techniques that have been used to study channel activity of VRAC by electrophysiology. Additionally, we will describe alternative methods such as radioactive osmolyte efflux and iodide quenching assays, direct measurements of the cell volume, and optical tools based on Förster resonance energy transfer (FRET).

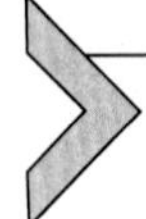

2. Animal models reveal the physiological functions of VRAC

After the leucine-rich repeat-containing protein 8A (LRRC8A) and its four paralogs were discovered to form heteromeric VRACs, numerous previously ascribed physiological roles were assessed with molecular biological tools and animal models. These studies have identified additional roles for VRAC, highlighting its physiological importance (Chen et al., 2019; Osei-Owusu, Yang, Vitery, & Qiu, 2018). Using cultured knockout cell lines helped to clarify the roles of different LRRC8 subunits in cell proliferation, migration, and apoptosis (Liu & Stauber, 2019; Lu et al., 2019; Planells-Cases et al., 2015; Sirianant et al., 2016). RNA interference-mediated knockdown of LRRC8 proteins was used to address cell-type-specific functions of VRAC such as in astrocytes and glioblastoma cells, macrophage purinergic signaling, insulin signaling, and myotube differentiation (Chen, Becker, Koch, & Stauber, 2019; Formaggio et al., 2019; Hyzinski-García, Rudkouskaya, & Mongin, 2014; Rubino, Bach, Schober, Lambert, & Mongin, 2018; Xie et al., 2017; Zahiri et al., 2021).

To study functions beyond the cellular level, genetically modified animal models, mostly mice, were used. The complete knockout of the essential VRAC component LRRC8A in mice leads to a drastic phenotype (Kumar et al., 2014). *Lrrc8a*$^{-/-}$ mice were born significantly below Mendelian ratio, with high postnatal lethality and lifespan up to only approximately 100 days. They exhibited growth retardation, curly hair, and weakness of the hind limbs. Male mutants were infertile. Many organs presented defective development: their kidneys displayed vacuolation of renal tubular cells and skeletal muscle bundles were thinned. Thymic development was also defective and peripheral T cell expansion was impaired (Kumar et al., 2014). A spontaneous mouse mutant with markedly reduced VRAC activity due to a 2-bp deletion in *Lrrc8a* that truncates 15 leucine-rich repeats of LRRC8A cytoplasmic C-terminal domain shared phenotypes with *Lrrc8a*$^{-/-}$ mice. These included reduced survival, curly hair (hence its name *ébouriffé* (*ebo*)), infertility, and kidney abnormalities (Kumar et al., 2014; Platt et al., 2017).

The pivotal importance of VRAC has also been highlighted in zebrafish, where it is crucial for embryonic development (Yamada, Wondergem, Morrison, Yin, & Strange, 2016). Morpholino-mediated knockdown of the functional orthologue of LRRC8A, *lrrc8aa*, resulted in embryos with pericardial edema and defects in trunk elongation, somatogenesis and brain ventricle development (Tseng et al., 2020; Yamada et al., 2016).

To circumvent the severity of the systemic gene depletion of LRRC8A and to study the function of VRAC in specific tissues or cell types, tissue-specific knockout mice for the essential VRAC component *Lrrc8a* have been generated using the Cre/LoxP system. LRRC8A was confirmed to be essential for VRAC function in skeletal myotubes (using *Myf5*^{Cre} or *Myl1*^{Cre} mice), where is involved in differentiation (A. Kumar et al., 2020), and adipocytes

(*Adiponectin*^{Cre}), where it was reported to play an important role in systemic glucose homeostasis (Zhang et al., 2017). In addition to adipocytes, *Lrrc8a* is responsible for glucose homeostasis mediated by pancreatic β-cells. When targeting murine β-cells (*Ins1*^{Cre} or *rat-Ins*^{Cre}), knockout of *Lrrc8a* caused impaired insulin secretion and provoked glucose intolerance in mutant animals (Kang et al., 2018; Stuhlmann et al., 2018). The analysis of endothelium-restricted *Lrrc8a* knockout mice ascribed a role in vascular function to VRAC as the mice were reported to develop hypertension in response to chronic angiotensin-II infusion and exhibit impaired retinal blood flow (Alghanem et al., 2021). The generation of macrophage-specific (*Cx3cr1*^{Cre}) *Lrrc8a* knockout mice demonstrated the involvement of VRAC in the hypotonicity-induced activation of the NLRP3 inflammasome complex contributing to its role in immune response (Green et al., 2020). Its further importance for antiviral immunity was demonstrated in *Lrrc8e*$^{-/-}$ mice, which exhibited impaired interferon responses and compromised immunity to the DNA virus HSV-1 (Zhou, Chen, et al., 2020).

Germ cell-specific (*Stra8*^{Cre}), but not Sertoli cell-specific (*AMH*^{Cre}) disruption of *Lrrc8a* in mice recapitulated the sperm defects observed in the pan-knockout mice, suggesting germ cell-autonomous roles of *Lrrc8a* in the testis and providing genetic evidence that LRRC8A-dependent VRAC activity is essential for sperm motility and spermatogenesis (Lück, Puchkov, Ullrich, & Jentsch, 2018).

Astrocyte-specific *Lrrc8a* knockout (*mGFAP*^{Cre}) has shown that VRAC contributes to increased glutamatergic input to hippocampal neurons and brain injury caused by an ischemic stroke (Yang et al., 2019). Mice with a brain-wide knockdown of *Lrrc8a* (*Nestin*^{Cre}) died at 5–9 weeks of age, indicating an essential role of LRRC8A in neural progenitors for survival (Wilson et al., 2021; Yang et al., 2019; Zhou et al., 2020).

3. Electrophysiological measurement of VRAC-mediated currents

3.1 First electrophysiological recordings of VRACs

Patch-clamp is the gold standard technique for the characterization of the electrical currents produced by ion channels (Hille, 2001). Electrophysiological recordings are performed with help of a glass micropipette, that forms a tight contact with a patch of the cellular membrane. Depending on the research interests, different patch-clamp configurations can be applied. Cell-attached, inside-out and outside-out patch configurations are used to study the behavior of individual ion channels in the section of membrane attached to the pipette. In the cell-attached configuration, the membrane patch is left intact allowing the recording of ion channels within the patch. The addition of a pore-forming agent, such as gramicidin, in the pipette solution, results in a perforated patch with established electrical continuity, which at the same time prevents the dialysis of intracellular proteins. The outside- and inside-out configurations or "excised patches" are formed upon removal of the membrane patch from the rest of the cellular membrane, permitting the recording of single channels and bath perfusion of the extracellular or cytoplasmic face of the membrane, respectively. Thus, it allows monitoring the effect of various compounds on the activity of the channels. Whole-cell is, by far, the most commonly used patch-clamp configuration. It is achieved from the cell-attached configuration by the membrane patch disruption with a briefly applied strong suction. Thereby electrical and molecular access to the intracellular space is established. The recordings can then be performed in one of the two modes: the voltage-clamp, in which the voltage is held constant allowing the study of membrane currents, and the current-clamp, in which, in turn, the current is controlled allowing the study of changes in membrane potential.

The activation of electrogenic anion and cation permeability in a response to osmotic cell swelling was first measured by whole-cell patch-clamp technique in Ehrlich ascites tumor cells (Hoffmann, 1978; Hoffmann, Simonsen, & Lambert, 1984; Hoffmann, Simonsen, & Sjoholm, 1979) and human lymphocytes (Grinstein, Clarke, Dupre, & Rothstein, 1982; Grinstein, Clarke, & Rothstein, 1982; Sarkadi, Attisano, Grinstein, Buchwald, & Rothstein, 1984). In 1988, the first whole-cell patch-clamp recordings of VRAC were described in human T-lymphocytes (Cahalan & Lewis, 1988) and intestinal epithelial cells (Hazama & Okada, 1988). The observed outwardly rectifying Cl^- currents were activated upon perfusion with a hypotonic solution and blocked upon the inhibition of RVD. To exclude the contribution of swelling-activated K^+ channels to these currents, K^+ was replaced by Cs^+ in the pipette solution. Activation was correlated solely with the increase in cell volume (Arreola, Melvin, & Begenisich, 1995; Ross, Garber, & Cahalan, 1994) and neither with changes in intracellular Ca^{2+} (Arreola et al., 1995; Nilius, Oike, Zahradnik, & Droogmans, 1994) nor with alterations of membrane capacitances (Ross et al., 1994). However, a permissive intracellular Ca^{2+} concentration of 50 nM was reported to be necessary for VRAC activation (Szücs, Heinke, Droogmans, & Nilius, 1996).

In the whole-cell mode, cell swelling is continuous due to intracellular dialysis from the patch pipette and the presence of a constant transmembrane osmotic gradient, which leads to a manifold increase of cell volume. Therefore, most whole-cell electrophysiology studies on VRAC are performed together with imaging cells by video-enhanced differential interference contrast (DIC) microscopy, which allows simultaneous quantification of cell volume changes and current activation (Bond, Basavappa, Christensen, & Strange, 1999; Cannon, Basavappa, & Strange, 1998). With dramatically increased cell volume, the cell membrane was observed to dissociate from the underlying cytoplasm. Several studies describe a swelling-induced formation of membrane blebs—domains that have separated from the cytoskeleton (Ikenouchi & Aoki, 2017). The appearance of membrane blebs was associated with a dramatic increase in the rate of VRAC activation (Strange et al., 2019). Swelling-induced currents were not accompanied by an increase in membrane capacitance, even the opposite effect was observed: upon the initial rise in VRAC currents the membrane capacitance gradually declined, therefore it seemed unlikely that vesicle fusion delivers VRACs to the plasma membrane (Ross et al., 1994). Recently, immune-labeling of the essential LRRC8A subunit followed by flow cytometry confirmed that the levels of VRAC at the cell surface did not change upon application of hypotonic solution or staurosporine (Yurinskaya et al., 2020).

Volume regulation is an important function for all the cell types mentioned above since they cope with osmotic perturbations in physiological conditions. In this regard, the discovery of volume-sensitive ion channels in chromaffin cells (Doroshenko, 1991) and neuroblastoma cells (Falke & Misler, 1989) was unexpected and gave the first hint towards much wider VRAC distribution and functions than previously thought. In the following years, besides the undisputed role of VRAC in cell volume regulation, it was suggested to be involved in the regulation of membrane potential, intracellular signaling, adiposity and glucose metabolism, cell proliferation, migration, apoptosis and necrosis, transepithelial ion transport, myogenesis, lysosomal ion homeostasis, and release of biologically active molecules (Akita & Okada, 2014; Chen, König, et al., 2019; Hoffmann, Lambert, & Pedersen, 2009; Jentsch, 2016; Kumar et al., 2020; Li et al., 2020; Stutzin & Hoffmann, 2006; Wehner, Olsen, Tinel, Kinne-Saffran, & Kinne, 2003).

Electrophysiological recordings remain a standard technique to measure the activity of VRACs; and since its molecular identification, this technique has also been used to study structure-function relationships in mutated LRRC8 proteins (see below). In addition to the common patch-clamp setup, VRAC currents were also measured by planar patch-clamp (Ehling et al., 2016), reconstituted in lipid bilayers (see below), and recently by patch-clamp of enlarged endo-lysosomes (Li et al., 2020).

3.2 Activation and inhibition of VRAC

Changes in the cellular volume may also be detected as mechanical stimuli through stretching of the membrane or its separation from the cytoskeleton. However, currents developed upon mechanical stimulation usually differed from VRAC currents in their characteristics, whereas typical VRAC currents were only observed in a response to osmotic stimulus (Christensen & Hoffmann, 1992). Therefore, a direct VRAC activation by mechanical stimuli has been mostly discarded. However, there may be a functional connection of VRAC to the cytoskeleton, since VRAC currents were potentiated upon disruption of the F-actin cytoskeleton in some cell types (Levitan, Almonte, Mollard, & Garber, 1995; Morishima, Shimizu, Kida, & Okada, 2000; Schwiebert, Mills, & Stanton, 1994). Another factor that could play a role in VRAC activation is the lipid composition of the plasma membrane, which is changing during cell swelling. Cholesterol, for instance, is known to mediate the membrane deformation energy and thereby possibly influence the opening of VRACs. Indeed, cholesterol depletion potentiated VRAC currents and even activated the currents in non-swollen cells (Klausen, Hougaard, Hoffmann, & Pedersen, 2006; König, Hao, Schwartz, Plested, & Stauber, 2019; Levitan, Christian, Tulenko, & Rothblat, 2000; Romanenko, Rothblat, & Levitan, 2004). The connection of the local cholesterol concentration with F-actin organization at the membrane supports the existence of a functional link between cytoskeleton and VRACs (Hoffmann et al., 2009).

Notably, in chromaffin cells, the currents were triggered by both intracellular perfusion with hypertonic solutions, and by inflating cells with a short pulse of slight pressure (5–10 s) applied through the patch pipette (Doroshenko & Neher, 1992). Both stimuli resulted in comparable volume changes and were sufficient to elicit the whole sequence of events. The current gradually increased to a peak value and subsequently decayed to its initial level within few minutes. However, the magnitude of these two kinds of stimuli is very different, since even 1 mOsm difference in osmolality was estimated to result in higher pressure than used in this work. The authors hypothesized, that the actual pressure achieved with osmotic stimulation is greatly attenuated by bulk flow through the cell-pipette connection.

This phenomenon should be considered when VRAC is measured in the whole-cell configuration (Doroshenko & Neher, 1992) and can be avoided in perforated patch whole-cell recordings. Gramicidin perforation of the membrane through the patch pipette is a particularly suitable method for studies of anionic channels since intracellular Cl^- concentration remained preserved (Kyrozis & Reichling, 1995).

Perforated patch recordings of VRAC currents from the pancreatic rat β-cells showed that increased cell volume, whether accompanied by raised intracellular osmolality or ionic strength, is a major determinant of VRAC activation in the β-cells. In this work, VRAC activation in perforated patches was achieved by isosmotic addition of the permeable osmolytes: urea, 3-O-methyl glucose, arginine, and NH_4Cl (Best & Brown, 2009).

Other isosmotic stimuli, known to activate VRAC include:

- GPCR stimulation by extracellular ATP (Akita, Fedorovich, & Okada, 2011; Burow et al., 2015), bradikynin (Akita et al., 2011; Liu, Akita, Shimizu, Sabirov, & Okada, 2009), glutamate (Akita & Okada, 2014), or sphingosine-1-phosphate (Furuya, Hirata, Kobayashi, & Sokabe, 2021; Zahiri et al., 2021); recently, G protein-coupled receptors GPRC5B and GPR37L1 were suggested to modulate VRAC activity (Alonso-Gardon et al., 2021)
- reactive oxygen species (Browe & Baumgarten, 2004; Gradogna, Gavazzo, et al., 2017; Shimizu, Numata, & Okada, 2004; Varela, Simon, Riveros, Jorgensen, & Stutzin, 2004)
- apoptosis-inducing drugs (Maeno, Ishizaki, Kanaseki, Hazama, & Okada, 2000; Planells-Cases et al., 2015; Shimizu et al., 2004)
- large reductions in intracellular ionic strength (Best & Brown, 2009; Cannon et al., 1998; Deneka et al., 2018; Nilius, Prenen, Voets, Eggermont, & Droogmans, 1998; Sabirov, Prenen, Tomita, Droogmans, & Nilius, 2000; Syeda et al., 2016; Voets, Droogmans, Raskin, Eggermont, & Nilius, 1999)
- intracellular GTP-γ-S (Estevez, Bond, & Strange, 2001; Nilius et al., 1999)
- shear stress in endothelial cells (Romanenko, Davies, & Levitan, 2002)
- membrane stretch in cardiomyocytes (Browe & Baumgarten, 2003).

Many of these activation pathways imply phosphorylation of VRAC as an essential event in the modulation of its activity (Eggermont, Trouet, Carton, & Nilius, 2001). The activation time of swelling-induced currents is taking up to several minutes, therefore the signaling cascade leading to VRAC opening may require multiple phosphorylation events. Inhibition of the protein tyrosine phosphatase PTP has a potentiating effect on

VRAC currents, while inhibition of the protein tyrosine kinase PTK prevents VRAC activation in lymphocytes (Lepple-Wienhues et al., 1998). However, it remains elusive whether VRAC itself or an interacting protein is phosphorylated in a regulatory manner (Bertelli et al., 2021).

Among suppressing factors for VRAC activity there are increased intracellular ionic strength (Best & Brown, 2009) and intracellular acidification (Mori, Morishima, Takasaki, & Okada, 2002; Nabekura et al., 2003; Sabirov, Prenen, Droogmans, & Nilius, 2000).

3.3 Biophysical properties of VRAC

The identification of VRAC currents aroused great interest in studying its properties. These properties have been extensively assessed with electrophysiology, even for decades before the molecular identity of VRAC was discovered. The early functional studies mainly described the correlation of VRAC-attributed anion currents with various (patho)physiological conditions and used VRAC blockers—such as NPPB, tamoxifen, phloretin, and DCPIB to single out its roles. A tremendous amount of experiments elicited strong similarities in VRAC properties, regardless of the studied cell type (König & Stauber, 2019; Nilius et al., 1997; Okada et al., 2009; Strange, Emma, & Jackson, 1996; Strange et al., 2019). First, the dependence of the channel activity on the intracellular ATP. A presence and possible binding of cytosolic ATP to a part of the channel or a regulatory subunit, but not its hydrolysis, is a prerequisite to the VRAC activation (Jackson, Morrison, & Strange, 1994; Oiki, Kubo, & Okada, 1994; Okada, 1997). Second, modest outward rectification (Fig. 2), reflecting voltage-dependent enhancement of the single-channel conductance in various cell types (Jackson & Strange, 1996; Nilius et al., 1997). In chromaffin cells and lymphocytes, the single-channel conductance of VRAC was defined to be about 2 pS, as estimated from the ratio of the current variance to the current mean value—so-called non-stationary fluctuation analysis (Cahalan & Lewis, 1988; Doroshenko, 1991; Doroshenko & Neher, 1992). It appeared, however, to be about 10-fold larger when directly measured in the inside-out patches from Ehrlich ascites (Hudson & Schultz, 1988). In human epithelial Intestine 407 cells, simultaneous whole-cell and cell-attached patch-clamp recordings (Okada et al., 1994) and the ensemble average current of inside-out records (Tsumura, Oiki, Ueda, Okuma, & Okada, 1996) provided clear evidence that single-channel events are responsible for the whole-cell current. Later works defined that the single-channel

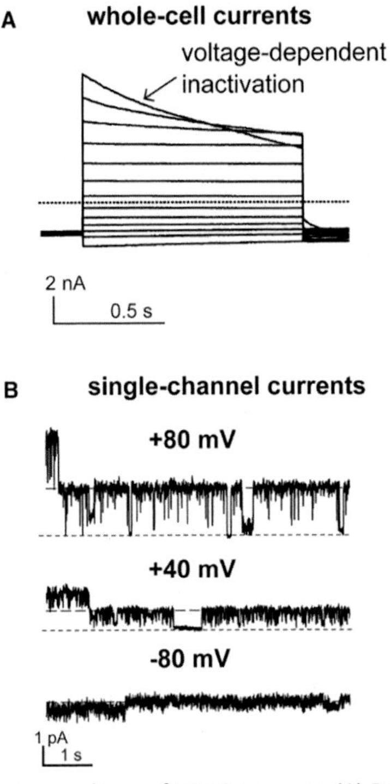

Fig. 2 Electrophysiological recordings of VRAC currents. (A) Representative traces (dotted line indicates zero current) of whole-cell currents mediated by native VRAC in human embryonic kidney cells. The hypotonicity-activated currents were measured in 20-mV steps between −120 and 120 mV. Note the outward rectification and voltage-dependent inactivation at positive potentials. (B) Representative single-channel currents from *Xenopus* oocytes expressing fluorescently tagged LRRC8A and LRRC8E. *Panel A: From Voss, F. K., Ullrich, F., Münch, J., Lazarow, K., Lutter, D., ..., Stauber, T., & Jentsch, T. J. (2014). Identification of LRRC8 heteromers as an essential component of the volume-regulated anion channel VRAC.* Science, *344(6184), 634-638. doi:10.1126/science.1252826. Panel B: From Gaitán-Peñas, H., Gradogna, A., Laparra-Cuervo, L., Solsona, C., Fernández-Dueñas, V., Barrallo-Gimeno, A., ... Estévez, R. (2016). Investigation of LRRC8-mediated volume-regulated anion currents in xenopus oocytes.* Biophysical Journal, *111(7), 1429-1443. doi:10.1016/j.bpj.2016.08.030.*

conductance of VRAC lies in the intermediate conductance range of approximately 50–80 pS at positive, and 10–20 pS at negative membrane potentials (Jackson & Strange, 1996; Nilius & Droogmans, 2003; Nilius et al., 1999; Okada, 2006). High variation of the early reported VRAC single-channel conductances could be explained by the technical difficulty of such recordings, which require swelling the cells while simultaneously or subsequently applying a patch pipette to the cell membrane surface or by a step-wise increase of the open probability (Okada, 2006; Strange et al., 2019).

Properties of the channel pore define not only the single-channel conductance but also the selectivity of the channel. As determined from the shifts in reversal potential, so-called low-field strength anion selectivity appeared to be another common feature of the VRACs (Ackerman, Wickman, & Clapham, 1994; Gosling, Smith, & Poyner, 1995; Kubo & Okada, 1992; Nilius et al., 1994; Verdon, Winpenny, Whitfield, Argent, & Gray, 1995; Viana et al., 1995; Voets et al., 1997). It results in the Eisenman's type I selectivity sequence for anionic molecules ($SCN^- > I^- > NO_3^- > Br^- > Cl^- > F^- > HCO_3^- >$ gluconate), suggesting that the anion channel contains weak binding sites (Pedersen et al., 2016). Due to the weak charge interactions of transported molecules with the amino acid moieties lining its vestibule, VRAC may poorly discriminate between various anions and neutral molecules. Fitting the relative permeabilities of the mentioned inorganic anions with known sizes as well as electrophysiological recordings in the presence of pore-permeating or pore-blocking sulphonic acid derivatives, calix[4]arenes, and polyethylene glycol molecules of various diameters (Droogmans, Maertens, Prenen, & Nilius, 1999; Ternovsky, Okada, & Sabirov, 2004) provided first estimations of VRAC channel pore size. The predicted pore size of around 11–12 Å would allow the conductance of relatively large molecules. The first evidence of transmembrane amino acid fluxes through anion channels was provided with the help of single-channel electrophysiological recordings in Madin-Darby canine kidney cells (Banderali & Roy, 1992). The recordings were performed in the presence of amino acids like aspartate or glutamate as charge carriers, which demonstrated their ability to permeate the membrane. Although in this particular study currents might not have been mediated by VRAC, as they were recorded from excised patches in ATP-free solutions, the subsequent whole-cell electrophysiological studies demonstrated VRAC was permeable to a variety of amino acids (Jackson et al., 1994; Jackson & Strange, 1993). Besides for small charged amino acids, organic osmolytes, and water (Gaitán-Peñas et al., 2016; Jackson & Strange, 1993; Liu, Tashmukhamedov, Inoue, Okada, & Sabirov, 2006; Lutter et al., 2017; Roy, 1995), VRAC permeability was also shown for larger molecules such as ATP (Hisadome et al., 2002), blasticidin S (Lee et al., 2014), and the natural antioxidant tripeptide glutathione (Friard et al., 2019; Sabirov, Kurbannazarova, Melanova, & Okada, 2013). Thus, VRACs seemed to intriguingly combine selectivity for inorganic anions and permeability for a wide variety of large molecules. Only after the molecular identity VRAC was uncovered, this phenomenon was explained by the variable heteromerization of LRRC8 paralogs (Gaitán-Peñas et al., 2016; Lee et al., 2014; Lutter et al., 2017; Planells-Cases et al., 2015; Schober et al., 2017).

Another "fingerprint" of VRAC current is an inactivation at inside-positive membrane potentials (Fig. 2A). Although likely of little physiological relevance, the comprehension of this phenomenon may shed light on the mechanisms of VRAC gating. The time course of VRAC inactivation varied among different cell types studied (Jackson & Strange, 1995; Leaney, Marsh, & Brown, 1997; Lepple-Wienhues et al., 1998). This could be due to the expression of different isoforms of the VRAC, and since the identification of LRRC8 heteromers as VRAC components also revealed that the subunit composition determines inactivation kinetics (Voss et al., 2014). In addition, the different experimental conditions, such as the composition of the extracellular medium affected this biophysical property. It has been shown that pH, extracellular concentrations of Mg^{2+}, Ca^{2+} and Cl^{-} as well as the size of the current, may influence the voltage- and time-dependence of VRAC current inactivation (Akita & Okada, 2014; Hernández-Carballo, De Santiago-Castillo, Rosales-Saavedra, Pérez-Cornejo, & Arreola, 2010; Jackson & Strange, 1995; Nilius et al., 1997; Nilius & Droogmans, 2001; Nilius, Sehrer, et al., 1994; Voets et al., 1997).

3.4 VRAC is formed by LRRC8 heteromers

Many attempts were made to discover the molecular identity of VRAC using combinations of different approaches, including electrophysiology. Among the false-positive candidates, there were P-glycoprotein, pIcln, ClC-3, Best1, and TMEM16F (ANO6) (Okada et al., 2019; Pedersen et al., 2015; Stauber, 2015). However, none of these candidates displayed properties fully compatible with those of VRAC. In 2014, the molecular identity of VRAC was discovered simultaneously in two studies using similar techniques (Qiu et al., 2014; Voss et al., 2014). They demonstrated that the multispan membrane protein LRRC8A is essential for VRAC pore formation. LRRC8A knockdown dramatically suppressed swelling-induced currents recorded from HEK cells, HeLa cells, and T lymphocytes (Qiu et al., 2014; Voss et al., 2014). These currents were completely absent in genome-edited, LRRC8A-deficient HEK and HCT116 cell lines, but were restored by heterologous expression of LRRC8A (Voss et al., 2014).

One line of evidence that LRRC8A is an integral part of the channel pore was provided by the identification of a critical threonine (T44) in its first transmembrane helix (Qiu et al., 2014). The substitution of this amino acid by cysteine (T44C) led to a strong suppression of VRAC currents upon extracellular application of the membrane-impermeable, thiol-reactive reagent 2-sulfonatoethyl methane-thiosulfate (MTSES), that was not observed in cells expressing wild-type LRRC8A. Accessibility of this amino acid to MTSES from the extracellular space indicates the participation of LRRC8A in VRAC pore formation. Moreover, direct evidence for the pore-forming ability of LRRC8 was provided by *in vitro* reconstitution of purified VRAC-containing protein complexes into droplet interface lipid bilayers (Syeda et al., 2016). In this system, osmotic gradient activated Cl^- currents, sensitive to the VRAC inhibitor DCPIB, demonstrating that purified LRRC8 proteins are sufficient to recapitulate the key properties of VRAC (Syeda et al., 2016).

Overexpression of LRRC8A, however, did not increase VRAC currents, and even suppressed them (Qiu et al., 2014; Voss et al., 2014), implying that LRRC8A is a part of a heteromeric complex. Other members of the LRRC8 family, LRRC8B, 8C, 8D, or 8E were shown to be required for functional VRAC activity (Gaitán-Peñas et al., 2016; Lutter et al., 2017; Sato-Numata, Numata, Inoue, Sabirov, & Okada, 2017; Syeda et al., 2016; Voss et al., 2014). Combined disruption of *LRRC8* genes in HCT116 cells revealed that LRRC8A alone is not sufficient for normal VRAC activity, whereas heteromers containing LRRC8A and at least one paralog are active (Voss et al., 2014). For instance, in contrast to overexpressing LRRC8A alone, its overexpression together with LRRC8C did not suppress swelling-induced currents. These currents were neither suppressed upon individually knocking out LRRC8B–E, indicating that none of them is essential for VRAC activity. However, when all of them were knocked out, cells did not exhibit any swelling-induced current, similarly to LRRC8A-deficient cells. Finally, in the cells where all five *LRRC8* genes were disrupted, VRAC currents could be restored only upon combined expression of LRRC8A with at least one of its paralogue (Voss et al., 2014). Interestingly, LRRC8C, D, or E subunits were reported to form functional homomeric channels if the intracellular loop connecting transmembrane domains 2 and 3 was replaced with that from LRRC8A (Yamada & Strange, 2018).

The subunit composition of LRRC8 heteromers was also shown to define depolarization-dependent inactivation of VRAC adding further evidence for pore formation by LRRC8 heteromers (Voss et al., 2014). HCT116 cells co-expressing LRRC8A with LRRC8E exhibited faster inactivation of swelling-induced currents and at less positive potentials in comparison to wild-type cells. In contrast, co-expression of LRRC8A with LRRC8C dramatically slowed down VRAC inactivation (Fig. 3). Consistently, LRRC8C-deficient cells demonstrated faster depolarization-dependent inactivation of VRAC current (Voss et al., 2014).

Both Qiu et al. and Voss et al. also showed that LRRC8 proteins are responsible for swelling-induced osmolyte release (Qiu et al., 2014; Voss et al., 2014). In addition to radiotracer experiments (see below), hypotonicity-induced whole-cell currents in HeLa cells decreased upon LRRC8A disruption when extracellular Cl^- was replaced by taurine (Qiu et al., 2014).

Although it is now well established that LRRC8 proteins form VRACs, several observations point to the existence of some auxiliary components. Firstly, overexpression of LRRC8A and LRRC8B/C/D/E caused no increase of the swelling-induced currents over the endogenous level in HEK293 and HCT116 cells (Voss et al., 2014) and HeLa cells (Okada, Islam, Tsiferova, Okada, & Sabirov, 2017). Moreover, overexpression did

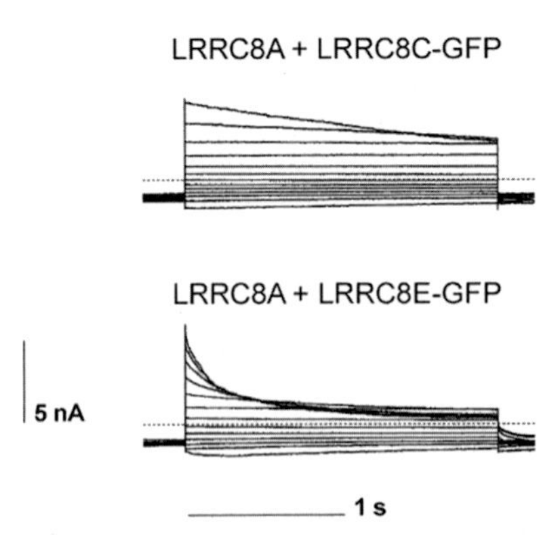

Fig. 3 Representative maximally activated swelling induced currents mediated by LRRC8A and either LRRC8C or LRRC8E expressed in HCT116 cells lacking all five *LRRC8* genes (20-mV steps from −120 to 120 mV). Currents with LRRC8E deactivate faster and at less positive potentials than with LRRC8C. *From Voss, F. K., Ullrich, F., Münch, J., Lazarow, K., Lutter, D., ..., Stauber, T., & Jentsch, T. J. (2014). Identification of LRRC8 heteromers as an essential component of the volume-regulated anion channel VRAC.* Science, *344(6184), 634-638. doi:10.1126/science.1252826.*

not restore swelling-induced currents in cisplatin-resistant KCP-4 cells, that do not exhibit VRAC activity up to the level in its parental cisplatin-sensitive KB cells (Okada et al., 2017). Second, while VRAC current density is in general relatively similar between cell lines, it may vary in cases despite similar LRRC8 expression levels (T. Okada et al., 2017). Thirdly, channels reconstituted with LRRC8A and LRRC8D/8E were not activated by pressure-induced inflation and exhibited VRAC currents independently of ATP (Syeda et al., 2016). Recently, despite clear evidence that LRRC8 proteins form VRAC also in astrocytes (Formaggio et al., 2019; Hyzinski-García et al., 2014; Yang et al., 2019), tweety homologs (TTYH1-3) were proposed as the VRAC forming molecules in mouse astrocytes (Han et al., 2019). However, the solved structure of this protein family revealed no evidence for channel function (Sukalskaia, Straub, Deneka, Sawicka, & Dutzler, 2021).

3.5 Gating of VRAC

3.5.1 Electrophysiological recordings in Xenopus *oocytes highlight a role for the C-terminal domain*

One of the proposed determinants of VRAC activation is low intracellular ionic strength. Supporting previous studies (Best & Brown, 2009; Cannon et al., 1998; Nilius et al., 1998; Sabirov, Prenen, Tomita, et al., 2000; Voets et al., 1999), incorporation of the recombinant LRRC8 proteins in lipid droplet bilayers also led to VRAC activation upon reduction of ionic strength (Syeda et al., 2016). Low ionic strength acted directly on the channel protein since a purified protein was employed in this system. The mechanism of this action may involve an alteration of the electrostatic interaction of the LRRDs, since its positively and negatively charged amino acids get less shielded by solvent ions, leading to channel opening.

However, activation of VRAC upon low intracellular ionic strength is compromised by several facts (König & Stauber, 2019; Strange et al., 2019): (i) the reductions of ionic strength needed for VRAC activation in patch-clamp recordings are much larger than what would be expected from activation by physiological osmolality changes (ii) the intracellular ion concentrations are "clamped" in the whole-cell configuration, allowing VRAC activation independently of intracellular ionic strength. (iii) VRAC can be activated in a variety of conditions not associated with alteration of ionic strength or extracellular osmolarity. Therefore, low ionic strength—while possibly contributing to VRAC activation—is unlikely the physiologically most relevant stimulus.

Since the structure of LRRC8 subunits resembles the ones of volume-insensitive connexins and pannexins except for the unique for LRRC8 C-terminal cytosolic LRRD (Abascal & Zardoya, 2012; Deneka et al., 2018; Kasuya et al., 2018; Kefauver et al., 2018; Kern et al., 2019), it is on the one hand highly possible that this domain is responsible for VRAC activation upon volume changes. On the other hand, pannexin-1 channel activation was also shown to be mediated by its C-terminus (Penuela, Gehi, & Laird, 2013). Both of these notions hint towards the involvement of C-terminal LRRD in the regulation of VRAC activation.

The importance of C-termini of LRRC8 proteins for VRAC activity was highlighted by using the *Xenopus* oocyte expression system, a complementary approach to the expression in knockout cell lines (Gaitán-Peñas et al., 2016). The *Xenopus* oocyte is a widely used expression system for the functional characterization of ion channels for two main reasons. First, their large size facilitates both injection of heterologous cRNA and subsequent electrophysiological recordings. Second, their efficient translation of cRNA results in a large number of ion channels in the plasma membrane. However, due to the large size of *Xenopus* oocytes, technical difficulties arise when measuring transmembrane currents in these cells. The current provided through a microelectrode is limited. Therefore, charging of the large membrane capacitance cannot be achieved immediately, and the voltage-clamp speed becomes thereby limited. Another problem is the excessively large current, which can result from high levels of ion channel expression and cause substantial voltage errors due to a potential drop in the series resistance in the bath solution. These problems have been addressed through the use of two microelectrode voltage-clamp technique with high-voltage amplifiers to force large currents through the current-passing microelectrode and by the use of electronic compensation for series resistance (Marmont, 1949; Stühmer & Parekh, 1995). Both microelectrodes form a contact with the membrane of the oocyte, one for voltage sensing and one for current injection. The transmembrane potential is measured by the voltage-sensing electrode connected to a high input impedance amplifier. The detected signal is compared with a command voltage, and the difference is brought to zero by a control amplifier. The injected current is monitored to provide a measure of the total membrane current. The advantage of the *Xenopus* oocyte system for VRAC studies is the possibility of performing a variety of electrophysiological, optical, and biochemical assays on single cells with the ability to control the expression of each LRRC8 subunit independently (Gaitán-Peñas et al., 2016; Gaitán-Peñas,

Pusch, & Estévez, 2018). Fusing a fluorescent protein such as GFP or mCherry to the C-terminus of each of the six LRRC8 subunits, resulted in a constitutively active channel in isotonically treated *Xenopus* oocytes, possibly by spatially separating the LRRDs (Gaitán-Peñas et al., 2016; Gaitán-Peñas et al., 2018). Expression of protein-tagged LRRC8 subunits did not open the channels completely and kept the mechanism of channel activation intact since VRAC currents were further enhanced upon hypotonic or other stimuli (Gaitán-Peñas et al., 2016; Gradogna, Gaitán-Peñas, Boccaccio, Estévez, & Pusch, 2017; Gradogna, Gavazzo, et al., 2017).

To study VRAC currents upon heterologous LRRC8 expression in *Xenopus* oocytes (Gaitán-Peñas et al., 2016), collagenase was used for complete removal of the enveloping follicular from the oocytes, resulting in the absence of endogenous VRAC currents. In contrast, oocytes defolliculated by forceps, thereby still partially surrounded by the follicular envelope, exhibited endogenous VRAC currents, suggesting that endogenous VRAC might be located in the follicular cell membrane (Decher et al., 2001; Voets et al., 1996). Although no currents were observed in collagenase-treated oocytes when fluorescently-tagged LRRC8A was expressed by alone, VRAC currents were detected upon expression of fluorescently-tagged LRRC8 accessory subunits (LRRC8C, D, and E). These currents were observed seldom and were comparably small, nevertheless suggesting that collagenase-digested oocytes may contain the *Xenopus* LRRC8A protein.

Although similar constitutive activation of the currents by adding fluorescent proteins was also observed when expressing LRRC8 proteins in HEK cells, the success rate of giga-seal formation in these cells was low and did not allow the analysis of the currents (Gaitán-Peñas et al., 2016).

Attaching mCherry to LRRC8A and LRRC8C-D, but not to LRRC8B, resulted in a similar current potentiation (Gaitán-Peñas et al., 2018). Tagging of the subunit of a single type was sufficient to activate the channel in isotonic conditions, whereas tagging of both subunits resulted in larger effects. The highest current was observed when LRRC8A and LRRC8E proteins were fused to mCherry. The size of the tag may be critical for VRAC activation since no activation was observed upon fusion with other tags (three copies of the hemagglutin (HA) tag or three copies of the flag tag). In addition, the kinetics of activation varied with the pair of tagged LRRC8 proteins expressed, proving that the observed currents were not endogenous to the oocyte, but resulted from the expressed LRRC8 subunits. The constitutive currents were reduced when applying a hyperosmotic solution, indicating that the fusion with the fluorescent proteins shifts the osmosensitivity of the channel (Gaitán-Peñas et al., 2016). It was concluded that the protein-tagging of the C-terminus changes the gating dependence of osmolarity, thereby acting as a "foot-in-the-door" mechanism.

In the future, the expression of the fluorescently-tagged subunits resulting in constitutively active VRACs could facilitate drug screening studies of VRAC channels and discover their novel structure-function relationships.

3.5.2 Chimeras between LRRC8 paralogs and cysteine modification reveal important regions for channel gating

Further regions important for VRAC activity are the intracellular loop connecting transmembrane helices 2 and 3, and the first extracellular loop connecting transmembrane helices 1 and 2 (Yamada & Strange, 2018). Exchanging either of these sequence stretches between LRRC8A and a paralog, resulted in constructs able to form functional homomeric channels. For example, homomers of LRR8C with the intracellular loop of LRRC8A mediated VRAC currents with properties of native VRACs such as sensitivity to cell swelling (Yamada & Strange, 2018), which are not entirely preserved in LRRC8A homomers (Yamada et al., 2021).

The N-termini of the LRRC8 proteins are also important for VRAC ion permeability and inactivation kinetics (Zhou, Polovitskaya, & Jentsch, 2018). This importance was revealed by the observations that (i) mutations of N-terminal amino acids and (ii) fusion of epitopes to their N-, but not their C-termini abolished VRAC currents (Zhou et al., 2018). Further investigation of LRRC8 N-terminal amino acids was performed by cysteine point mutagenesis—a useful technique for revealing the functional importance of the respective residue and its exposure to a hydrophilic environment (Holmgren, Liu, Xu, & Yellen, 1996). Cysteine point mutations allow observing changes in channel properties upon application of cysteine-modifying reagents such as 2-aminoethyl methanethiosulfonate (MTSEA) and Cd^{2+}. Single-cysteine replacements obliterated currents when inserted into both subunits of LRRC8A/C channels. Replacing Arg8 by cysteine in both LRRC8A and LRRC8C resulted in functional channels. These channels demonstrated reduction or enhancement of VRAC currents upon application of MTSEA, indicating that the mutated residues are both accessible from the aqueous phase. Additionally, conserved Glu6 was shown to mediate anion selectivity of the LRRC8A/C, and, to a lesser extent, LRRC8A/D, and LRRC8A/E channels. Thus, the sequence diversity at the N-terminal region might be important for substrate permeability, which likely affects the reactivity of MTSES to the mutations at Glu6. Taken together, these findings strongly suggest that the LRRC8 N-termini line the VRAC pore (Zhou et al., 2018). In consistence with this idea, structure determination of human LRRC8D in a homo-hexamer revealed that Glu6 faces the channel pore (Nakamura et al., 2020). It was shown that the N-terminal helix of each LRRC8D subunit protruded into the channel pore from the intracellular side to modulate gating. Electrophysiological analyses based on the described structure using a cysteine modifier confirmed that the N-terminal helix observed in the LRRC8D structure also exists in the LRRC8A/D hetero-hexamer, with channel entry from the cytoplasmic side (Nakamura et al., 2020).

Finally, VRAC inactivation was shown to be determined by a highly conserved C-terminal part of the first extracellular loop of LRRC8 proteins (Ullrich et al., 2016). Since the isoforms LRRC8C and LRRC8E exhibit a large difference in their depolarization-dependent inactivation kinetics when expressed with LRRC8A (Voss et al., 2014), creating the chimeras assembled from these isoforms elucidated the molecular basis for this variability. Point mutations identified residues in the first extracellular loop as the main determinants for differential inactivation properties (Ullrich et al., 2016). Charge reversal of these residues did not only dramatically alter the kinetics and voltage dependence of VRAC inactivation, but also reduced its $I^- > Cl^-$ selectivity. Moreover, the latter effect was facilitated when both LRRC8A and the co-expressed subunit, LRR8C or LRR8CE, carried such mutations, suggesting that the C-terminal part of the first extracellular loop of LRRC8 proteins participates in forming the outer pore of VRAC (Ullrich et al., 2016). Therefore, inactivation of VRAC likely involves a conformational change of the external pore constriction, either by a direct effect of positive voltages on charged segments of the protein or indirectly, for instance by the interaction of the protein with substrates.

Membrane stretch associated with osmotic cell swelling is also known to activate TRPM7 - a nonselective cation channel. It was reported that TRPM7 affected VRAC activity, thereby implying a functional coupling of these two swelling-activated channels (Numata, Sato-Numata, Hermosura, Mori, & Okada, 2021). VRAC currents were suppressed by knockdown of TRPM7 in HeLa cells or knockout in chicken DT40 cells. The currents were rescued by heterologous expression of TRPM7 in TRPM7-deficient DT40 cells. When the α-kinase domain of TRPM7 was deleted, VRAC activity was abolished. However, it was not affected upon expression of mutant TRPM7, in which the α-kinase was rendered inactive by a point mutation. Therefore, it is postulated that the physical interaction of the C-terminal kinase domain of TRPM7 after osmotic cell swelling rather than the enzyme activity of the α-kinase domain regulates VRAC activity (Numata et al., 2021).

3.6 VRAC reconstituted in lipid bilayers

A purified ion channel can be studied after reconstitution into model membrane systems, such as lipid vesicles or planar bilayers (Morera, Vargas, González, Rosenmann, & Latorre, 2007). As opposed to working with cells, the incorporation of ion channels in planar lipid bilayers allows the measurement and detection of ion channel activity in a well-controlled manner. Planar lipid bilayers form a unique controllable environment in which specific regulatory components such as lipids, ligands, inhibitors, specific ions, and proteins, as well as the temperature that modulates the activity of many ion channels can be maintained (Zakharian, 2021). It also eliminates the interference by the activities of endogenous ion channels. Several methods have been developed to incorporate ion channels into planar bilayers (Braha et al., 1997; Heginbotham, LeMasurier, Kolmakova-Partensky, & Miller, 1999; Schindler & Rosenbusch, 1978; Slatin, Qiu, Jakes, & Finkelstein, 1994). Syeda and colleagues studied VRAC for the first time in a droplet interface bilayer (DIB) (Syeda et al., 2016), in which two droplets under an oil/lipid mixture first become encased within lipid monolayers and are then joined to form bilayers (Funakoshi, Suzuki, & Takeuchi, 2006; Holden, Needham, & Bayley, 2007). Electrodes placed within the droplets enabled the electrical recording of the channel activity at the single-molecule level (Holden et al., 2007). Using the DIB, the authors investigated the activity of different LRRC8A-containing reconstituted heteromers by mechanical manipulations of the membrane or a decrease in ionic strength. A 20% increase in droplet volume and subsequent mechanical changes in lipid bilayers activated the mechanosensitive channel (MscS) (Battle, Petrov, Pal, & Martinac, 2009) but not the LRRC8-containing VRAC (Syeda et al., 2016). Instead, a decrease in ionic strength activated LRRC8A-containing VRACs (for VRAC activation by reduced ionic strength, see above).

Recently, several studies showed cryo-EM structures of LRRC8A or LRRC8D homomers, or LRRC8A/C heteromers (Deneka et al., 2018; Kasuya et al., 2018; Kefauver et al., 2018; Kern et al., 2019; Nakamura et al., 2020). In one study, *Mus musculus* LRRC8A homomer in complex with the VRAC inhibitor DCPIB was reconstituted in lipid nanodiscs, providing insight into LRRC8 gating and inhibition (Kern et al., 2019), while properties of LRRC8A homomers clearly differ from endogenous VRACs (Yamada et al., 2021). Purified LRRC8A was reconstituted into phosphatidylcholine lipids and recorded from proteoliposome patches (Kern et al., 2019). LRRC8A homomer activity was only observed in low ionic strength solutions, as with other reconstituted preparations (Kasuya et al., 2018; Syeda et al., 2016).

Similar to the *Xenopus* oocyte system, the disadvantage of planar lipid bilayers is their large capacitance due to the large size of the bilayer. As a result, the voltage response time is slower. The second disadvantage of this technique is the generation of high amplitude noise due to the large area of bilayers in traditional systems. Intensive filtering may be used to circumvent this. Therefore, low pass filtering of the single-channel signal dampens time resolution and at some point hinders detection of fast-gating events (Zakharian, 2013).

3.7 Pharmacological VRAC modulation

The investigation of VRAC function was—especially before its molecular identification—largely based on its pharmacological modulation, but the applied reagents were mostly not very specific (Friard et al., 2017; Sato-Numata, Numata, Inoue, & Okada, 2016). The lack of high-affinity inhibitors of VRAC has in fact been an obstacle for the identification of its molecular identity and the investigation of its physiological importance with electrophysiology and other approaches. Classical Cl^- channel blockers such as DIDS and NPPB can inhibit VRAC currents at micro- to millimolar concentrations, however, their effect is rather unspecific, since they also affect many other channels and transporters (Jentsch, Stein, Weinreich, & Zdebik, 2002). The swelling-induced anion current is also inhibited by the estrogen receptor antagonists tamoxifen, cliphen, and nafoxidine at micromolar concentrations but the effectiveness of these compounds differs depending on the cell type studied (Maertens, Droogmans, Chakraborty, & Nilius, 2001). To some extent, a blocking effect was also reported for niflumic, flufenamic and arachidonic acids, glibenclamide, and some antimalarials such as mefloquine (Jentsch et al., 2002; Nilius & Droogmans, 2003). Carbenoxolone, known to inhibit gap junctions, was also capable of inhibiting VRAC currents at similar concentrations

(Benfenati et al., 2009). The most specific inhibitors of VRAC so far are the acidic di-aryl-urea NS3728 and the indanone compound DCPIB (Decher et al., 2001; Helix, Strobaek, Dahl, & Christophersen, 2003). DCPIB was found to bind to an arginine in a cryo-EM structure of LRRC8A (Kern et al., 2019), but replacing this amino acid with leucine did not drastically alter VRAC inhibition by the compound (Yamada et al., 2021). DCPIB was also shown to exert a promiscuous effect on the glutamate transporter GLT-1, connexin hemichannels, the H^+,K^+-ATPase and potassium channels (Bowens, Dohare, Kuo, & Mongin, 2013; Deng, Mahajan, Baumgarten, & Logothetis, 2016; Fujii et al., 2015; Lv et al., 2019; Minieri et al., 2013).

Pranlukast and zafirlukast, antagonists of the cysteinyl leukotriene receptor 1 (CysLT1R) were identified as VRAC inhibitors in a high-throughput screen study (Figueroa, Kramer, Strange, & Denton, 2019). Interestingly, as the study was performed on HEK293 cells, that are lacking CYSLTR1 mRNA, pranlukast and zafirlukast seem to be direct channel inhibitors that work independently of the CysLT1R through direct interactions with the channel protein. Thus these two chemical scaffolds can be used for the development of more potent and specific VRAC inhibitors.

Oxidation was also shown to modulate LRRC8 channels in a subunit-dependent manner (Friard, Laurain, Rubera, & Duranton, 2021; Gradogna, Gaitán-Peñas, et al., 2017). By fluorescently tagging LRRC8 proteins that yield large constitutive currents direct effects of oxidation were demonstrated. The currents mediated by LRRC8A/LRRC8E heteromers were more than 10-fold potentiated upon the oxidation of intracellular cysteine residues. However, hypotonicity-induced activation of these heteromers channels appeared to be not affected by oxidation. In contrast, LRRC8A/LRRC8C- and LRRC8A/LRRC8D-mediated currents were strongly inhibited by oxidation (Gradogna, Gaitán-Peñas, et al., 2017).

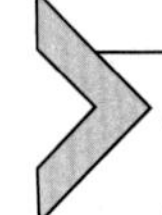

4. Measurement of transported substrate

4.1 Measurement of halide flux

Instead of measuring the electrical current mediated by VRACs, it is possible to measure concentration changes or flux of the transport substrate, both for uncharged as well as charged compounds. As VRAC mediates anion transport, irrespective of its subunit composition, its opening will allow for Cl^- flux along its electrochemical gradient. A decrease in cytosolic [Cl^-] concomitant with VRAC activation has been observed using the Cl^- indicator

6-methoxy-N-3-sulfopropyl quinolinium (SPQ) during myoblast differentiation (L. Chen, König, & Stauber, 2020). Another method to measure halide transport relies on the I^- permeability of VRAC and a genetically encoded halide sensor. The influx of externally applied I^- upon VRAC activation will quench an iodide-sensitive mutant of the yellow-fluorescent protein (YFP) expressed in this cell after transient or stable transfection (Fig. 4), so the extent of YFP quenching correlates with VRAC activity. In fact, both studies that identified LRRC8 proteins as essential VRAC compounds used the $YFP^{H148Q,I152L}$ (Galietta, Haggie, & Verkman, 2001) in their assay for the genome-wide siRNA screen (Qiu et al., 2014; Voss et al., 2014) and YFP^{H148Q} had been used to identify Bestrophin 1 as the swelling-activated anion channel in *Drosophila* (Stotz & Clapham, 2012).

For the siRNA screens for the mammalian VRAC, HEK293 cells lines with stable (inducible) $YFP^{H148Q,I152L}$ expression were used. Downregulation of LRRC8A expression, as well as pharmacological inhibition of VRAC with DCPIB or carbenoxolone, significantly reduced fluorescence quenching as measured with a plate reader after a fixed time or in a time course after application of extracellular hypotonicity and iodide (Qiu et al., 2014; Voss et al., 2014). Quenching of halide-sensitive YFP was subsequently used in multiple studies to determine VRAC activity, for example, to confirm VRAC activity upon hypotonic cell swelling in various cell lines like HeLa, HEK293 and keratinocytes, to study VRAC activation by apoptosis-stimulating agents, to characterize VRAC inhibitors (Figueroa et al., 2019; Ghosh, Khandelwal,

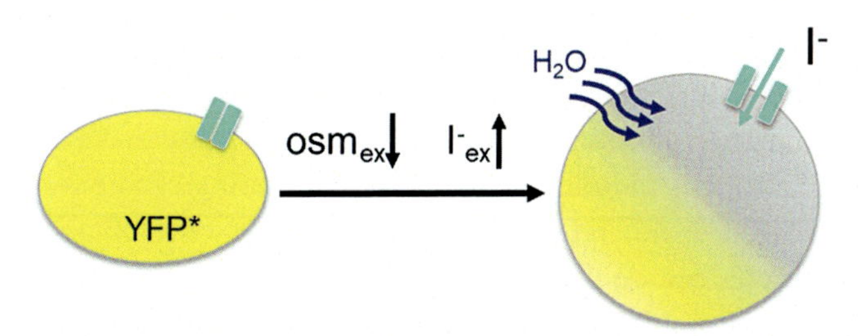

Fig. 4 Schematic representation of the assay based on YFP quenching by VRAC-mediated I^- influx. A cell expressing an I^--sensitive YFP variant faces extracellular hypotonicity and I^-. Cell swelling-induced opening of VRAC will allow the influx of I^- that quenches the YFP. *From Voss, F. K., Ullrich, F., Münch, J., Lazarow, K., Lutter, D., ..., Stauber, T., & Jentsch, T. J. (2014). Identification of LRRC8 heteromers as an essential component of the volume-regulated anion channel VRAC.* Science, *344(6184), 634-638. doi:10.1126/science.1252826.*

Kumar, & Bera, 2017; Green et al., 2020; Planells-Cases et al., 2015; Serra et al., 2021; Sirianant et al., 2016; Trothe et al., 2018).

4.2 Measuring the flux of larger osmolytes

As VRAC does not only conduct halides but also various organic osmolytes, it has alternatively been named volume-sensitive osmolyte/anion channel (VSOAC) (Qiu et al., 2014; Strange et al., 1996; Voss et al., 2014). Its permeability for organic osmolytes ranges from small negatively charged molecules such as amino acids, taurine to large uncharged polyols such as *myo*-inositol and sorbitol (Chen, König, et al., 2019; Jentsch, 2016).

Release of the osmolytes via VRAC has been investigated extensively using a radiotracer osmolyte efflux assay (Bowens et al., 2013; Hyzinski-García et al., 2014; Mongin & Kimelberg, 2005). Radiotracer variants of a specific osmolyte can be loaded into cells and after stimulation, radioactivity can be subsequently measured in the supernatant. Usually, osmolytes labeled with tritium [^{3}H] or the carbon isotope [^{14}C] are used. Examples comprise [^{3}H]-labeled taurine (Bach, Sørensen, & Lambert, 2018; Hyzinski-García et al., 2014; Lutter et al., 2017; Planells-Cases et al., 2015; Qiu et al., 2014; Schober et al., 2017; Sørensen, Nielsen, Thorsteinsdottir, Hoffmann, & Lambert, 2016; Voss et al., 2014), ^{14}[C]-taurine (Hyzinski-García et al., 2014; Kirk, Ellory, & Young, 1992; Mongin & Kimelberg, 2005), D-[^{14}C]-aspartate (Schober et al., 2017), D-[^{3}H]-aspartate (Hyzinski-García et al., 2014; Lutter et al., 2017; Mongin & Kimelberg, 2002) and a range of further [^{3}H]-labeled amino acids, [^{3}H]-GABA and [^{3}H]-*myo*-inositol (Hyzinski-García et al., 2014; Lutter et al., 2017; Planells-Cases et al., 2015; Schober et al., 2017; Voss et al., 2014). Osmolytes are taken up by various cellular transport systems during several hours of incubation (Bröer, 2002). Following this uptake, VRAC can be activated by different stimuli and scintillation counting can be used to evaluate radiolabeled osmolytes release (Fig. 5). Otherwise, uptake of radiolabeled osmolytes can be measured, as, for example, with *Xenopus* oocytes (Gaitán-Peñas et al., 2016). In addition to providing direct evidence of VRAC conductivity for the various osmolytes, the radiotracer studies provided insight into the role of specific LRRC8 family members in the swelling-activated release of organic osmolytes, for example, the critical role for LRRC8D in taurine permeability (Planells-Cases et al., 2015) and of subunits LRRC8C and LRRC8E for different amino acids (Lutter et al., 2017).

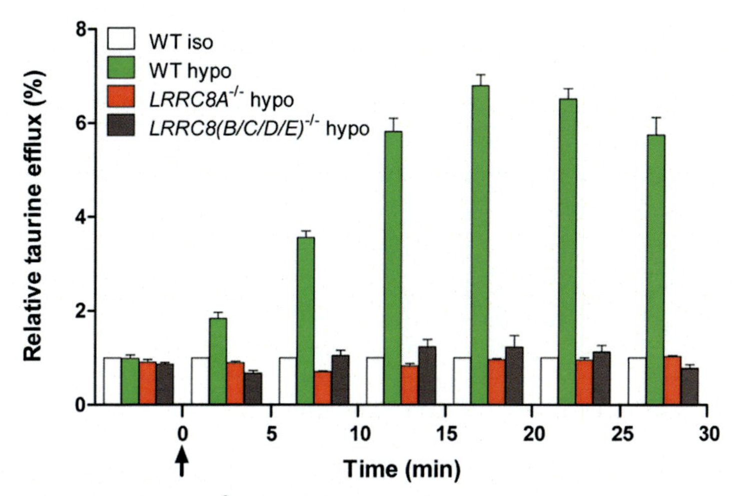

Fig. 5 Efflux of pre-loaded [^{3}H]-labeled taurine after a hypotonic stimulus (*arrow*) from wildtype HEK293 cells (*green*), but not under isosmotic control condition (*white*). Hypotonicity-induced taurine-efflux is impaired by knockout of LRRC8A (*red*) or combined knockout of LRRC8B-E (*gray*). *From Voss, F. K., Ullrich, F., Münch, J., Lazarow, K., Lutter, D., ..., Stauber, T., & Jentsch, T. J. (2014). Identification of LRRC8 heteromers as an essential component of the volume-regulated anion channel VRAC.* Science, *344(6184), 634-638. doi:10.1126/science.1252826.*

In the absence of exogenously applied radiotracers, cellular amino acid content can be measured using a separation method such as high-performance liquid chromatography (HPLC). Cells incubated with and without hypotonic medium are lysed, and their protein content is denatured, resulting in soluble amino acids that can be measured, as, for example, for taurine levels in cultured astrocytes challenged with hypotonicity (Kimelberg, Goderie, Higman, Pang, & Waniewski, 1990). Likewise, analyzing the supernatant of swollen cells by enzymatic assay kits can provide information about VRAC-mediated osmolyte efflux. For instance, hypotonicity-induced cell swelling resulted in a VRAC-dependent increase in glutamate levels in HEK293 cell supernatant, as determined by glutamate oxidase assay kit analysis (Lutter et al., 2017). ATP released into the culture medium by RAW macrophages upon isosmotic sphingosine-1-phosphate-induced VRAC activation or by various LRRC8 combinations expressed in *Xenopus* oocytes was analyzed by a luciferase assay (Burow et al., 2015; Gaitán-Peñas et al., 2016). cGAMP transport by VRAC was measured by an enzyme assay for the dinucleotide released from cells (Lahey et al., 2020) or by assessing the amount of imported cGAMP by liquid chromatography–mass spectrometry (LC-MS) (Chen et al., 2021; Zhou, Chen, et al., 2020). As with taurine efflux, uptake of the antibiotic blasticidin S and anti-cancer drugs cisplatin and carboplatin were strongly dependent on the presence of LRRC8D in VRAC heteromers (Lee et al., 2014; Planells-Cases et al., 2015). Cell survival may be used as an indirect read-out of LRRC8D-containing VRAC activity, or measurement of the drugs' uptake—for example by inductively coupled plasma-mass spectrometry (ICP-MS) in the case of platinum-containing drugs (Planells-Cases et al., 2015)—may serve this purpose.

5. Measurement of cell volume

VRAC is critically involved in regulatory volume decrease (RVD) upon hypotonic cell swelling, but also in isosmotic volume decrease such as during apoptotic volume decrease (AVD) (Chen, König, et al., 2019). Therefore, the measurement of cell volume following VRAC activation is a simple method to estimate VRAC activity. Measurement of cell volume can be performed with a wide variety of methods, ranging from simple observation under a light microscope to optical computer tomography (Model, 2018). The Coulter method has been regarded as a standard method. A typical Coulter sizer has two compartments, each containing an electrode. One of the compartments is filled with a solution of a standard electrolyte, and the second has a suspension of cells in an electro-conductive fluid, i.e. any typical cell medium with salts. The two compartments are separated by a small (50–100 μm) aperture. After voltage application between the electrodes, the amount of electric current is limited by this narrow passage so when a dielectric cell passes through the aperture, the resistance between the compartments is further increased. The increase in resistance is recorded by an electric circuit as a voltage spike and can be converted into the cell volume using calibration with standard beads (Model, 2018). The Coulter sizing method was applied in one of the studies that identified LRRC8A as a VRAC component to assess its role in the RVD of HeLa cells (Qiu et al., 2014). This method has since been applied in various studies on VRACs in different cell lines, for example, glioblastoma cells and chondrocytes (Kittl et al., 2020; Rubino et al., 2018; Serra et al., 2021). Similarly, volume changes with VRAC activity can be measured by flow cytometry (Sirianant et al., 2016).

A method to measure volume changes of attached cells relies on fluorophores that undergo self-quenching at high concentrations, i.e. the fluorescence intensity declines with increasing fluorophore concentration. This variation in the fluorescence indirectly reflects the volume of the cell. The fluorophore calcein, a derivative of fluorescein, provided a better signal-to-noise ratio compared to fluorescein and exhibited stable fluorescence, which was not affected by physiological changes in pH, Ca^{2+}, or Mg^{2+} (Crowe, Altamirano, Huerto, & Alvarez-Leefmans, 1995). Further refinement of this approach was achieved by calcein self-quenching at higher concentrations (Capo-Aponte, Iserovich, & Reinach, 2005). The calcein approach was used in the other study that identified LRRC8 members as VRAC components. With this method, a defective RVD was shown for LRRC8A-deficient HEK293 cells (Voss et al., 2014). Later on, it was used for studying the role of the different LRRC8 subunits, RVD in various cell types, including pancreatic β-cells and keratinocytes and VRAC inhibitors (Friard et al., 2017; Planells-Cases et al., 2015; Stuhlmann et al., 2018; Trothe et al., 2018). Cells are loaded with calcein, usually as cell-permeant AM-calcein that becomes fluorescent and loses membrane permeability upon cleavage by intracellular esterases. Upon application of extracellular hypotonicity, calcein fluorescence rapidly increases, indicating cell swelling. A subsequent rapid fluorescence decline, in turn, indicates cell volume reduction caused by RVD that is proportional to the applied hypotonicity.

Fluorescence self-quenching requires relatively high (mM) and often toxic loading concentrations. The calcein-based method is also less accurate than the Coulter sizing and volume estimations with calcein are usually semiquantitative (Model, 2018). Another, less commonly used method to determine relative cell volume changes involves optical sectioning (Bond et al., 1999; Cannon et al., 1998; Figueroa & Denton, 2021). The diameter and area of the cross-section of detached, round cells are measured using differential interference contrast (DIC) images. Using the ratio of cross-sectional areas at two different time points, the relative volume change is approximated. Each time a cell is measured, the same plane of the cell must be used. Besides being cumbersome, this approach is time-consuming and its potential inaccuracy renders it a less-suited method to generally assess VRAC-mediated cell volume changes (Model, 2018).

6. Monitoring VRAC activity by optical imaging

In the past few decades, we have developed a thorough understanding of ion channel biology through electrophysiological approaches, owing to their high temporal resolution and single-molecule sensitivity. However, recent, rapid advancements in imaging techniques have made it possible to test long-standing hypotheses regarding ion channel location, interactions, dynamics, and composition in living cells. To this end, the evolution of fluorescence microscopy has increased both the sensitivity and the resolution of microscopy. In contrast to classical approaches, optical methods offer several advantages for studying the conformational dynamics of ion channels in living cells. Optical methods provide spatio-temporal information about channels and are less invasive than electrophysiological measurements. As such, Förster resonance energy transfer (FRET) offers a powerful method. It is a distance-dependent physical process in which energy is transferred non-radiatively from an excited molecule (donor) to another molecule (acceptor) having overlapping emission and absorption spectra respectively. An angstrom-level measurement of molecular proximity can be accomplished with this method (10–100 Å) and it is highly efficient when donor and acceptor are positioned within the Förster radius—the distance at which half of the excitation energy of the donor is transferred to the acceptor (Sekar & Periasamy, 2003). Furthermore, FRET measurements provide both high-throughput and non-invasive long-term recording of live cells. Such fluorescence microscopy has been successfully applied to investigate conformational changes of various channels such as cyclic nucleotide-gated (CNG) channels, the ClC-0 Cl^- channel, BK K^+ channels and glutamate receptors (Bykova, Zhang, Chen, & Zheng, 2006; Miranda et al., 2013; Zachariassen et al., 2016; Zheng & Zagotta, 2000)

VRAC activity can also be tracked by FRET in live cells with subcellular resolution (König et al., 2019). Intra-complex FRET between fluorophores attached to the C-termini of LRRC8 subunits (specifically CFP/YFP variants Cerulean or Venus to LRRC8A and LRRC8E, respectively) was used to monitor the activity of VRAC in HeLa and HEK293 cells. As the C-termini rearranged conformation during VRAC activation, which is in accordance with certain flexibility suggested by structural studies (Deneka et al., 2018; Kasuya et al., 2018; Kefauver et al., 2018; Kern et al., 2019), the FRET signal decreased (Fig. 6). Patch-clamp fluorometry, i.e. simultaneous measurement of FRET and electrophysiological current recording confirmed that the reversible drop in FRET mirrored channel opening. The subcellular resolution allowed distinguishing between VRAC activation at the plasma membrane and the lack of activation at the endomembrane system. This would be hard to reconcile with VRAC activation by reduced ionic strength, as this was observed equally over the entire cell (König et al., 2019). Instead, plasma membrane diacylglycerol (DAG) seemed to be involved in

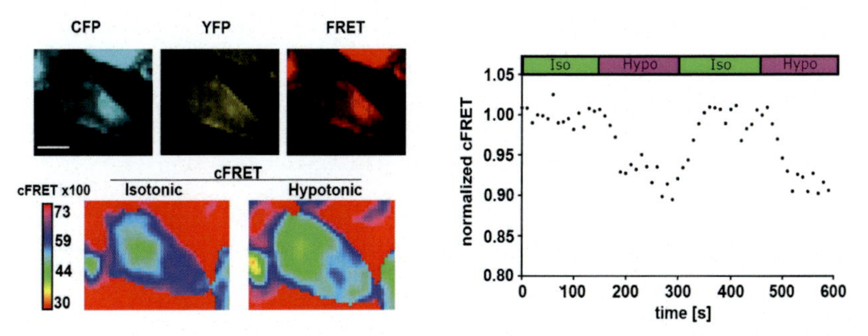

Fig. 6 Sensitized-emission FRET experiment with an LRRC8A-CFP/LRRC8E-YFP-expressing cell. *Left*: images of the acquired channels (donor CFP, acceptor YFP, FRET) with the calculated cFRET map for two time points in isotonic and hypotonic buffer below. Scale bar, 10 μm. *Right*: normalized cFRET values during the repeated exchange between isotonic and hypotonic buffer. *From König, B., Hao, Y., Schwartz, S., Plested, A. J., & Stauber, T. (2019). A FRET sensor of C-terminal movement reveals VRAC activation by plasma membrane DAG signaling rather than ionic strength.* eLife, *8, e45421. doi:10.7554/eLife.45421, under CC BY 4.0 license.*

VRAC activation. In addition, the non-invasive measurement of VRAC activation by FRET allowed investigation of the signaling that was observable by whole-cell patch clamp only directly after membrane breakthrough, but was lost within minutes of dialysis (König et al., 2019). Subsequently, the FRET sensor was used to monitor VRAC activity during myogenic differentiation of C2C12 myoblast, where it was temporarily activated during the first two hours concomitant with a reduction in intracellular chloride (L. Chen et al., 2020).

In summary, there is an increasing repertoire of methods to investigate the function of VRACs by biochemical, structural and biophysical means. The combination of these tools unequivocally brings forward our insight into the structure-function relationships and cell biological and physiological roles of this anion and osmolyte channel.

References

Abascal, F., & Zardoya, R. (2012). LRRC8 proteins share a common ancestor with pannexins, and may form hexameric channels involved in cell-cell communication. *BioEssays*, *34*(7), 551–560. https://doi.org/10.1002/bies.201100173.

Ackerman, M. J., Wickman, K. D., & Clapham, D. E. (1994). Hypotonicity activates a native chloride current in Xenopus oocytes. *The Journal of General Physiology*, *103*(2), 153–179.

Akita, T., Fedorovich, S. V., & Okada, Y. (2011). Ca^{2+} nanodomain-mediated component of swelling-induced volume-sensitive outwardly rectifying anion current triggered by autocrine action of ATP in mouse astrocytes. *Cellular Physiology and Biochemistry*, *28*(6), 1181–1190. https://doi.org/10.1159/000335867.

Akita, T., & Okada, Y. (2014). Characteristics and roles of the volume-sensitive outwardly rectifying (VSOR) anion channel in the central nervous system. *Neuroscience, 275C*, 211–231. https://doi.org/10.1016/j.neuroscience.2014.06.015.

Alghanem, A. F., Abello, J., Maurer, J. M., Kumar, A., Ta, C. M., Gunasekar, S. K., et al. (2021). The SWELL1-LRRC8 complex regulates endothelial AKT-eNOS signaling and vascular function. *eLife, 10*. https://doi.org/10.7554/eLife.61313.

Alonso-Gardon, M., Elorza-Vidal, X., Castellanos, A., La Sala, G., Armand-Ugon, M., Gilbert, A., et al. (2021). Identification of the GlialCAM interactome: The G protein-coupled receptors GPRC5B and GPR37L1 modulate Megalencephalic leukoencephalopathy proteins. *Human Molecular Genetics*. https://doi.org/10.1093/hmg/ddab155.

Arreola, J., Melvin, J. E., & Begenisich, T. (1995). Volume-activated chloride channels in rat parotid acinar cells. *The Journal of Physiology, 484*, 677–687.

Bach, M. D., Sørensen, B. H., & Lambert, I. H. (2018). Stress-induced modulation of volume-regulated anions channels in human alveolar carcinoma cells. *Physiological Reports, 6*(19), e13869. https://doi.org/10.14814/phy2.13869.

Banderali, U., & Roy, G. (1992). Anion channels for amino acids in MDCK cells. *The American Journal of Physiology, 263*(6 Pt 1), C1200–C1207.

Battle, A. R., Petrov, E., Pal, P., & Martinac, B. (2009). Rapid and improved reconstitution of bacterial mechanosensitive ion channel proteins MscS and MscL into liposomes using a modified sucrose method. *FEBS Letters, 583*(2), 407–412. https://doi.org/10.1016/j.febslet.2008.12.033.

Benfenati, V., Caprini, M., Nicchia, G. P., Rossi, A., Dovizio, M., Cervetto, C., et al. (2009). Carbenoxolone inhibits volume-regulated anion conductance in cultured rat cortical astroglia. *Channels (Austin, Tex.), 3*(5), 323–336.

Bertelli, S., Remigante, A., Zuccolini, P., Barbieri, R., Ferrera, L., Picco, C., et al. (2021). Mechanisms of activation of LRRC8 volume regulated anion channels. *Cellular Physiology and Biochemistry, 55*(S1), 41–56. https://doi.org/10.33594/000000329.

Best, L., & Brown, P. D. (2009). Studies of the mechanism of activation of the volume-regulated anion channel in rat pancreatic β-cells. *The Journal of Membrane Biology, 230*(2), 83–91. https://doi.org/10.1007/s00232-009-9189-x.

Best, L., Brown, P. D., Sener, A., & Malaisse, W. J. (2010). Electrical activity in pancreatic islet cells: The VRAC hypothesis. *Islets, 2*(2), 59–64. https://doi.org/10.4161/isl.2.2.11171.

Bond, T., Basavappa, S., Christensen, M., & Strange, K. (1999). ATP dependence of the ICl, swell channel varies with rate of cell swelling. Evidence for two modes of channel activation. *The Journal of General Physiology, 113*(3), 441–456. https://doi.org/10.1085/jgp.113.3.441.

Bowens, N. H., Dohare, P., Kuo, Y. H., & Mongin, A. A. (2013). DCPIB, the proposed selective blocker of volume-regulated anion channels, inhibits several glutamate transport pathways in glial cells. *Molecular Pharmacology, 83*(1), 22–32. https://doi.org/10.1124/mol.112.080457.

Braha, O., Walker, B., Cheley, S., Kasianowicz, J. J., Song, L., Gouaux, J. E., et al. (1997). Designed protein pores as components for biosensors. *Chemistry & Biology, 4*(7), 497–505. https://doi.org/10.1016/s1074-5521(97)90321-5.

Bröer, S. (2002). Adaptation of plasma membrane amino acid transport mechanisms to physiological demands. Pflügers Archiv, 444(4), 457-466. doi:https://doi.org/10.1007/s00424-002-0840-y.

Browe, D. M., & Baumgarten, C. M. (2003). Stretch of beta 1 integrin activates an outwardly rectifying chloride current via FAK and Src in rabbit ventricular myocytes. *The Journal of General Physiology, 122*(6), 689–702. https://doi.org/10.1085/jgp.200308899.

Browe, D. M., & Baumgarten, C. M. (2004). Angiotensin II (AT1) receptors and NADPH oxidase regulate Cl^- current elicited by beta1 integrin stretch in rabbit ventricular myocytes. *The Journal of General Physiology*, *124*(3), 273–287. https://doi.org/10.1085/jgp.200409040.
Burow, P., Klapperstück, M., & Markwardt, F. (2015). Activation of ATP secretion via volume-regulated anion channels by sphingosine-1-phosphate in RAW macrophages. *Pflügers Archiv*, *467*(6), 1215–1226. https://doi.org/10.1007/s00424-014-1561-8.
Bykova, E. A., Zhang, X. D., Chen, T. Y., & Zheng, J. (2006). Large movement in the C terminus of CLC-0 chloride channel during slow gating. *Nature Structural & Molecular Biology*, *13*(12), 1115–1119.
Cahalan, M. D., & Lewis, R. S. (1988). Role of potassium and chloride channels in volume regulation by T lymphocytes. *Society of General Physiologists Series*, *43*, 281–301.
Cannon, C. L., Basavappa, S., & Strange, K. (1998). Intracellular ionic strength regulates the volume sensitivity of a swelling-activated anion channel. *The American Journal of Physiology*, *275*(2 Pt 1), C416–C422.
Capo-Aponte, J. E., Iserovich, P., & Reinach, P. S. (2005). Characterization of regulatory volume behavior by fluorescence quenching in human corneal epithelial cells. *The Journal of Membrane Biology*, *207*(1), 11–22. https://doi.org/10.1007/s00232-005-0800-5.
Chen, L., Becker, T. M., Koch, U., & Stauber, T. (2019). The LRRC8/VRAC anion channel facilitates myogenic differentiation of murine myoblasts by promoting membrane hyperpolarization. *The Journal of Biological Chemistry*, *294*(39), 14279–14288. https://doi.org/10.1074/jbc.RA119.008840.
Chen, L., König, B., Liu, T., Pervaiz, S., Razzaque, Y. S., & Stauber, T. (2019). More than just a pressure relief valve: physiological roles of volume-regulated LRRC8 anion channels. *Biological Chemistry*. https://doi.org/10.1515/hsz-2019-0189.
Chen, L., König, B., & Stauber, T. (2020). LRRC8 channel activation and reduction in cytosolic chloride concentration during early differentiation of C2C12 myoblasts. *Biochemical and Biophysical Research Communications*, *532*, 482–488. https://doi.org/10.1016/j.bbrc.2020.08.080.
Chen, X., Wang, L., Cao, L., Li, T., Li, Z., Sun, Y., et al. (2021). Regulation of anion channel LRRC8 volume-regulated anion channels in transport of 2'3'-cyclic GMP-AMP and cisplatin under steady state and inflammation. *Journal of Immunology*, *206*(9), 2061–2074. https://doi.org/10.4049/jimmunol.2000989.
Christensen, O., & Hoffmann, E. K. (1992). Cell swelling activates K^+ and Cl^- channels as well as nonselective, stretch-activated cation channels in Ehrlich ascites tumor cells. *The Journal of Membrane Biology*, *129*(1), 13–36.
Crowe, W. E., Altamirano, J., Huerto, L., & Alvarez-Leefmans, F. J. (1995). Volume changes in single N1E-115 neuroblastoma cells measured with a fluorescent probe. *Neuroscience*, *69*(1), 283–296. https://doi.org/10.1016/0306-4522(95)00219-9.
Decher, N., Lang, H. J., Nilius, B., Bruggemann, A., Busch, A. E., & Steinmeyer, K. (2001). DCPIB is a novel selective blocker of $I_{Cl,swell}$ and prevents swelling-induced shortening of guinea-pig atrial action potential duration. *British Journal of Pharmacology*, *134*(7), 1467–1479. https://doi.org/10.1038/sj.bjp.0704413.
Deneka, D., Sawicka, M., Lam, A. K. M., Paulino, C., & Dutzler, R. (2018). Structure of a volume-regulated anion channel of the LRRC8 family. *Nature*, *558*(7709), 254–259. https://doi.org/10.1038/s41586-018-0134-y.
Deng, W., Mahajan, R., Baumgarten, C. M., & Logothetis, D. E. (2016). The $I_{Cl,swell}$ inhibitor DCPIB blocks Kir channels that possess weak affinity for PIP2. *Pflügers Archiv*, *468*(5), 817–824. https://doi.org/10.1007/s00424-016-1794-9.
Doroshenko, P. (1991). Second messengers mediating activation of chloride current by intracellular GTP gamma S in bovine chromaffin cells. *The Journal of Physiology*, *436*, 725–738.

Doroshenko, P., & Neher, E. (1992). Volume-sensitive chloride conductance in bovine chromaffin cell membrane. *The Journal of Physiology, 449*, 197–218.

Droogmans, G., Maertens, C., Prenen, J., & Nilius, B. (1999). Sulphonic acid derivatives as probes of pore properties of volume-regulated anion channels in endothelial cells. *British Journal of Pharmacology, 128*(1), 35–40. https://doi.org/10.1038/sj.bjp.0702770.

Eggermont, J., Trouet, D., Carton, I., & Nilius, B. (2001). Cellular function and control of volume-regulated anion channels. *Cell Biochemistry and Biophysics, 35*(3), 263–274. https://doi.org/10.1385/CBB:35:3:263.

Ehling, P., Meuth, P., Eichinger, P., Herrmann, A. M., Bittner, S., Pawlowski, M., et al. (2016). Human T cells in silico: Modelling their electrophysiological behaviour in health and disease. *Journal of Theoretical Biology, 404*, 236–250. https://doi.org/10.1016/j.jtbi.2016.06.001.

Estevez, A. Y., Bond, T., & Strange, K. (2001). Regulation of $I_{Cl,swell}$ in neuroblastoma cells by G protein signaling pathways. *American Journal of Physiology. Cell Physiology, 281*(1), C89–C98. https://doi.org/10.1152/ajpcell.2001.281.1.C89.

Falke, L., & Misler, S. (1989). Activity of ion channels during volume regulation by clonal N1E115 neuroblastoma cells. *Proceedings of the National Academy of Sciences of the United States of America, 86*(10), 3919–3923.

Feustel, P. J., Jin, Y., & Kimelberg, H. K. (2004). Volume-regulated anion channels are the predominant contributors to release of excitatory amino acids in the ischemic cortical penumbra. *Stroke, 35*(5), 1164–1168. https://doi.org/10.1161/01.STR.0000124127.57946.a1.

Figueroa, E. E., & Denton, J. S. (2021). Zinc pyrithione activates the volume-regulated anion channel through an antioxidant-sensitive mechanism. *American Journal of Physiology-Cell Physiology, 320*(6), C1088–C1098. https://doi.org/10.1152/ajpcell.00070.2021.

Figueroa, E. E., Kramer, M., Strange, K., & Denton, J. S. (2019). CysLT1 receptor antagonists pranlukast and zafirlukast inhibit LRRC8-mediated volume regulated anion channels independently of the receptor. *American Journal of Physiology-Cell Physiology, 317*(4), C857–C866. https://doi.org/10.1152/ajpcell.00281.2019.

Formaggio, F., Saracino, E., Mola, M. G., Rao, S. B., Amiry-Moghaddam, M., Muccini, M., et al. (2019). LRRC8A is essential for swelling-activated chloride current and for regulatory volume decrease in astrocytes. *The FASEB Journal, 33*(1), 101–113. https://doi.org/10.1096/fj.201701397RR.

Friard, J., Corinus, A., Cougnon, M., Tauc, M., Pisani, D. F., Duranton, C., et al. (2019). LRRC8/VRAC channels exhibit a noncanonical permeability to glutathione, which modulates epithelial-to-mesenchymal transition (EMT). *Cell Death & Disease, 10*(12), 925. https://doi.org/10.1038/s41419-019-2167-z.

Friard, J., Laurain, A., Rubera, I., & Duranton, C. (2021). LRRC8/VRAC channels and the Redox balance: A complex relationship. *Cellular Physiology and Biochemistry, 55*(S1), 106–118. https://doi.org/10.33594/000000341.

Friard, J., Tauc, M., Cougnon, M., Compan, V., Duranton, C., & Rubera, I. (2017). Comparative effects of chloride channel inhibitors on LRRC8/VRAC-mediated chloride conductance. *Frontiers in Pharmacology, 8*, 328. https://doi.org/10.3389/fphar.2017.00328.

Fujii, T., Takahashi, Y., Takeshima, H., Saitoh, C., Shimizu, T., Takeguchi, N., et al. (2015). Inhibition of gastric H^+,K^+-ATPase by 4-(2-butyl-6,7-dichloro-2-cyclopentylindan-1-on-5-yl)oxybutyric acid (DCPIB), an inhibitor of volume-regulated anion channel. *European Journal of Pharmacology, 765*, 34–41. https://doi.org/10.1016/j.ejphar.2015.08.011.

Funakoshi, K., Suzuki, H., & Takeuchi, S. (2006). Lipid bilayer formation by contacting monolayers in a microfluidic device for membrane protein analysis. *Analytical Chemistry, 78*(24), 8169–8174. https://doi.org/10.1021/ac0613479.

Furuya, K., Hirata, H., Kobayashi, T., & Sokabe, M. (2021). Sphingosine-1-phosphate induces ATP release via volume-regulated anion channels in breast cell lines. *Life, 11*(8). https://doi.org/10.3390/life11080851.

Gaitán-Peñas, H., Gradogna, A., Laparra-Cuervo, L., Solsona, C., Fernández-Dueñas, V., Barrallo-Gimeno, A., et al. (2016). Investigation of LRRC8-mediated volume-regulated anion currents in xenopus oocytes. *Biophysical Journal, 111*(7), 1429–1443. https://doi.org/10.1016/j.bpj.2016.08.030.

Gaitán-Peñas, H., Pusch, M., & Estévez, R. (2018). Expression of LRRC8/VRAC currents in xenopus oocytes: Advantages and caveats. *International Journal of Molecular Sciences, 19*(3). https://doi.org/10.3390/ijms19030719.

Galietta, L. J., Haggie, P. M., & Verkman, A. S. (2001). Green fluorescent protein-based halide indicators with improved chloride and iodide affinities. *FEBS Letters, 499*(3), 220–224.

Ghosh, A., Khandelwal, N., Kumar, A., & Bera, A. K. (2017). Leucine-rich repeat-containing 8B protein is associated with the endoplasmic reticulum Ca^{2+} leak in HEK293 cells. *Journal of Cell Science, 130*(22), 3818–3828. https://doi.org/10.1242/jcs.203646.

Gosling, M., Smith, J. W., & Poyner, D. R. (1995). Characterization of a volume-sensitive chloride current in rat osteoblast-like (ROS 17/2.8) cells. *The Journal of Physiology, 485*(Pt 3), 671–682.

Gradogna, A., Gaitán-Peñas, H., Boccaccio, A., Estévez, R., & Pusch, M. (2017). Cisplatin activates volume sensitive LRRC8 channel mediated currents in Xenopus oocytes. *Channels (Austin, Tex.), 11*(3), 254–260. https://doi.org/10.1080/19336950.2017.1284717.

Gradogna, A., Gavazzo, P., Boccaccio, A., & Pusch, M. (2017). Subunit-dependent oxidative stress sensitivity of LRRC8 volume-regulated anion channels. *The Journal of Physiology, 595*(21), 6719–6733. https://doi.org/10.1113/JP274795.

Green, J. P., Swanton, T., Morris, L. V., El-Sharkawy, L. Y., Cook, J., Yu, S., et al. (2020). LRRC8A is essential for hypotonicity-, but not for DAMP-induced NLRP3 inflammasome activation. *eLife, 9*. https://doi.org/10.7554/eLife.59704.

Grinstein, S., Clarke, C. A., Dupre, A., & Rothstein, A. (1982). Volume-induced increase of anion permeability in human lymphocytes. *The Journal of General Physiology, 80*(6), 801–823.

Grinstein, S., Clarke, C. A., & Rothstein, A. (1982). Increased anion permeability during volume regulation in human lymphocytes. *Philosophical Transactions of the Royal Society of London. Series B, Biological Sciences, 299*(1097), 509–518.

Han, Y. E., Kwon, J., Won, J., An, H., Jang, M. W., Woo, J., et al. (2019). Tweety-homolog (Ttyh) family encodes the pore-forming subunits of the swelling-dependent volume-regulated anion channel ($VRAC_{swell}$) in the brain. *Experimental Neurobiology, 28*(2), 183–215. https://doi.org/10.5607/en.2019.28.2.183.

Hazama, A., & Okada, Y. (1988). Ca^{2+} sensitivity of volume-regulatory K^+ and Cl^- channels in cultured human epithelial cells. *The Journal of Physiology, 402*, 687–702.

Heginbotham, L., LeMasurier, M., Kolmakova-Partensky, L., & Miller, C. (1999). Single streptomyces lividans K^+ channels: Functional asymmetries and sidedness of proton activation. *The Journal of General Physiology, 114*(4), 551–560. https://doi.org/10.1085/jgp.114.4.551.

Helix, N., Strobaek, D., Dahl, B. H., & Christophersen, P. (2003). Inhibition of the endogenous volume-regulated anion channel (VRAC) in HEK293 cells by acidic di-aryl-ureas. *The Journal of Membrane Biology, 196*(2), 83–94.

Hernández-Carballo, C. Y., De Santiago-Castillo, J. A., Rosales-Saavedra, T., Pérez-Cornejo, P., & Arreola, J. (2010). Control of volume-sensitive chloride channel inactivation by the coupled action of intracellular chloride and extracellular protons. *Pflügers Archiv, 460*(3), 633–644. https://doi.org/10.1007/s00424-010-0842-0.

Hille, B. (2001). *Ion channels of excitable membranes* (3rd ed.). Sunderland: Sinauer.
Hisadome, K., Koyama, T., Kimura, C., Droogmans, G., Ito, Y., & Oike, M. (2002). Volume-regulated anion channels serve as an auto/paracrine nucleotide release pathway in aortic endothelial cells. *The Journal of General Physiology, 119*(6), 511–520.
Hoffmann, E. K. (1978). *Regulation of cell volume by selective changes in the leak permeabilities of Ehrlich ascites tumor cells* (pp. 397–417). Alfred Benzon Symp, XI.
Hoffmann, E. K., Lambert, I. H., & Pedersen, S. F. (2009). Physiology of cell volume regulation in vertebrates. *Physiological Reviews, 89*(1), 193–277.
Hoffmann, E. K., Schettino, T., & Marshall, W. S. (2007). The role of volume-sensitive ion transport systems in regulation of epithelial transport. *Comparative Biochemistry and Physiology. Part A, Molecular & Integrative Physiology, 148*(1), 29–43. https://doi.org/10.1016/j.cbpa.2006.11.023.
Hoffmann, E. K., Simonsen, L. O., & Lambert, I. H. (1984). Volume-induced increase of K^+ and Cl^- permeabilities in Ehrlich ascites tumor cells. Role of internal Ca^{2+}. *The Journal of Membrane Biology, 78*, 211–222.
Hoffmann, E. K., Simonsen, L. O., & Sjoholm, C. (1979). Membrane potential, chloride exchange, and chloride conductance in Ehrlich mouse ascites tumour cells. *The Journal of Physiology, 296*, 61–84.
Holden, M. A., Needham, D., & Bayley, H. (2007). Functional bionetworks from nanoliter water droplets. *Journal of the American Chemical Society, 129*(27), 8650–8655. https://doi.org/10.1021/ja072292a.
Holmgren, M., Liu, Y., Xu, Y., & Yellen, G. (1996). On the use of thiol-modifying agents to determine channel topology. *Neuropharmacology, 35*(7), 797–804.
Hudson, R. L., & Schultz, S. G. (1988). Sodium-coupled glycine uptake by Ehrlich ascites tumor cells results in an increase in cell volume and plasma membrane channel activities. *Proceedings of the National Academy of Sciences of the United States of America, 85*(1), 279–283. https://doi.org/10.1073/pnas.85.1.279.
Hyzinski-García, M. C., Rudkouskaya, A., & Mongin, A. A. (2014). LRRC8A protein is indispensable for swelling-activated and ATP-induced release of excitatory amino acids in rat astrocytes. *The Journal of Physiology, 592*(Pt 22), 4855–4862. https://doi.org/10.1113/jphysiol.2014.278887.
Ikenouchi, J., & Aoki, K. (2017). Membrane bleb: A seesaw game of two small GTPases. *Small GTPases, 8*(2), 85–89. https://doi.org/10.1080/21541248.2016.1199266.
Jackson, P. S., Morrison, R., & Strange, K. (1994). The volume-sensitive organic osmolyte-anion channel VSOAC is regulated by nonhydrolytic ATP binding. *The American Journal of Physiology, 267*(5 Pt 1), C1203–C1209. https://doi.org/10.1152/ajpcell.1994.267.5.C1203.
Jackson, P. S., & Strange, K. (1993). Volume-sensitive anion channels mediate swelling-activated inositol and taurine efflux. *The American Journal of Physiology, 265*(6 Pt 1), C1489–C1500.
Jackson, P. S., & Strange, K. (1995). Characterization of the voltage-dependent properties of a volume-sensitive anion conductance. *The Journal of General Physiology, 105*(5), 661–676.
Jackson, P. S., & Strange, K. (1996). Single channel properties of a volume sensitive anion channel: lessons from noise analysis. *Kidney International, 49*(6), 1695–1699. https://doi.org/10.1038/ki.1996.250.
Jentsch, T. J. (2016). VRACs and other ion channels and transporters in the regulation of cell volume and beyond. *Nature Reviews. Molecular Cell Biology, 17*(5), 293–307. https://doi.org/10.1038/nrm.2016.29.
Jentsch, T. J., Lutter, D., Planells-Cases, R., Ullrich, F., & Voss, F. K. (2016). VRAC: molecular identification as LRRC8 heteromers with differential functions. *Pflügers Archiv, 468*(3), 385–393. https://doi.org/10.1007/s00424-015-1766-5.

Jentsch, T. J., Stein, V., Weinreich, F., & Zdebik, A. A. (2002). Molecular structure and physiological function of chloride channels. *Physiological Reviews*, *82*(2), 503–568.

Kang, C., Xie, L., Gunasekar, S. K., Mishra, A., Zhang, Y., Pai, S., et al. (2018). SWELL1 is a glucose sensor regulating beta-cell excitability and systemic glycaemia. *Nature Communications*, *9*(1), 367. https://doi.org/10.1038/s41467-017-02664-0.

Kasuya, G., Nakane, T., Yokoyama, T., Jia, Y., Inoue, M., Watanabe, K., et al. (2018). Cryo-EM structures of the human volume-regulated anion channel LRRC8. *Nature Structural & Molecular Biology*, *25*(9), 797–804. https://doi.org/10.1038/s41594-018-0109-6.

Kefauver, J. M., Saotome, K., Dubin, A. E., Pallesen, J., Cottrell, C. A., Cahalan, S. M., et al. (2018). Structure of the human volume regulated anion channel. *eLife*, 7, e38461. https://doi.org/10.7554/eLife.38461.

Kern, D. M., Oh, S., Hite, R. K., & Brohawn, S. G. (2019). Cryo-EM structures of the DCPIB-inhibited volume-regulated anion channel LRRC8A in lipid nanodiscs. *eLife*, *8*, e42636. https://doi.org/10.7554/eLife.42636.

Kimelberg, H. K., Goderie, S. K., Higman, S., Pang, S., & Waniewski, R. A. (1990). Swelling-induced release of glutamate, aspartate, and taurine from astrocyte cultures. *The Journal of Neuroscience*, *10*(5), 1583–1591.

Kirk, K., Ellory, J. C., & Young, J. D. (1992). Transport of organic substrates via a volume-activated channel. *The Journal of Biological Chemistry*, *267*(33), 23475–23478.

Kittl, M., Winklmayr, M., Helm, K., Lettner, J., Gaisberger, M., Ritter, M., et al. (2020). Acid- and volume-sensitive chloride currents in human chondrocytes. *Frontiers in Cell and Development Biology*, *8*, 583131. https://doi.org/10.3389/fcell.2020.583131.

Klausen, T. K., Hougaard, C., Hoffmann, E. K., & Pedersen, S. F. (2006). Cholesterol modulates the volume-regulated anion current in Ehrlich-Lettre ascites cells via effects on Rho and F-actin. *American Journal of Physiology. Cell Physiology*, *291*(4), C757–C771. https://doi.org/10.1152/ajpcell.00029.2006.

König, B., Hao, Y., Schwartz, S., Plested, A. J., & Stauber, T. (2019). A FRET sensor of C-terminal movement reveals VRAC activation by plasma membrane DAG signaling rather than ionic strength. *eLife*, *8*, e45421. https://doi.org/10.7554/eLife.45421.

König, B., & Stauber, T. (2019). Biophysics and structure-function relationships of LRRC8-formed volume-regulated anion channels. *Biophysical Journal*, *116*(7), 1185–1193. https://doi.org/10.1016/j.bpj.2019.02.014.

Kubo, M., & Okada, Y. (1992). Volume-regulatory Cl- channel currents in cultured human epithelial cells. *The Journal of Physiology*, *456*, 351–371.

Kumar, A., Xie, L., Ta, C. M., Hinton, A. O., Gunasekar, S. K., Minerath, R. A., et al. (2020). SWELL1 regulates skeletal muscle cell size, intracellular signaling, adiposity and glucose metabolism. *eLife*, *9*. https://doi.org/10.7554/eLife.58941.

Kumar, L., Chou, J., Yee, C. S., Borzutzky, A., Vollmann, E. H., von Andrian, U. H., et al. (2014). Leucine-rich repeat containing 8A (LRRC8A) is essential for T lymphocyte development and function. *The Journal of Experimental Medicine*, *211*(5), 929–942. https://doi.org/10.1084/jem.20131379.

Kyrozis, A., & Reichling, D. B. (1995). Perforated-patch recording with gramicidin avoids artifactual changes in intracellular chloride concentration. *Journal of Neuroscience Methods*, *57*(1), 27–35.

Lahey, L. J., Mardjuki, R. E., Wen, X., Hess, G. T., Ritchie, C., Carozza, J. A., et al. (2020). LRRC8A:C/E Heteromeric Channels Are Ubiquitous Transporters of cGAMP. *Molecular Cell*, *80*(4), 578–591. e575. https://doi.org/10.1016/j.molcel.2020.10.021.

Lang, F., & Hoffmann, E. K. (2012). Role of ion transport in control of apoptotic cell death. *Comprehensive Physiology*, *2*(3), 2037–2061. https://doi.org/10.1002/cphy.c110046.

Leaney, J. L., Marsh, S. J., & Brown, D. A. (1997). A swelling-activated chloride current in rat sympathetic neurones. *The Journal of Physiology*, *501*, 555–564.

Lee, C. C., Freinkman, E., Sabatini, D. M., & Ploegh, H. L. (2014). The protein synthesis inhibitor blasticidin s enters mammalian cells via leucine-rich repeat-containing protein 8D. *The Journal of Biological Chemistry*, *289*(24), 17124–17131. https://doi.org/10.1074/jbc.M114.571257.

Lepple-Wienhues, A., Szabo, I., Laun, T., Kaba, N. K., Gulbins, E., & Lang, F. (1998). The tyrosine kinase p56lck mediates activation of swelling-induced chloride channels in lymphocytes. *The Journal of Cell Biology*, *141*(1), 281–286.

Levitan, I., Almonte, C., Mollard, P., & Garber, S. S. (1995). Modulation of a volume-regulated chloride current by F-actin. *The Journal of Membrane Biology*, *147*(3), 283–294.

Levitan, I., Christian, A. E., Tulenko, T. N., & Rothblat, G. H. (2000). Membrane cholesterol content modulates activation of volume-regulated anion current in bovine endothelial cells. *The Journal of General Physiology*, *115*(4), 405–416.

Li, P., Hu, M., Wang, C., Feng, X., Zhao, Z., Yang, Y., et al. (2020). LRRC8 family proteins within lysosomes regulate cellular osmoregulation and enhance cell survival to multiple physiological stresses. *Proceedings of the National Academy of Sciences of the United States of America*, *117*(46), 29155–29165. https://doi.org/10.1073/pnas.2016539117.

Liu, H. T., Akita, T., Shimizu, T., Sabirov, R. Z., & Okada, Y. (2009). Bradykinin-induced astrocyte-neuron signalling: glutamate release is mediated by ROS-activated volume-sensitive outwardly rectifying anion channels. *The Journal of Physiology*, *587*(Pt 10), 2197–2209. https://doi.org/10.1113/jphysiol.2008.165084.

Liu, H. T., Tashmukhamedov, B. A., Inoue, H., Okada, Y., & Sabirov, R. Z. (2006). Roles of two types of anion channels in glutamate release from mouse astrocytes under ischemic or osmotic stress. *Glia*, *54*(5), 343–357. https://doi.org/10.1002/glia.20400.

Liu, T., & Stauber, T. (2019). The volume-regulated anion channel LRRC8/VRAC is dispensable for cell proliferation and migration. *International Journal of Molecular Sciences*, *20*(11), E2663. https://doi.org/10.3390/ijms20112663.

Lu, P., Ding, Q., Li, X., Ji, X., Li, L., Fan, Y., et al. (2019). SWELL1 promotes cell growth and metastasis of hepatocellular carcinoma in vitro and in vivo. *eBioMedicine*, *48*, 100–116. https://doi.org/10.1016/j.ebiom.2019.09.007.

Lück, J. C., Puchkov, D., Ullrich, F., & Jentsch, T. J. (2018). LRRC8/VRAC anion channels are required for late stages of spermatid development in mice. *The Journal of Biological Chemistry*, *293*(30), 11796–11808. https://doi.org/10.1074/jbc.RA118.003853.

Lutter, D., Ullrich, F., Lueck, J. C., Kempa, S., & Jentsch, T. J. (2017). Selective transport of neurotransmitters and modulators by distinct volume-regulated LRRC8 anion channels. *Journal of Cell Science*, *130*(6), 1122–1133. https://doi.org/10.1242/jcs.196253.

Lv, J., Liang, Y., Zhang, S., Lan, Q., Xu, Z., Wu, X., et al. (2019). DCPIB, an Inhibitor of Volume-Regulated Anion Channels, Distinctly Modulates K2P Channels. *ACS Chemical Neuroscience*, *10*(6), 2786–2793. https://doi.org/10.1021/acschemneuro.9b00010.

Maeno, E., Ishizaki, Y., Kanaseki, T., Hazama, A., & Okada, Y. (2000). Normotonic cell shrinkage because of disordered volume regulation is an early prerequisite to apoptosis. *Proceedings of the National Academy of Sciences of the United States of America*, *97*(17), 9487–9492. https://doi.org/10.1073/pnas.140216197.

Maertens, C., Droogmans, G., Chakraborty, P., & Nilius, B. (2001). Inhibition of volume-regulated anion channels in cultured endothelial cells by the anti-oestrogens clomiphene and nafoxidine. *British Journal of Pharmacology*, *132*(1), 135–142. https://doi.org/10.1038/sj.bjp.0703786.

Marmont, G. (1949). Studies on the axon membrane; a new method. *Journal of Cellular and Comparative Physiology*, *34*(3), 351–382. https://doi.org/10.1002/jcp.1030340303.

Minieri, L., Pivonkova, H., Caprini, M., Harantova, L., Anderova, M., & Ferroni, S. (2013). The inhibitor of volume-regulated anion channels DCPIB activates TREK potassium channels in cultured astrocytes. *British Journal of Pharmacology, 168*(5), 1240–1254. https://doi.org/10.1111/bph.12011.

Miranda, P., Contreras, J. E., Plested, A. J., Sigworth, F. J., Holmgren, M., & Giraldez, T. (2013). State-dependent FRET reports calcium- and voltage-dependent gating-ring motions in BK channels. *Proceedings of the National Academy of Sciences of the United States of America, 110*(13), 5217–5222. https://doi.org/10.1073/pnas.1219611110.

Model, M. A. (2018). Methods for cell volume measurement. *Cytometry. Part A, 93*(3), 281–296. https://doi.org/10.1002/cyto.a.23152.

Mongin, A. A. (2016). Volume-regulated anion channel--a frenemy within the brain. *Pflügers Archiv, 468*(3), 421–441. https://doi.org/10.1007/s00424-015-1765-6.

Mongin, A. A., & Kimelberg, H. K. (2002). ATP potently modulates anion channel-mediated excitatory amino acid release from cultured astrocytes. *American Journal of Physiology. Cell Physiology, 283*(2), C569–C578. https://doi.org/10.1152/ajpcell.00438.2001.

Mongin, A. A., & Kimelberg, H. K. (2005). ATP regulates anion channel-mediated organic osmolyte release from cultured rat astrocytes via multiple Ca^{2+}-sensitive mechanisms. *American Journal of Physiology. Cell Physiology, 288*(1), C204–C213. https://doi.org/10.1152/ajpcell.00330.2004.

Morera, F. J., Vargas, G., González, C., Rosenmann, E., & Latorre, R. (2007). Ion-channel reconstitution. *Methods in Molecular Biology (Clifton, N.J.), 400*, 571–585. https://doi.org/10.1007/978-1-59745-519-0_38.

Mori, S., Morishima, S., Takasaki, M., & Okada, Y. (2002). Impaired activity of volume-sensitive anion channel during lactacidosis-induced swelling in neuronally differentiated NG108-15 cells. *Brain Research, 957*(1), 1–11.

Morishima, S., Shimizu, T., Kida, H., & Okada, Y. (2000). Volume expansion sensitivity of swelling-activated Cl^- channel in human epithelial cells. *The Japanese Journal of Physiology, 50*(2), 277–280.

Nabekura, T., Morishima, S., Cover, T. L., Mori, S.-I., Kannan, H., Komune, S., et al. (2003). Recovery from lactacidosis-induced glial cell swelling with the aid of exogenous anion channels. *Glia, 41*(3), 247–259. https://doi.org/10.1002/glia.10190.

Nakamura, R., Numata, T., Kasuya, G., Yokoyama, T., Nishizawa, T., Kusakizako, T., et al. (2020). Cryo-EM structure of the volume-regulated anion channel LRRC8D isoform identifies features important for substrate permeation. *Communications Biology, 3*(1), 240. https://doi.org/10.1038/s42003-020-0951-z.

Nilius, B., & Droogmans, G. (2001). Ion channels and their functional role in vascular endothelium. *Physiological Reviews, 81*(4), 1415–1459. https://doi.org/10.1152/physrev.2001.81.4.1415.

Nilius, B., & Droogmans, G. (2003). Amazing chloride channels: an overview. *Acta Physiologica Scandinavica, 177*(2), 119–147.

Nilius, B., Eggermont, J., Voets, T., Buyse, G., Manolopoulos, V., & Droogmans, G. (1997). Properties of volume-regulated anion channels in mammalian cells. *Progress in Biophysics and Molecular Biology, 68*(1), 69–119.

Nilius, B., Oike, M., Zahradnik, I., & Droogmans, G. (1994). Activation of a Cl^- current by hypotonic volume increase in human endothelial cells. *The Journal of General Physiology, 103*(5), 787–805.

Nilius, B., Prenen, J., Voets, T., Eggermont, J., & Droogmans, G. (1998). Activation of volume-regulated chloride currents by reduction of intracellular ionic strength in bovine endothelial cells. *The Journal of Physiology, 506*(Pt 2), 353–361.

Nilius, B., Sehrer, J., Viana, F., De Greef, C., Raeymaekers, L., Eggermont, J., et al. (1994). Volume-activated Cl^- currents in different mammalian non-excitable cell types. *Pflügers Archiv, 428*(3–4), 364–371.

Nilius, B., Voets, T., Prenen, J., Barth, H., Aktories, K., Kaibuchi, K., et al. (1999). Role of Rho and Rho kinase in the activation of volume-regulated anion channels in bovine endothelial cells. *The Journal of Physiology*, *516*(Pt 1), 67–74.

Numata, T., Sato-Numata, K., Hermosura, M. C., Mori, Y., & Okada, Y. (2021). TRPM7 is an essential regulator for volume-sensitive outwardly rectifying anion channel. *Communications Biology*, *4*(1), 599. https://doi.org/10.1038/s42003-021-02127-9.

Oiki, S., Kubo, M., & Okada, Y. (1994). Mg^{2+} and ATP-dependence of volume-sensitive Cl^- channels in human epithelial cells. *The Japanese Journal of Physiology*, *44*(Suppl 2), S77–S79.

Okada, T., Islam, M. R., Tsiferova, N. A., Okada, Y., & Sabirov, R. Z. (2017). Specific and essential but not sufficient roles of LRRC8A in the activity of volume-sensitive outwardly rectifying anion channel (VSOR). *Channels (Austin, Tex.)*, *11*(2), 109–120. https://doi.org/10.1080/19336950.2016.1247133.

Okada, Y. (1997). Volume expansion-sensing outward-rectifier Cl^- channel: fresh start to the molecular identity and volume sensor. *The American Journal of Physiology*, *273*(3 Pt 1), C755–C789.

Okada, Y. (2006). Cell volume-sensitive chloride channels: phenotypic properties and molecular identity. *Contributions to Nephrology*, *152*, 9–24.

Okada, Y., Kubo, M., Oiki, S., Petersen, C. C., Tominaga, M., Hazama, A., et al. (1994). Properties of volume-sensitive Cl- channels in a human epithelial cell line. *The Japanese Journal of Physiology*, *44*(Suppl 2), S31–S35.

Okada, Y., Okada, T., Sato-Numata, K., Islam, M. R., Ando-Akatsuka, Y., Numata, T., et al. (2019). Cell volume-activated and volume-correlated anion channels in mammalian cells: Their biophysical, molecular, and pharmacological properties. *Pharmacological Reviews*, *71*(1), 49–88. https://doi.org/10.1124/pr.118.015917.

Okada, Y., Sato, K., & Numata, T. (2009). Pathophysiology and puzzles of the volume-sensitive outwardly rectifying anion channel. *The Journal of Physiology*, *587*(10), 2141–2149.

Osei-Owusu, J., Yang, J., Vitery, M. D. C., & Qiu, Z. (2018). Molecular biology and physiology of Volume-Regulated Anion Channel (VRAC). *Current Topics in Membranes*, *81*, 177–203. https://doi.org/10.1016/bs.ctm.2018.07.005.

Pedersen, S. F., Hoffmann, E. K., & Novak, I. (2013). Cell volume regulation in epithelial physiology and cancer. *Frontiers in Physiology*, *4*, 233. https://doi.org/10.3389/fphys.2013.00233.

Pedersen, S. F., Klausen, T. K., & Nilius, B. (2015). The identification of a volume-regulated anion channel: an amazing Odyssey. *Acta Physiologica (Oxford, England)*, *213*(4), 868–881. https://doi.org/10.1111/apha.12450.

Pedersen, S. F., Okada, Y., & Nilius, B. (2016). Biophysics and physiology of the volume-regulated anion channel (VRAC)/volume-sensitive outwardly rectifying anion channel (VSOR). *Pflügers Archiv*, *468*(3), 371–383. https://doi.org/10.1007/s00424-015-1781-6.

Penuela, S., Gehi, R., & Laird, D. W. (2013). The biochemistry and function of pannexin channels. *Biochimica et Biophysica Acta*, *1828*(1), 15–22. https://doi.org/10.1016/j.bbamem.2012.01.017.

Pervaiz, S., Kopp, A., von Kleist, L., & Stauber, T. (2019). Absolute protein amounts and relative abundance of volume-regulated anion channel (VRAC) LRRC8 subunits in cells and tissues revealed by quantitative immunoblotting. *International Journal of Molecular Sciences*, *20*(23). https://doi.org/10.3390/ijms20235879.

Planells-Cases, R., Lutter, D., Guyader, C., Gerhards, N. M., Ullrich, F., Elger, D. A., et al. (2015). Subunit composition of VRAC channels determines substrate specificity and cellular resistance to Pt-based anti-cancer drugs. *The EMBO Journal*, *34*(24), 2993–3008. https://doi.org/10.15252/embj.201592409.

Platt, C. D., Chou, J., Houlihan, P., Badran, Y. R., Kumar, L., Bainter, W., et al. (2017). Leucine-rich repeat containing 8A (LRRC8A)-dependent volume-regulated anion channel activity is dispensable for T-cell development and function. *The Journal of Allergy and Clinical Immunology*, *140*(6), 1651–1659. e1651. https://doi.org/10.1016/j.jaci.2016.12.974.

Poletto Chaves, L. A., & Varanda, W. A. (2008). Volume-activated chloride channels in mice Leydig cells. *Pflügers Archiv*, *457*(2), 493–504. https://doi.org/10.1007/s00424-008-0525-2.

Qiu, Z., Dubin, A. E., Mathur, J., Tu, B., Reddy, K., Miraglia, L. J., et al. (2014). SWELL1, a plasma membrane protein, is an essential component of volume-regulated anion channel. *Cell*, *157*(2), 447–458. https://doi.org/10.1016/j.cell.2014.03.024.

Romanenko, V. G., Davies, P. F., & Levitan, I. (2002). Dual effect of fluid shear stress on volume-regulated anion current in bovine aortic endothelial cells. *American Journal of Physiology. Cell Physiology*, *282*(4), C708–C718.

Romanenko, V. G., Rothblat, G. H., & Levitan, I. (2004). Sensitivity of volume-regulated anion current to cholesterol structural analogues. *The Journal of General Physiology*, *123*(1), 77–87. https://doi.org/10.1085/jgp.200308882.

Ross, P. E., Garber, S. S., & Cahalan, M. D. (1994). Membrane chloride conductance and capacitance in Jurkat T lymphocytes during osmotic swelling. *Biophysical Journal*, *66*, 169–178.

Roy, G. (1995). Amino acid current through anion channels in cultured human glial cells. *The Journal of Membrane Biology*, *147*(1), 35–44.

Rubino, S., Bach, M. D., Schober, A. L., Lambert, I. H., & Mongin, A. A. (2018). Downregulation of leucine-rich repeat-containing 8A limits proliferation and increases sensitivity of glioblastoma to temozolomide and carmustine. *Frontiers in Oncology*, *8*, 142. https://doi.org/10.3389/fonc.2018.00142.

Sabirov, R. Z., Kurbannazarova, R. S., Melanova, N. R., & Okada, Y. (2013). Volume-sensitive anion channels mediate osmosensitive glutathione release from rat thymocytes. *PLoS One*, *8*(1), e55646. https://doi.org/10.1371/journal.pone.0055646.

Sabirov, R. Z., Prenen, J., Droogmans, G., & Nilius, B. (2000). Extra- and intracellular proton-binding sites of volume-regulated anion channels. *The Journal of Membrane Biology*, *177*(1), 13–22. https://doi.org/10.1007/s002320001090.

Sabirov, R. Z., Prenen, J., Tomita, T., Droogmans, G., & Nilius, B. (2000). Reduction of ionic strength activates single volume-regulated anion channels (VRAC) in endothelial cells. *Pflügers Archiv*, *439*(3), 315–320.

Sarkadi, B., Attisano, L., Grinstein, S., Buchwald, M., & Rothstein, A. (1984). Volume regulation of Chinese hamster ovary cells in anisoosmotic media. *Biochimica et Biophysica Acta*, *774*, 159–168.

Sato-Numata, K., Numata, T., Inoue, R., & Okada, Y. (2016). Distinct pharmacological and molecular properties of the acid-sensitive outwardly rectifying (ASOR) anion channel from those of the volume-sensitive outwardly rectifying (VSOR) anion channel. *Pflügers Archiv*, *468*(5), 795–803. https://doi.org/10.1007/s00424-015-1786-1.

Sato-Numata, K., Numata, T., Inoue, R., Sabirov, R. Z., & Okada, Y. (2017). Distinct contributions of LRRC8A and its paralogs to the VSOR anion channel from those of the ASOR anion channel. *Channels (Austin, Tex.)*, *11*(2), 167–172. https://doi.org/10.1080/19336950.2016.1230574.

Schindler, H., & Rosenbusch, J. P. (1978). Matrix protein from Escherichia coli outer membranes forms voltage-controlled channels in lipid bilayers. *Proceedings of the National Academy of Sciences*, *75*(8), 3751. https://doi.org/10.1073/pnas.75.8.3751.

Schober, A. L., Wilson, C. S., & Mongin, A. A. (2017). Molecular composition and heterogeneity of the LRRC8-containing swelling-activated osmolyte channels in primary rat astrocytes. *The Journal of Physiology*. https://doi.org/10.1113/JP275053.

Schwiebert, E. M., Mills, J. W., & Stanton, B. A. (1994). Actin-based cytoskeleton regulates a chloride channel and cell-volume in a renal cortical collecting duct cell-line. *The Journal of Biological Chemistry, 269*(10), 7081–7089.

Sekar, R. B., & Periasamy, A. (2003). Fluorescence resonance energy transfer (FRET) microscopy imaging of live cell protein localizations. *The Journal of Cell Biology, 160*(5), 629–633. https://doi.org/10.1083/jcb.200210140.

Serra, S. A., Stojakovic, P., Amat, R., Rubio-Moscardo, F., Latorre, P., Seisenbacher, G., et al. (2021). LRRC8A-containing chloride channel is crucial for cell volume recovery and survival under hypertonic conditions. *Proceedings of the National Academy of Sciences of the United States of America, 118*(23). https://doi.org/10.1073/pnas.2025013118.

Shimizu, T., Numata, T., & Okada, Y. (2004). A role of reactive oxygen species in apoptotic activation of volume-sensitive Cl^- channel. *Proceedings of the National Academy of Sciences of the United States of America, 101*(17), 6770–6773. https://doi.org/10.1073/pnas.0401604101.

Sirianant, L., Wanitchakool, P., Ousingsawat, J., Benedetto, R., Zormpa, A., Cabrita, I., et al. (2016). Non-essential contribution of LRRC8A to volume regulation. *Pflügers Archiv, 468*(5), 805–816. https://doi.org/10.1007/s00424-016-1789-6.

Slatin, S. L., Qiu, X.-Q., Jakes, K. S., & Finkelstein, A. (1994). Identification of a translocated protein segment in a voltage-dependent channel. *Nature, 371*(6493), 158–161. https://doi.org/10.1038/371158a0.

Sørensen, B. H., Nielsen, D., Thorsteinsdottir, U. A., Hoffmann, E. K., & Lambert, I. H. (2016). Downregulation of LRRC8A protects human ovarian and alveolar carcinoma cells against Cisplatin-induced expression of p53, MDM2, p21Waf1/Cip1, and Caspase-9/-3 activation. *American Journal of Physiology. Cell Physiology, 310*(11), C857–C873. https://doi.org/10.1152/ajpcell.00256.2015.

Stauber, T. (2015). The volume-regulated anion channel is formed by LRRC8 heteromers - molecular identification and roles in membrane transport and physiology. *Biological Chemistry, 396*(9–10), 975–990. https://doi.org/10.1515/hsz-2015-0127.

Stotz, S. C., & Clapham, D. E. (2012). Anion-sensitive fluorophore identifies the Drosophila swell-activated chloride channel in a genome-wide RNA interference screen. *PLoS One,* 7(10), e46865. https://doi.org/10.1371/journal.pone.0046865.

Strange, K., Emma, F., & Jackson, P. S. (1996). Cellular and molecular physiology of volume-sensitive anion channels. *The American Journal of Physiology, 270*(3 Pt 1), C711–C730.

Strange, K., Yamada, T., & Denton, J. S. (2019). A 30-year journey from volume-regulated anion currents to molecular structure of the LRRC8 channel. *The Journal of General Physiology, 151*(2), 100–117. https://doi.org/10.1085/jgp.201812138.

Stuhlmann, T., Planells-Cases, R., & Jentsch, T. J. (2018). LRRC8/VRAC anion channels enhance beta-cell glucose sensing and insulin secretion. *Nature Communications, 9*(1), 1974. https://doi.org/10.1038/s41467-018-04353-y.

Stühmer, W., Parekh, A. B., Sakmann, B., & Neher, E. (1995). Electrophysiological recordings from Xenopus oocytes. Single-channel recording. In *Single-Channel Recording* (pp. 341–356).

Stutzin, A., & Hoffmann, E. K. (2006). Swelling-activated ion channels: functional regulation in cell-swelling, proliferation and apoptosis. *Acta Physiologica (Oxford, England), 187*(1-2), 27–42.

Sukalskaia, A., Straub, M. S., Deneka, D., Sawicka, M., & Dutzler, R. (2021). Cryo-EM structures of the TTYH family reveal a novel architecture for lipid interactions. *Nature Communications, 12*(1), 4893. https://doi.org/10.1038/s41467-021-25106-4.

Syeda, R., Qiu, Z., Dubin, A. E., Murthy, S. E., Florendo, M. N., Mason, D. E., et al. (2016). LRRC8 proteins form volume-regulated anion channels that sense ionic strength. *Cell, 164*(3), 499–511. https://doi.org/10.1016/j.cell.2015.12.031.

Szücs, G., Heinke, S., Droogmans, G., & Nilius, B. (1996). Activation of the volume-sensitive chloride current in vascular endothelial cells requires a permissive intracellular Ca^{2+} concentration. *Pflügers Archiv*, *431*(3), 467–469.

Ternovsky, V. I., Okada, Y., & Sabirov, R. Z. (2004). Sizing the pore of the volume-sensitive anion channel by differential polymer partitioning. *FEBS Letters*, *576*(3), 433–436. https://doi.org/10.1016/j.febslet.2004.09.051.

Trothe, J., Ritzmann, D., Lang, V., Scholz, P., Pul, U., Kaufmann, R., et al. (2018). Hypotonic stress response of human keratinocytes involves LRRC8A as component of volume-regulated anion channels. *Experimental Dermatology*. https://doi.org/10.1111/exd.13789.

Tseng, Y. T., Ko, C. L., Chang, C. T., Lee, Y. H., Huang Fu, W. C., & Liu, I. H. (2020). Leucine-rich repeat containing 8A contributes to the expansion of brain ventricles in zebrafish embryos. *Biology Open*, *9*(1). https://doi.org/10.1242/bio.048264.

Tsumura, T., Oiki, S., Ueda, S., Okuma, M., & Okada, Y. (1996). Sensitivity of volume-sensitive Cl^- conductance in human epithelial cells to extracellular nucleotides. *The American Journal of Physiology*, *271*(6 Pt 1), C1872–C1878.

Ullrich, F., Reincke, S. M., Voss, F. K., Stauber, T., & Jentsch, T. J. (2016). Inactivation and anion selectivity of volume-regulated anion channels (VRACs) depend on C-terminal residues of the first extracellular loop. *The Journal of Biological Chemistry*, *291*(33), 17040–17048. https://doi.org/10.1074/jbc.M116.739342.

Varela, D., Simon, F., Riveros, A., Jorgensen, F., & Stutzin, A. (2004). NAD(P)H oxidase-derived H_2O_2 signals chloride channel activation in cell volume regulation and cell proliferation. *The Journal of Biological Chemistry*, *279*(14), 13301–13304.

Verdon, B., Winpenny, J. P., Whitfield, K. J., Argent, B. E., & Gray, M. A. (1995). Volume-activated chloride currents in pancreatic duct cells. *The Journal of Membrane Biology*, *147*(2), 173–183.

Viana, F., Van Acker, K., De Greef, C., Eggermont, J., Raeymaekers, L., Droogmans, G., et al. (1995). Drug-transport and volume-activated chloride channel functions in human erythroleukemia cells: relation to expression level of P-glycoprotein. *The Journal of Membrane Biology*, *145*(1), 87–98.

Voets, T., Buyse, G., Tytgat, J., Droogmans, G., Eggermont, J., & Nilius, B. (1996). The chloride current induced by expression of the protein pI_{Cln} in Xenopus oocytes differs from the endogenous volume-sensitive chloride current. *The Journal of Physiology*, *495*, 441–447.

Voets, T., Droogmans, G., Raskin, G., Eggermont, J., & Nilius, B. (1999). Reduced intracellular ionic strength as the initial trigger for activation of endothelial volume-regulated anion channels. *Proceedings of the National Academy of Sciences of the United States of America*, *96*(9), 5298–5303.

Voets, T., Wei, L., De Smet, P., Van Driessche, W., Eggermont, J., Droogmans, G., et al. (1997). Downregulation of volume-activated Cl^- currents during muscle differentiation. *The American Journal of Physiology*, *272*(2 Pt 1), C667–C674. https://doi.org/10.1152/ajpcell.1997.272.2.C667.

Voss, F. K., Ullrich, F., Münch, J., Lazarow, K., Lutter, D., Stauber, T., et al. (2014). Identification of LRRC8 heteromers as an essential component of the volume-regulated anion channel VRAC. *Science*, *344*(6184), 634–638. https://doi.org/10.1126/science.1252826.

Wehner, F. (2006). Cell volume-regulated cation channels. *Contributions to Nephrology*, *152*, 25–53. https://doi.org/10.1159/000096315.

Wehner, F., Olsen, H., Tinel, H., Kinne-Saffran, E., & Kinne, R. K. (2003). Cell volume regulation: osmolytes, osmolyte transport, and signal transduction. *Reviews of Physiology, Biochemistry and Pharmacology*, *148*, 1–80. https://doi.org/10.1007/s10254-003-0009-x.

Wilson, C. S., Dohare, P., Orbeta, S., Nalwalk, J. W., Huang, Y., Ferland, R. J., et al. (2021). Late adolescence mortality in mice with brain-specific deletion of the volume-regulated anion channel subunit LRRC8A. *The FASEB Journal*, *35*(10), e21869. https://doi.org/10.1096/fj.202002745R.

Xie, L., Zhang, Y., Gunasekar, S. K., Mishra, A., Cao, L., & Sah, R. (2017). Induction of adipose and hepatic SWELL1 expression is required for maintaining systemic insulin-sensitivity in obesity. *Channels (Austin, Tex.)*, 1–5. https://doi.org/10.1080/19336950.2017.1373225.

Yamada, T., Figueroa, E. E., Denton, J. S., & Strange, K. (2021). LRRC8A homohexameric channels poorly recapitulate VRAC regulation and pharmacology. *American Journal of Physiology. Cell Physiology*, *320*(3), C293–C303. https://doi.org/10.1152/ajpcell.00454.2020.

Yamada, T., & Strange, K. (2018). Intracellular and extracellular loops of LRRC8 are essential for volume-regulated anion channel function. *The Journal of General Physiology*, *150*(7), 1003–1015. https://doi.org/10.1085/jgp.201812016.

Yamada, T., Wondergem, R., Morrison, R., Yin, V. P., & Strange, K. (2016). Leucine-rich repeat containing protein LRRC8A is essential for swelling-activated Cl^- currents and embryonic development in zebrafish. *Physiological Reports*, *4*(19). https://doi.org/10.14814/phy2.12940.

Yang, J., Vitery, M. D. C., Chen, J., Osei-Owusu, J., Chu, J., & Qiu, Z. (2019). Glutamate-releasing SWELL1 channel in astrocytes modulates synaptic transmission and promotes brain damage in stroke. *Neuron*. https://doi.org/10.1016/j.neuron.2019.03.029.

Yurinskaya, V., Aksenov, N., Moshkov, A., Goryachaya, T., Shemery, A., & Vereninov, A. (2020). Flow fluorometry quantification of anion channel VRAC subunit LRRC8A at the membrane of living U937 cells. *Channels (Austin, Tex.)*, *14*(1), 45–52. https://doi.org/10.1080/19336950.2020.1730535.

Zachariassen, L. G., Katchan, L., Jensen, A. G., Pickering, D. S., Plested, A. J., & Kristensen, A. S. (2016). Structural rearrangement of the intracellular domains during AMPA receptor activation. *Proceedings of the National Academy of Sciences of the United States of America*, *113*(27), E3950–E3959. https://doi.org/10.1073/pnas.1601747113.

Zahiri, D., Burow, P., Grossmann, C., Müller, C. E., Klapperstück, M., & Markwardt, F. (2021). Sphingosine-1-phosphate induces migration of microglial cells via activation of volume-sensitive anion channels, ATP secretion and activation of purinergic receptors. *Biochimica et Biophysica Acta, Molecular Cell Research*, *1868*(2), 118915. https://doi.org/10.1016/j.bbamcr.2020.118915.

Zakharian, E. (2013). Recording of ion channel activity in planar lipid bilayer experiments. *Methods in Molecular Biology (Clifton, N.J.)*, *998*, 109–118. https://doi.org/10.1007/978-1-62703-351-0_8.

Zakharian, E. (2021). Chapter Eleven - Ion channel reconstitution in lipid bilayers. In D. L. Minor, & H. M. Colecraft (Eds.), *Vol. 652. Methods in Enzymology* (pp. 273–291). Academic Press.

Zhang, Y., Xie, L., Gunasekar, S. K., Tong, D., Mishra, A., Gibson, W. J., et al. (2017). SWELL1 is a regulator of adipocyte size, insulin signalling and glucose homeostasis. *Nature Cell Biology*, *19*(5), 504–517. https://doi.org/10.1038/ncb3514.

Zheng, J., & Zagotta, W. N. (2000). Gating rearrangements in cyclic nucleotide-gated channels revealed by patch-clamp fluorometry. *Neuron*, *28*(2), 369–374.

Zhou, C., Chen, X., Planells-Cases, R., Chu, J., Wang, L., Cao, L., et al. (2020). Transfer of cGAMP into Bystander Cells via LRRC8 Volume-Regulated Anion Channels Augments STING-Mediated Interferon Responses and Anti-viral Immunity. *Immunity*, *52*(5), 767–781. e766 https://doi.org/10.1016/j.immuni.2020.03.016.

Zhou, J. J., Luo, Y., Chen, S. R., Shao, J. Y., Sah, R., & Pan, H. L. (2020). LRRC8A-dependent volume-regulated anion channels contribute to ischemia-induced brain injury and glutamatergic input to hippocampal neurons. *Experimental Neurology*, *332*, 113391. https://doi.org/10.1016/j.expneurol.2020.113391.

Zhou, P., Polovitskaya, M. M., & Jentsch, T. J. (2018). LRRC8 N termini influence pore properties and gating of volume-regulated anion channels (VRACs). *The Journal of Biological Chemistry*, *293*(35), 13440–13451. https://doi.org/10.1074/jbc.RA118.002853.

Studying cell volume beyond cell volume

Michael A. Model*
Kent State University, Kent, Ohio, United States
*Corresponding author: e-mail address: mmodel@kent.edu

Contents

Abstract

The first part of the paper describes two simple microscopic techniques that we use in our laboratory. One measures cell volumes in adherent cultures and the other measures cell dry mass; both measurements are done on the same instrument (a standard bright-field transmission microscope with only one or two narrow-band color filters added) and on the same cells. The reason for combining cell volume with dry mass is that the ratio of the two—dry mass concentration (MC)—is an important and insufficiently utilized biological parameter. We then describe a few applications of MC. The available experimental data strongly suggest its critical role in biological processes, including cell volume regulation. For example, most eukaryotic cells have surprisingly similar values of MC. Moreover, MC (and not cell volume) is tightly controlled in growing cell cultures at highly variable external osmolarities. We review the results showing that elevation of MC is a direct cause of shrinkage-induced apoptosis. Also, by focusing on MC, one can study heterogenous processes, such as necrotic swelling, or discriminate between apoptotic dehydration and the loss of cell fragments.

Current Topics in Membranes, Volume 88
ISSN 1063-5823
https://doi.org/10.1016/bs.ctm.2021.08.001

Abbreviations

AB9 Acid Blue 9
AVD apoptotic volume decrease
CV cell volume
MC cell dry mass
QPI quantitative phase imaging
RVD regulatory volume decrease
RVI regulatory volume increase
TIE transport-of-intensity microscopy
TTD transmission-through-dye microscopy

1. Introduction

Cell volume regulation has become a classical subject covered in standard physiology textbooks. As a result of the work of many researchers (reviewed by Kay & Blaustein, 2019), we know that cell volume behavior can be reasonably well described by a model in which water fluxes are driven by osmolarity differences between the cell interior and exterior. In a stable state, extracellular osmolarity is equal to intracellular. Intracellular osmolytes are mostly represented by inorganic ions, such as K^+, Cl^-, and Na^+, and by small organic molecules, such as amino acids, sugars, taurine, and urea. When the external osmolarity exceeds the internal osmolarity, water comes out of the cell, causing shrinkage; when the external osmolarity falls below internal, water enters the cell, causing swelling. Because the extent of active water transport (Zeuthen, 2010) is uncertain, the cell must maintain its volume through the regulation of internal osmolytes.

The typical volume regulation experiment involves exposing the cells to an abrupt change in external osmolarity. As has been known since the 1960s, after the first inevitable volume change, cells often take steps to restore their initial volume. Regulatory volume decrease (RVD) and increase (RVI) are the processes of cell volume restoration that follow the initial swelling in a hypoosmotic solution or shrinkage in a hyperosmotic solution (Okada et al., 2001). The other phenomenon that has received much scrutiny is the apoptotic volume decrease (AVD)—a characteristic isosmotic shrinkage of cells that are induced to apoptosis by various factors (Bortner & Cidlowski, 2002; Model, 2014; Okada et al., 2001; Rana & Model, 2020). The opposite of the

AVD is the necrotic volume increase—a spontaneous swelling resulting from severe cell damage (Barros, Hermosilla, & Castro, 2001).

To study cell volume regulation, one needs, at the very least, to have a way to measure it. The diverse methods of cell volume measurements have been recently reviewed (Model, 2018). In our opinion, transmission-through dye (TTD) microscopy is technically the simplest of all microscopic methods and is also one of the most accurate. The first part of this paper describes its principle and some practical aspects.

Unlike cell volume, the regulation of cell dry mass concentration (MC) has been relatively neglected. Recently, the interest in MC has been stimulated by the development of the branch of microscopy known as quantitative phase imaging (QPI). QPI measures the phase delay difference $\Delta\phi$ accumulated upon passage of light through a sample. If the refractive indices of the sample and of the surrounding medium are n and n', respectively, and h is the sample thickness, the relationship between these quantities is

$$\Delta\phi = \frac{2\pi h}{\lambda_0}(n - n') \tag{1}$$

where λ_0 is the wavelength in the vacuum. This relationship can be extended to cell volume (V) by using the integral quantities over the cell area A:

$$A \cdot \Delta\bar{\phi} = \frac{2\pi V}{\lambda_0}(n - n') \tag{2}$$

where the bar over ϕ denotes the average; it is also assumed that the cell is uniform and can be characterized by a single n. The latter is clearly not quite correct (to be more precise, it is never correct except for erythrocytes), but the significance of spatial heterogeneity for the subject of this paper is not well understood at present. One exception is local water accumulation that will be described below.

The usefulness of QPI comes from the fact that n is proportional to MC. The main contributions to the cell dry mass come from macromolecules: proteins (60–75%) and, to a lesser degree, nucleic acids (5–10%) (Model & Petruccelli, 2018). (Because of the abundance of proteins and following the convention, we will be occasionally referring to macromolecules as "proteins"). Although the refractive increments (the dependence of n on the concentration) of different proteins and other macromolecules may differ slightly, a reasonable approximation can be made:

$$n - n' \approx 0.185(c - c') \tag{3}$$

(Theisen, 2000), where c and c′ stand for the intracellular and extracellular MC, respectively, and are expressed in g/mL. Thus, if V is known from independent measurements, MC can be found:

$$c = c' + 0.86\lambda_0 \frac{\Delta\overline{\phi}}{V} \tag{4}$$

If, as is usually the case, c′ is zero or negligible, then

$$c = 0.86\lambda_0 \frac{\Delta\overline{\phi}}{CV} \tag{5}$$

Several commercial QPI microscopes of varying cost and complexity are available, but the popular digital holographic instruments do not provide an easy way to compute the refractive index. However, one type of QPI, known as the transport-of-intensity-equation (TIE) imaging, does not require any dedicated or unusual equipment and, like TTD, can be realized on almost any standard optical microscope. Thus, despite its intimidating name, the TIE method is not only practical but compatible with volume measurements by TTD, as well as with fluorescence. The second part of the paper is devoted to TIE.

In the third part, we turn to the biological significance of MC. Despite the long history of successful research in cell volume regulation, it is not equally applicable to all situations. First, the concept of volume regulation best applies to short time periods or to terminally differentiated cells to avoid complications resulting from cell growth and division. Indeed, RVI and RVD are usually observed within 10–30 min of a step change in external osmolarity. In many cases, these rapid responses are absent or incomplete, and if a cell continues to grow, the meaning of volume regulation becomes uncertain. In this case, it would be natural to substitute cell volume with MC. Unlike cell volume, MC is an intrinsic property, which is independent, or almost independent, of the phase of the cell cycle. Thus, by focusing on MC instead, or in addition, to CV, one can extend the study of osmotic responses to much longer times.

The other reason to include MC in cell volume studies is that some processes affect CV and MC differently. For example, cell volume can increase either from the normal growth or from the accumulation of water.

Apoptotic shrinkage can be caused by cell fragmentation or by water loss. Only by complementing cell volume measurements with MC, these cases can be easily differentiated.

An additional practical benefit for microscopic observation is that MC values are much more consistent, and sometimes, conclusions can be made based on the analysis of only several cells, whereas hundreds of cells may be required to compare cell volumes between different populations.

Third, MC has a biological significance of its own. When proteins are present at concentrations as high as those observed in the cells—the situation known as macromolecular crowding—small changes in MC (which can also be understood as an acronym for macromolecular crowding) are expected to have disproportionately strong, nonlinear effects on biochemical reactions. The best-studied MC effects are of three types: (1) an increase in MC favors aggregation of proteins into complexes; (2) an increase in MC favors more compact protein conformations, and several fluorescent MC sensors utilize this property [e.g., Boersma, Zuhorn, & Poolman, 2015; Germond, Fujita, Ichimura, & Watanabe, 2016; Mouton, Veenhoff, & Boersma, 2020]; (3) reactions rates are also affected, but those are less predictable due to the opposing effects of increased proximity and increased viscosity (Minton, 2001). These effects do not depend, at least in theory, on the chemical nature of macromolecules but result from reduced free space. When studied *in vitro*, macromolecular crowding is often mimicked by synthetic polymers, such as polyethylene glycol or Ficoll.

While these consequences of macromolecular crowding are well understood theoretically and have been thoroughly studied *in vitro*, observations on living cells are very few. Regulation and the role of MC in living cells and organisms present a large and promising area for research.

2. CV measurement by dye exclusion

In TTD imaging, live cells (which can be attached to a coverglass) are kept in their normal medium that additionally contains a membrane-impermeant nontoxic dye. The sample is placed in a shallow space with depth h_0, which is typically 30–40 μm but can be made smaller or larger, depending on the cell size. The sample is illuminated in transmission at a wavelength corresponding to the peak of the dye absorbance. The dye attenuates the light passing through the sample according to the Beer-Lambert law:

$$I_0 = I_i e^{-\alpha h_0} \tag{6}$$

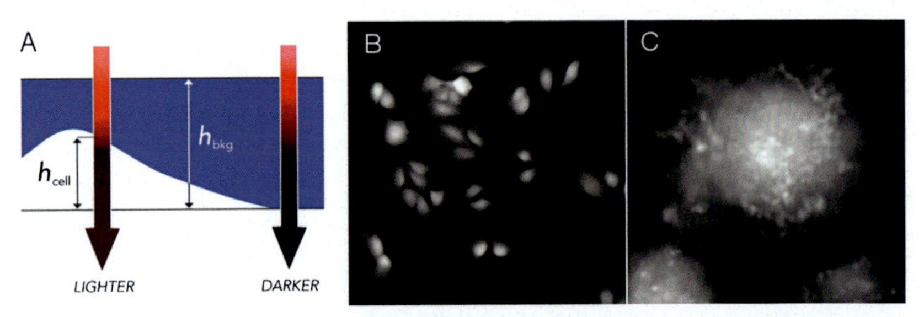

Fig. 1 (A) The principle of contrast generation in TTD. (B) A low-resolution (×10/0.4) image of live T24 cells. (C) A high-resolution (×60/1.2) image of the membrane of a formaldehyde fixed T24 cell. A high vertical resolution was achieved by increasing the dye concentration. *The pictures are taken panel (A) from Gregg, J. L., McGuire, K. M., Focht, D. C., and Model, M. A. (2010). Measurement of the thickness and volume of adherent cells using transmission-through-dye microscopy. Pflügers Archiv / European Journal of Physiology, 460, 1097–1104 and Panels (B and C) from Pelts, M., Pandya, S. M., Oh, C. J., & Model, M. A. (2011). Thickness profiling of formaldehyde-fixed cells by transmission-through-dye microscopy. BioTechniques, 50, 389–396, with permission.*

where I_i is the incident intensity, I_0 is the intensity upon exiting the sample, and α is the absorbance of the dye-containing solution. If an intact cell with thickness h_c is present, it excludes the dye and thereby reduces the attenuation path from h_0 to $h_0 - h_c$, resulting in the local intensity

$$I_c = I_i e^{-\alpha(h_0 - h_c)} \tag{7}$$

Thus, the cells appear brighter than the background by the factor

$$\frac{I_c}{I_0} = e^{\alpha h_c} \tag{8}$$

(Fig. 1A). This simple relationship allows a straightforward conversion of image contrast to cell thickness and volume. In practice, it is convenient to convert image intensity into logarithmic units normalized by α:

$$h_{ij} = \frac{\ln I_{ij}}{\alpha} \tag{9}$$

so that gray values at every pixel ij directly report cell height. Cell volume is found by summation of heights over the cell area A:

$$V = \sum_{A} \left(h_{ij} - h_{bkg}\right) = A\left(\overline{h} - h_{bkg}\right) \tag{10}$$

Note that the result does not depend on h_0 or on the absolute value of incident intensity: h_c and CV are found only from the relative contrast between the cell and background. The absorption coefficient α can be easily measured on an inverted microscope by imaging a half-ball lens immersed in a small droplet of the dye, which effectively creates a well-defined depth profile around the contact point (Model, Khitrin, & Blank, 2008).

Fig. 1B shows the typical appearance of cultured cells viewed in TTD at moderate magnification. Quite often, the local profile is slightly affected by the contrast of bright-field origin, especially around vesicles, nucleoli, or steep cell edges (the so-called Becke lines; Fig. 3A), but these intensity variations have minimal effect on cell volume, as they mostly cancel each other during summation (Eq. 10) and can be additionally corrected if desired (Model, 2012). The accuracy of TTD has been confirmed by comparison with confocal scanning, which produced less than a 1% discrepancy (Model, 2020), and by imaging a spherical bead in a refractive index-matched medium (Model, 2012). Only when the bright-field contrast is very strong (more than 20% of the contrast due to dye exclusion, which may happen when cells are small and dehydrated), the results become less reliable.

For the contrast dye, we use the food colorant Acid Blue 9 (AB9). It is inexpensive and nontoxic to cells at concentrations necessary to achieve high contrast (~7 mg/mL for eukaryotic cells, giving $\alpha \approx 0.155\ \mu m^{-1}$ and an osmolarity increase by 14 mM). Its peak absorbance lies in the red range, around 630 nm, and the dye is completely transparent below 500 nm. Most cell types can be kept in the AB9-containing medium for many hours without any noticeable effect on their viability, locomotion, or proliferation.

Several other features of TTD are worth noting:

- TTD can be realized on any widefield microscope (or even on a laser scanning confocal microscope using transmission detector). Red illumination can be achieved by placing a bandpass 630 nm filter on the condenser wheel or anywhere between the light source and detector. By switching the filter from red to blue (e.g., 485/10), a brightfield image unaffected by the dye can be obtained, which is very useful for combining TTD with TIE; see below). The presence of AB9 does not preclude fluorescence imaging using excitation between UV and green.
- As already stated, the result does not depend on the illumination intensity; therefore, fluctuations in the light source do not affect the results. Likewise, the depth of the chamber does not need to be tightly controlled. It only must be not too small to avoid squeezing the cells or

too large, so the sample is sufficiently transparent for red light. As a rule of thumb, the product of AB9 concentration in mg/mL and the chamber depth in μm should be within 300. The chamber can be prepared using commercial slides with 40 μm strips (Model, 2015) or just by placing a coverglass over small spots of silicone grease and gently pressing it down. With some practice, the depth can be adjusted by the eye, going by the sample color.

- The vertical resolution of TTD depends on α. By increasing the AB9 concentration 10-fold and reducing the chamber depth, the volumes of live bacteria can be accurately measured (Lababidi, Pelts, Moitra, Leff, & Model, 2011). For profiling of material surfaces, a depth resolution down to a nanometer can be achieved (Model et al., 2008). For live cells, the depth resolution is only limited by the amount of the dye that can be tolerated; also, higher resolution comes at the expense of the resolvable range of depths.
- Only live cells with intact membranes exclude the dye; therefore, TTD is well suited for detecting protozoa in contaminated environmental samples (Model & Davis, 2016) or parasites in feces [Submitted]. In such samples, dead matter and debris remain dark, but live organisms appear bright red.

Additional technical details of TTD imaging can be found in Model (2015).

The main inconvenience of TTD is the use of a strongly colored (though nontoxic and washable) stain and the need to keep cells in a shallow compartment. During long experiments, one must watch that the cells do not dry up. It would help to have a water-immersion objective with sufficiently short working distance, so that the observation could be carried out directly in a cell culture dish under an upright microscope; unfortunately, such objectives are not available at present.

A similar method based on fluorescence exclusion has been developed (Bottier et al., 2011). Its advantage is that the images are not affected by the Becke lines and that the contrast agent (e.g., fluorescently stained dextran) is much less concentrated than AB9. However, the calibration procedure for fluorescence measurements is more involved, and, for reasons not entirely clear, the fluorescence method failed to achieve comparable reproducibility and accuracy in our hands (Model, 2020). The other notable techniques for cell volume measurement are confocal scanning followed by summation of the areas of all cell slices, a combination of front- and side-viewing, some microfluidic-based methods and a qualitative method based on the quenching of intracellular trapped calcein (reviewed in (Model, 2018)).

3. MC measurement by TIE/TTD

Visibility of unstained cells under bright-field illumination depends on their refractive index and the plane of focus. According to A. K. Khitrin (https://physicstoday.scitation.org/do/10.1063/PT.6.4o.20180921a/full/), this fact can be understood from the following reasoning (Khitrin, Petruccelli, & Model, 2017).

A microscopic image replicates the distribution of intensity (real or virtual) at the front focal plane of the objective. In the absence of a refractive sample, the field is uniformly illuminated by the condenser, producing a featureless image. A refractive sample deflects the rays from a straight path, making them denser in some areas (resulting in an increased brightness, or bright Becke lines) or sparser, producing darker areas (Fig. 2). We have used this simple geometrical model to accurately predict the distribution of intensity in bright-field images of complex objects (Clements, Davidson, & Model, 2021).

A mathematical analysis of Khitrin's model leads to a differential equation equivalent to TIE (which has been initially derived from different considerations (Teague, 1983)). TIE imaging is based on two or more mutually defocused brightfield images (Fig. 3), which are then processed by a

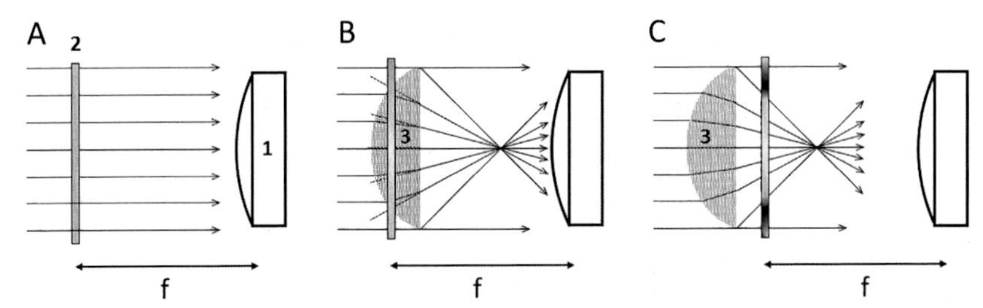

Fig. 2 In the absence of a specimen, a uniform illumination at the focal plane (2) of the objective (1) produces no contrast (A). A refractive specimen (3), which, for the sake of simplicity, is depicted here as a lens, alters the distribution of intensity at the focal plane of the objective; the latter may lie inside (B) or outside (C) the sample. The intensity pattern at the focal plane is replicated by the optical system and generates an image. In the case (B), one expects a brighter area in the central part, where the density of back-projected rays is higher. In the case (C), the area immediately outside the cone of light formed by the specimen is completely dark. *The figure is reproduced from Khitrin, A. K., Petruccelli, J. C., & Model, M. A. (2017). Bright-field microscopy of transparent objects: A ray tracing approach. Microscopy and Microanalysis, 23, 1116–1120, with permission.*

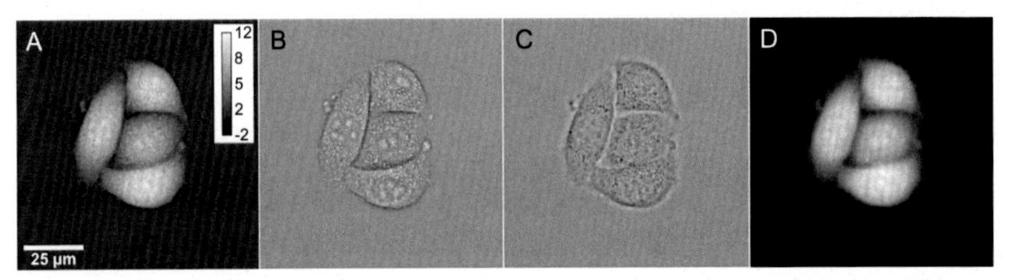

Fig. 3 Images of a group of HeLa cells. (A) TTD with a calibration bar showing cell height. The brighter nucleoli result from the Becke effect and represent an artifact. (B) In-focus (i.e., focused on the plane of the coverslip) bright-field image BF1. (C) A second bright-field image BF2, with the objective brought closer to the sample by 5 μm. (D) The result of TIE computation based on images B and C. The ratio of TIE/TTD integrated over the cell area is proportional to the average MC.

software, producing a map of the object's "optical thickness"—the product of physical thickness and relative refractive index (n_{cell}/n_{medium}).

TIE literature tends to be highly mathematical; but the method is simple in realization (besides being zero-cost) and, if used correctly, produces smooth and easy-to analyze images (e.g., Fig. 3D). In the Appendix, some of its practical aspects are discussed.

The other methods of MC measurements have been reviewed (Model & Petruccelli, 2018). One characteristic closely related to MC is the cell buoyant density. The relationship between the density and MC comes from the fact that proteins and especially nucleic acids have a higher density than water. According to various estimates, the average density of proteins is between 1.22 and 1.43 g/cm^3 (Fischer, Polikarpov, & Craievich, 2004), and the density of DNA is around 1.7–2 g/mL^3 (Neurohr & Amon, 2020; Panijpan, 1977). (An interesting discussion of the buoyancy of marine phytoplankton can be found in Boyd & Gradmann, 2002). MC also correlates with viscosity and diffusion rates of fluorescent probes (Luby-Phelps, 1999), but the latter strongly depends on the probe size. Of the QPI methods, TIE has the advantage that it is easily compatible with CV determination.

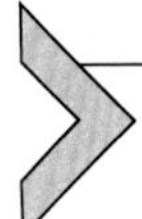

4. MC at work

4.1 The law of 0.2?

The available data on the typical MC values are limited, and a partial list is given in Table 1. Human red blood cells have a higher MC (over 0.3 g/mL), and the density of bacteria is variable. With these and perhaps some other

Table 1 Experimental values of MC.

Species	Cell type	MC (g/mL)	Source
Human	HeLa, adherent	0.20–0.21	Mudrak, Rana, and Model (2018) and Model, Mudrak, Rana, and Clements (2018)
	HeLa, suspension	0.24 0.25	Model et al. (2018) and Lue et al. (2006)
	Prostate cancer cell line	0.19	Mudrak et al. (2018)
	Lymphocytes	0.19	Zipursky, Bow, Seshadri, and Brown (1976) and Grover et al. (2011)
	Spleen cells	0.19	Bosslet, Ruffmann, Altevogt, and Schirrmacher (1981)
	Smooth muscle	0.12	Curl et al. (2005)
	Pancreatic tumor	0.22	Kemper et al. (2006)
Dog	Kidney cell line	0.19 0.19 0.26	Mudrak et al. (2018), Model et al. (2018), and Song et al. (2006)
Mouse	Fibroblasts	0.18	Model, Hollembeak, and Kurokawa (2021)
	Cortical neurons	0.18	Model et al. (2021)
	Neurons	0.20–0.21	Rappaz et al. (2005)
	Chondrocytes	0.07–0.18	Cooper et al. (2013)
	Leukemic cell line	0.28	Bryan et al. (2014)
	Various lines	0.19–0.24	Loken and Kubitschek (1984)
Rat	Kidney cell line	0.09	Choi et al. (2010)
	Heart muscle	0.19	Aliev et al. (2002)
	Cortical neurons	0.11–0.21	Maric, Maric, and Barker (1998)
Frog	Various	0.15–0.25	Ling (2004)
Yeast		0.20 0.40	White (1952) and Baldwin and Kubitschek (1984)

Some of the original numbers have been converted from refractive index according to Eq. 3, others from buoyant cell density ρ as $C = 2.5\ (\rho - 1)$ or from water mass fraction F as $C = (1 - F)/(1 + 0.4F)$ (Model & Petruccelli, 2018). Partly reused from Model M. A, Hollembeak J. E, & Kurokawa M. (2021). Macromolecular crowding: A hidden link between cell volume and everything else. Cellular Physiology and Biochemistry 55, 25–40.

exceptions (some of which may be due to technical issues), most healthy eukaryotic cells under stable conditions have a protein concentration close to 0.2 g/mL.

Is it a universal requirement that cells must have their cytoplasmic MC close to 0.2 g/mL? Some thoughts on this subject can be found in the literature (Guigas, Kalla, & Weiss, 2007; Neurohr & Amon, 2020), but more measurements on cells from diverse origins will help answer this question. For examples, some freshwater mollusks have an extremely low blood osmolarity, down to 30 mOsm/kg (Medeiros, Faria, & Souza, 2020), and it would be interesting to measure their MC.

4.2 Slow MC regulation

From our experiments on HeLa cells, MC appears to be extremely resilient to osmotic disturbances, whether hyperosmotic or hypoosmotic. Although HeLa lack RVI during the first hour (in agreement with other authors (Barros, 1999; Tivey, Simmons, & Aiton, 1985)), a more prolonged incubation results in a complete restoration of MC (Fig. 4). This observation underscores the difference between the rapid RVI and the slow MC regulation.

We have extended these observations to a much wider range of osmolarities. Hypoosmotic solutions were prepared by diluting cell growth medium with water, and hyperosmotic solutions—by adding sucrose. Except for the most extreme osmolarity of ¼ normal, MC was highly resistant to osmotic challenges (Fig. 5).

A similar stability was observed in response to antibiotic gramicidin, which forms channels for Na^+ and K^+ and thus alters ionic and water balance. Like hypotonicity, gramicidin causes rapid cell swelling and a corresponding decrease in MC within the first 0.5–1 h. It appears that at moderate gramicidin concentrations, MC recovers after 24 h, but at high concentrations, the initial swelling eventually gives way to shrinkage, much like with hypoosmolarity. These are all preliminary results, and we plan to publish a detailed study of how cells adapt to osmotic and ionic disturbances and what happens if they don't.

4.3 A closer look at the necrotic volume increase

The high concentration of membrane-impermeant negatively charged macromolecules in the cytosol requires an extra amount of counterions, with the result that true equilibrium between a cell and extracellular fluid cannot

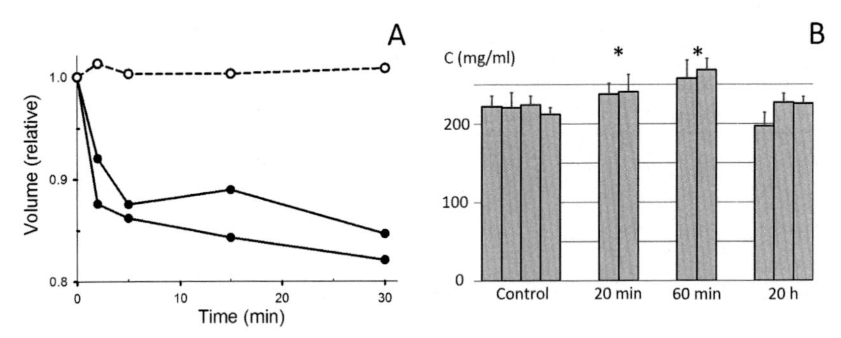

Fig. 4 (A) Volume changes in HeLa cells following replacement of a 360 mOsm medium (DMEM with AB9) with a 440 mOsm medium. Solid lines show the results of two separate experiments, and the broken line is the control. (B) Protein concentration following a transition to the 440 mOsm medium. *From Mudrak, N. J., Rana, P. S., & Model, M. A. (2018). Calibrated brightfield-based imaging for measuring intracellular protein concentration. Cytometry Part A, 93, 297–304, with permission.*

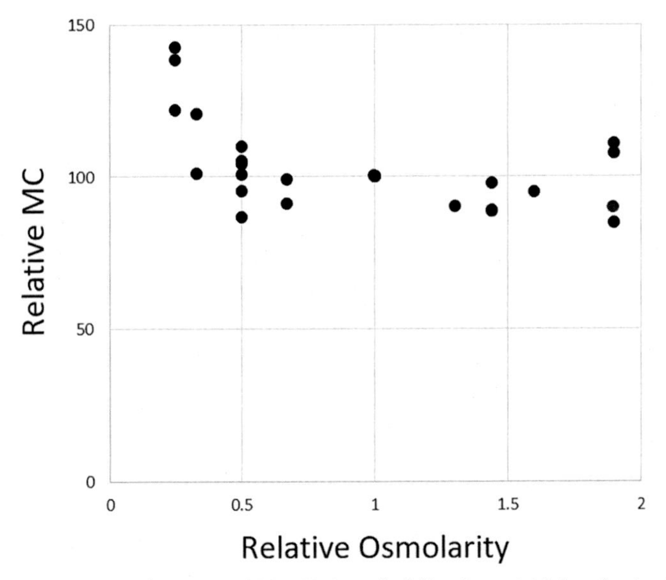

Fig. 5 Relative percent changes in MC in HeLa cells following a 24-h incubation in media (DMEM) of different osmolarities.

be achieved. Dynamic equilibrium can only be maintained due to energy-consuming active transport of ions. A damage to the plasma membrane or cessation of the pump activity are expected to cause water accumulation and cell swelling (Armstrong, 2003), called sometimes the necrotic volume increase (Barros et al., 2001).

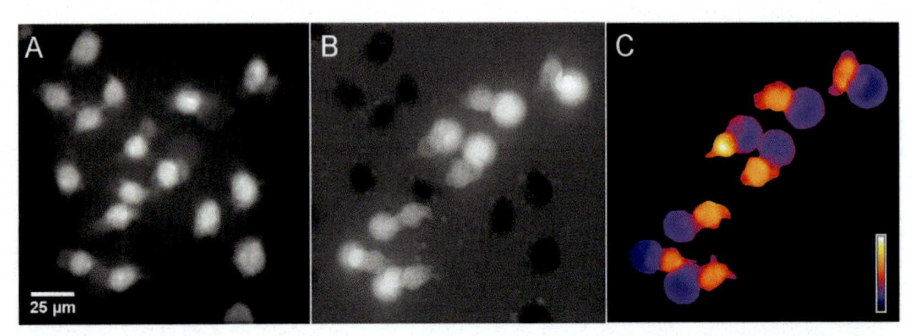

Fig. 6 Necrotic blebs in HeLa cells exposed to free radical producer menadione. (A) TIE image showing protein mass; (B) TTD image showing cell thickness (broken cells are dark); (C) A semiquantitative map of protein concentration obtained by dividing A by B. High concentrations are indicated with yellow-white and low concentrations with purple color. *Based on Rana, P. S., Mudrak, N. J., Lopez, R., Lynn, M., Kershner, L., & Model, M. A. (2017). Phase separation in necrotic cells. Biochemical and Biophysical Research Communications, 492, 300–303, with permission.*

However, a TTD/TIE analysis of HeLa cells exposed to free radicals reveals a more complex picture (Rana et al., 2017). Water accumulation does not occur uniformly throughout the cell but is limited to a bleb—an organelle-free protrusion under the membrane (Fig. 6); in the rest of the cytosol, MC remains at a normal level.

This observation leads to the next question: if the protein concentration within the bleb is less than in the cytosol and no membrane separates them (as we have shown by fluorescence photobleaching, diffusion in and out of the bleb is unimpeded), how can osmotic equilibrium exist between these two compartments? Ions and small molecules are expected to be distributed evenly between the bleb and the cytosol, and the excess of proteins in the cytosol should create a small but significant osmotic imbalance. We hypothesized that since large watery blebs develop only at advanced stages of necrosis, most cellular proteins by that time have lost their native conformation and aggregated, resulting in a decrease in their effective molar concentration. Indeed, a biochemical analysis showed extensive protein aggregation at this stage.

Interestingly, a similar looking water-filled vacuoles are observed in cells exposed to extremely low osmolarity (not shown) or even at a moderate 50% hypoosmolarity in the presence of cytoskeleton inhibitors. It has yet to be determined if these vacuoles are enclosed in a membrane or represent a phase separation.

4.4 A closer look at the apoptotic volume decrease

A smaller size is a characteristic feature of apoptotic cells; however, the "size" can be a vague quantity, as three very different processes contribute to apoptotic shrinkage. One is the loss of attachment to the substrate and the resulting cell rounding, making them appear smaller under the microscope. The other is the loss of water, and the term "apoptotic volume decrease," or AVD, often applies to this aspect of shrinkage. The third is detachment of cell fragments. Clearly, cell volume measurements cannot distinguish between the last two types of shrinkage.

The following example illustrates the problem (Model et al., 2018). Staurosporine is a protein kinase inhibitor and a classic apoptosis inducer. It causes characteristic morphological changes that look very similar in adherent HeLa and Madin-Darby Canine Kidney (MDCK) cells (Fig. 7). However, a TTD/TIE analysis showed a significant loss of water (a 50–70% increase in MC) only in MDCK cells, but not in adherent HeLa. (However, a similar in magnitude water loss was observed in suspended HeLa cells). This shows that equating AVD to water loss cannot be done based on morphology or volume measurements alone. We suspect that in some of the published reports of a 50–70% shrinkage, fragmentation was a significant but unnoticed component of the observed volume decrease.

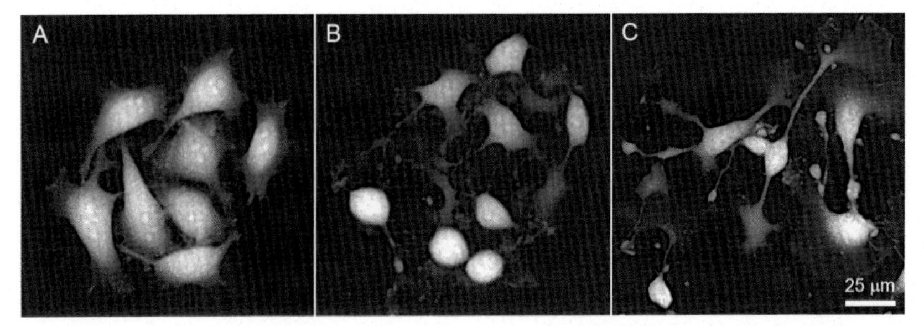

Fig. 7 The effect of staurosporine treatment on cell morphology. (A and B) TTD images of the same group of HeLa cells before and 1 h after addition of staurosporine. (C) MDCK cells treated with staurosporine. Despite the development of similar morphology characterized by flattened projections and bead-like structures, only MDCK cells have experienced water loss. *Reproduced from Model, M. A., Mudrak, N. J., Rana, P. S., & Clements, R. J. (2018). Staurosporine-induced apoptotic water loss is cell-and attachment-specific. Apoptosis, 23, 449–455, with permission.*

4.5 Death by MC

It is well known that not only is apoptosis accompanied by a water loss, but persistent osmotic compression by itself leads to apoptosis (Bortner & Cidlowski, 1996). However, the very initial trigger of the shrinkage-induced apoptosis has not been identified. Obviously, it must be some aspect of shrinkage, and there are four of them: (1) a decrease in the volume itself; (2) an increase in the concentrations of small molecules and ions (including their cumulative characteristic—the ionic strength); (3) an altered tension in the membrane and/or the cytoskeleton; (4) an increase in MC. All other shrinkage-related events (phosphorylation, activation of channels, generation of second messengers, etc.) can only be mediated by changes in mechanical forces or in concentrations. As there are no established mechanisms by which the volume can be sensed directly[a] (Leslie, 2011), we have focused on the mechanisms (2) and (3) as the alternatives to MC (mechanism 4) (Rana, Kurokawa, & Model, 2020).

Of thousands of molecular species whose concentrations change with volume, we chose Na^+ and K^+ that are traditionally associated with apoptosis (Bortner & Cidlowski, 2003; McCarthy & Cotter, 1997; Yu, 2003). Their concentrations were clamped by a combination of ionophores as previously described (Rana et al., 2019), and shrinkage was induced by membrane-impermeant sucrose. In this way, we have been able to control ion composition and MC independently. We found that when the cells were allowed to swell (in response to ionophores in the absence of sucrose), they maintained healthy morphology and developed no signs of apoptosis regardless of ion composition; however, when shrinkage was imposed on them by the addition of sucrose, it produced apoptosis with all its classical features: caspase activation, translocation of phosphatidylserine, release of cytochrome *c*, and dependence on Bcl-2 (Fig. 8). Although apoptosis was somewhat enhanced at low K^+, that was mostly limited to swollen cells (in experiments where apoptosis was induced by staurosporine in ionophore-treated cells). Disruption of the cytoskeleton by cytochalasin had no effect on the apoptosis development. We did not attempt to alter the membrane tension (which could be done, for example, by local anesthetics) but it should not be a major factor in shrinkage. Thus, by verifying that intracellular ion composition (Na^+ vs K^+) and preservation of actin

[a] Nevertheless, cells are clearly aware of their volume, as their average size is somehow conserved.

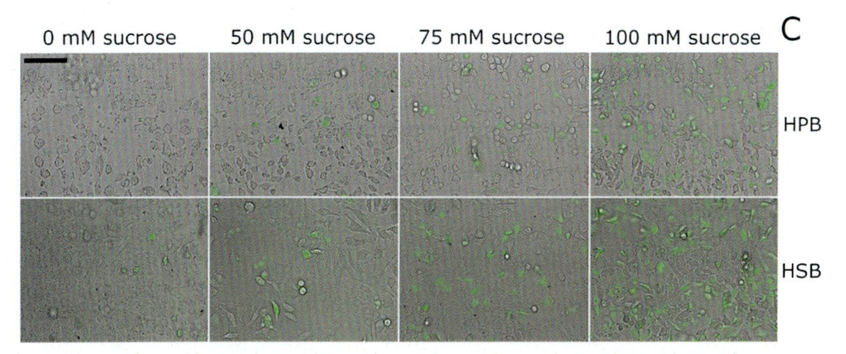

Fig. 8 Overlay of brightfield images of HeLa cells and NucView 488 fluorescence (green), indicating caspase-3 activation. The incubation was carried out either in a high-potassium (HPB) or a high-sodium buffer (HSB). Scale bar: 100 μm. *From Rana, P. S., Kurokawa, M., & Model, M. A. (2020). Evidence for macromolecular crowding as a direct apoptotic stimulus. Journal of Cell Science, 133(9), with permission.*

cytoskeleton had no effect on apoptosis, we had to conclude that MC is the only remaining viable candidate for this role.

However, this example also illustrates the general difficulty of studying macromolecular crowding and its role in living cells. While the role of an enzyme or a channel in a biological process can be verified using a specific inhibitor or genetic manipulation, there is no such resource in the case of MC. We dealt with this problem by excluding the alternative mechanisms. Proof by exclusion may be uncommon in cell biology, but in this case, the number of possibilities is limited, and the approach seems justified.

5. Conclusions

The indisputable fact is that macromolecular crowding is an inherent feature of the cytosol, and life apparently cannot exist in an excessively concentrated or excessively dilute state. Unfailing restoration of MC in a strongly anisosmotic environment (Fig. 5) is a clear sign that this parameter is important enough to be worth maintaining. Presumably, when MC is very high, too little room is left for the exchange of metabolites, and only metabolically dormant spores can survive extreme dehydration. The exact consequences of a too low MC have yet to be investigated. But we know from in vitro data that macromolecular crowding can control transcription, translation, activity of numerous enzymes, assembly of the cytoskeleton, and at least some ion channels (Ge, Luo, & Xu, 2011; Vibhute et al., 2020; Zimmerman & Minton, 1993), and there is no reason to believe that all these effects would become irrelevant in intact cells. Moreover, the idea that macromolecular crowding may be critical in cell volume regulation has been occasionally put forward by other authors (e.g., Al-Habori, 2001; Burg, 2000; Minton, 1994; Parker, 1993). It can be hoped that with the advent of effective experimental techniques for MC measurement, future work will reveal many important facts about the workings of macromolecular crowding in the living cell.

Acknowledgments

The author is grateful to J. Petruccelli (University at Albany) for discussing TIE imaging. Some data for Fig. 5 have been collected by J. Hollembeak (Kent State University). The work was supported by the University Research Council and the Department of Biological Sciences of Kent State University.

Appendix

Most TIE codes are written in MATLAB language; however, a TIE plugin for ImageJ is being currently developed by R. Clements (Kent State University) and J. Petruccelli (University at Albany), which should simplify the use of TIE by biologists.

We use the code written by Gorthi and Schonbrun (2012); it inputs two square bright-field images BF1 and BF2 collected at a specified wavelength and vertical separation and generates a phase map. TIE is notorious for being sensitive to noise and producing uneven background; therefore, we developed a few simple tricks that significantly improve the quality of the resultant phase images. Bright-field images are collected through a 485 nm bandpass filter, which allows simultaneous acquisition of TTD images through a 630 nm filter. The optimal vertical separation Δz between BF1 and BF2 is determined empirically; we usually use $\Delta z = 5$ μm with a 20× objective. There is always a small difference in the average intensities of BF1 and BF2, but we ensure that their intensities are equal to a fraction of a grayscale unit before processing. That can be accomplished either manually or, better, using a simple ImageJ plugin (http://drosophila.biology.kent.edu/users/rclement/extras/equalizer.html). Failure to equalize the intensities results in a cave-in or bulge-out background (Fig. 9B and C). We also crop images to relatively small square areas, trying to avoid cutting through any cells.

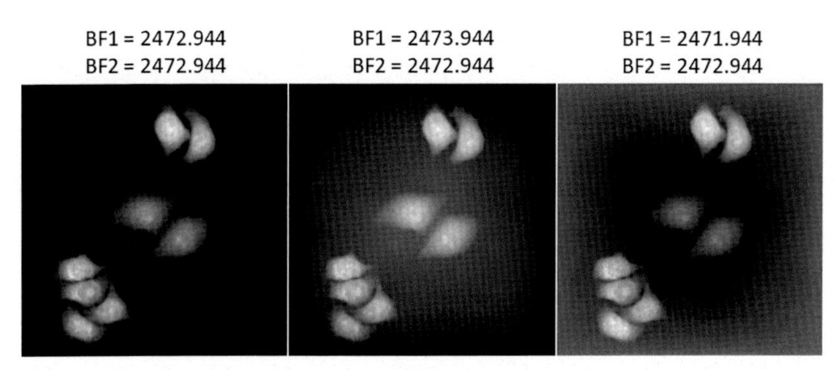

Fig. 9 (A) A TIE image based on bright-field images that had exactly the same average intensity (2472.944 on a 0–4095 scale). (B) The intensity of BF1 was purposely increased by 1 unit. (C) The intensity of BF1 was decreased by 1 unit.

The TIE output depends on the accurate knowledge of Δz; specifically, it is inversely proportional to its presumed value. Thus, if the true Δz was 4 μm and the parameter entered was 5 μm, the phase will be underestimated by 20%.

Unfortunately, the accuracy of the vertical travel of the stage (or of the objective) is not routinely tested and cannot be easily corrected. One way to get around it is to calibrate TIE directly in terms of protein concentration, bypassing the explicit determination of the phase. To do that, we assume that the result of TIE computation is not necessarily equal but only proportional to $\Delta\phi$. Then Eq. 4 can be replaced with

$$c - c' = p\frac{T}{V} \tag{11}$$

where T is the output of TIE integrated over the cell volume and p may differ from the theoretical $0.86\lambda_0 = 0.417$. The problem then becomes equivalent to finding the coefficient p.

To measure p experimentally without having a calibration object of defined size and shape, we perform duplicate measurements of T and V on the same cells in two AB9-containing buffers with different concentrations of bovine serum albumin (BSA) and use the fact that within a short time period, the total amount of intracellular protein must remain unchanged. If V_1 and V_2 are the volumes of the same cell in two solutions with BSA concentrations $c_1'$and c_1' and T_1 and T_2 are the TIE results integrated over the same cell, we obtain three linear equations with three unknowns c_1, c_2, and p:

$$T_1 = \frac{1}{p} V_1 \left(c_1 - c_1' \right) \tag{12}$$
$$T_2 = \frac{1}{p} V_2 \left(c_2 - c_2' \right)$$
$$V_1 c_1 = V_2 c_2$$

where the latter equation represents the condition of dry mass conservation. We solve these equations for p:

$$p = \frac{V_2 c_2' - V_1 c_1'}{T_1 - T_2} \tag{13}$$

This result becomes particularly simple if the first measurement is performed in a protein-free buffer buffer ($c_1' = 0$). Then p can be found from the slope of the line

$$\frac{T_1 - T_2}{V_2} = \frac{1}{p} c_2' \tag{14}$$

Representative results are shown in Fig. 10. There is no need to ensure isosmolarity of the two solutions, which is especially convenient since the osmometer is not a standard piece of equipment in every lab.

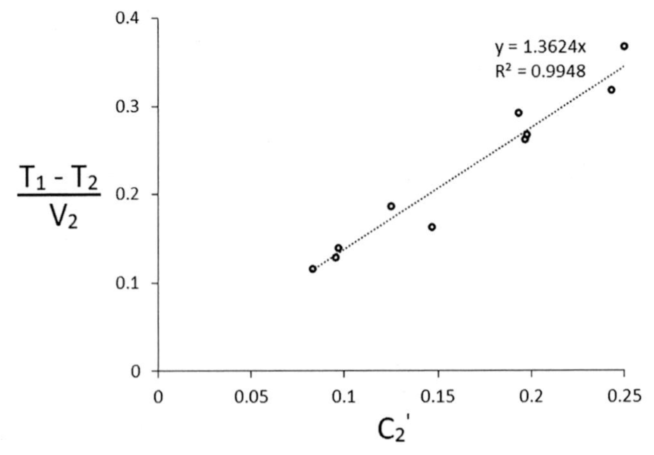

Fig. 10 The results of TIE calibration, where each point represents the average value obtained in one experiment. There is a large difference between the empirical slope 1.36 and the theoretical 1/0.417 = 2.4. We did not attempt to determine if this was entirely due to systematic errors in the vertical travel or if other factors have also played a role.

References

Al-Habori, M. (2001). Macromolecular crowding and its role as intracellular signalling of cell volume regulation. *The International Journal of Biochemistry & Cell Biology*, *33*, 844–864.

Aliev, M. K., Dos Santos, P., Hoerter, J. A., Soboll, S., Tikhonov, A. N., & Saks, V. A. (2002). Water content and its intracellular distribution in intact and saline perfused rat hearts revisited. *Cardiovascular Research*, *53*, 48–58.

Armstrong, C. M. (2003). The Na/K pump, Cl ion, and osmotic stabilization of cells. *Proceedings of the National Academy of Sciences of the United States of America*, *100*, 6257–6262.

Baldwin, W. W., & Kubitschek, H. E. (1984). Buoyant density variation during the cell cycle of Saccharomyces cerevisiae. *Journal of Bacteriology*, *158*, 701–704.

Barros, L. F. (1999). Measurement of sugar transport in single living cells. *Pflügers Archiv*, *437*, 763–770.

Barros, L. F., Hermosilla, T., & Castro, J. (2001). Necrotic volume increase and the early physiology of necrosis. *Comparative Biochemistry and Physiology Part A: Molecular & Integrative Physiology*, *130*, 401–409.

Boersma, A. J., Zuhorn, I. S., & Poolman, B. (2015). A sensor for quantification of macromolecular crowding in living cells. *Nature Methods*, *12*, 227–229.

Bortner, C. D., & Cidlowski, J. A. (1996). Absence of volume regulatory mechanisms contributes to the rapid activation of apoptosis in thymocytes. *American Journal of Physiology-Cell Physiology*, *271*, C950–C961.

Bortner, C. D., & Cidlowski, J. A. (2002). Apoptotic volume decrease and the incredible shrinking cell. *Cell Death and Differentiation*, *9*, 1307–1310.

Bortner, C. D., & Cidlowski, J. A. (2003). Uncoupling cell shrinkage from apoptosis reveals that Na^+ influx is required for volume loss during programmed cell death. *Journal of Biological Chemistry*, *278*, 39176–39184.

Bosslet, K., Ruffmann, R., Altevogt, P., & Schirrmacher, V. (1981). A rapid method for the isolation of metastasizing tumour cells from internal organs with the help of isopycnic density-gradient centrifugation in Percoll. *British Journal of Cancer*, *44*, 356–362.

Bottier, C., Gabella, C., Vianay, B., Buscemi, L., Sbalzarini, I. F., Meister, J. J., et al. (2011). Dynamic measurement of the height and volume of migrating cells by a novel fluorescence microscopy technique. *Lab on a Chip*, *11*, 3855–3863.

Boyd, C., & Gradmann, D. (2002). Impact of osmolytes on buoyancy of marine phytoplankton. *Marine Biology*, *141*, 605–618.

Bryan, A. K., Hecht, V. C., Shen, W., Payer, K., Grover, W. H., & Manalis, S. R. (2014). Measuring single cell mass, volume, and density with dual suspended microchannel resonators. *Lab on a Chip*, *14*, 569–576.

Burg, M. B. (2000). Macromolecular crowding as a cell VolumeSensor. *Cellular Physiology and Biochemistry*, *10*, 251–256.

Choi, W. J., Jeon, D. I., Ahn, S. G., Yoon, J. H., Kim, S., & Lee, B. H. (2010). Full-field optical coherence microscopy for identifying live cancer cells by quantitative measurement of refractive index distribution. *Optics Express*, *18*, 23285–23295.

Clements, R. J., Davidson, M., & Model, M. A. (2021). Experimental test of the geometric model of image formation in bright-field microscopy. *Journal of Microscopy*, *283*, 3–8.

Cooper, K. L., Oh, S., Sung, Y., Dasari, R. R., Kirschner, M. W., & Tabin, C. J. (2013). Multiple phases of chondrocyte enlargement underlie differences in skeletal proportions. *Nature*, *495*, 375–378.

Curl, C. L., Bellair, C. J., Harris, T., Allman, B. E., Harris, P. J., Stewart, A. G., et al. (2005). Refractive index measurement in viable cells using quantitative phase-amplitude microscopy and confocal microscopy. *Cytometry Part A: The Journal of the International Society for Analytical Cytology*, *65*, 88–92.

Fischer, H., Polikarpov, I., & Craievich, A. F. (2004). Average protein density is a molecular-weight-dependent function. *Protein Science*, *13*, 2825–2828.
Ge, X., Luo, D., & Xu, J. (2011). Cell-free protein expression under macromolecular crowding conditions. *PLoS One*, *6*, e28707.
Germond, A., Fujita, H., Ichimura, T., & Watanabe, T. M. (2016). Design and development of genetically encoded fluorescent sensors to monitor intracellular chemical and physical parameters. *Biophysical Reviews*, *8*, 121–138.
Gorthi, S. S., & Schonbrun, E. (2012). Phase imaging flow cytometry using a focus-stack collecting microscope. *Optics Letters*, *37*, 707–709.
Grover, W. H., Bryan, A. K., Diez-Silva, M., Suresh, S., Higgins, J. M., & Manalis, S. R. (2011). Measuring single-cell density. *Proceedings of the National Academy of Sciences of the United States of America*, *108*, 10992–10996.
Guigas, G., Kalla, C., & Weiss, M. (2007). The degree of macromolecular crowding in the cytoplasm and nucleoplasm of mammalian cells is conserved. *FEBS Letters*, *581*, 5094–5098.
Kay, A. R., & Blaustein, M. P. (2019). Evolution of our understanding of cell volume regulation by the pump-leak mechanism. *Journal of General Physiology*, *151*, 407–416.
Kemper, B., Carl, D. D., Schnekenburger, J., Bredebusch, I., Schäfer, M., Domschke, W., et al. (2006). Investigation of living pancreas tumor cells by digital holographic microscopy. *Journal of Biomedical Optics*, *11*, 034005.
Khitrin, A. K., Petruccelli, J. C., & Model, M. A. (2017). Bright-field microscopy of transparent objects: A ray tracing approach. *Microscopy and Microanalysis*, *23*, 1116–1120.
Lababidi, S. L., Pelts, M., Moitra, M., Leff, L. G., & Model, M. A. (2011). Measurement of bacterial volume by transmission-through-dye imaging. *Journal of Microbiological Methods*, *87*, 375–377.
Leslie, M. (2011). How does a cell know its size. *Science*, *334*, 1047–1048.
Ling, G. N. (2004). What determines the normal water content of a living cell? *Physiological Chemistry and Physics and Medical NMR*, *36*, 1–20.
Loken, M. R., & Kubitschek, H. E. (1984). Constancy of cell buoyant density for cultured murine cells. *Journal of Cellular Physiology*, *118*(1), 22–26.
Luby-Phelps, K. (1999). Cytoarchitecture and physical properties of cytoplasm: Volume, viscosity, diffusion, intracellular surface area. *International Review of Cytology*, *192*, 189–221.
Lue, N., Popescu, G., Ikeda, T., Dasari, R. R., Badizadegan, K., & Feld, M. S. (2006). Live cell refractometry using microfluidic devices. *Optics Letters*, *31*, 2759–2761.
Maric, D., Maric, I., & Barker, J. L. (1998). Buoyant density gradient fractionation and flow cytometric analysis of embryonic rat cortical neurons and progenitor cells. *Methods*, *16*, 247–259.
McCarthy, J. V., & Cotter, T. G. (1997). Cell shrinkage and apoptosis: A role for potassium and sodium ion efflux. *Cell Death and Differentiation*, *4*, 756–770.
Medeiros, I. P. M., Faria, S. C., & Souza, M. M. (2020). Osmoionic homeostasis in bivalve mollusks from different osmotic niches: Physiological patterns and evolutionary perspectives. *Comparative Biochemistry and Physiology Part A: Molecular & Integrative Physiology*, *240*, 110582.
Minton, A. P. (1994). Influence of macromolecular crowding on intracellular association reactions: Possible role in volume regulation. In *Cellular and molecular physiology of cell volume regulation* (pp. 181–190). Boca Raton: CRC Press (K. strange, editor).
Minton, A. P. (2001). The influence of macromolecular crowding and macromolecular confinement on biochemical reactions in physiological media. *Journal of Biological Chemistry*, *276*, 10577–10580.
Model, M. A. (2012). Imaging the cell's third dimension. *Microscopy Today*, *20*, 32–37.
Model, M. A. (2014). Possible causes of apoptotic volume decrease: An attempt at quantitative review. *American Journal of Physiology-Cell Physiology*, *306*, C417–C424.

Model, M. A. (2015). Cell volume measurements by optical transmission microscopy. *Current Protocols in Cytometry*, *72*, 12–39.
Model, M. A. (2018). Methods for cell volume measurement. *Cytometry Part A*, *93*, 281–296.
Model, M. A. (2020). Comparison of cell volume measurements by fluorescence and absorption exclusion microscopy. *Journal of Microscopy*, *280*, 12–18.
Model, M. A., & Davis, M. A. (2016). Observation of living organisms in environmental samples by transmission-through-dye microscopy. *Microscopy Today*, *24*, 46–51.
Model, M. A., Hollembeak, J. E., & Kurokawa, M. (2021). Macromolecular crowding: A hidden link between cell volume and everything else. *Cellular Physiology and Biochemistry*, *55*, 25–40.
Model, M. A., Khitrin, A. K., & Blank, J. L. (2008). Measurement of the absorption of concentrated dyes and their use for quantitative imaging of surface topography. *Journal of Microscopy*, *231*, 156–167.
Model, M. A., Mudrak, N. J., Rana, P. S., & Clements, R. J. (2018). Staurosporine-induced apoptotic water loss is cell-and attachment-specific. *Apoptosis*, *23*, 449–455.
Model, M. A., & Petruccelli, J. C. (2018). Intracellular macromolecules in cell volume control and methods of their quantification. *Current Topics in Membranes*, *81*, 237–289.
Mouton, S. N., Veenhoff, L. M., & Boersma, A. J. (2020). Macromolecular crowding measurements with genetically encoded probes based on Förster resonance energy transfer in living cells. In R. Hancock (Ed.), *The nucleus* (pp. 169–180). New York: Humana Press.
Mudrak, N. J., Rana, P. S., & Model, M. A. (2018). Calibrated brightfield-based imaging for measuring intracellular protein concentration. *Cytometry Part A*, *93*, 297–304.
Neurohr, G. E., & Amon, A. (2020). Relevance and regulation of cell density. *Trends in Cell Biology*, *30*, 213–225.
Okada, Y., Maeno, E., Shimizu, T., Dezaki, K., Wang, J., & Morishima, S. (2001). Receptor-mediated control of regulatory volume decrease (RVD) and apoptotic volume decrease (AVD). *The Journal of Physiology*, *532*, 3–16.
Panijpan, B. (1977). The buoyant density of DNA and the G + C content. *Journal of Chemical Education*, *54*, 172–173.
Parker, J. C. (1993). In defense of cell volume? *American Journal of Physiology-Cell Physiology*, *265*, C1191–C1200.
Rana, P. S., Gibbons, B. A., Vereninov, A. A., Yurinskaya, V. E., Clements, R. J., Model, T. A., et al. (2019). Calibration and characterization of intracellular Asante potassium green probes, APG-2 and APG-4. *Analytical Biochemistry*, *567*, 8–13.
Rana, P. S., Kurokawa, M., & Model, M. A. (2020). Evidence for macromolecular crowding as a direct apoptotic stimulus. *Journal of Cell Science*, *133*(9), jcs243931.
Rana, P. S., & Model, M. A. (2020). A reverse-osmosis model of apoptotic shrinkage. *Frontiers in Cell and Development Biology*, *8*, 588721.
Rana, P. S., Mudrak, N. J., Lopez, R., Lynn, M., Kershner, L., & Model, M. A. (2017). Phase separation in necrotic cells. *Biochemical and Biophysical Research Communications*, *492*, 300–303.
Rappaz, B., Marquet, P., Cuche, E., Emery, Y., Depeursinge, C., & Magistretti, P. J. (2005). Measurement of the integral refractive index and dynamic cell morphometry of living cells with digital holographic microscopy. *Optics Express*, *13*, 9361–9373.
Song, W. Z., Zhang, X. M., Liu, A. Q., Lim, C. S., Yap, P. H., & Hosseini, H. M. M. (2006). Refractive index measurement of single living cells using on-chip Fabry-Pérot cavity. *Applied Physics Letters*, *89*, 203901.
Teague, M. R. (1983). Deterministic phase retrieval: A Green's function solution. *JOSA*, *73*, 1434–1441.
Theisen, A. (2000). *Refractive increment data-book for polymer and biomolecular scientists*. Nottingham University Press.

Tivey, D. R., Simmons, N. L., & Aiton, J. F. (1985). Role of passive potassium fluxes in cell volume regulation in cultured HeLa cells. *The Journal of Membrane Biology*, *87*, 93–105.
Vibhute, M. A., Schaap, M. H., Maas, R. J., Nelissen, F. H., Spruijt, E., Heus, H. A., et al. (2020). Transcription and translation in cytomimetic protocells perform most efficiently at distinct macromolecular crowding conditions. *ACS Synthetic Biology*, *9*, 2797–2807.
White, J. (1952). Variation in water content of yeast cells caused by varying temperatures of growth and by other cultural conditions. *Journal of the Institute of Brewing*, *58*, 47–50.
Yu, S. P. (2003). Regulation and critical role of potassium homeostasis in apoptosis. *Progress in Neurobiology*, *70*, 363–386.
Zeuthen, T. (2010). Water-transporting proteins. *Journal of Membrane Biology*, *234*, 57–73.
Zimmerman, S. B., & Minton, A. P. (1993). Macromolecular crowding: Biochemical, biophysical, and physiological consequences. *Annual Review of Biophysics and Biomolecular Structure*, *22*, 27–65.
Zipursky, A., Bow, E., Seshadri, R. S., & Brown, E. J. (1976). Leukocyte density and volume in normal subjects and in patients with acute lymphoblastic leukemia. *Blood*, *48*, 361–371.

CHAPTER SIX

Membrane tension

Pei-Chuan Chao[a] and Frederick Sachs[b,*]
[a]Department of Civil, Structural and Environmental Engineering, University at Buffalo, The State University of New York, Buffalo, NY, United States
[b]Department of Physiology and Biophysics, University at Buffalo, The State University of New York, Buffalo, NY, United States
*Corresponding author: e-mail address: sachs@buffalo.edu

Abstract

The cell membrane serves as a barrier that restricts the rate of exchange of diffusible molecules. Tension in the membrane regulates many crucial cell functions involving shape changes and motility, cell signaling, endocytosis, and mechanosensation. Tension reflects the forces contributed by the lipid bilayer, the cytoskeleton, and the extracellular matrix. With a fluid-like bilayer model, membrane tension is presumed uniform and hence propagated instantaneously. In this review, we discuss techniques to measure the mean membrane tension and how to resolve the stresses in different components and consider the role of bilayer heterogeneity.

Besides known as a participant in writing The Declaration of Independence, Benjamin Franklin was also a pioneer scientist studying lipid monolayers (Franklin & Brownrigg, 1774; Tanford, 2004; Wang, Stieglitz, Marden, & Tamm, 2013). Inspired by his experiments of oil on water, scientists calculated the thickness of a monolayer (Rayleigh, 1889) and a bilayer (Gorter & Grendel, 1925), Tanford later invented the concept of a hydrophobic force to describe the forces (Tanford, 1980).

Cells are covered with a membrane that limits the rate of exchange of diffusible molecules using a bilayer of phospholipids. However, since mechanical stresses from cell motility and forces from the environment could stretch the bilayer beyond its lytic limit, nature added reinforcing protein polymers along the inside and the outside. As a two-dimensional structure, membrane mechanics are well described by the tension dependence. Membrane tension affects many cell functions including morphogenesis and motility (Clark, Wartlick, Salbreux, & Paluch, 2014; Diz-Muñoz, Fletcher, & Weiner, 2013; Houk et al., 2012; Keren, 2011; Keren et al., 2008), vesicle trafficking and fusion (Apodaca, 2002; Masters, Pontes, Viasnoff, Li, & Gauthier, 2013; Shin et al., 2018; Wen et al., 2016), cell

Current Topics in Membranes, Volume 88
ISSN 1063-5823
https://doi.org/10.1016/bs.ctm.2021.09.002

signaling (Groves & Kuriyan, 2010; Houk et al., 2012; Huse, 2017; Masters et al., 2013), endocytosis (Raucher & Sheetz, 1999; Shi, Graber, Baumgart, Stone, & Cohen, 2018), and mechanosensation (Cox et al., 2016; Cox, Bavi, & Martinac, 2019; He et al., 2018; Sachs, 2010).

As we begin to study membrane mechanics, we must realize that the term "membrane" is not precise when referring to mechanics. Biological membranes are heterogeneous and the structure begs to have the mean stress resolved at least into bilayer, cytoskeletal and extracellular matrix (ECM) stresses (Akinlaja & Sachs, 1998). To study tension in the cell membrane, we will start by discussing changes in surface tension in the bilayer.

The fluid-mosaic membrane model proposed in 1972 (Singer & Nicolson, 1972) depicted the membrane as a two-dimensional viscous fluid film composed of two lipid monolayers containing dissolved proteins. The lipid bilayer has phospholipids with the aliphatic chains facing each other and the hydrophilic heads facing the extracellular and intracellular aqueous phases (Fig. 1). This idealized model has since been amended to include interaction with cytoskeletal and extracellular proteins, aggregates of segregated soluble proteins and multiple types of lipids (Edidin, 2003; Engelman, 2005; Goñi, 2014; Nicolson, 2014; Zalba & Ten Hagen, 2017). Different types of lipids may be distributed asymmetrically across the bilayer and can also aggregate into domains often called "lipid rafts." These rafts are rich in cholesterol and sphingolipids (Simons & Vaz, 2004; Van Meer, Voelker, & Feigenson, 2008) and are formed in a Liquid-ordered (l_o) phase and are able to coexist with the surrounding lipids in a Liquid-disordered (l_d) phase depending on the lipid composition and temperature (Baumgart, Hess, & Webb, 2003;

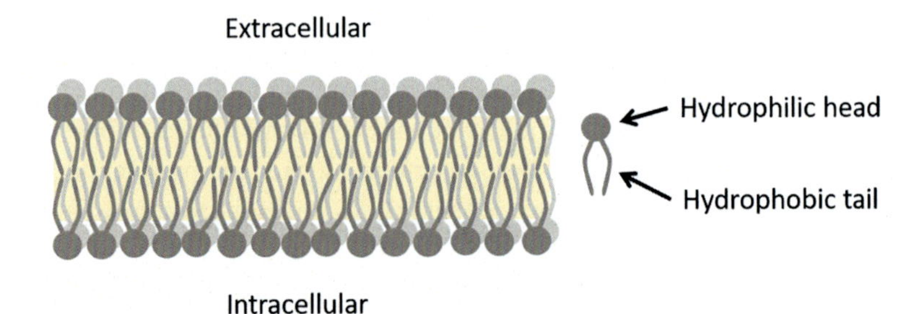

Fig. 1 Schematic illustration of the structure of lipid bilayer. Phospholipids form a bilayer with the aliphatic chains facing the interior and the hydrophilic heads facing the extracellular and intracellular aqueous phases.

de la Serna, Perez-Gil, Simonsen, & Bagatolli, 2004; Dent et al., 2015). The group of Kusumi et al. (2005) used high-speed single-particle tracking to observe heterogeneity and diffusion within the membrane. Although liquid-liquid phase separations and their coexistence in biological membranes have been observed, the concept of specific lipid rafts remains controversial due to the lack of direct microscopic visualizations and conclusive domain properties in living cells (Klotzsch & Schütz, 2013; Levental & Veatch, 2016; Sezgin, Levental, Mayor, & Eggeling, 2017).

For simplicity, lipid bilayer mechanics are often studied using giant unilamellar vesicles (GUVs) (Dimova et al., 2006; Rideau, Dimova, Schwille, Wurm, & Landfester, 2018). GUVs are aqueous vesicles covered with lipid bilayers and comparable in size to eukaryotic cells (1–10 μm). They can be visualized with an optical microscope that allows for a direct measurement of membrane strain. Stress can be applied using pipette aspiration (Fig. 2) (Dimova et al., 2006; Roux et al., 2010), optical tweezers (Fig. 6) (Lieber, Yehudai-Resheff, Barnhart, Theriot, & Keren, 2013; Tian & Baumgart, 2009), or magnetic probes (Neuman & Nagy, 2008).

Micropipette aspiration (Hochmuth, 2000) uses suction applied to a pipette while contacting a cell. This draws in a bleb of cell membrane (Fig. 2). Laplace's law relates patch tension (σ) to the radius of curvature (r) and the transmembrane pressure (P): $\sigma = 0.5\,Pr$, and provides a first order

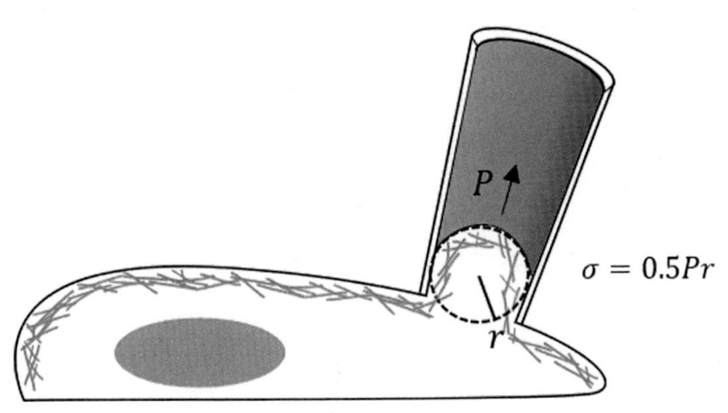

Fig. 2 Schematic representation of micropipette aspiration experiment to measure the mechanical properties of cell membranes. A portion of membrane is aspirated into a micropipette. The mean membrane tension, σ, can be approximated using suction pressure, P, and the radius of patch membrane curvature, r, by the law of Laplace, $\sigma = 0.5Pr$.

approximation to the mean tension. But note that for cells it is not the bilayer tension since the biological bilayer has tension sharing with cortical and ECM proteins (Hochmuth, Mohandas, Mohandas, & Blackshear, 1973; Pontes, Monzo, & Gauthier, 2017; Sens & Plastino, 2015). Aspiration makes the implicit assumption that pulling a piece of cell cortex into a pipette does not disrupt any of its mechanical properties. Fig. 3 shows video micrographs of an area expansion measurement for a GUV (Rawicz et al., 2000). Fig. 4 shows images of both flaccid and swollen red blood cells aspirated into a micropipette (Hochmuth, 2000).

One experiment attempting to resolve the sharing of lipid with protein stress examined how mean membrane tension affected the voltage required for lysis (Akinlaja & Sachs, 1998). Since the lipid bilayer is the only membrane component with a significant voltage drop, voltage stress is only applied to the bilayer (see Fig. 5). Bilayers will lyse at distinct voltages

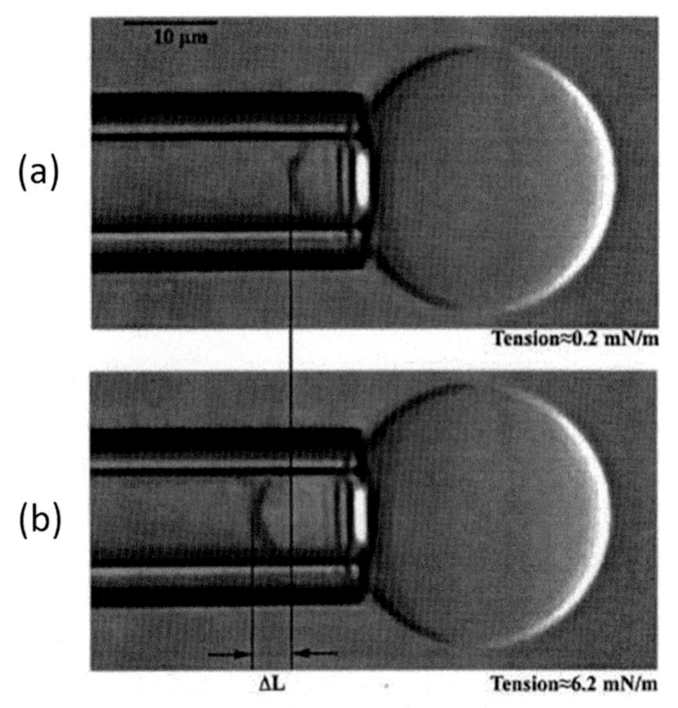

Fig. 3 Micropipette aspiration of a giant vesicle. Video micrographs of a GUV aspirated into a micropipette at (A) low tension and (B) high tension. *Adapted with permission from Rawicz, W., Olbrich, K. C., McIntosh, T., Needham, D., & Evans, E. (2000). Effect of chain length and unsaturation on elasticity of lipid bilayers.* Biophysical Journal *79(1), 328–339.*

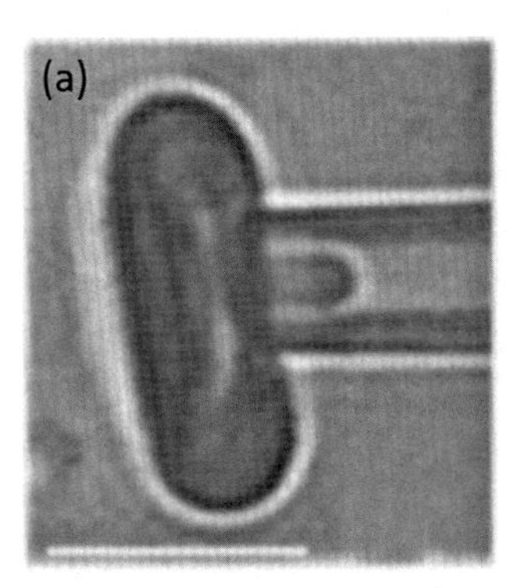

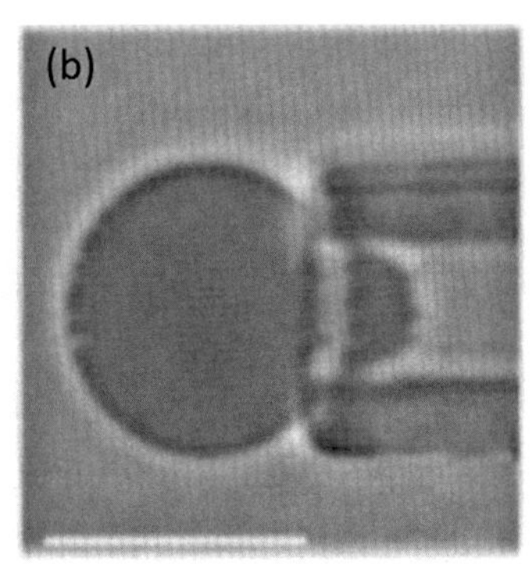

Fig. 4 Micropipette aspiration *of red blood cells.* (A) A flaccid and (B) a swollen red cell aspirated into a micropipette. *Adapted with permission from Hochmuth, R. M. (2000). Micropipette aspiration of living cells.* Journal of Biomechanics 33*(1), 15–22.*

and the higher the tension in the bilayer the less voltage is required to poke a hole; lysis occurs at a characteristic strain energy. Those experiments suggested that in a patch of HEK membrane, about half of the mean membrane tension is born by the bilayer.

Another technique to probe bilayer tension is to pull lipid tethers from a cell using optical tweezers or an atomic force microscope (Fig. 6). To form a tether, a sticky bead is bound to a membrane protein, and the bead is then pulled away from the cell. Fig. 7A shows a tether pulled away from the bilayer by an attached bead. Fig. 7B shows a typical tether extraction force curve showing the maximum and steady-state tether forces (Pontes et al., 2013). With sufficient applied force, lipids separate from the structural proteins and flow into the tether. By measuring the force needed to pull the tether (f) and knowing the bending stiffness of the bilayer (κ), the membrane tension (σ) can be calculated by $\sigma = f^2/8\pi^2\kappa$ (Dai & Sheetz, 1995; Hormeno & Arias-Gonzalez, 2006; Lieber et al., 2013; Sens & Plastino, 2015; Sheetz & Dai, 1996). The values of the bending modulus for various cell types from the literature range from 0.14 to 0.27 pN μm (Evans, 1983; Hochmuth, Shao, Dai, & Sheetz, 1996; Lieber et al., 2013). The membrane bending moment can be derived from the tether force and the tether radius (r_t), that tends to be in the range of 100 nm: $\kappa = fr_t/2\pi$ (Hochmuth et al., 1996, Lieber et al., 2013). Literature values for cell bilayer tension span a wide range: 5–276 pN/μm (0.005–0.276 dyne/cm) (Dai & Sheetz, 1995; Lieber et al., 2013; Sens & Plastino, 2015; Sun et al., 2005), but the best data for resting cells suggests they resting cells have very little bilayer tension

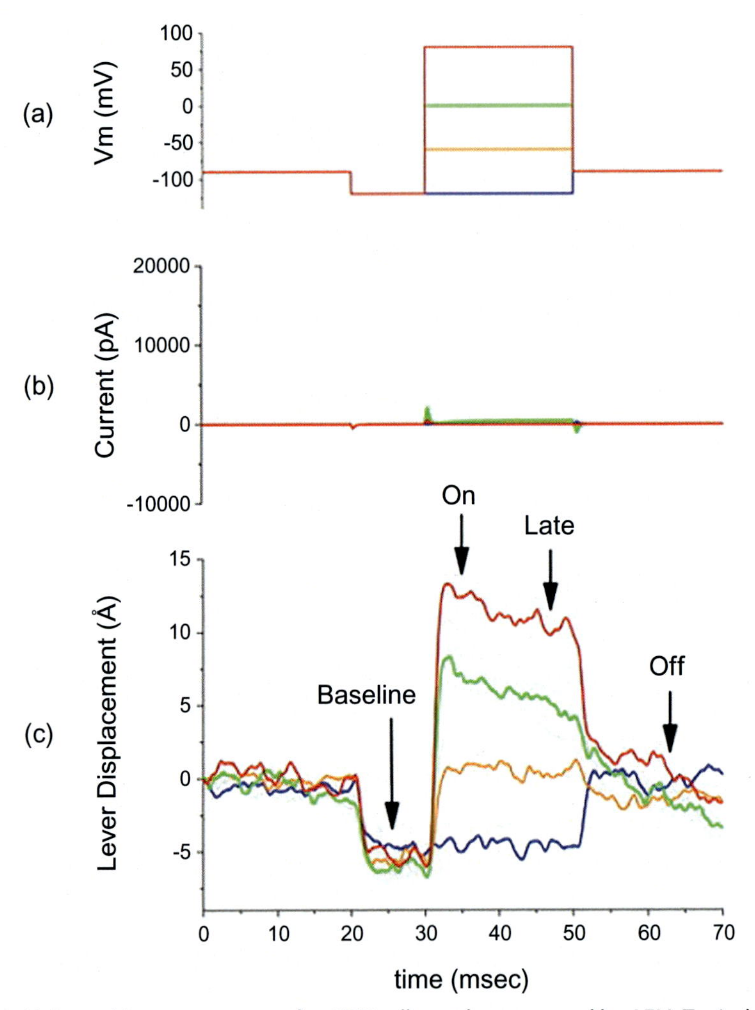

Fig. 5 Voltage driven movement of an HEK cell membrane sensed by AFM. Typical data for wild type wtHEK cells. (A) Voltage step protocol with whole-cell patch clamp at a holding potential of $V_m = -80$ mV. At 20 ms into the record, V_m is prepulsed to -120 mV and then in 10 mV increments to +80 mV. (B) Whole-cell ionic currents are negligible in these cells. (C) voltage-induced membrane displacement. A voltage step produces a jump in membrane tension reported by the cantilever parked on the cell membrane. For wtHEKs, depolarization results in an increase in membrane tension and thus an upward cantilever deflection (positive). Hyperpolarization decreases membrane tension, resulting in cantilever sinking into the cell (negative). Displacements are calculated from a baseline at 120 mV, averaged between 25 and 30 ms. After the voltage-induced displacement (average 33–35 ms), the cantilever drifts toward baseline probably because of cortical and lipid relaxation with a time constant $\gg$10mS. *Adapted with the permission from Beyder, A., & Sachs, F. (2009). Electromechanical coupling in the membranes of shaker-transfected HEK cells.* Proceedings of the National Academy of Sciences of the United States of America 106*(16), 6626–6631.*

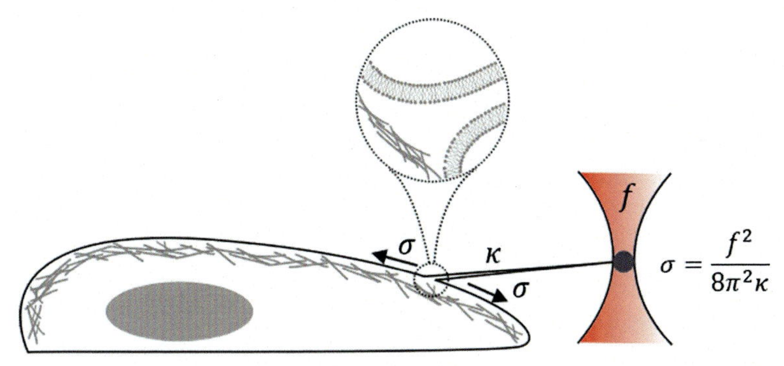

Fig. 6 Schematic representation of tether-pulling experiment with optical tweezers to measure mechanical properties of the cell bilayer. A reactive bead is attached to the cell membrane and the bead, trapped by optical tweezer, is pulled away. Mobile lipids flow into the tether. By measuring the force needed to pull the tether (f) and the bending stiffness of the bilayer (κ), the membrane tension (σ) is obtained with $\sigma = f^2/8\pi^2\kappa$.

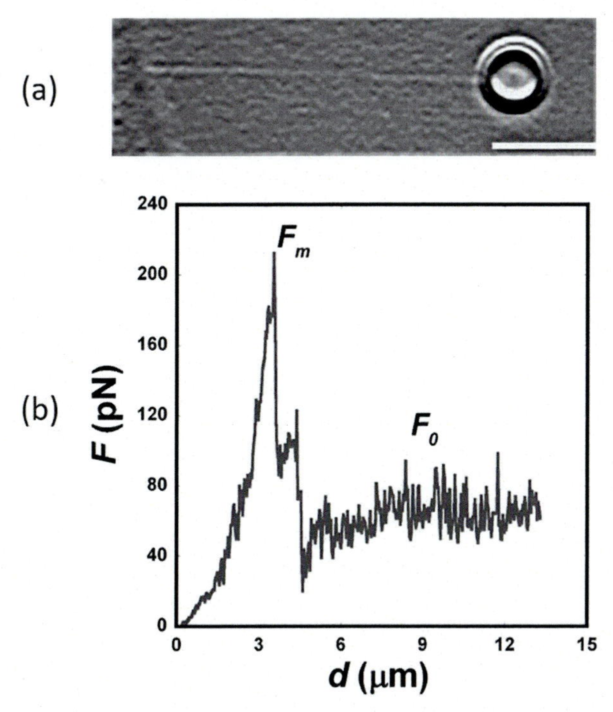

Fig. 7 Tether extraction from microglial cell. (A) A tether pulled away from a cell lipid bilayer by the attached bead. (B) Tether extraction force curve where F_m is the maximum force and F_0 is the steady-state force. That reflects the energy to pull lipids from the cytoskeleton. *Adapted with the permission from Pontes, B., Ayala, Y., Fonseca, A. C. C., Romão, L. F., Amaral, R. F., Salgado, L. T., et al. (2013). Membrane elastic properties and cell function.* PLoS One 8*(7), e67708.*

(Sheetz & Dai, 1996). We recommend the review by Cohen and Shi for a discussion of these issues (Cohen & Shi, 2020).

While relating bilayer tension to tether force, a correction factor is the force contributed by the lipid-cytoskeleton adhesion that tends to pull the tether back into the cell. That varies with cell type (Dai & Sheetz, 1999; Sens & Plastino, 2015). The contribution of cytoskeleton adhesion can be reduced by eliminating the cytoskeleton with drugs or by pulling tethers from blebs instead of the resting membrane. Nascent blebs contain minimal cytoskeleton (Cox et al., 2016; Lieber et al., 2013; Sens & Plastino, 2015).

To report membrane tension in the structural proteins of the membrane, genetically coded FRET sensors (fluorescence resonance energy transfer) can measure local forces in real time in specific proteins. These probes have been expressed in living cultured cells and transgenic organisms (Meng, 2011; Meng & Sachs, 2012; Meng, Suchyna, & Sachs, 2008). The probes work by measuring the distance between the donor and acceptor fluorophores separated by an elastic coupling. They are sensitive to forces in the range of 10 pN (Markin & Sachs, 2015) and can be genetically encoded into host proteins. These probes are unfortunately not capable of measuring tension in lipids since there is no fixed location to pull against.

However, tension pulls lipids apart and reduces lipid viscosity. Measurement of the rotational correlation time of dissolved probes can infer tension changes (Chambers et al., 2018; Steinmark et al., 2020; Yang et al., 2013). Colom et al. (2018) proposed using fluorescent membrane tension probes. These mechanosensitive "flipper" probes respond to the changes of membrane tension (via local density) through changes in fluorescent lifetime. Tension decreases the density of acyl chains yielding a red-shifted excitation maxima and longer lifetimes (Goujon et al., 2019). Tension gradients can also be observed by observing with FLIM (Fluorescence Lifetime Imaging Microscopy (Colom et al., 2018)). Since spatial resolution is limited by the optics, any observed properties may represent an average across local domains with different properties (Carquin, D'Auria, Pollet, Bongarzone, & Tyteca, 2016; Lindblom & Orädd, 2009).

Tension in a bilayer stretches it according to the two-dimensional analogy of Hooke's law:

$$T = K\left(\frac{\Delta A}{A}\right) \tag{1}$$

where K is the elastic area stretch modulus, A is the surface area and ΔA is the change of area induced by T (Diz-Muñoz et al., 2013). One way to create a gradient of tension is by hydrodynamics. Following the literature (Bussell, Koch, & Hammer, 1995; Shi et al., 2018), lipid flow in a membrane containing randomly dispersed immobile proteins can be described by a modified Stokes equation with drag:

$$\nabla\sigma = -\eta\nabla^2\vec{v} + \frac{\eta}{k}\vec{v} \tag{2}$$

where σ is the membrane tension, η is the two-dimensional membrane viscosity, $\vec{v}$ is the velocity field of lipid flow, and k is the Darcy permeability of the protein array (Brinkman, 1949; Bussell et al., 1995; Howells, 1974; Shi et al., 2018). The Darcy permeability represents the ability of the fluid to flow through a porous mass (lipid bilayer with embedded immobile obstacles) (Brinkman, 1949), and is affected by the radius, a, of the immobile particles and area fraction of fixed particles, ϕ, by $k = a^2 f(\phi)$, where $f(\phi)$ is a dimensionless function of immobile area (Bussell et al., 1995; Cohen & Shi, 2020; Howells, 1974).

Considering only pure lipid bilayers, the drag term in Eq. (2) can be neglected,

$$\nabla\sigma = -\eta\nabla^2\vec{v} \tag{3}$$

If a small perturbation to lipid density is assumed, $\rho = \rho_0 + \delta\rho$, conservation of mass is satisfied by

$$\frac{\partial\rho}{\partial t} + \rho_0\nabla\cdot\vec{v} = 0 \tag{4}$$

Assuming membrane tension follows a linear stress-strain relation (Hochmuth, Mohandas, Mohandas, & Blackshear, 1973)

$$\delta\sigma = -\frac{E_m}{\rho_0}\delta\rho \tag{5}$$

where E_m is the area expansion modulus. Eqs. (3)–(5) can describe the hydrodynamics of lipid flow in bilayer without immobile proteins or lipid domains (Shi et al., 2018). The relation for tension relaxation is,

$$\frac{\partial\nabla^2\sigma}{\partial t} = -\frac{E_m}{\eta}\nabla^2\sigma \tag{6}$$

where η/E_m is identified as a relaxation time. With an estimated value of E_m(40 pN/μm; Hochmuth, 2000) and η (3×10^{-3}pN · s/μm), the relaxation time is less than 0.1 ms (Keren et al., 2008; Shi et al., 2018).

To measure the viscosity of a lipid membrane, a small fluorescence probe termed a "molecular rotor" has proved a useful tool (Haidekker et al., 2001; Haidekker & Theodorakis, 2007; Kuimova, 2012; Páez-Pérez, López-Duarte, Vyšniauskas, Brooks, & Kuimova, 2021; Wu et al., 2013). Molecular rotors are a group of fluorescent molecules where their quantum yield and lifetime are dependent on the local viscosity. FLIM microscopy can provide spatiotemporally-resolved maps of microviscosity. An interesting observation is that when molecular rotors are applied as viscosity sensors in cells, their reported measurement of intracellular viscosity (80–600 cP) (Dent et al., 2015; Kuimova, 2012; Levitt et al., 2009) is much higher than the values probed by aqueous probes (1–3 cP) (Fogelson & Mogilner, 2014; Saffman & Delbrück, 1975; Shi et al., 2018). This is because the molecular rotors are hydrophobic, and are located in the aliphatic region of the bilayer, and hence are insensitive to aqueous phase viscosity.

While membrane viscosity is often considered a uniform property, the "microviscosity" can vary in time and space. The heterogeneous composition of biologic bilayers includes items such as rafts or caveolae, and they have a much different viscosity than the surrounding liquid regions (Carquin et al., 2016; Lingwood & Simons, 2010; Sherin et al., 2017; Steinkühler, Sezgin, Urbančič, Eggeling, & Dimova, 2019). The variations of microviscosity can lead to time dependent gradients of tension, which may lead to localized mechano-signaling in addition to stress on the cytoskeleton (Fogelson & Mogilner, 2014; Lieber, Schweitzer, Kozlov, & Keren, 2015) and immobile proteins (Shi et al., 2018). Membrane tension also affects the physical properties of a lipid bilayer, including thickness (Hamill & Martinac, 2001), lipid structures (Colom et al., 2018; Muddana, Gullapalli, Manias, & Butler, 2011), polarity (Zhang, Frangos, & Chachisvilis, 2006), and fluidity (Butler, Norwich, Weinbaum, & Chien, 2001; Haidekker et al., 2001; Muddana et al., 2011; Reddy, Warshaviak, & Chachisvilis, 2012). The area per lipid increases with increasing membrane tension, which leads to the increase in fluidity (Butler et al., 2001; Reddy et al., 2012). The fact that the physical properties of the bilayer are also a function of tension makes numerical analysis a challenging task.

The time dependence of stress relaxation of a cell membrane affects what is observed at different timescales. Some literature suggests that local perturbations propagate globally in milliseconds leading to nearly isotropic membrane tension (Huse, 2017; Keren et al., 2008; Pontes et al., 2017; Sens & Plastino, 2015). Cohen and Shi analyzed the time scale of tension relaxation in cell membranes (Cohen & Shi, 2020) and showed that long-range gradients of membrane tension may last for more than 10 min (Shi et al., 2018). They felt that the slow relaxation rate was due to small values of the lipid diffusion coefficient controlled by the size and density of immobile obstacles, the bilayer area expansion modulus, and the bilayer viscosity. These results are bound to vary across different cell types reminding us to be careful about presuming a homogeneous membrane tension.

References

Akinlaja, J., & Sachs, F. (1998). The breakdown of cell membranes by electrical and mechanical stress. *Biophysical Journal, 75*, 247–254.

Apodaca, G. (2002). Modulation of membrane traffic by mechanical stimuli. *American Journal of Physiology-Renal Physiology, 282*, F179–F190.

Baumgart, T., Hess, S. T., & Webb, W. W. (2003). Imaging coexisting fluid domains in biomembrane models coupling curvature and line tension. *Nature, 425*(6960), 821–824.

Brinkman, H. (1949). A calculation of the viscous force exerted by a flowing fluid on a dense swarm of particles. *Flow, Turbulence and Combustion, 1*(1), 27–34.

Bussell, S. J., Koch, D. L., & Hammer, D. A. (1995). Effect of hydrodynamic interactions on the diffusion of integral membrane proteins: Diffusion in plasma membranes. *Biophysical Journal, 68*(5), 1836–1849.

Butler, P. J., Norwich, G., Weinbaum, S., & Chien, S. (2001). Shear stress induces a time-and position-dependent increase in endothelial cell membrane fluidity. *American Journal of Physiology-Cell Physiology, 280*(4), C962–C969.

Carquin, M., D'Auria, L., Pollet, H., Bongarzone, E. R., & Tyteca, D. (2016). Recent progress on lipid lateral heterogeneity in plasma membranes: From rafts to submicrometric domains. *Progress in Lipid Research, 62*, 1–24.

Chambers, J. E., Kubánková, M.t., Huber, R. G., López-Duarte, I., Avezov, E., Bond, P. J., et al. (2018). An optical technique for mapping microviscosity dynamics in cellular organelles. *ACS Nano, 12*(5), 4398–4407.

Clark, A. G., Wartlick, O., Salbreux, G., & Paluch, E. K. (2014). Stresses at the cell surface during animal cell morphogenesis. *Current Biology, 24*, R484–R494.

Cohen, A. E., & Shi, Z. (2020). Do cell membranes flow like honey or jiggle like jello? *BioEssays, 42*(1), 1900142.

Colom, A., Derivery, E., Soleimanpour, S., Tomba, C., Dal Molin, M., Sakai, N., et al. (2018). A fluorescent membrane tension probe. *Nature Chemistry, 10*, 1118–1125.

Cox, C. D., Bae, C., Ziegler, L., Hartley, S., Nikolova-Krstevski, V., Rohde, P. R., et al. (2016). Removal of the mechanoprotective influence of the cytoskeleton reveals PIEZO1 is gated by bilayer tension. *Nature Communications, 7*, 1–13.

Cox, C. D., Bavi, N., & Martinac, B. (2019). Biophysical principles of ion-channel-mediated mechanosensory transduction. *Cell Reports, 29*, 1–12.

Dai, J., & Sheetz, M. P. (1995). Mechanical properties of neuronal growth cone membranes studied by tether formation with laser optical tweezers. *Biophysical Journal, 68*, 988–996.

Dai, J., & Sheetz, M. P. (1999). Membrane tether formation from blebbing cells. *Biophysical Journal, 77*, 3363–3370.

de la Serna, J. B., Perez-Gil, J., Simonsen, A. C., & Bagatolli, L. A. (2004). Cholesterol rules: Direct observation of the coexistence of two fluid phases in native pulmonary surfactant membranes at physiological temperatures. *Journal of Biological Chemistry*, *279*(39), 40715–40722.

Dent, M. R., López-Duarte, I., Dickson, C. J., Geoghegan, N. D., Cooper, J. M., Gould, I. R., et al. (2015). Imaging phase separation in model lipid membranes through the use of BODIPY based molecular rotors. *Physical Chemistry Chemical Physics*, *17*(28), 18393–18402.

Dimova, R., Aranda, S., Bezlyepkina, N., Nikolov, V., Riske, K. A., & Lipowsky, R. (2006). A practical guide to giant vesicles. Probing the membrane nanoregime via optical microscopy. *Journal of Physics: Condensed Matter*, *18*(28), S1151.

Diz-Muñoz, A., Fletcher, D. A., & Weiner, O. D. (2013). Use the force: Membrane tension as an organizer of cell shape and motility. *Trends in Cell Biology*, *23*, 47–53.

Edidin, M. (2003). Lipids on the frontier: A century of cell-membrane bilayers. *Nature Reviews. Molecular Cell Biology*, *4*(5), 414–418.

Engelman, D. M. (2005). *Membranes are more mosaic than fluid*. Nature Publishing Group.

Evans, E. A. (1983). Bending elastic modulus of red blood cell membrane derived from buckling instability in micropipet aspiration tests. *Biophysical Journal*, *43*, 27–30.

Fogelson, B., & Mogilner, A. (2014). Computational estimates of membrane flow and tension gradient in motile cells. *PLoS One*, *9*(1), e84524.

Franklin, B., & Brownrigg, W. (1774). XLIV. Of the stilling of waves by means of oil. Extracted from sundry letters between Benjamin Franklin, LL. DFRS William Brownrigg, MDFRS and the Reverend Mr. Farish. *Philosophical Transactions of the Royal Society of London*, *64*, 445–460.

Goñi, F. M. (2014). The basic structure and dynamics of cell membranes: An update of the Singer–Nicolson model. *Biochimica et Biophysica Acta (BBA)-Biomembranes*, *1838*(6), 1467–1476.

Gorter, E., & Grendel, F. (1925). On bimolecular layers of lipoids on the chromocytes of the blood. *The Journal of Experimental Medicine*, *41*(4), 439.

Goujon, A., Colom, A., Strakova, K., Mercier, V., Mahecic, D., Manley, S., et al. (2019). Mechanosensitive fluorescent probes to image membrane tension in mitochondria, endoplasmic reticulum, and lysosomes. *Journal of the American Chemical Society*, *141*, 3380–3384.

Groves, J. T., & Kuriyan, J. (2010). Molecular mechanisms in signal transduction at the membrane. *Nature Structural & Molecular Biology*, *17*, 659–665.

Haidekker, M. A., Ling, T., Anglo, M., Stevens, H. Y., Frangos, J. A., & Theodorakis, E. A. (2001). New fluorescent probes for the measurement of cell membrane viscosity. *Chemistry & Biology*, *8*(2), 123–131.

Haidekker, M. A., & Theodorakis, E. A. (2007). Molecular rotors—Fluorescent biosensors for viscosity and flow. *Organic & Biomolecular Chemistry*, *5*(11), 1669–1678.

Hamill, O. P., & Martinac, B. (2001). Molecular basis of mechanotransduction in living cells. *Physiological Reviews*, *81*(2), 685–740.

He, L., Tao, J., Maity, D., Si, F., Wu, Y., Wu, T., et al. (2018). Role of membrane-tension gated Ca2 + flux in cell mechanosensation. *Journal of Cell Science*, *131*, jcs208470.

Hochmuth, R. M. (2000). Micropipette aspiration of living cells. *Journal of Biomechanics*, *33*(1), 15–22.

Hochmuth, R., Mohandas, N., Mohandas, P., & Blackshear. (1973). Measurement of the elastic modulus for red cell membrane using a fluid mechanical technique. *Biophysical Journal*, *13*, 747–762.

Hochmuth, F. M., Shao, J.-Y., Dai, J., & Sheetz, M. P. (1996). Deformation and flow of membrane into tethers extracted from neuronal growth cones. *Biophysical Journal*, *70*, 358–369.

Hormeno, S., & Arias-Gonzalez, J. R. (2006). Exploring mechanochemical processes in the cell with optical tweezers. *Biology of the Cell, 98*, 679–695.

Houk, A. R., Jilkine, A., Mejean, C. O., Boltyanskiy, R., Dufresne, E. R., Angenent, S. B., et al. (2012). Membrane tension maintains cell polarity by confining signals to the leading edge during neutrophil migration. *Cell, 148*, 175–188.

Howells, I. (1974). Drag due to the motion of a Newtonian fluid through a sparse random array of small fixed rigid objects. *Journal of Fluid Mechanics, 64*(3), 449–476.

Huse, M. (2017). Mechanical forces in the immune system. *Nature Reviews. Immunology, 17*, 679.

Keren, K. (2011). Cell motility: The integrating role of the plasma membrane. *European Biophysics Journal, 40*, 1013.

Keren, K., Pincus, Z., Allen, G. M., Barnhart, E. L., Marriott, G., Mogilner, A., et al. (2008). Mechanism of shape determination in motile cells. *Nature, 453*, 475–480.

Klotzsch, E., & Schütz, G. J. (2013). A critical survey of methods to detect plasma membrane rafts. *Philosophical Transactions of the Royal Society B: Biological Sciences, 368*(1611), 20120033.

Kuimova, M. K. (2012). Mapping viscosity in cells using molecular rotors. *Physical Chemistry Chemical Physics, 14*(37), 12671–12686.

Kusumi, A., Nakada, C., Ritchie, K., Murase, K., Suzuki, K., Murakoshi, H., et al. (2005). Paradigm shift of the plasma membrane concept from the two-dimensional continuum fluid to the partitioned fluid: High-speed single-molecule tracking of membrane molecules. *Annual Review of Biophysics and Biomolecular Structure, 34*, 351–378.

Levental, I., & Veatch, S. L. (2016). The continuing mystery of lipid rafts. *Journal of Molecular Biology, 428*(24), 4749–4764.

Levitt, J. A., Kuimova, M. K., Yahioglu, G., Chung, P.-H., Suhling, K., & Phillips, D. (2009). Membrane-bound molecular rotors measure viscosity in live cells via fluorescence lifetime imaging. *The Journal of Physical Chemistry C, 113*(27), 11634–11642.

Lieber, A. D., Schweitzer, Y., Kozlov, M. M., & Keren, K. (2015). Front-to-rear membrane tension gradient in rapidly moving cells. *Biophysical Journal, 108*(7), 1599–1603.

Lieber, A. D., Yehudai-Resheff, S., Barnhart, E. L., Theriot, J. A., & Keren, K. (2013). Membrane tension in rapidly moving cells is determined by cytoskeletal forces. *Current Biology, 23*, 1409–1417.

Lindblom, G., & Orädd, G. (2009). Lipid lateral diffusion and membrane heterogeneity. *Biochimica et Biophysica Acta (BBA)-Biomembranes, 1788*(1), 234–244.

Lingwood, D., & Simons, K. (2010). Lipid rafts as a membrane-organizing principle. *Science, 327*(5961), 46–50.

Markin, V. S., & Sachs, F. (2015). Free volume in membranes: Viscosity or tension? *Open Journal of Biophysics, 05*(03), 80–83.

Masters, T. A., Pontes, B., Viasnoff, V., Li, Y., & Gauthier, N. C. (2013). Plasma membrane tension orchestrates membrane trafficking, cytoskeletal remodeling, and biochemical signaling during phagocytosis. *Proceedings of the National Academy of Sciences of the United States of America, 110*, 11875–11880.

Meng, F. (2011). Real time FRET based detection of mechanical stress in cytoskeletal and extracellular matrix proteins. *Cellular and Molecular Bioengineering, 4*, 148–159.

Meng, F., & Sachs, F. (2012). Orientation-based FRET sensor for real-time imaging of cellular forces. *Journal of Cell Science, 125*, 743–750.

Meng, F., Suchyna, T., & Sachs, F. (2008). A fluorescence energy transfer-based mechanical stress sensor for specific proteins in situ. *The FEBS Journal, 275*, 3072–3087.

Muddana, H. S., Gullapalli, R. R., Manias, E., & Butler, P. J. (2011). Atomistic simulation of lipid and DiI dynamics in membrane bilayers under tension. *Physical Chemistry Chemical Physics, 13*(4), 1368–1378.

Neuman, K. C., & Nagy, A. (2008). Single-molecule force spectroscopy: Optical tweezers, magnetic tweezers and atomic force microscopy. *Nature Methods*, *5*(6), 491–505.
Nicolson, G. L. (2014). The fluid—Mosaic model of membrane structure: Still relevant to understanding the structure, function and dynamics of biological membranes after more than 40 years. *Biochimica et Biophysica Acta (BBA)-Biomembranes*, *1838*(6), 1451–1466.
Páez-Pérez, M., López-Duarte, I., Vyšniauskas, A., Brooks, N. J., & Kuimova, M. K. (2021). Imaging non-classical mechanical responses of lipid membranes using molecular rotors. *Chemical Science*, *12*(7), 2604–2613.
Pontes, B., Ayala, Y., Fonseca, A. C. C., Romão, L. F., Amaral, R. F., Salgado, L. T., et al. (2013). Membrane elastic properties and cell function. *PLoS One*, *8*(7), e67708.
Pontes, B., Monzo, P., & Gauthier, N. C. (2017). Membrane tension: A challenging but universal physical parameter in cell biology. *Seminars in Cell & Developmental Biology*.
Raucher, D., & Sheetz, M. P. (1999). Membrane expansion increases endocytosis rate during mitosis. *Journal of Cell Biology*, *144*, 497–506.
Rawicz, W., Olbrich, K. C., McIntosh, T., Needham, D., & Evans, E. (2000). Effect of chain length and unsaturation on elasticity of lipid bilayers. *Biophysical Journal*, *79*(1), 328–339.
Rayleigh, L. (1889). Measurements of the amount of oil necessary in order to check the motions of camphor upon water. *Proceedings of the Royal Society of London*, *47*, 364–367.
Reddy, A. S., Warshaviak, D. T., & Chachisvilis, M. (2012). Effect of membrane tension on the physical properties of DOPC lipid bilayer membrane. *Biochimica et Biophysica Acta (BBA)-Biomembranes*, *1818*(9), 2271–2281.
Rideau, E., Dimova, R., Schwille, P., Wurm, F. R., & Landfester, K. (2018). Liposomes and polymersomes: A comparative review towards cell mimicking. *Chemical Society Reviews*, *47*(23), 8572–8610.
Roux, A., Koster, G., Lenz, M., Sorre, B., Manneville, J.-B., Nassoy, P., et al. (2010). Membrane curvature controls dynamin polymerization. *Proceedings of the National Academy of Sciences of the United Sates of America*, *107*(9), 4141–4146.
Sachs, F. (2010). Stretch-activated ion channels: What are they? *Physiology*, *25*, 50–56.
Saffman, P., & Delbrück, M. (1975). Brownian motion in biological membranes. *Proceedings of the National Academy of Sciences of the United Sates of America*, *72*(8), 3111–3113.
Sens, P., & Plastino, J. (2015). Membrane tension and cytoskeleton organization in cell motility. *Journal of Physics: Condensed Matter*, *27*, 273103.
Sezgin, E., Levental, I., Mayor, S., & Eggeling, C. (2017). The mystery of membrane organization: Composition, regulation and roles of lipid rafts. *Nature Reviews. Molecular Cell Biology*, *18*(6), 361–374.
Sheetz, M. P., & Dai, J. (1996). Modulation of membrane dynamics and cell motility by membrane tension. *Trends in Cell Biology*, *6*, 85–89.
Sherin, P. S., López-Duarte, I., Dent, M. R., Kubánková, M., Vyšniauskas, A., Bull, J. A., et al. (2017). Visualising the membrane viscosity of porcine eye lens cells using molecular rotors. *Chemical Science*, *8*(5), 3523–3528.
Shi, Z., Graber, Z. T., Baumgart, T., Stone, H. A., & Cohen, A. E. (2018). Cell membranes resist flow. *Cell*, *175*, 1769–1779.
Shin, W., Ge, L., Arpino, G., Villarreal, S. A., Hamid, E., Liu, H., et al. (2018). Visualization of membrane pore in live cells reveals a dynamic-pore theory governing fusion and endocytosis. *Cell*, *173*, 934–945.
Simons, K., & Vaz, W. L. (2004). Model systems, lipid rafts, and cell membranes. *Annual Review of Biophysics and Biomolecular Structure*, *33*, 269–295.
Singer, S. J., & Nicolson, G. L. (1972). The fluid mosaic model of the structure of cell membranes. *Science*, *175*(4023), 720–731.
Steinkühler, J., Sezgin, E., Urbančič, I., Eggeling, C., & Dimova, R. (2019). Mechanical properties of plasma membrane vesicles correlate with lipid order, viscosity and cell density. *Communications Biology*, *2*(1), 1–8.

Steinmark, I. E., Chung, P. H., Ziolek, R. M., Cornell, B., Smith, P., Levitt, J. A., et al. (2020). Time-resolved fluorescence anisotropy of a molecular rotor resolves microscopic viscosity parameters in complex environments. *Small, 16*(22), 1907139.

Sun, M., Graham, J. S., Hegedüs, B., Marga, F., Zhang, Y., Forgacs, G., et al. (2005). Multiple membrane tethers probed by atomic force microscopy. *Biophysical Journal, 89*, 4320–4329.

Tanford, C. (1980). *The hydrophobic effect: Formation of micelles and biological membrane* (2d ed.). J. Wiley.

Tanford, C. (2004). *Ben Franklin stilled the waves: An informal history of pouring oil on water with reflections on the ups and downs of scientific life in general.* OUP Oxford.

Tian, A., & Baumgart, T. (2009). Sorting of lipids and proteins in membrane curvature gradients. *Biophysical Journal, 96*, 2676–2688.

Van Meer, G., Voelker, D. R., & Feigenson, G. W. (2008). Membrane lipids: Where they are and how they behave. *Nature Reviews. Molecular Cell Biology, 9*(2), 112–124.

Wang, D.-N., Stieglitz, H., Marden, J., & Tamm, L. K. (2013). Benjamin Franklin, Philadelphia's favorite son, was a membrane biophysicist. *Biophysical Journal, 104*(2), 287–291.

Wen, P. J., Grenklo, S., Arpino, G., Tan, X., Liao, H.-S., Heureaux, J., et al. (2016). Actin dynamics provides membrane tension to merge fusing vesicles into the plasma membrane. *Nature Communications*, 7, 1–14.

Wu, Y., Štefl, M., Olzyńska, A., Hof, M., Yahioglu, G., Yip, P., et al. (2013). Molecular rheometry: Direct determination of viscosity in L o and L d lipid phases via fluorescence lifetime imaging. *Physical Chemistry Chemical Physics, 15*(36), 14986–14993.

Yang, Z., He, Y., Lee, J.-H., Park, N., Suh, M., Chae, W.-S., et al. (2013). A self-calibrating bipartite viscosity sensor for mitochondria. *Journal of the American Chemical Society, 135*(24), 9181–9185.

Zalba, S., & Ten Hagen, T. L. (2017). Cell membrane modulation as adjuvant in cancer therapy. *Cancer Treatment Reviews, 52*, 48–57.

Zhang, Y.-L., Frangos, J. A., & Chachisvilis, M. (2006). Laurdan fluorescence senses mechanical strain in the lipid bilayer membrane. *Biochemical and Biophysical Research Communications, 347*(3), 838–841.

Methods for assessment of membrane protrusion dynamics

Jordan Fauser[a], Martin Brennan[a], Denis Tsygankov[b], and Andrei V. Karginov[a,*]

[a]University of Illinois at Chicago, Department of Cellular and Molecular Pharmacology and Regenerative Medicine, Chicago, IL, United States
[b]Georgia Institute of Technology, Wallace H. Coulter Department of Biomedical Engineering, Atlanta, GA, United States
*Corresponding author: e-mail address: karginov@uic.edu

Contents

Abstract

Membrane protrusions are a critical facet of cell function. Mediating fundamental processes such as cell migration, cell-cell interactions, phagocytosis, as well as assessment and remodeling of the cell environment. Different protrusion types and morphologies can promote different cellular functions and occur downstream of distinct signaling pathways. As such, techniques to quantify and understand the inner workings of protrusion dynamics are critical for a comprehensive understanding of cell biology. In this chapter, we describe approaches to analyze cellular protrusions and correlate physical changes in cell morphology with biochemical signaling processes. We address methods to quantify and characterize protrusion types and velocity, mathematical approaches to

Current Topics in Membranes, Volume 88
ISSN 1063-5823
https://doi.org/10.1016/bs.ctm.2021.09.005

predictive models of cytoskeletal changes, and implementation of protein engineering and biosensor design to dissect cell signaling driving protrusive activity. Combining these approaches allows cell biologists to develop a comprehensive understanding of the dynamics of membrane protrusions.

1. Introduction

Fundamental physiological processes such as cell migration or differentiation are characterized by profound changes in cell morphology. These changes are driven by highly dynamic reorganization of the plasma membrane in response to physiological cues. In human pathologies, these processes are often dysregulated. Thus, it is important to determine how membrane dynamics are regulated during physiological and pathological cell behavior. The dynamic reorganization of the plasma membrane occurs through formation of different types of membrane protrusions in response to extracellular signals. The type and the dynamics of membrane protrusions can characterize physiological responses to specific stimuli or indicate a pathological change in internal cell signaling. Protrusions allow the cell to probe the surrounding environment and sense mechanical properties, such as matrix stiffness, or biochemical composition (Jacquemet, Hamidi, & Ivaska, 2015; Krause & Gautreau, 2014; Mattila & Lappalainen, 2008). This can indicate to the cell a direction for migration and invasion or induce changes in morphology and cell function. Thus, studying the dynamic changes in membrane protrusions provides vital information on cell signaling and cellular response to physiological and pathological stimuli.

Researchers have categorized membrane protrusion types that drive distinct changes in cell membrane morphologies. Each protrusion type is involved in specific cellular processes due to specific cytoskeletal changes and signaling. These protrusion types are characterized based on their cytoskeletal structure and components and their physical characteristics (Fig. 1). Filopodia are thin, actin-rich protrusions involved in cell migration, cell-cell adhesion, and environment sensing (Fig. 1A) (Krause & Gautreau, 2014; Mattila & Lappalainen, 2008). Filopodia often precede the large, flat lamellipodium in cell migration. Changes in filopodia density and dynamics have also been correlated with metastatic behavior in cancer (Jacquemet et al., 2013, 2015, 2017). Unlike filopodia, which contain tight bundles of filamentous actin (Mattila & Lappalainen, 2008), the sheet-like structure of lamellipodia is supported by branched networks of actin (Fig. 1B) (Schaub, Meister, & Verkhovsky, 2007). There are several other highly

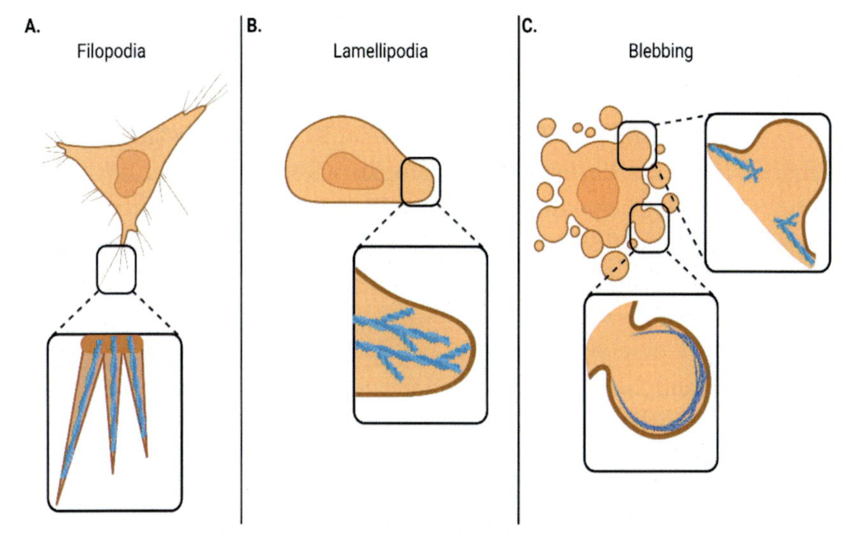

Fig. 1 Cartoon representation of Actin cytoskeleton in different cell protrusions. Created with biorender.com. (A) Representation of filamentous actin (cyan) in filopodia. (B) Representation of branched actin network (cyan) in lamellipodia. (C) Representation of disrupted actin (cyan) in an expanding bleb (top right) and actin assembly in the bleb cortex (bottom).

specific cell membrane protrusions, such as invadopodia and podosomes involved in cell invasion, migration, and metastasis (Jacquemet et al., 2015; Murphy & Courtneidge, 2011). These structures are characterized by a dense actin core surrounded by scaffolding and adhesion proteins. While both invadopodia and podosomes contain similar features, they exhibit distinct dynamics and actin organization (Murphy & Courtneidge, 2011). Another important dynamic protrusion type is cell blebbing (Fig. 1C). Blebs can form both during cell migration and during apoptosis in a small G-protein and Rho-associated Kinase-dependent manner (Aoki et al., 2016, 2020). Investigation into the biophysical and biochemical differences in blebs during cell migration versus during cell death is necessary for distinguishing these two important pathways.

Changes in cell morphology and the dynamics of membrane protrusions can be investigated from several different perspectives. First, protrusion dynamics can be assessed from the physical changes in cell shape and morphology. This approach allows for categorization of the types of protrusions as well as quantification of the cell edge dynamics. The different types of protrusions are indicative of different types of signaling and morphological changes within cells. Second, protrusions can be modeled mathematically to predict the effect of dynamic changes on cell behavior (Dickinson, 2009; Lee, 2018; Rossier et al., 2010; Ryan et al., 2017; Shemesh, Alexander, & Kozlov, 2012; Vavylonis, Yang, & O'Shaughnessy, 2005). As the first approach can provide information on the velocity of the cell edge, the mathematical model can correlate morphologies with biochemical changes in the cytoskeleton. The predictive nature of these models can further enhance our understanding of cell behavior in systems that are challenging to study experimentally. Third, these changes can be evaluated from a biochemical perspective; investigating the signaling and cytoskeletal rearrangements involved in the different protrusion behaviors. Biochemical approaches to better understand the signaling behind the changes in protrusions are critical for revealing which pathways and stimuli result in each morphology. This is of particular importance in dissecting pathological signaling to find new ways to prevent morphological changes accompanying a disease.

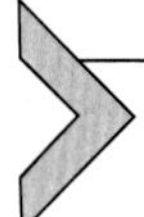

2. Approaches to assess physical changes in cell morphodynamics

Analysis of membrane protrusion changes often involves analyzing the physical changes in the cell membrane structure. These physical attributes include changes in cell area, filopodia number or size, increased protrusive or retractive activity, cell edge velocity, or other morphodynamic changes. Often these dynamic changes are tracked in real-time using live cell imaging to evaluate the physical changes in cell structure. Through the implementation of fluorescent markers or phase-contrast imaging, cell biologists can distinguish these changes in an unbiased and quantifiable manner. Several software suites have been developed to aid in these quantifications and provide methods for analysis of cell protrusion dynamics.

2.1 Cell spreading

Changes in cell spreading can provide vital information about initial cell adhesion, interaction with specific substrates, and the mechanosensitivity of the cell to matrix rigidity. Cells can spread isotropically, in a uniform manner resulting in evenly spread round cells, or anisotropically resulting in elongated or asymmetrically spread cells. The type of spreading can be influenced by matrix rigidity, its composition and distribution of adhesive molecules, and cell type (Zemel et al., 2010). Distinguishing the subtle differences in cell spreading, depending on environment, requires precise identification of the cell boundary and reliable, replicable analysis of the changes in the cell perimeter and area over time.

Different approaches have been developed to identify and track changes in the cell boundary over time. The MovThresh segmentation module from the CellGeo platform produced by Tsygankov et al. determines the cell boundary for each frame of a time-lapse fluorescent imaging experiment (Fig. 2A) (Tsygankov et al., 2014a). The interactive Graphical User Interface (GUI) for this platform allows the user to control the thresholding used to establish the cell boundary, limiting artifacts in the definition of cell boundary. This yields a cell boundary representation with well resolved fine details, such as filopodia, which can be applied throughout the CellGeo platform to analyze specific protrusion morphologies and changes in cell geometry. The processed cell images can then be analyzed by the BisectoGraph cell geometry module of CellGeo, which generates a tree graph of the cell shape (Fig. 2B–D). Using a tree graph to represent the cell shape allows for detailed description of the more subtle changes in cell edge morphology. This approach allows researchers to define and quantitatively analyze fine

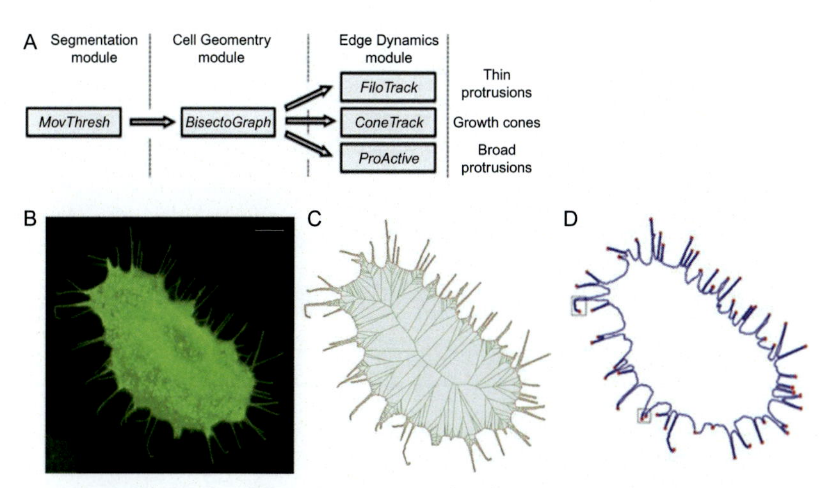

Fig. 2 (A) Workflow of CellGeo package to define, detect and track thin protrusions, broad protrusions, and growth cones. (B) Fluorescence microscopy image of a cell exhibiting complex protrusions (C) Tree graph representation of the cell in (B), showing accurate mapping of the boundary. (D) Boundary profile of the distance measurements from (C) with the local maxim, protrusion tips, in red. *Reproduction of figure from Tsygankov, D., et al. (2014). CellGeo: A computational platform for the analysis of shape changes in cells with complex geometries.* Journal of Cell Biology, 204*(3), 443–460.*

details in the shape of a cell boundary. Furthermore, these approaches can be coupled with newly developed optogenetic tools capable of inducing transient changes in cell area to quantify the effect of specific cell signaling on cell area (Shaaya et al., 2020).

Other approaches to segmentation of cell images allow for the use of phase-contrast microscopy to evaluate changes in cell morphology in the absence of fluorescent markers. Ersoy et al. demonstrated the use of ridge-based cell detection to overcome some of the limitations of the more commonly used edge-based detection methods (Ersoy et al., 2008). When using phase-contrast microscopy, cells often exhibit a halo which can result in imprecise detection of the cell edge for analysis. Edge detection of the cell boundary is also very sensitive to background noise or lower resolution images. Ersoy et al. exploit the presence of the halo in phase-contrast microscopy to develop their cell edge detection. They initially employ ridge-based detection, resulting in a cell mask larger than the actual cell, encompassing the phase halo (Fig. 3). They then refine this detection using an evolving

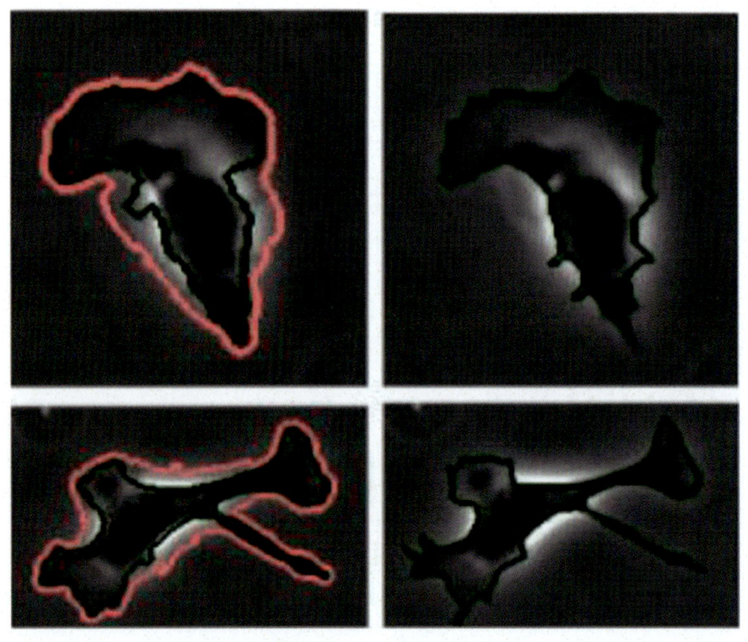

Fig. 3 Images from Ersoy et al. (2008) demonstrating automatic detection of cell edge in phase contrast images. Pink is ridge based detection and green is final segmentation. The left two images are using their algorithm to detect the cell edge while the right two images are manual detection in green.

active contour that passes through the phase halo to create an accurate depiction of the cell edge. This approach avoids boundary leak or inclusion of the phase halo, both of which can cause artifact cell area analysis and skew the statistics.

2.2 Protrusive and retractive activity

While cell area can provide initial information on the spreading behavior of the cell, more detailed analysis of protrusion types, cell edge velocity, persistence of protrusions, and changes in retraction of the cell can further our understanding of cell morphological changes. Fortunately, the methods used to detect the cell boundary for cell area analysis can also be employed in the analysis of these protrusive activities.

2.2.1 Lamellipodia persistence

Utilizing the cell boundary established using the MovThresh module of the CellGeo platform, changes in the shape of the cell body can be tracked over time (Tsygankov et al., 2014a). The ProActive module separates lamellipodium from filopodia by focusing on the cell body. Thresholding of the velocity of the cell edge can also be used to exclude small changes in the cell area to only indicate large protrusive changes. The MATLAB GUI used here again allows the user to easily control the parameters used to quantify the dynamic changes in cell protrusions in a reproducible manner. Furthermore, the user can easily adjust the time-lapse used to analyze protrusions. These protrusive changes can be normalized relative to the cell perimeter or cell area to control for differences in cell sizes and allow for more direct comparisons of the overall changes in cell protrusive activity. This strategy has been employed to demonstrate the dynamic effects of Src kinase signaling using optogenetic or chemogenetic tools to interrogate cell protrusive activity (Klomp et al., 2019; Shaaya et al., 2020). The CellGeo approach to analyzing protrusions and persistence of lamellipodia over the course of a live imaging experiment allows for dissection of large protrusions as well as highly dynamic small protrusions that occur during cell edge ruffling.

Lamellipodia can exhibit more complex boundary displacements such as ruffles along the edge of the protrusions. These deformations in the boundary are indicative of different cytoskeletal changes and cannot be efficiently detected by only looking at cell area changes. Using a segmentation approach to dissect local boundary perturbations, Machacek and Danuser developed a method to track protrusions and retractions within distinct segments along the cell boundary (Machacek & Danuser, 2006). Utilizing this segmentation approach in combination with multiplexed Rho GTPase biosensors, Machacek et al. determined the coordinated Rho GTPase signaling involved in protrusive activity (Machacek et al., 2009; Machacek & Danuser, 2006).

2.2.2 Kymograph analysis of protrusion dynamics

Another approach to analyzing membrane protrusion dynamics is to use a kymograph analysis to evaluate cell edge velocity within protrusions. This approach has been widely implemented as a method to analyze protrusion velocity for several decades (Bear et al., 2002; Zimmerman et al., 2017). Image analysis software such as ImageJ make this type of analysis possible. Kymographs allow for generation of a single, 2D image, containing information on the position of the cell edge over time (Fig. 4) By drawing a single line region of interest intersecting the cell edge, researchers can determine how the position of the cell edge changes along that line over time (Fig. 4A). Due to the positional and time information provided by this method, velocity of the cell edge at different time points can be determined (Fig. 4B). This approach can be used to evaluate the effect of specific signaling complexes on protrusion velocity. For example, Hartman et al. were able to quantitatively demonstrate the importance of Shp2 tyrosine phosphatase and Focal Adhesion Kinase signaling in growth factor-induced lamellipodia persistence, important for cell migration (Hartman, Schaller, & Agazie, 2013). This approach can also be combined with temporal

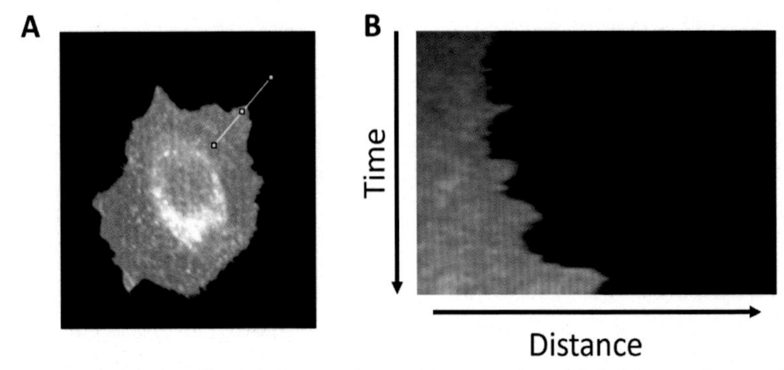

Fig. 4 HeLa cell with cell mask deep red membrane marker. (A) Cell image from a single frame of a time lapsed movie with a slice intersecting the cell edge (yellow). (B) 2D re-sliced image showing the change in the cell edge position along the yellow line from (A) over the course of the time-lapsed movie.

control of signaling utilizing optogenetic regulation of a protein of interest, such as Cdc42, to evaluate temporal regulation of morphological changes (Zimmerman et al., 2017).

However, manual analysis of kymographs can prove challenging and cause difficulties with reproduction of the analysis. This issue was addressed by Barry et al. through the creation of an open-source and freely available ImageJ plugin for protrusion dynamics analysis, ADAPT (Automated Detection and Analysis of Protrusions) (Barry et al., 2015). ADAPT further advances cell edge velocity detection by allowing for velocity determination for the entire cell perimeter, not just a slice of the cell edge as is typically done with kymographic analysis. This approach offers an additional benefit over several other cell edge velocity measurement plug-ins, ADAPT can correlate fluorescent intensity, thus protein recruitment, with cell edge velocity. ADAPT was adeptly implemented in combination with GTPase biosensors to demonstrate the spatial and temporal coordination of Rho, Rac, and Cdc42 to regulate membrane protrusions (Martin et al., 2016). Similarly, Zhurikhina et al. developed an addition to the CellGeo suite called EdgeProps which also allows for correlation of fluorescent intensity and cell edge velocity (Zhurikhina et al., 2018). As EdgeProps allows for determination of velocity across the entire cell boundary, this module can be implemented to show local changes in cell edge velocity over time. The power of this approach was demonstrated with analysis of optogenetically regulated changes in cell edge velocity (Shaaya et al., 2020). Shaaya et al. utilized a light-regulated Src kinase (LightR-Src) to stimulate Src activity within a single section of a cell membrane (Fig. 5). Analysis of this local stimulation revealed pulsatile increases in cell edge velocity and specific signaling involved in these protrusions.

2.3 Filopodia analysis

Filopodia act as sensors for the cell environment, investigating adjacent cells as well as the soluble cues and extracellular matrix composition and mechanical properties. These finger-like protrusions are involved in both physiological and pathological cell migration and invasion in a 3D environment (Heckman & Plummer, 2013; Jacquemet et al., 2015). The importance of filopodia in processes such as wound healing, neurite extension, and embryonic development, as well as pathological functions such as solid tumor metastasis make these a critical protrusion to understand (Jacquemet et al., 2015). Unlike large protrusive events such as lamellipodia, filopodia

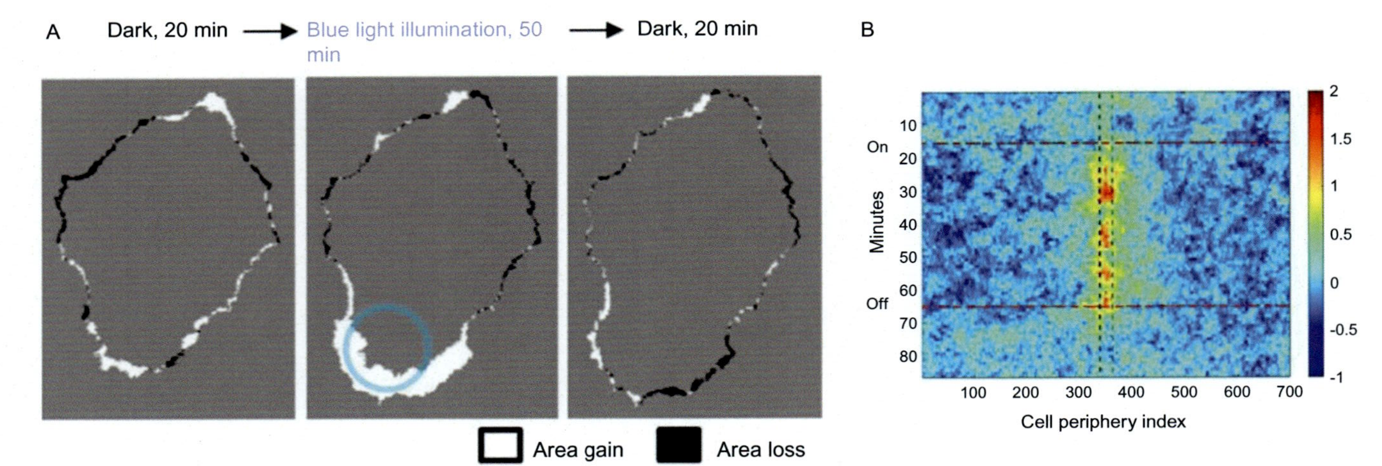

Fig. 5 Local activation of an engineered optogenetic Src (Shaaya et al., 2020) (A) HeLa cells expressing light-regulated Src (LightR-Src) are stimulated with a laser at sub-cellular resolution resulting in local protrusive activity. Representative cell mask from CellGeo showing area gained (white) and area lost (black) over the given time intervals. Blue circle indicates region of Src activation. (B) Kymograph of cell velocity for 700 points along the cell boundary (y axis) over the duration of imaging (x axis). Red colors indicate positive increase in cell edge velocity, protrusions, while blue colors indicate negative changes in cell edge velocity, retraction. Teal color indicates no change in velocity.

are small needle-like protrusions requiring specific approaches to analyze. Many of the cell morphology analysis suites contain methods to quantify filopodia using cell segmentation and masking described above. There have also been software suites developed specifically for the analysis of filopodia dynamics.

FiloTrack is an extension of the CellGeo suite that allows for quantification of filopodia dynamics. This method allows researchers to quantify length, duration, and number of filopodia over time as well as protrusion and retraction rates (Tsygankov et al., 2014a). The FiloTrack approach enables the identification and tracking of individual filopodia through detection of both the filopodia tip and the core of the filopodia. This ensures that all filopodia remain counted even if they are very dynamic. This approach was applied to quantify the differences in cellular protrusion types following stimulation of distinct pathways mediated by Src kinase; revealing that Src interaction with p130Cas results in filopodia formation while Src interaction with FAK causes general cell spreading (Karginov et al., 2014). This method also allows researchers to distinguish filopodia protrusions from morphologically similar cell retraction fibers that are formed when cells contract but retain some points of attachment to extracellular matrix in the retracted area (Fig. 6).

Another method for filopodia quantification correlates fluorescent intensity with filopodia dynamics. Filopodyan was created to evaluate fluorescent intensity at the tip and base of individual filopodia concurrently with filopodia morphodynamics (Urbančič et al., 2017). This approach can be used to quantify dynamics in open-source, freely available software such as Fiji (ImageJ) or R. Through their correlation of fluorescent intensity of ENA/VASP proteins with changes in filopodia extension or retraction, they were able to identify morphologically similar but biochemically different sub-populations of filopodia (Urbančič et al., 2017). This is important in creating a comprehensive understanding of the driving forces of filopodia formation.

Several other ImageJ plugins were created for the purpose of quantifying filopodia. Both FiloQuant and ADAPT offer facile methods to quantify filopodia in live-cell imaging. FiloQuant takes advantage of the skeletonize function in ImageJ to aid in the quantification of filopodia density and length (Jacquemet et al., 2017). Skeletonization of the filopodia uses a thinning algorithm to identify the center-line, or skeleton, of the filopodia to then use for quantification of their number and length. By skeletonizing the cell edge as well, the noise from the cell body can be subtracted leaving only the

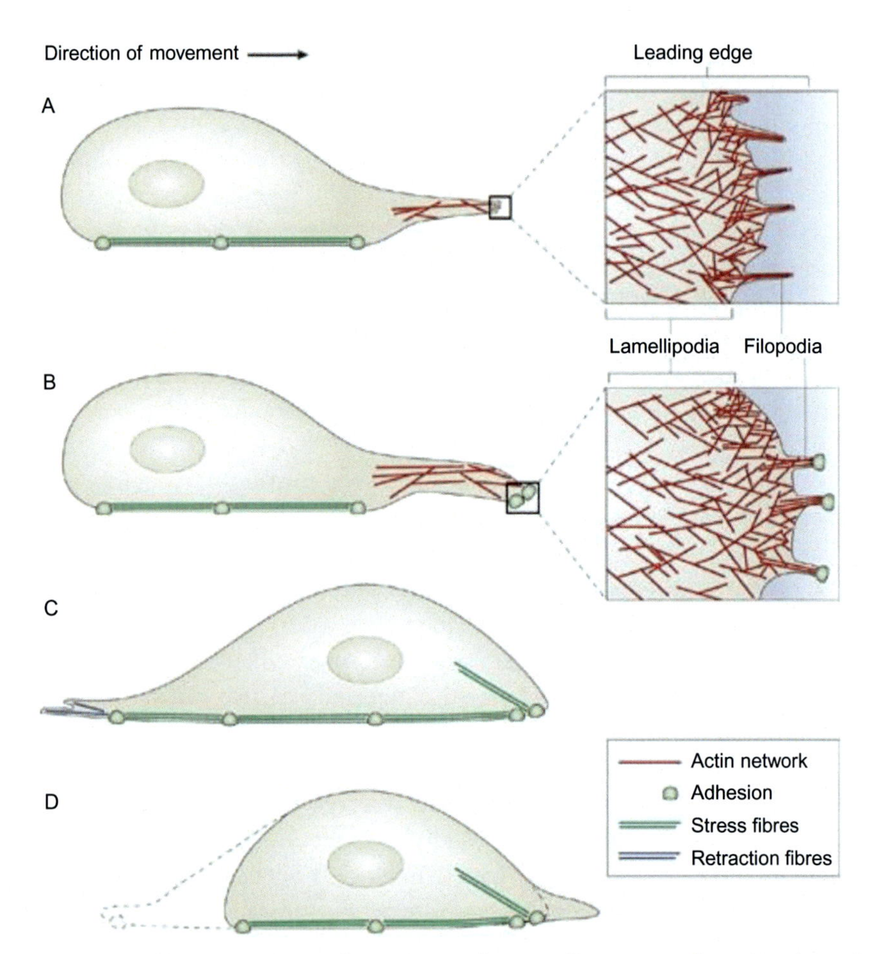

Fig. 6 Cartoon representation of protrusions during cell migration from Mattila and Lappalainen (2008). (A) Membrane protrusion supported by the actin cytoskeleton. Callout shows filopodia on the edge of the protruding lamella. (B) Filopodia on the leading-edge form adhesions to the matrix to anchor the cell for migration. (C) As the cell is pulled forward retraction fibers form at the trailing edge of the cell. (D) Adhesions at the rear of the cell are disassembled as the cell migrates forward.

skeletons of the filopodia at the cell edge to be quantified (Fig. 7). This skeletonized cell edge can also be used to determine number of filopodia relative to the perimeter of the cell. Furthermore, this approach can be used to quantify filopodia extension and retraction through comparison of filopodia length from frame to frame. As this method relies on skeletonization of the individual filopodia, high density filopodia can overlap and result in incorrect assessment of filopodia length. However, this approach provides

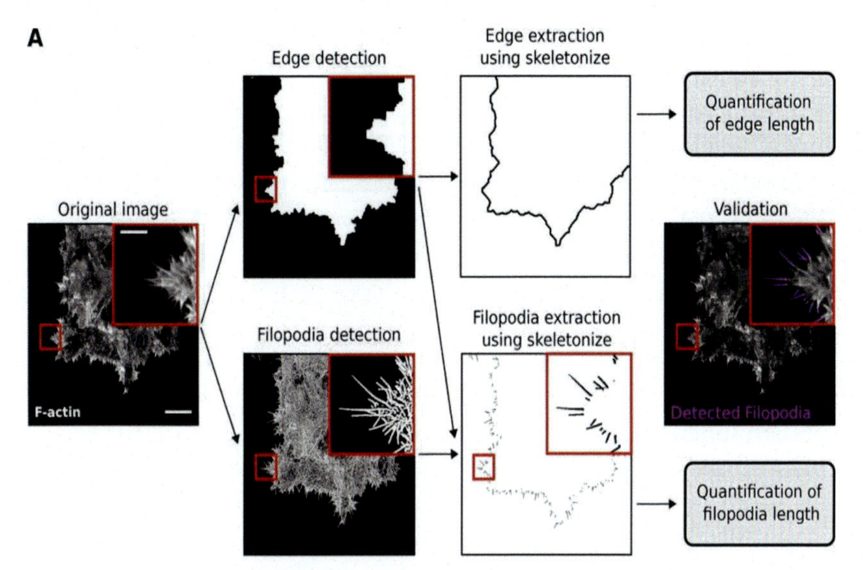

Fig. 7 Workflow for FiloQuant analysis of filopodia length and density. Original image of MCF10A cells invading through a collagen gel input in ImageJ for analysis. The image undergoes two processing steps. The first (top) defines the edge of the cell and erases the filopodia. Then using the Seletonize3D plugin extracts the edge of the cell. The second processing step uses a contrast enhanced image to detect faint filopodia. The cell edge is subtracted from the detected filopodia to isolate filopodia along the cell edge. The length and number of filopodia are automatically calculated using the Skeletonize3D and AnalyzeSkeleton plugins in ImageJ. Filopodia density can be determined from the ratio of the number of filopodia to the edge length. *Figure reproduced from Jacquemet, G., et al. (2017). FiloQuant reveals increased filopodia density during breast cancer progression.* Journal of Cell Biology, 216*(10), 3387–3403.*

a simple and quick analysis of filopodia length with comparable accuracy to manually quantified filopodia. The authors also demonstrated the robustness of this approach through analysis of in vivo filopodia formation both in angiogenesis and tumor spheroids (Jacquemet et al., 2017). Meanwhile, ADAPT uses a different approach to identify and track filopodia. The binary cell image has the periphery eroded, resulting in a binary image of the cell core. This cell core is then subtracted from the original cell image leaving only the filopodia on the periphery of the cell (Barry et al., 2015). This approach allows for comparison of filopodia from one cell type to another. Filopodia analysis can also be combined with fluorescent intensity correlation already established in the ADAPT plugin.

Filopodia identification can be challenging to accurately determine in cases where filopodia density is great and filopodia intersect one another.

Furthermore, some filopodia can be highly dynamic and thus challenging to track from frame to frame automatically. All methods require well-resolved signal to accurately quantify filopodia, thus the use of weakly labeled filopodia markers should be avoided. The methods used to skeletonize or identify the tip and base of each filopodium aim to address these issues as best as possible, though the challenges of quantifying filopodia should be kept in mind when deciding which approach is best suited to the needs of individual imaging experiments.

2.4 Analysis of blebbing dynamics

Blebs are spherical membrane protrusions that occur as the plasma membrane separates from the filamentous actin cytoskeleton. These blebs can drive cell migration and serve as an important indicator of apoptosis. Analysis of the morphologies and velocity of extension and retraction in blebs can aid in distinguishing these different cellular fate outputs. If these morphological changes can be correlated with different protein recruitment or cellular signaling, researchers can better understand the interplay between biophysical and biochemical characteristics of cell blebbing.

Aoki et al. have used a kymograph approach to evaluate the velocity of blebs in both migration and apoptosis. Using a membrane localizing fluorescent molecule they were able to quantify the protrusion and retraction of blebs as previously described in the section on kymograph analysis of membrane protrusions. Through correlation of localization of specific proteins, RhoA or Rnd3, and bleb velocity and size, they were able to identify important signaling in bleb formation and distinguish between bleb morphologies in migration versus apoptosis at early and late stages (Aoki et al., 2016, 2020).

Using the ADAPT plugin again allows for automated identification and tracking of bleb velocity. Here the user can establish thresholds for membrane curvature to improve the accuracy of bleb detection. After identifying the curvature extreme of an individual bleb over the period of time specified by the user, the plugin will track the bleb and determine cell edge velocity (Barry et al., 2015). As with other applications of ADAPT, bleb dynamics can be correlated with fluorescent intensity of a protein of interest.

Cellular blebbing can occur across the entire dorsal surface of adherent cells, as such, utilization of microscopy techniques capable of capturing dynamic protrusions in 3D is essential. Lattice Light-Sheet or Light-Sheet Fluorescence Microscopy is particularly advantageous in 3D imaging. Due to the method of illumination, the entire imaging plane is detected simultaneously, resulting in a reduction in phototoxicity and photobleaching and faster acquisition (Fischer et al., 2011; Mimori-Kiyosue, 2021). Applications of light-sheet microscopy for imaging morphological changes have been described by several groups which highlight the combined advantage of light-sheet microscopy with biomimetic microenvironments (Albert-Smet et al., 2019; Welf et al., 2016). This approach is particularly useful when analyzing cellular blebbing, allowing for fast acquisition of dynamic cellular blebs in 3D. Dean et al. capitalized on these properties of light-sheet microscopy for rapid imaging of live cells (Dean et al., 2016). They furthered the 3D advantages of light-sheet microscopy through implementation of diagonally scanned light-sheet microscopy (DiaSLM) (Dean et al., 2016). This unique approach allowed for more comprehensive analysis of 3D bleb dynamics through volumetric analysis. Visualization and analysis of 3D bleb motion dynamics have been advanced by Manandhar et al. (2020). Their image processing pipeline can be generally applied to estimate the motion of live cells in 3D from light-sheet microscopy images. Lai et al. also utilized a new tool to detect bud emergence in *Saccharomyces cerevisiae* in 3D (Lai et al., 2018). Imaging z-stacks of yeast budding, the "Automated Fitting" function of *BudTrack* created a continuous 3-Dimensional surface rendering of the budding yeast. This allowed them to track bud emergence from any angle. Bud emergence in yeast is morphologically similar to blebbing in mammalian cells and involves similar signaling pathways. The automated *BudTrack* used in this analysis could be applied to 3D detection of cellular blebbing.

2.5 Analysis of cell morphodynamics

Cells exhibiting dynamic changes in cellular signaling can exhibit several different protrusive morphologies over the duration of an experiment. To evaluate these transient and dynamic transitions in cellular behavior, researchers needed a reproducible and automated approach to categorizing these changes on a single-cell level. The development of a MATLAB-based system to evaluate overall changes in rate of cell area change as well as polarity index, allowed for categorization of individual cell behaviors over time (Tsygankov et al., 2014b). This program, SquigglyMorph, overlays the perimeter of the cell from one time frame to the next. Then, using polarity and rate of area change parameters established by the user, the software sorts the cell behavior into one of six categories for each time frame (Fig. 8). The six categories of morphologies are: uniform spreading, polarized spreading,

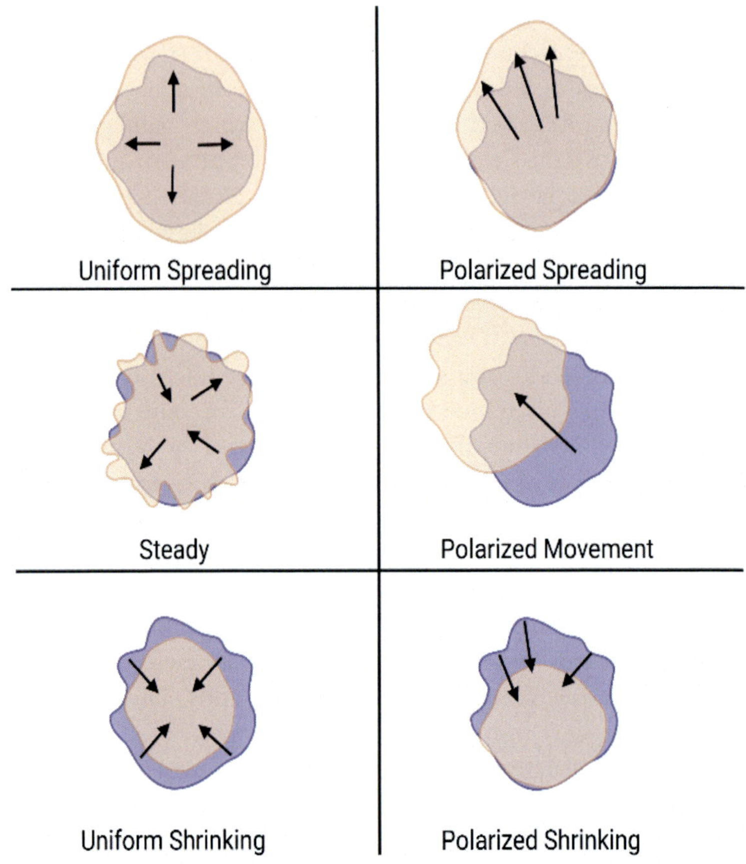

Fig. 8 Cartoon representation of the six categories of cell morphodynamic changes determined by SquigglyMorph software. Modified representation from Tsygankov et al. (2014b) created with BioRender.com. Original cell shape represented in purple and cell change represented in peach with directional arrows. Top to bottom arranged by cell area change and right to left are organized by polarity, the two parameters which can be adjusted in Squigglymorph. *Modified representation from Tsygankov, D., et al. (2014). User-friendly tools for quantifying the dynamics of cellular morphology and intracellular protein clusters.* Methods in cell biology. *Elsevier. pp. 409–427.*

polarized migration, uniform shrinking, polarized shrinking, or steady (Tsygankov et al., 2014b). The user can define the parameters based on visual analysis of the cell behavior over time, aided by the simplified representation of protrusion magnitude and polarity provided in the GUI. Once the parameters have been established, these same parameters can be applied to all cells using batch processing. While categorization of cell behavior is inherently qualitative, the ability to apply set parameters to automatically sort the cell behavior allows for reproducible analysis with minimized bias.

Application of this approach to synthetically regulated signaling proteins allows for analysis of dynamic changes in cell protrusion and morphology downstream of specific cellular signaling. This was demonstrated by Chu et al. who applied this approach to study the effects of specific activation of different Src family kinases (Chu et al., 2014). Using batch processing of cells before and after activation of specific kinases, they were able to determine temporal changes in cell spreading and migration, revealing coordinated induction of specific behaviors after activation of each kinase.

3. Approaches to theoretical modeling of cell protrusion activity

Cell protrusion morphology and dynamics are dependent on the dynamic and mechanical properties of the cytoskeleton. As a result, many researchers have found analysis of the polymerization rate and mechanical behaviors of the cytoskeleton to be a beneficial approach to understanding the morphological changes in cell protrusions. Polymerization and branching behaviors of actin have been thoroughly studied over the years, in vitro, in cellulo, and in silico (Carlsson, 2006; Mogilner & Oster, 1996; Sept, Elcock, & Mccammon, 1999; Vavylonis et al., 2005). This well-understood pathway can be used to build mathematical models of cell protrusion behaviors. Computational models allow for prediction of different cellular behaviors due to changes to environment or specific signaling. While cell protrusion and migration has been well characterized on 2D surfaces, cells very rarely encounter such environment in vivo. Thus, many of these studies lack the effects of the 3D environment on cell protrusion and migration. Utilization of 3D matrices can introduce additional challenges to live-cell imaging, making these types of analysis more complex. Combining mathematical approaches to model and predict cell behavior with biological experiments creates a more comprehensive picture of the complex dynamics of cell signaling processes, as well as providing a method to understand cell protrusions and migration in a 3D space. Here we will discuss some of the mathematical models used to predict and evaluate cellular protrusions as well as some of the complementary biophysical experiments.

3.1 Models of actin polymerization

To design these mathematical models, computational scientists must first decide the best model for actin polymerization to fit their system.

There are a variety of well described models of actin polymerization which have been covered in depth in other reviews (Dickinson, 2009). Here we will highlight some of the key components of several of these models and the ways they have been implemented to evaluate cell protrusions.

Two models often used to describe the extension of filamentous actin are the Brownian Ratchet and tethered ratchet models (Fig. 9). The Brownian ratchet model determines the force applied by the polymerization of filamentous actin that is not attached to the membrane surface (Dickinson, 2009; Ryan et al., 2017). This model implements a force on the cell edge determined by the elongation of the actin filament. The actin filament pushes against the cell edge, due to the stochastic fluctuations of the particles, gaps will appear between the cell edge and the filament. If the gap is sufficiently large for an actin monomer to attach to the filament, then the filament will elongate resulting in a perturbation of the membrane. The tethered ratchet model is a variation upon the Brownian ratchet model in

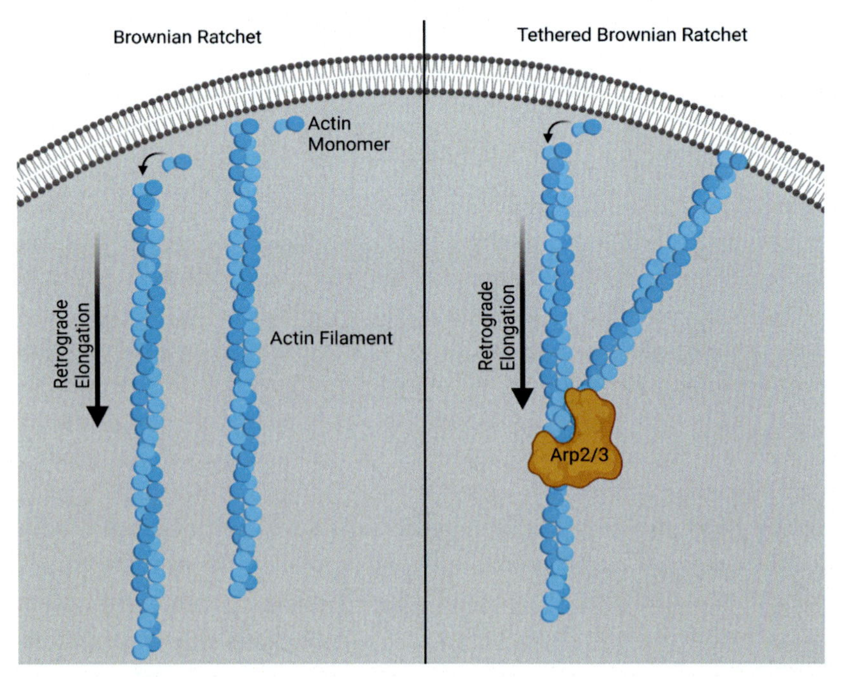

Fig. 9 Representation of Brownian ratchet and tethered Brownian ratchet models of actin polymerization. Brownian ratchet model relies on the stochastic motion of the particles creating a sufficient gap between the actin filament and the cell edge to allow for elongation of the filament. Tethered Brownian ratchet theory adds an additional branch of the actin filament which is anchored to the membrane and exerts a constant force on the cell edge.

which the unattached elongating filaments push upon the cell edge through a particle propulsion mechanism, however, it also accounts for a second population of filaments that are tethered to the cell edge and allow for transient binding of actin-associated proteins such as Arp2/3 or VASP and actin branching. These secondary filaments are not elongating, but rather acting as a constant spring force on the cell edge allowing for propulsion driven by the filaments behaving like the Brownian ratchet (Dickinson, 2009; Enculescu, Danuser, & Falcke, 2010).

Another model, the stick-slip model, accounts for the force of adhesion molecules acting on the elongating actin filaments. As actin filaments exert force on the cell edge through elongation, there is a retrograde flow of actin away from the cell edge which is influenced by actomyosin contraction and the tension force of the membrane itself (Ennomani et al., 2016; Rossier et al., 2010). Furthermore, there is a friction-like force exerted by adhesion molecules interacting with the actin filament. The stick-slip model aims to account for these additional forces on the dynamics of membrane protrusions while providing the added advantage of mechanosensitivity in the model (Sens, 2020). In this model, slow retrograde actin motion correlates with protrusive activities (sticking) while fast retrograde motion (slipping) correlates with retraction. This model can be used in combination with other models to enhance the understanding of the role of adhesions in protrusion dynamics (Shemesh et al., 2012).

Treatment of the actin network as a viscoelastic gel is another commonly used approach to demonstrate the mechanical properties of actin networks and their influence on cell edge dynamics (Heck et al., 2020; Shemesh et al., 2012). The viscoelastic properties of cells are tightly controlled by the biophysical and mechanical structure of the actin cytoskeleton, these biophysical properties can be important in the morphodynamic behaviors exhibited by cells. Tabatabaei et al. revealed the different viscoelastic properties of invasive versus non-invasive breast cancer cells, demonstrating the physiological importance of improved ways to model and evaluate these cellular properties (Tabatabaei, Tafazzoli-Shadpour, & Khani, 2021). Furthermore, application of the viscoelastic model improves models of lamellipodium dynamics and thus provides enhanced methods to model cell migration (Shemesh et al., 2012). All of these models benefit from established measured biochemical and biophysical properties of cells, however, the true power of mathematical modeling is the iterative application of experimental and theoretical approaches to develop a highly comprehensive model of dynamic cell behavior (Lee, 2018; Tabatabaei et al., 2021).

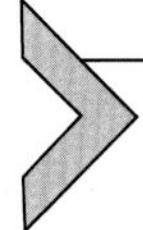

4. Biochemical assessment of membrane protrusion dynamics

Cells regulate morphological and structural changes through intricate and complex signaling pathways. As scientists develop increasingly elegant approaches to probe cell signaling dynamics in real-time, these signaling pathways have been illuminated to better understand the way the cell interacts with and moves within its environment. Several approaches can provide highly specific information on the biochemistry of cellular protrusions. The first of these techniques involves the use of biosensors for analysis of the dynamic changes in biomolecules during protrusive events within the cell. The second approach takes advantage of the rapidly developing field of optogenetics. Genetically encoding a light-sensitive protein that can affect specific cell signaling pathways allows researchers tight spatiotemporal control over the biochemical pathways involved in protrusions. This leads to a better understanding of the way these biomolecules regulate various aspects of protrusive behavior.

4.1 Biosensors for protrusion signaling and cytoskeletal changes

Biosensors can provide detailed spatial and temporal information about cell signaling dynamics and the correlation between signaling and cytoskeletal rearrangements and cell morphology. The diversity of biosensors allows researchers to probe a variety of biomolecules involved in protrusive activity; from upstream Rho GTPase signaling, the cytoskeletal components themselves, or the membrane components and physical properties.

4.1.1 Rho GTPases

While a multitude of proteins is involved in the regulation of protrusions and cell migration, a few key components of initiating these changes have come to light. Small GTPases, such as RhoA, Rac1, and Cdc42 are of particular interest in cellular protrusion dynamics. Rho GTPases have been established to play an important role in protrusions and polarized migration (Lawson & Ridley, 2018; Ridley, 2015). Furthermore, the activation dynamics and localization of these proteins within the cell create critical signaling gradients for the initiation of polarized migration. Researchers often implement Förster Resonance Energy Transfer (FRET) based biosensors to indicate the activity levels of a protein of interest. The design and characterization

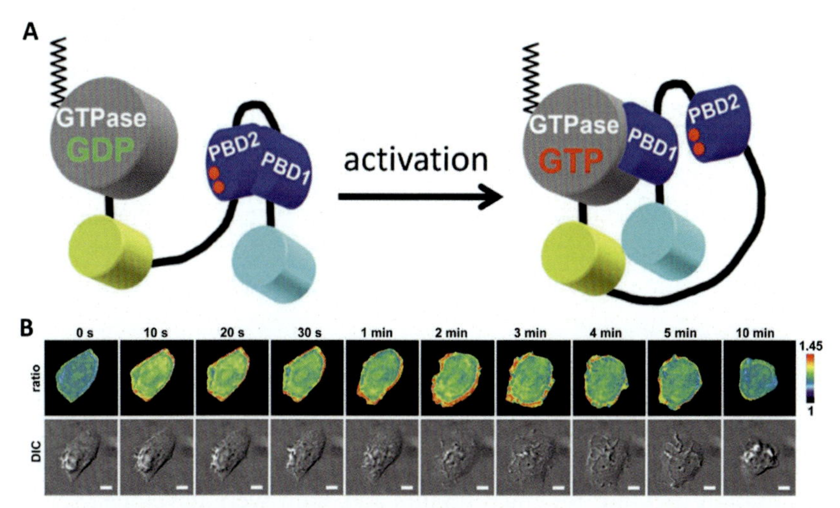

Fig. 10 FRET-based GTPase sensors. (A) A FRET pair and a binding domain (BD) is fused to a GTPase to enable visualization of the GTPase's activity. In the inactive, GDP bound state, the FRET signal is low. When the GTPase is active, in the GTP bound state, the BD binds to the GTPase bringing the cerulean and yellow florescent protein closer together thereby enhancing the FRET signal. (B) Live-cell imaging of a Rac1 FRET sensor expressed in cells that were stimulated with a chemokine. Rac1 activity correlates with the induced membrane dynamics that are observed in the differential interference contrast (DIC) images. *Figure reproduced from Donnelly, S. K., et al. (2018). Characterization of genetically encoded FRET biosensors for rho-family GTPases,* Methods in molecular biology. *New York: Springer. pp. 87–106.*

of these biosensors for Rho GTPases have been well described (Fig. 10) (Donnelly et al., 2018; Hodgson, Shen, & Hahn, 2010). Using a variety of biosensors, researchers can concurrently gain information on the activity and localization of these proteins. The utility of biosensors to small GTPases was artfully implemented by Machacek et al. to show the spatiotemporal coordination of these proteins in cell edge dynamics (Machacek et al., 2009) While RhoA, Rac1, and Cdc42 are all known to be active at the leading edge of a cell during migration, computational multiplexing of biosensors for each of these GTPases revealed spatial and temporal segregation of their activities. RhoA was found to be activated as the cell edge advanced, indicating RhoA initiates protrusive events, while Cdc42 and Rac1 exhibited a spatial and temporal delay in activation behind RhoA. The use of these biosensors allowed researchers to dissect the role of each of these proteins in protrusion initiation and stabilization. These multiplexing studies with various biosensors allow for dissection and signaling network analysis, furthering our understanding of the complex signaling involved with Rho

GTPase signaling in protrusions (Machacek et al., 2009; Marston et al., 2020). Application of these tools to pathogenic cells, such as cancers, can provide detailed information on the role of specific proteins in metastatic behaviors (Al Haddad et al., 2020). FRET-based biosensors are not limited to GTPases specifically and can also be implemented to study other effectors of cytoskeletal changes. Lorenz et al. evaluated the role of a protein involved in actin polymerization, N-WASP, in invadopodia and lamellipodia formation in cancer cells (Lorenz et al., 2004).

4.1.2 Changes in cytoskeletal components

Sustained changes in the cell edge morphology are dictated by the physical properties of cytoskeletal structure and membrane. As previously discussed, different protrusion types exhibit different cytoskeletal structures. Visualization methods for microtubules and actin were developed several decades ago, though there have been advances in their implementations (Melak, Plessner, & Grosse, 2017). Initially, actin organization could be observed in fixed cells with the use of immunofluorescent staining which was eventually supplanted with the use of phalloidin in fixed cells (Lazarides & Weber, 1974; Vandekerckhove et al., 1985). The use of these stains in fixed cells allow researchers to evaluate the static structures and correlate those structures with observed behaviors, such as the formation of podosomes with cell invasion and matrix degradation (Collins et al., 2020). However, both of these techniques were unable to capture the dynamic behaviors of actin in live cells due to the inability to efficiently cross the cell membrane. The use of fluorescently tagged actin allows for analysis of dynamics of actin assembly in live cells (Clark, Dierkes, & Paluch, 2013). Additionally, genetically encoded fluorescently tagged actin-binding domains can be utilized to evaluate actin dynamics in live cells (Lee & Knecht, 2002). Lifeact is a thoroughly validated actin-binding probe with specificity to filamentous actin that has been demonstrated both in cells and in vivo (Melak et al., 2017; Schachtner et al., 2012). These probes may provide additional benefits to evaluating actin dynamics without disrupting them. The development of a far-red, membrane permeable, fluorogenic probe (SiR-actin and SiR-tubulin) capable of binding actin or tubulin without disrupting the dynamics of these cytoskeletal components has further enhanced the ability of researchers to evaluate cytoskeletal dynamics (Lukinavičius et al., 2014). Furthermore, the use of a far-red probe allows researchers to evaluate additional proteins in the yellow-green-blue ranges, which enhances the amount of information provided by each experiment.

Focal Adhesions are an important anchor for cytoskeletal components. Focal adhesion turnover and maturation can be correlated with cell spreading, protrusions, and migration. As such, correlation of focal adhesion dynamics with morphological changes in the cell edge can provide useful information on the role of adhesions in protrusions and migration. Berginski et al. sought to automate detection and tracking of focal adhesions from TIRF imaging to then correlate focal adhesion dynamics with changes to the cell edge (Berginski et al., 2011) This focal adhesion tracking software is freely available on the Gomez Lab website.

4.1.3 Membrane tension sensors/membrane composition and curvature

The physical and chemical properties of the cell membrane can also influence the protrusive activity of a cell. The composition of the cell membrane affects the signaling occurring at the membrane as well as the physical properties of the membrane and how it responds to internal and external stimuli. To fully understand the role of the plasma membrane itself in cell migration, researchers must probe the dynamic changes in various membrane components and the mechanical properties of the membrane within certain environments.

The plasma membrane is composed of an asymmetric lipid bilayer with embedded proteins. The lipid and protein composition of the membrane influence both its fluidity and flexibility as well as the biochemical signaling occurring at the membrane. Biosensors capable of evaluating lipid composition within the membrane, particularly those capable of distinguishing between lipid concentrations in the inner and outer membrane, can provide critical information on how the cell may respond to internal and external stimuli. Lipids such as cholesterol have been established as important mediators of signaling and membrane fluidity. Recently a probe capable of quantifying available cholesterol levels on the inner and outer leaflet in living cells was reported (Buwaneka et al., 2021; Liu et al., 2017). This will allow researchers to correlate cholesterol levels with cell behaviors such as protrusion and migration following specific stimuli. Similarly, genetically encoded lipid biosensors can provide information about the dynamics of different lipid signaling as has been demonstrated with phosphoinositides and phosphatidic acid (Goulden et al., 2019; Hertel et al., 2020; Nishioka et al., 2010). All of these various lipid sensors take advantage of protein-binding domains with highly specific lipid binding in combination with a dye sensitive to the chemical environment or FRET-based changes (Buwaneka et al., 2021; Goulden et al., 2019; Hertel et al., 2020; Liu et al., 2017; Nishioka et al., 2010). While the membrane composition can influence signaling adjacent to the membrane, it can also influence the physical properties of the membrane such as curvature, fluidity, and flexibility. There are several sensors capable of probing the changes in membrane tension under specific conditions. The use of a FRET sensor with a spring-like domain allows for quantification of the stress experienced by the membrane under conditions such as shear stress (Li et al., 2018).

4.2 Optogenetic and chemogenetic tools to control membrane protrusions

Thus far, we have discussed a variety of sensors that can provide dynamic information on the activity levels of various biomolecules; however, these approaches still leave some unanswered questions on the isolated roles of each of these proteins in protrusive activity. Recent advances in optogenetic and chemogenetic approaches to regulate enzymes provide a unique set of tools to dissect cell signaling involved in protrusive activity. Cell signaling can be regulated either through allosteric control of an enzyme or through localization of a protein to a specific region of the cell.

4.2.1 Allosteric chemogenetic and optogenetic regulation of protrusive signaling

Engineered allosteric regulation of a specific enzyme allows researchers tight spatial and temporal control over the activity of a protein of interest. Utilizing this approach to study proteins involved in cell protrusion and migration, the unique role of individual proteins can be dissected. This has been demonstrated with Src family kinases, both with chemogenetic and optogenetic control of their activity to show the importance of Src kinases in cell protrusions and morphodynamic changes (Fig. 5) (Chu et al., 2014; Karginov et al., 2010; Shaaya et al., 2020). Furthermore, these approaches can be used to evaluate the effect of a specific protein complex using the Rapamycin regulated targeted activation of a protein (RapR-TAP) system (Fig. 11) (Karginov et al., 2014). This system combines allosteric rapamycin regulation of a protein of interest with rapamycin-induced dimerization of two proteins to exhibit control of both activation and localization simultaneously. Specific targeting of c-Src to FAK or p130Cas resulted in different protrusions, further unraveling the role of Src in regulation of cell morphodynamics (Karginov et al., 2014).

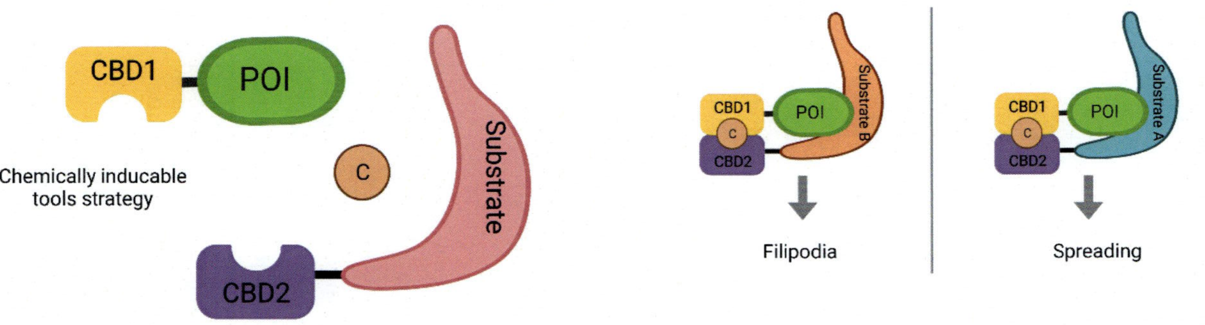

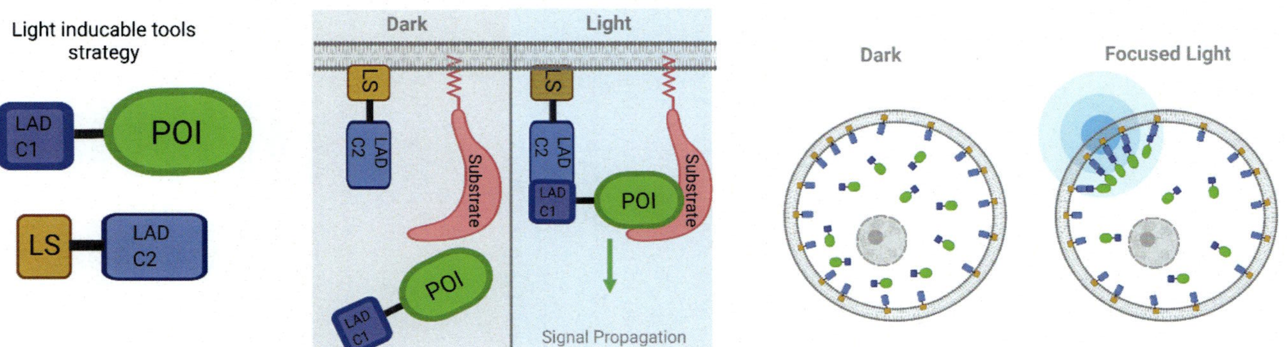

Fig. 11 Tool based strategies for dissecting specific pathways involved in protrusion. (A) Chemically inducible binding domains (CBD) are fused to a protein of interest (POI) and substrate. These components are expressed in the cell and then targeted and activated upon the addition of a chemical (C). Individual nodes of signaling pathways can then be selectively activated to reveal their role in membrane dynamics (Karginov et al., 2014). (B) Light activated dimers (LAD) are fused to a protein of interest and a localization sequence (LS). In this example the localization sequence targets the plasma membrane of the cell. In the dark the expressed POI is diffuse in the cell, but by exposing the cell to focused light the POI can be selectively targeted to a portion of the membrane allowing the study of local signaling (Mühlhäuser, Weber, & Radziwill, 2019). Figures created with biorender.com.

4.2.2 Optogenetic control of protein localization

Protein activation and activity are often highly dependent on localization of the protein within the cell. Bearing this in mind protein engineers implemented the intrinsic dimerization and oligomerization properties of many photoresponsive proteins to target specific proteins of interest to a designated region of the cell, including the plasma membrane (Fig. 11) (Mühlhäuser et al., 2017, 2019; Shaaya, Fauser, & Karginov, 2021; Weitzman & Hahn, 2014). This approach also has the added advantage of creating asymmetric signaling through focused illumination of one side of the cell. Implementation of this approach has been particularly useful for studying Rho GTPase signaling (Shaaya et al., 2021; Valon et al., 2015; Van Geel, Cheung, & Gadella, 2020). Combining optogenetic control with FRET biosensors, such as those previously mentioned, can add an additional level of information from an experiment to parse the complex signaling involved in cell protrusions (Van Geel et al., 2020). These targeting approaches can also be implemented to control the level of a specific lipid in the plasma membrane with spatiotemporal control to reveal the signaling effects of these lipids. Optogenetic control of PIP3 levels in the plasma membrane demonstrated an important role for PIP3 in cytoskeletal changes and lamellipodia and filipodia formation (Kakumoto & Nakata, 2013).

5. Conclusion

Here we presented a variety of approaches to reliably describe and quantify membrane morphological changes and dynamics. Cellular protrusion dynamics and velocity can be determined by using a selection of open-source software suites. Many of these suites can also be used to correlate fluorescent intensity with cell edge dynamics. Taking these functions in conjunction with recent advances in biosensors to indicate specific protein activity, it is possible to correlate specific signaling events with morphodynamic changes in the membrane. In addition, we highlighted several predictive computational models that can be used to enhance the understanding of cytoskeletal rearrangements in protrusions. These advances have improved the quality and reproducibility of protrusion analysis while also filling in some of the gaps in our understanding of the signaling dynamics. We look forward to advancements in the implementation of these techniques to parse out the signaling involved in protrusions as well as the development of new tools to further dissect these pathways.

Acknowledgment

This work was supported by the National Institute of Health (NIGMS grant R01GM118582 to A.V.K.) and (T32 Training grant HL007829-22 to J.F. and M.B.).

References

Al Haddad, M., et al. (2020). Differential regulation of rho GTPases during lung adenocarcinoma migration and invasion reveals a novel role of the tumor suppressor StarD13 in invadopodia regulation. *Cell Communication and Signaling: CCS*, *18*(1).

Albert-Smet, I., et al. (2019). Applications of light-sheet microscopy in microdevices. *Frontiers in Neuroanatomy*, *13*(1).

Aoki, K., et al. (2016). A RhoA and Rnd3 cycle regulates actin reassembly during membrane blebbing. *Proceedings of the National Academy of Sciences*, *113*(13), E1863–E1871.

Aoki, K., et al. (2020). Coordinated changes in cell membrane and cytoplasm during maturation of apoptotic bleb. *Molecular Biology of the Cell*, *31*(8), 833–844.

Barry, D. J., et al. (2015). Open source software for quantification of cell migration, protrusions, and fluorescence intensities. *Journal of Cell Biology*, *209*(1), 163–180.

Bear, J. E., et al. (2002). Antagonism between Ena/VASP proteins and actin filament capping regulates fibroblast motility. *Cell*, *109*(4), 509–521.

Berginski, M. E., et al. (2011). High-resolution quantification of focal adhesion spatiotemporal dynamics in living cells. *PLoS One*, *6*(7), e22025.

Buwaneka, P., et al. (2021). Evaluation of the available cholesterol concentration in the inner leaflet of the plasma membrane of mammalian cells. *Journal of Lipid Research*, *62*, 100084.

Carlsson, A. E. (2006). Stimulation of actin polymerization by filament severing. *Biophysical Journal*, *90*(2), 413–422.

Chu, P.-H., et al. (2014). Engineered kinase activation reveals unique morphodynamic phenotypes and associated trafficking for Src family isoforms. *Proceedings of the National Academy of Sciences*, *111*(34), 12420–12425.

Clark, A. G., Dierkes, K., & Paluch, E. K. (2013). Monitoring actin cortex thickness in live cells. *Biophysical Journal*, *105*(3), 570–580.

Collins, K. B., et al. (2020). Septin2 mediates podosome maturation and endothelial cell invasion associated with angiogenesis. *Journal of Cell Biology*, *219*(2).

Dean, M., Kevin, et al. (2016). Diagonally scanned light-sheet microscopy for fast volumetric imaging of adherent cells. *Biophysical Journal*, *110*(6), 1456–1465.

Dickinson, R. B. (2009). Models for actin polymerization motors. *Journal of Mathematical Biology*, *58*(1–2), 81–103.

Donnelly, S. K., et al. (2018). Characterization of genetically encoded FRET biosensors for rho-family GTPases. In *Methods in molecular biology* (pp. 87–106). New York: Springer.

Enculescu, M. S.-G. M., Danuser, G., & Falcke, M. (2010). Modeling of protrusion phenotypes driven by the actin-membrane interaction. *Biophysical Journal*, *98*(8), 1571–1581.

Ennomani, H., et al. (2016). Architecture and connectivity govern actin network contractility. *Current Biology*, *26*(5), 616–626.

Ersoy, I., et al. (2008). Cell spreading analysis with directed edge profile-guided level set active contours. In *Medical image computing and computer-assisted intervention—MICCAI 2008* (pp. 376–383). Berlin, Heidelberg: Springer.

Fischer, R. S., et al. (2011). Microscopy in 3D: A biologist's toolbox. *Trends in Cell Biology*, *21*(12), 682–691.

Goulden, B. D., et al. (2019). A high-avidity biosensor reveals plasma membrane PI(3,4)P2 is predominantly a class I PI3K signaling product. *Journal of Cell Biology*, *218*(3), 1066–1079.

Hartman, Z. R., Schaller, M. D., & Agazie, Y. M. (2013). The tyrosine phosphatase SHP2 regulates focal adhesion kinase to promote EGF-induced Lamellipodia persistence and cell migration. *Molecular Cancer Research*, *11*(6), 651–664.
Heck, T., et al. (2020). The role of actin protrusion dynamics in cell migration through a degradable viscoelastic extracellular matrix: Insights from a computational model. *PLoS Computational Biology*, *16*(1), e1007250.
Heckman, C. A., & Plummer, H. K. (2013). Filopodia as sensors. *Cellular Signalling*, *25*(11), 2298–2311.
Hertel, F., et al. (2020). Fluorescent biosensors for multiplexed imaging of phosphoinositide dynamics. *ACS Chemical Biology*, *15*(1), 33–38.
Hodgson, L., Shen, F., & Hahn, K. (2010). Biosensors for characterizing the dynamics of rho family GTPases in living cells. *Current Protocols in Cell Biology*, *46*(1).
Jacquemet, G., Hamidi, H., & Ivaska, J. (2015). Filopodia in cell adhesion, 3D migration and cancer cell invasion. *Current Opinion in Cell Biology*, *36*, 23–31.
Jacquemet, G., et al. (2013). RCP-driven α5β1 recycling suppresses Rac and promotes RhoA activity via the RacGAP1–IQGAP1 complex. *Journal of Cell Biology*, *202*(6), 917–935.
Jacquemet, G., et al. (2017). FiloQuant reveals increased filopodia density during breast cancer progression. *Journal of Cell Biology*, *216*(10), 3387–3403.
Kakumoto, T., & Nakata, T. (2013). Optogenetic control of PIP3: PIP3 is sufficient to induce the actin-based active part of growth cones and is regulated via endocytosis. *PLoS One*, *8*(8), e70861.
Karginov, A. V., et al. (2010). Engineered allosteric activation of kinases in living cells. *Nature Biotechnology*, *28*(7), 743–747.
Karginov, A. V., et al. (2014). Dissecting motility signaling through activation of specific Src-effector complexes. *Nature Chemical Biology*, *10*(4), 286–290.
Klomp, J. E., et al. (2019). Time-variant SRC kinase activation determines endothelial permeability response. *Cell Chemical Biology*, *26*(8), 1081–1094.e6.
Krause, M., & Gautreau, A. (2014). Steering cell migration: Lamellipodium dynamics and the regulation of directional persistence. *Nature Reviews Molecular Cell Biology*, *15*(9), 577–590.
Lai, H., et al. (2018). Temporal regulation of morphogenetic events in Saccharomyces cerevisiae. *Molecular Biology of the Cell*, *29*(17), 2069–2083.
Lawson, C. D., & Ridley, A. J. (2018). Rho GTPase signaling complexes in cell migration and invasion. *Journal of Cell Biology*, *217*(2), 447–457.
Lazarides, E., & Weber, K. (1974). Actin antibody: The specific visualization of actin filaments in non-muscle cells. *Proceedings of the National Academy of Sciences*, *71*(6), 2268–2272.
Lee, J. (2018). Insights into cell motility provided by the iterative use of mathematical modeling and experimentation. *AIMS Biophysics*, *5*(2), 97–124.
Lee, E., & Knecht, D. A. (2002). Visualization of actin dynamics during macropinocytosis and exocytosis. *Traffic*, *3*(3), 186–192.
Li, W., et al. (2018). A membrane-bound biosensor visualizes shear stress-induced inhomogeneous alteration of cell membrane tension. *iScience*, *7*, 180–190.
Liu, S.-L., et al. (2017). Orthogonal lipid sensors identify transbilayer asymmetry of plasma membrane cholesterol. *Nature Chemical Biology*, *13*(3), 268–274.
Lorenz, M., et al. (2004). Imaging sites of N-WASP activity in Lamellipodia and Invadopodia of carcinoma cells. *Current Biology*, *14*(8), 697–703.
Lukinavičius, G., et al. (2014). Fluorogenic probes for live-cell imaging of the cytoskeleton. *Nature Methods*, *11*(7), 731–733.
Machacek, M., & Danuser, G. (2006). Morphodynamic profiling of protrusion phenotypes. *Biophysical Journal*, *90*(4), 1439–1452.

Machacek, M., et al. (2009). Coordination of rho GTPase activities during cell protrusion. *Nature*, *461*(7260), 99–103.

Manandhar, S., et al. (2020). 3D flow field estimation and assessment for live cell fluorescence microscopy. *Bioinformatics*, *36*(5), 1317–1325.

Marston, D. J., et al. (2020). Multiplexed GTPase and GEF biosensor imaging enables network connectivity analysis. *Nature Chemical Biology*, *16*(8), 826–833.

Martin, K., et al. (2016). Spatio-temporal co-ordination of RhoA, Rac1 and Cdc42 activation during prototypical edge protrusion and retraction dynamics. *Scientific Reports*, *6*(1), 21901.

Mattila, P. K., & Lappalainen, P. (2008). Filopodia: Molecular architecture and cellular functions. *Nature Reviews Molecular Cell Biology*, *9*(6), 446–454.

Melak, M., Plessner, M., & Grosse, R. (2017). Actin visualization at a glance. *Journal of Cell Science*, *130*(3), 525–530.

Mimori-Kiyosue, Y. (2021). Imaging mitotic processes in three dimensions with lattice light-sheet microscopy. *Chromosome Research*, *29*(1), 37–50.

Mogilner, A., & Oster, G. (1996). Cell motility driven by actin polymerization. *Biophysical Journal*, *71*(6), 3030–3045.

Mühlhäuser, W. W. D., Weber, W., & Radziwill, G. (2019). OpEn-tag—A customizable optogenetic toolbox to dissect subcellular signaling. *ACS Synthetic Biology*, *8*(7), 1679–1684.

Mühlhäuser, W. W. D., et al. (2017). Optogenetics—Bringing light into the darkness of mammalian signal transduction. *Biochimica et Biophysica Acta (BBA) – Molecular Cell Research*, *1864*(2), 280–292.

Murphy, D. A., & Courtneidge, S. A. (2011). The 'ins' and 'outs' of podosomes and invadopodia: Characteristics, formation and function. *Nature Reviews Molecular Cell Biology*, *12*(7), 413–426.

Nishioka, T., et al. (2010). Heterogeneity of phosphatidic acid levels and distribution at the plasma membrane in living cells as visualized by a Förster resonance energy transfer (FRET) biosensor. *Journal of Biological Chemistry*, *285*(46), 35979–35987.

Ridley, A. J. (2015). Rho GTPase signalling in cell migration. *Current Opinion in Cell Biology*, *36*, 103–112.

Rossier, O. M., et al. (2010). Force generated by actomyosin contraction builds bridges between adhesive contacts. *The EMBO Journal*, *29*(6), 1055–1068.

Ryan, G. L., et al. (2017). Cell protrusion and retraction driven by fluctuations in actin polymerization: A two-dimensional model. *Cytoskeleton*, *74*(12), 490–503.

Schachtner, H., et al. (2012). Tissue inducible Lifeact expression allows visualization of actin dynamics in vivo and ex vivo. *European Journal of Cell Biology*, *91*(11), 923–929.

Schaub, S. B., Meister, J.-J., & Verkhovsky, A. B. (2007). Analysis of actin filament network organization in lamellipodia by comparing experimental and simulated images. *Journal of Cell Science*, *120*(8), 1491–1500.

Sens, P. (2020). Stick–slip model for actin-driven cell protrusions, cell polarization, and crawling. *Proceedings of the National Academy of Sciences*, *117*(40), 24670–24678.

Sept, D., Elcock, A. H., & Mccammon, J. A. (1999). Computer simulations of actin polymerization can explain the barbed-pointed end asymmetry. *Journal of Molecular Biology*, *294*(5), 1181–1189.

Shaaya, M., Fauser, J., & Karginov, A. V. (2021). Optogenetics: The art of illuminating complex signaling pathways. *Physiology*, *36*(1), 52–60.

Shaaya, M., et al. (2020). Light-regulated allosteric switch enables temporal and subcellular control of enzyme activity. *eLife*, *9*.

Shemesh, T., Alexander, D. B., & Kozlov, M. M. (2012). Physical model for self-organization of actin cytoskeleton and adhesion complexes at the cell front. *Biophysical Journal*, *102*(8), 1746–1756.

Tabatabaei, M., Tafazzoli-Shadpour, M., & Khani, M. M. (2021). Altered mechanical properties of actin fibers due to breast cancer invasion: Parameter identification based on micropipette aspiration and multiscale tensegrity modeling. *Medical & Biological Engineering & Computing*, *59*(3), 547–560.

Tsygankov, D., et al. (2014a). CellGeo: A computational platform for the analysis of shape changes in cells with complex geometries. *Journal of Cell Biology*, *204*(3), 443–460.

Tsygankov, D., et al. (2014b). User-friendly tools for quantifying the dynamics of cellular morphology and intracellular protein clusters. In *Methods in cell biology* (pp. 409–427). Elsevier.

Urbančič, V., et al. (2017). Filopodyan: An open-source pipeline for the analysis of filopodia. *Journal of Cell Biology*, *216*(10), 3405–3422.

Valon, L., et al. (2015). Predictive spatiotemporal manipulation of signaling perturbations using Optogenetics. *Biophysical Journal*, *109*(9), 1785–1797.

Van Geel, O., Cheung, S., & Gadella, T. W. J. (2020). Combining optogenetics with sensitive FRET imaging to monitor local microtubule manipulations. *Scientific Reports*, *10*(1).

Vandekerckhove, J., et al. (1985). The phalloidin binding site of F-actin. *The EMBO Journal*, *4*(11), 2815–2818.

Vavylonis, D., Yang, Q., & O'Shaughnessy, B. (2005). Actin polymerization kinetics, cap structure, and fluctuations. *Proceedings of the National Academy of Sciences*, *102*(24), 8543–8548.

Weitzman, M., & Hahn, K. M. (2014). Optogenetic approaches to cell migration and beyond. *Current Opinion in Cell Biology*, *30*, 112–120.

Welf, S. E., et al. (2016). Quantitative multiscale cell imaging in controlled 3D microenvironments. *Developmental Cell*, *36*(4), 462–475.

Zemel, A., et al. (2010). Cell shape, spreading symmetry, and the polarization of stress-fibers in cells. *Journal of Physics: Condensed Matter*, *22*(19), 194110.

Zhurikhina, A., et al. (2018). EdgeProps: A computational platform for correlative analysis of cell dynamics and near-edge protein activity. In *Methods in molecular biology* (pp. 47–56). New York: Springer.

Zimmerman, S. P., et al. (2017). Cells lay their own tracks: Optogenetic Cdc42 activation stimulates fibronectin deposition supporting directed migration. *Journal of Cell Science*, *130*(18), 2971–2983.

Evaluating membrane structure by Laurdan imaging: Disruption of lipid packing by oxidized lipids

Irena Levitan*
Division of Pulmonary and Critical Care, Department of Medicine, University of Illinois at Chicago, Chicago, IL, United States
*Corresponding author: e-mail address: levitan@uic.edu

Contents

Abstract

Impact of different lipids on membrane structure/lipid order is critical for multiple biological processes. Laurdan microscopy provides a unique tool to assess this property in heterogeneous biological membranes. This review describes the general principles of the approach and its application in model membranes and cells. It also provides an in-depth discussion of the insights obtained using Laurdan microscopy to evaluate the differential effects of cholesterol, oxysterols and oxidized phospholipids on lipid packing of ordered and disordered domains in vascular endothelial cells.

Current Topics in Membranes, Volume 88
ISSN 1063-5823
https://doi.org/10.1016/bs.ctm.2021.10.003

1. Introduction

Lipid packing or order is a fundamental property of biological membranes that has a major impact on the function of membrane proteins, formation of lipid-protein signaling platforms and membrane recycling (Cooper, 1977; Yeagle, 1991). In earlier studies, changes in membrane microstructure were detected primarily by measuring the degree of movement of different fluorescent probes, which in most cases inversely correlates with an increase in lipid order. Currently, increasing number of studies turn to a fluorescent probe, Laurdan, that has spectral sensitivity to the membrane phase, its excitation/emission spectra depending on the gel-like/liquid crystalline-like phases of the phospholipids (Gaus, Zech, & Harder, 2006; Jay & Hamilton, 2017; Gunther, Malacrida, Jameson, Gratton, & Sanchez, 2021). This property allows imaging the local lipid order in model membranes (Dietrich et al., 2001) and cells (Gaus et al., 2003). Laurdan imaging can be used in both live and fixed cells (Gaus et al., 2003; Gaus, Le Lay, Balasubramanian, & Schwartz, 2006), as well as in intact tissues (Owen, Rentero, Magenau, Abu-Siniyeh, & Gaus, 2011). Most importantly, it allows to assess local membrane properties and obtain space-resolved maps of lipid order heterogeneity in live cells and tissues.

2. Basic principles of Laurdan fluorescence

LAURDAN (6-dodecanoyl-2-(dimethylamino) naphthalene), was synthesized and first characterized as an environmentally sensitive fluorescent probe by Gregorio Weber and Fay Farris in 1979 (Weber & Farris, 1979; Gunther et al., 2021). It is composed of naphthalene hydrophilic head linked to a hydrophobic tail composed of lauric fatty acid and it partitions to the hydrophilic-hydrophobic interface of the phospholipid bilayer with its lauric acid tail aligning with the phospholipid acyl chains (Fig. 1).

When it is fluorescently excited, a charge separation creates a strong dipole between the amino and the carbonyl groups. This strong dipole, in turn, aligns the surrounding dipole molecules, such as water, to restrict their free rotational motions. The energy dissipation for aligning water molecules or restricting the free rotational motions of water molecules results in a longer emission wavelength. Therefore, the spectra of the probe is sensitive to the water contents in the membrane. A more disordered membrane (liquid crystalline-like) should has more defects (or voids), which

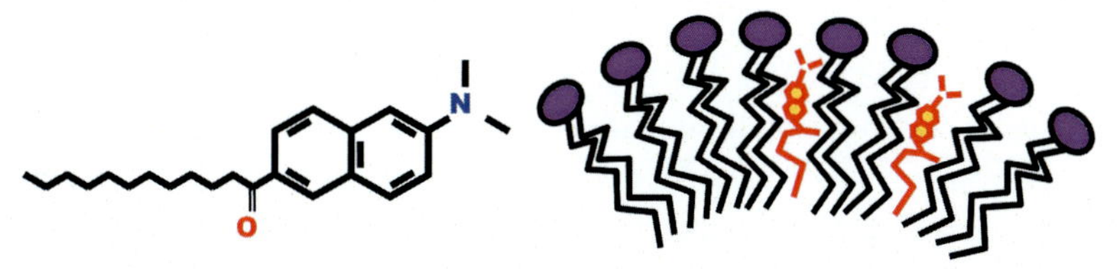

Fig. 1 Chemical structure and incorporation of Laurdan in membrane. (A) Laurdan (6-dodecanoyl-2-dimethylaminonaphthalene, C24H35NO) has tswo aromatic rings and lauric fatty acid tail. When it is fluorescently excited, a charge separation between the amino and the carbonyl groups create a dipole moment sensitive to the presence of water dipoles in its environment. (B) Incorporation of Laurdan into the membrane bilayer (a single leaflet is shown). Purple indicates hydrophilic head, black line indicates hydrophobic acyl chain, and red line indicates Laurdan.

allow more water molecules to partition into the bilayer. Consistently, studies showed that Laurdan undergoes a red spectral shift during the membrane phase transition from gel-to liquid crystalline-like, which was attributed to dipole relaxation and the sensitivity to the polarity of the environment (Parasassi, De Stasio, Ravagnan, Rusch, & Gratton, 1991). Furthermore, the sensitivity of Laurdan to the lipid order of the membrane is a result of its sensitivity to the presence and mobility of water dipoles within the lipid bilayer and as more mobile water dipoles surround the Laurdan ring, it causes a shift in the emission spectrum. Thus, as water penetration into the bilayer decreases with an increase in the lipid order of the membrane, Laurdan emission spectrum shifts accordingly (Parasassi et al., 1991; Parasassi & Gratton, 1995). It is also important to note that Laurdan partitions equally to solid and liquid lipid phases, does not bind to specific fatty acids and is not soluble in water (Bagatolli, Sanchez, Hazlett, & Gratton, 2003). Another parameter that might be important to take into the account is the sub-cellular distribution of the Laurdan dye between the apical and basal membranes, which might affect the polarization properties.

Since Laurdan undergoes a 50-nm red shift in the emission maximum in polar vs no-polar environments, measuring fluorescence intensity at the two specific wavelengths provides a ratiometric method to assess the local membrane properties. This method defines the general polarization (GP) function to quantify the emission shift (Parasassi, De Stasio, d'Ubaldo, & Gratton, 1990; Parasassi et al., 1991). Specifically, the emission peaks of Laurdan immersed into lipids in the gel phase is 440 nm and in liquid crystalline phase is 490 nm. Thus, the GP ratio is defined as

$$GP = (I_{440} - I_{490})/(I_{440} + I_{490}),$$

where I represents the emission intensities at the given wavelengths. The excitation is done at 340–360 nm. At very low concentration, variations in Laurdan concentration do not affect the GP measurement. The GP values can vary from −1.0 to +1.0 with the more positive values indicating increase in lipid order. Experimentally though, the values typically range between −0.6 and +0.6 with the liquid phase yielding values between −0.3 and +0.3 and gel phase typically yielding values between 0.5 and 0.6.

In early studies, Laurdan GP values of lipid vesicles with different lipid compositions were measured using conventional fluorometer in a "cuvette" to determine the homogeneity and the presence of domains in lipid vesicles (Parasassi, Di Stefano, Loiero, Ravagnan, & Gratton, 1994; Parasassi & Gratton, 1995). Interestingly, Laurdan measurements led to a surprising observation about the effect of cholesterol on membrane fluidity: while it was considered to be well known that increase in membrane cholesterol decreases membrane fluidity, GP measurements suggested that the effect is more complex. Surprisingly, it was found that cholesterol has a "homogenizing" effect on the membrane, decreasing the fluidity of the liquid-crystalline phase, whereas also increasing the fluidity of the gel phase (Parasassi et al., 1994). Specifically, they found that the addition of 30 mol%, cholesterol, a physiological level of cholesterol in biological membranes, decreases the fluctuation rate between the liquid and gel phases of the bilayer, while at the same time decreasing water concentration on hydrophobic-hydrophilic interface of due to higher lipid packing. These observations provided significant insights into the impact of cholesterol on the physical properties of lipid bilayers.

Clearly, the use of Laurdan is not limited to the studies of cholesterol effects on the lipid bilayers. Multiple studies applied this approach to evaluate effects of various lipids or other hydrophobic molecules that can be incorporated into the bilayers on lipid packing of the membrane. For example, using Laurdan, Amyloid-β peptide was found to alter membrane phase properties in astrocytes through its direct insertion and indirectly through the phospholipase A2 pathway (Hicks et al., 2008). More recently, Morandi et al. (2021), investigated the impact of plastic pollutants found in sea water that can enter marine organisms and consequently, the human food chain. Morandi et al found that incorporation of polystyrene, one of the widely used plastics, into liposomes resulted in a strong shift in Laurdan emission indicating a shift in the gel-to-liquid transition and lipid packing of the membrane suggesting that plastic nanopollutants may disrupt the function of biological membranes. Another very interesting recent study by Salvador-Castell, Brooks, Winter, Peters, and Oger (2021) used Laurdan fluorescence to evaluate the effects of unique phospholipids of archaeal membranes on the stability of the membrane under different temperatures to get insights into the ability of these membranes to keep their integrity under extreme temperatures. Thus, Laurdan GP provides a versatile approach to evaluate the physical properties of lipid bilayers.

Measuring Laurdan fluorescence in a cuvette, however, provides only the average behavior of a vesicle population without the information about spatial distribution of the domains.

The main constraint of using Laurdan probe in microscopy was limited by its rapid photobleaching, which made it non-reliable for conventional microscopy (Bagatolli et al., 2003).

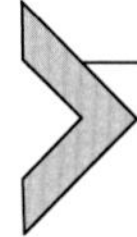

3. Laurdan two photon imaging: Visualizing domains in membrane vesicles

Laurdan imaging was developed with the advance of the two-photon fluorescence microscopy, an approach in which a fluorescent probe absorbs two photons simultaneously with each photon contributing half of the required energy and thus shifting excitation to the wavelengths longer than the emission wavelengths. Typically, a probe is excited by two photons in near-infrared range, substantially reducing photobleaching and scattering. Most of the current Laurdan studies use two-photon microscopy.

One of the earliest studies to shift from "cuvette" to two-photon microscopy by Parasassi, Gratton, Yu, Wilson, and Levi (1997), suggested that while different domains co-exist in membrane vesicles, their size is smaller than the microscope resolution of 200 nm, an important observation in light of subsequent debate about the dimensions of lipid ordered domains in biological membranes. They also found that cholesterol induces heterogeneity in the gel phase: in contrast to previous observations that cholesterol increases the fluidity of the gel phase, Parasassi et al. (1997), did not see a disordering effect of cholesterol and instead observed a general shift of the GP values towards the more ordered state. A more recent study by Aguilar et al. (2012) showed a clear spatial separation of Laurdan GP values in ordered vs. disordered domains in giant unilamellar vesicles (GUVs) containing cholesterol in temperature below the transition temperature.

This separation was not observed above the transition temperatures of 37–38 °C and the membrane becomes more homogenous. The authors concluded that an increase in the cholesterol content in the vesicles results in the stabilization of a *liquid-ordered* (*Lo*) phase, an intermediate between *gel* (So) and *liquid-crystalline* (*Ld*) phases.

Sanchez, Tricerri, and Gratton (2007) applied Laurdan GP imaging of GUVs to study the kinetics of cholesterol removal by high density lipoproteins (HDL) from different membrane domains (pools). As expected, the GP values decreased as a function of methyl-β-cyclodextrin (MβCD), a well-known cholesterol-depleting agent. Imaging of the GUVs show that at temperatures below the phase transition (30 °C), small islands/domains of membrane with high GP values floating in a larger membrane pool with lower GP. These domains vary significantly in size among the vesicles. Above 30 °C, the two phases merged, and separate domains were not visible. Most interestingly, exposing the GUVs containing cholesterol to HDL particles resulted in a decrease in the GP values in the liquid but not from the ordered state suggesting that HDL particles selectively removes cholesterol from the liquid phase of the membrane. A similar approach was used by Sanchez, Gunther, Tricerri, and Gratton (2011) to study the kinetics and the pool/domain specificity of cholesterol removal by MβCD. Consistent with the observations described above for HDL particles, MβCD was also found to preferentially remove cholesterol from the liquid/disordered state in spite of the fact that it is the ordered phase that is more rich in cholesterol. Thus, Laurdan imaging allows distinguishing between cholesterol kinetics of two separate membrane pools.

These observations have major implications both for understanding the nature of cholesterol interactions with the surrounding phospholipids and for the use of MβCD to assess the physiological roles of cholesterol rich membrane domains. A molecular mechanism proposed for the preferential removal of cholesterol from the disordered phase was that the kinks in the structure of unsaturated phospholipids present in the disordered state prevent cholesterol from penetrating deeper into the bilayer and thus making it more accessible to the acceptor molecules. In contrast, increased hydrophobic thickness and packing of the ordered phase impedes its removal by MβCD or other acceptors. It is most important to note that based on these observations, MβCD should not be used as a selective tool for the disruption of the ordered domains. In fact, not only MβCD removes cholesterol from the disordered domains, it appears to preferentially target this phase.

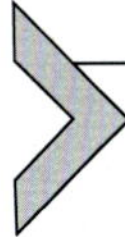

4. Laurdan two photon imaging: Visualizing membrane domains in living cells

Laurdan imaging of living cells was pioneered by Gaus et al. (2003) and Gaus, Le Lay, et al. (2006). First, Gaus et al. (2003) showed that Laurdan-stained macrophages have highly heterogenous distributions of the GP values with a punctate distribution of areas of low and high GPs corresponding to ordered "gel" and disordered "fluid phase" domains throughout the entire surface of the cells. The phase separation is not observed but rather the cell surfaces appeared to have a continuum of GP values varying from low to high. Multiple studies including from our group showed a similar pattern of highly heterogenous punctate staining through a cell with the ordered domains typically concentrated at cell edges (Ayee, LeMaster et al. 2017; Levitan and Shentu 2011; Shentu et al., 2010) (see a typical image of GP distribution in Fig. 2A).

The areas of the high GP puncta were also very heterogenous in size varying from hundreds of nanometers to micrometers. This is larger than the estimated size of lipid rafts, cholesterol-rich membrane domains that play major roles in cell signaling (e.g., Lingwood and Simons, 2010; Sonnino and Prinetti, 2013). Indeed, the size, the exact composition, the dynamics of lipid rafts have been investigated in numerous studies with different

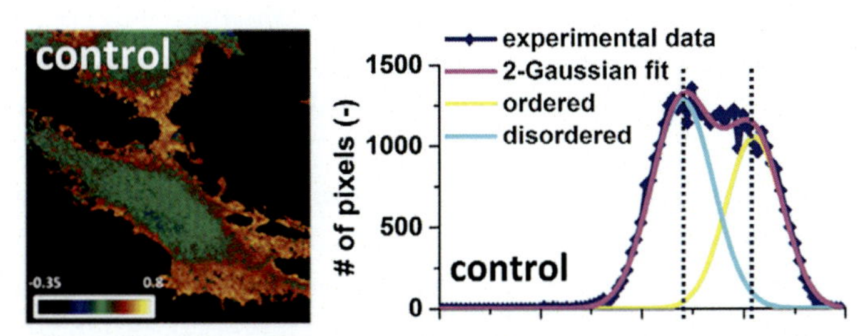

Fig. 2 Typical Laurdan image and the distribution of GP values in a living cell. (Left) A typical GP image of an endothelial cells (human aortic endothelium) shown in pseudo-color, as indicated in the gradient bar: the disordered regions are shown in blue and green and the ordered regions in orange and yellow. (Right) GP histograms (navy dots) fitted by two-Gaussian distributions, with right-shifted curve (yellow) representing ordered domains and left-shifted curve (blue) representing fluid domains. The sum of the Gaussians is shown as a solid, mauve curve. GP distribution is obtained from the region –0.35 to + 0.8. *Adapted from Ayee, M. A., LeMaster, E., Shentu, T. P., Singh, D. K., Barbera, N., Soni, D., et al. (2017). Molecular-scale biophysical modulation of an endothelial membrane by oxidized phospholipids.* Biophysical Journal 112*(2), 325–338.*

approaches but still remain controversial. A working definition of a lipid raft was adopted during the Keystone Symposium on Lipid Rafts and Cell Function in 2006. The question "What is a raft?" was answered with the following statement: "Membrane rafts are small (10–200 nm), heterogeneous, highly dynamic, sterol- and sphingolipid-enriched do-mains that compartmentalize cellular processes. Small rafts can sometimes be stabilized to form larger platforms through protein-protein and protein-lipid interactions." (Pike, 2006). This is below the resolution of the two-photon microscopy at 800 nm (~180 nm) and thus, the puncta are unlikely to represent individual rafts. Instead, they may represent areas where rafts coalesce into larger domains or areas with a higher fraction of ordered domains. Notably, a subsequent study by Gaus, Le Lay et al. (2006) established that in endothelial cells and in embryonic fibroblasts, the puncta with high GP values appear to co-localize with the markers of focal adhesions, phosphorylated FAK (pFAK) and phosphorylated caveolin-1 (pYCav1), whereas Tranferrin, a marker of disordered domains was shown to co-localize with low GP puncta.

Typically, GP histograms for single cells show double peak distributions with the two peaks overlapping to different degrees depending on the degree of the separation between the areas with low and high GP values (see a typical histogram in Fig. 2B). The distributions are fitted by a sum of two Gaussian curves, which are interpreted to represent ordered (peak with higher GP values) and fluid (lower GP values) membrane domains. It is important to note that the distribution of the GP values is continuous and the fit to a sum of two Gaussian is an approximation. The specific positions of the two peaks represent the degree of lipid order for each population of the domains and shift to higher or lower values depending on the experimental conditions.

As described in more detail below, increasing or decreasing membrane cholesterol also shifts the peaks appropriately. The areas under the two curves represent the coverage of the cells surface by ordered or disordered domains. Both values vary between different cell types or other parameters. For example, in macrophages, Gaus, Gratton et al. (2003) found the peaks corresponding to the two phases at 0.14 GP and at 0.55 GP for the fluid and the gel phase respectively. The areas covered by ordered and disordered domains were comparable with disordered area being somewhat larger. In endothelial cells (porcine aorta) (Gaus, Le Lay, et al., 2006), found that the separation between the two peaks was a little less pronounced (peaks at 0.17 and 0.51, respectively) and the % coverage was significantly different than what was found in the previous study with macrophages. In contrast to macrophages, where % coverage of the membrane by the ordered domains was only slightly smaller than % coverage by the disordered domains, in endothelial cells, the difference was clearly very significant with the ordered domains being a relatively minor component of the membrane. Our later studies (Ayee, LeMaster, et al., 2017; Shentu et al., 2010) also showed somewhat smaller separation between the peaks in human aortic endothelial cells but still two clear peaks with the disordered area being higher than the ordered one. Thus, this method constitutes a unique approach to assess the impact of different experimental conditions on distinct membrane domains in living cells.

5. Impact of cholesterol depletion on Laurdan GP values

A decrease in membrane cholesterol is well known to have profound effects on membrane fluidity and lipid packing. Numerous studies, using a variety of techniques, showed that cholesterol removal results in increased membrane fluidity and decreased lipid packing (Zidovetzki and Levitan, 2007). As such, testing the impact of cholesterol depletion on Laurdan signal serves a valuable control to validate this approach. Indeed, as expected cholesterol depletion using MβCD, the most commonly used tool to remove membrane cholesterol from biological membranes (Zidovetzki and Levitan, 2007), was shown to decrease the GP values in several preparations indicating that the probe is sensitive to experimental perturbations of membrane cholesterol in living cells (e.g., Gaus, Gratton, et al., 2003; Shentu et al., 2010).

A much more controversial question is what is the dynamics of cholesterol depletion from ordered vs. disordered domains. This question is critical for the interpretation of studies that use MβCD as a tool to disrupt lipid rafts, a widely used strategy that frequently assumes that MβCD specifically affects lipid rafts and disregards its possible effect on the disordered domains. This assumption was previously questioned in studies that used biochemical techniques to separate between low-density (cholesterol-rich) and high density (cholesterol-poor) membrane fractions and which showed that cholesterol can be depleted from both pools. The degree to which MβCD takes cholesterol from one pool vs. another varied between different studies, as described in detail in our earlier review (Zidovetzki and Levitan, 2007). Clearly, though, biochemical separation of membrane domains may introduce multiple artifacts for their lipid composition and is not the most reliable approach to address this question.

In this regard, Laurdan imaging that discriminates between ordered and disordered domains in living cells presents a much more powerful tool to establish whether MβCD has preferential effects on ordered vs disordered domains. The results, however, reported by different studies are still controversial. As mentioned above, studies in GUVs suggested that MβCD actually preferentially removes cholesterol from fluid rather than ordered domains (Sanchez, Gunther, Tricerri, & Gratton, 2011). In living cells, the results are more complex and more controversial. Gaus, Gratton, et al. (2003) showed that in macrophages, cholesterol depletion with MβCD resulted in a significant decrease in the cell coverage by the ordered domains but the shifts in the GP values were not analyzed in detail.

Our study, Shentu et al. (2010) provided a detailed quantitative analysis of the impact of MβCD on the distribution of the GP values in aortic endothelial cells and established that both ordered and disordered domains are affected. As shown previously, punctate distribution of ordered and disordered domains was observed throughout the cell surface, with ordered domains concentrating on the cells periphery (Fig. 3). Exposure to MβCD resulted in a shift of the GP values of *both* types of the domains to less ordered and more fluid membrane structure and the magnitudes of both shifts were comparable (Fig. 3, Table 1). In contrast, the relative coverage of the cell surface by the two types of the domains remained unchanged with approximately 30% of the cell surface covered by the ordered and 70% by the disordered domains. Clearly, these observations further challenge the general belief that MβCD-induced cholesterol depletion is a specific tool for the disruption of lipid rafts. There is no doubt that exposing cells to MβCD results in a significant decrease in membrane cholesterol but it appears that a decrease occurs both in rafts and in non-rafts fraction. The significance of this observation is in the notion that at least in this case, MβCD-induced cholesterol depletion does not eliminate lipid rafts but instead alters the lipid order within the domains making them less ordered and that a similar change develops in the disordered domains in parallel. We also found that a decrease in the GP values or disruption of lipid packing is associated with an increase in the elastic modulus of endothelial cells, an effect dependent on the activation of RhoA/ROCK signaling cascade (Oh et al., 2016). The mechanism underlying the connection between lipid packing and activation of RhoA is currently under the investigation.

Interestingly, the loss of caveolin-1, the structural component of caveolae, known as cholesterol-rich membrane invaginations, also led to significant shifts in the GP values in both ordered and disordered domains to more fluid

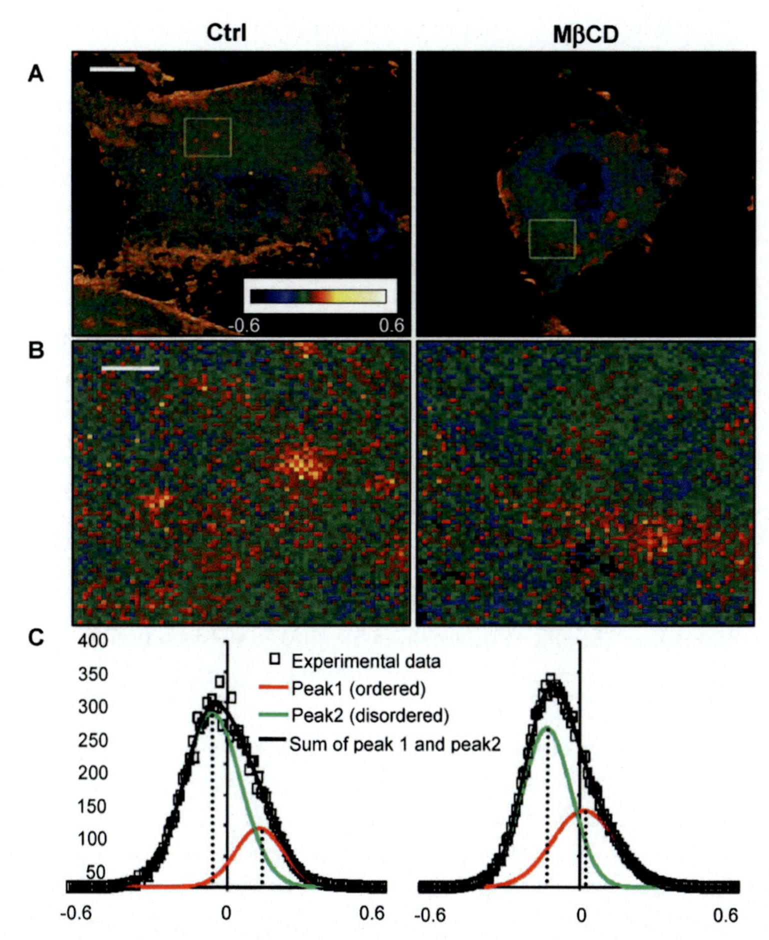

Fig. 3 MβCD-induced cholesterol depletion shifts Laurdan GP to more fluid values in both ordered and disordered domains. (A) Typical GP images for control cells and MβCD-treated cells. All images are shown in pseudocolor, as described above (green/blue—disordered, yellow/orange—ordered). (B) Zoomed images showing the puncta in more details. (C) GP histograms showing the experimental values fitted with a sum of two Gaussians, as described above. *Adapted from Shentu, T. P., Titushkin, I., Singh, D. K., Gooch, K. J., Subbaiah, P. V., Cho, M., et al. (2010). oxLDL-induced decrease in lipid order of membrane domains is inversely correlated with endothelial stiffness and network formation.* American Journal of Physiology Cell Physiology 299*(2), C218–229.*

values in embryonic fibroblasts (Gaus, Le Lay, et al., 2006). A lack of caveolin-1 also resulted in a significant decrease in membrane coverage by the ordered domains and both effects were reversible upon reconstitution of caveolin-1. Since caveolin-1 is known to regulate cholesterol trafficking

Table 1 Peak GP values and % coverage for control and MβCD-treated cells.

	Control	MβCD
GP value		
Peak 1	0.09 ± 0.01	0.03 ± 0.01*
Peak 2	−0.06 ± 0.01	−0.12 ± 0.01*
General GP	−0.02 ± 0.01	−0.08 ± 0.01*
% Coverage		
Peak 1	31 ± 4	33 ± 3
Peak 2	69 ± 4	67 ± 3

Adapted from Shentu, T. P., Titushkin, I., Singh, D. K., Gooch, K. J., Subbaiah, P. V., Cho, M., et al. (2010). oxLDL-induced decrease in lipid order of membrane domains is inversely correlated with endothelial stiffness and network formation. *American Journal of Physiology Cell Physiology 299*(2), C218–229.

to the membrane (Frank, Woodman, et al., 2003), the observed shift in the GP values could be the result of lower cholesterol content of the membrane in Cav-1 knock-out cells. It is also possible that Cav-1 itself may play a role in regulating lipid order of the membrane. It is also interesting to note that the GP values differ significantly between the different studies, which might be due to differences in the lipid composition in different cellular sub-types or culture conditions. It would be interesting to explore the heterogeneity of membrane fluidity on the single cell basis.

A new Laurdan imaging approach was developed more recently to allow discriminating between changes in membrane cholesterol and changes in membrane fluidity (Bonaventura, Barcellona, et al., 2014; Golfetto, Hinde, et al., 2013). This approach is based on the dual dependence of Laurdan fluorescence on the polarity of the environment and on dynamics of dipolar relaxation and quantifies the decay of the Laurdan fluorescence using phasor analysis. Done in living cells, the phasor approach allows monitoring dynamic changes in local membrane fluidity/lipid packing during physiological processes, such as cell migration, independently of changes in membrane cholesterol. Notably, using time-resolved Laurdan fluorescence analysis of live HeLa cells, Ma, Benda, et al. (2018) showed that the time-resolved GP data is consistent with 20–30 mol% of cholesterol and suggested that most of the membrane heterogeneity comes from the local variations of membrane cholesterol.

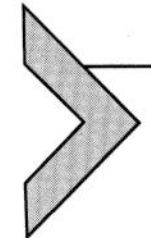

6. Opposite effects of low-density lipoproteins (LDL) and oxidized low-density lipoproteins (oxLDL) on the lipid order of endothelial cells

The major cholesterol carrier in blood is low-density lipoprotein (LDL), a lipoprotein particle that emulsificates cholesterol and other lipids allowing their transport in blood stream (Baigent, Blackwell, et al. 2010; Castelli, Anderson, et al. 1992). It is well known that elevated level blood LDL correlates the development of fatty streaks in the vessel walls of the arteries, which progress to atherosclerotic lesions and clogging of the arteries. LDL particles are internalized by cells via LDL receptor-mediated endocytotic pathway, which is then degraded and free unesterified cholesterol incorporated into the plasma membrane. Indeed, numerous studies showed that plasma hypercholesterolemia leads to significant cholesterol loading of vascular tissues in vivo, particularly macrophages that become cholesterol-loaded foam cells. However, a controversy arose about whether LDL is directly responsible for cholesterol loading of cells or whether the loading requires LDL to be oxidized. The source of this controversy was the discovery that macrophages exposed to high levels of LDL *in vitro,* do not show significant cholesterol loading. It was then proposed that a lack of cholesterol loading in macrophages exposed to LDL is due to the downregulation of the LDL receptor and that the loading is the result of exposing the cells to oxidative modifications of LDL, oxLDL (Steinberg, 2002). Indeed, oxLDL, is also found in blood concurrently with LDL although at much lower levels was shows to be highly pro-inflammatory and cytotoxic further supporting the idea that it is a major proatherogenic form of LDL (Levitan, Volkov, et al., 2010).

We have studied the effects of LDL and oxLDL on membrane cholesterol and lipid packing of endothelial cells using Laurdan two-photon microscopy imaging and found that the two types of lipoproteins have opposite effects on lipid packing: while exposure to LDL resulted in increased lipid order of the membrane, exposure to oxLDL resulted in the fluidization of the membrane.

LDL: Exposure of human aortic endothelial cells to clinically-relevant high levels of LDL resulted in a relatively mild but significant cholesterol loading (~20% increase in free cholesterol) and a pronounced shift in Laurdan GP values towards more ordered lipid structure (Bogachkov, Chen, et al., 2020). Fig. 4 shows typical Laurdan images of endothelial cells (HAECs) exposed to normal (50 mg/dL) or hypercholesterolemic (250 mg/dL) levels of LDL.

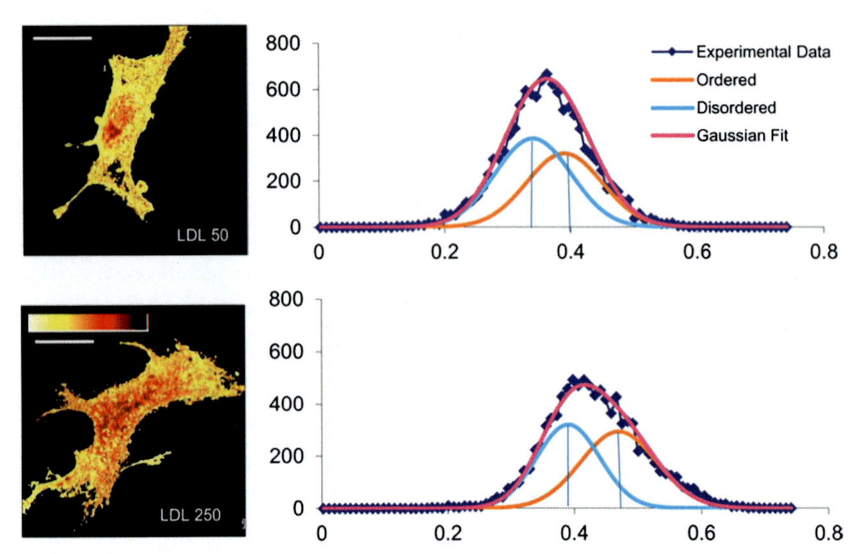

Fig. 4 Exposure to high levels of LDL increases lipid packing of endothelial cells in both ordered and disordered domains. Left: *typical* GP images for endothelial cells exposed to 50 vs 250 mg/dL (pseudocolor, as described above). Right: GP histograms showing the experimental values fitted with a sum of two Gaussians, as described above. *Adapted from Bogachkov, Y. Y., Chen, L., Le Master, E., Fancher, I. S., Zhao, Y., Aguilar, V., et al. (2020). LDL induces cholesterol loading and inhibits endothelial proliferation and angiogenesis in Matrigels: Correlation with impaired angiogenesis during wound healing.* American Journal of Physiology. Cell Physiology 318*(4), C762–C776.*

As described above, the images are pseudo-colored with higher GP values shown as red and lower as yellow. The shift to higher GP values in cells exposed to hypercholesterolemic LDL level is apparent both in the representative images and in the corresponding histograms of the GP values. Notably, the rightward shift is observed in both Gaussians, the one corresponding to the fluid domains and the one corresponding to the ordered domains indicating that both domains shift to the more ordered structure. The relative abundance of the two types of domains did not change, however, suggesting that there is no increase in the membrane coverage by the ordered domains. This is important because it allows discriminating between two conceptually different models for the impact of cholesterol on membrane domains in living cells: one model is to suggest that increase in membrane cholesterol results in increased abundance of lipid rafts/ordered domains, whereas the alternative is that the main changes occur within the domains. Our observations clearly support the second possibility.

OxLDL: We found that in contrast to LDL, exposing endothelial cells to oxLDL does not load the cells with cholesterol but results in changes in the elastic modulus of the cells similar to those induced by cholesterol depletion, rather than cholesterol enrichment (Byfield, Tikku, et al., 2006). Furthermore, cholesterol depletion-like changes in the cellular elastic modulus in oxLDL treated cells were not accompanied with a decrease in cellular cholesterol. To understand better this phenomenon, we analyzed the impact of oxLDL on lipid packing using Laurdan and found that exposure to oxLDL shifts the GP values to lower values, an effect that is indeed similar to the effect of cholesterol depletion describe above (Shentu et al., 2010). Furthermore, providing surplus of cholesterol to oxLDL treated cells by exposing them to MβCD saturated with cholesterol reverses the effect of oxLDL on lipid packing. The reversal is particularly apparent in a strong shift of the ordered domains towards higher GP values in cells exposed to oxLDL and then to MβCD-cholesterol, as compared to oxLDL alone (Shentu et al., 2010).

These observations indicate that, in contrast to previous belief, exposure to oxLDL is not a more efficient method to load cell with cholesterol but has a completely different and opposite effect on the biophysical properties of the lipid bilayer. Our next question, therefore, was to investigate the mechanism responsible for this dichotomy. We found that this phenomenon can be attributed to the incorporation of oxidized lipids (Fig. 5).

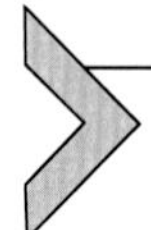

7. Disruption of lipid packing of endothelial membrane by oxidized lipids: Oxysterols and oxidized phospholipids

7.1 Oxysterols

Earlier studies showed that in contrast to cholesterol, incorporation of oxysterols into liposomes disrupt the formation of the ordered domains (Wang, Megha, & London, 2004). We hypothesized, therefore, that a decrease in lipid packing we observed in endothelial cells exposed to oxLDL could be due to the incorporation of oxidized lipids. To test this idea, we exposed cells to 7ketocholesterol, one of the major oxysterol components of oxLDL (Garcia-Cruset, Carpenter, Guardiola, Stein, & Mitchinson, 2001; Oh et al., 2016). Our observations showed that exposure to 7ketocholesterol resulted in a significant shift in Laurdan GP to more fluid values, an effect similar to that of oxLDL and cholesterol depletion (Fig. 6, (Shentu et al., 2010)). Interestingly, this effect was observed mostly in the

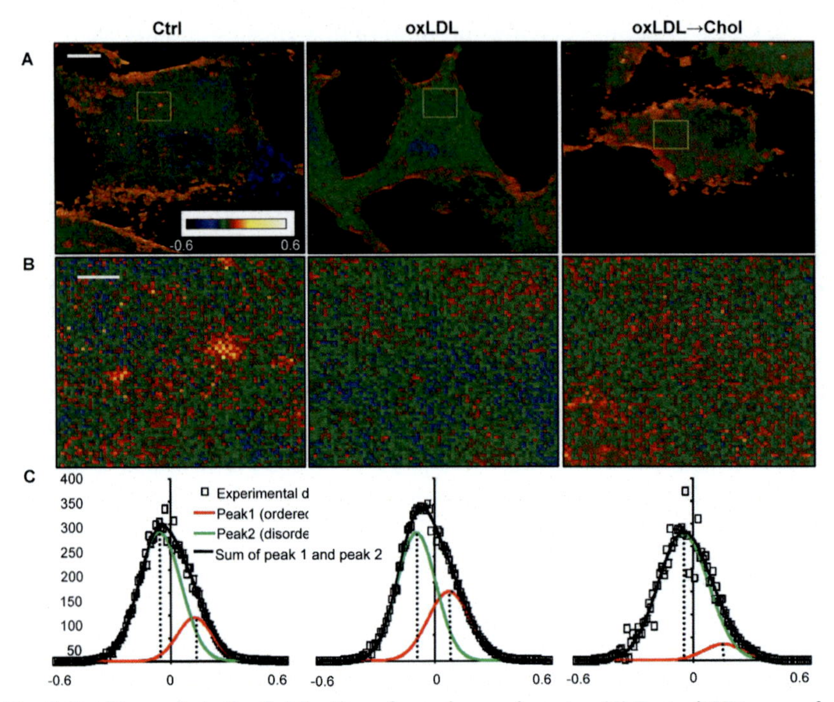

Fig. 5 OxLDL results in the fluidization of membrane domains. (A) Typical GP images for control cells and MβCD-treated cells. All images are shown in pseudocolor, as described above (green/blue—disordered, yellow/orange—ordered). (B) Zoomed images showing the puncta in more details. (C) GP histograms showing the experimental values fitted with a sum of two Gaussians, as described above. *Adapted from Shentu, T. P., Titushkin, I., Singh, D. K., Gooch, K. J., Subbaiah, P. V., Cho, M., et al. (2010). oxLDL-induced decrease in lipid order of membrane domains is inversely correlated with endothelial stiffness and network formation.* American Journal of Physiology Cell Physiology 299*(2), C218–229.*

ordered domains. Furthermore, a decrease in lipid packing was also observed in cells exposed to anderstenol, another oxysterol that was shown previously to disrupt membrane the formation of lipid ordered domains in liposomes (Xu and London, 2000).

More recently, we provided a theoretical explanation for the differential effects of cholesterol and 7ketocholesterol on lipid order of the membrane bilayer (Ayee & Levitan, 2021). Briefly, we used Molecular Dynamics simulation to analyze structural and dynamic changes in phospholipid bilayers that result from the incorporation of physiological levels of cholesterol or 7ketocholesterol, the amounts of both sterols were based on mass spectrometry analysis of cells exposed to LDL or oxLDL respectively. Our analysis

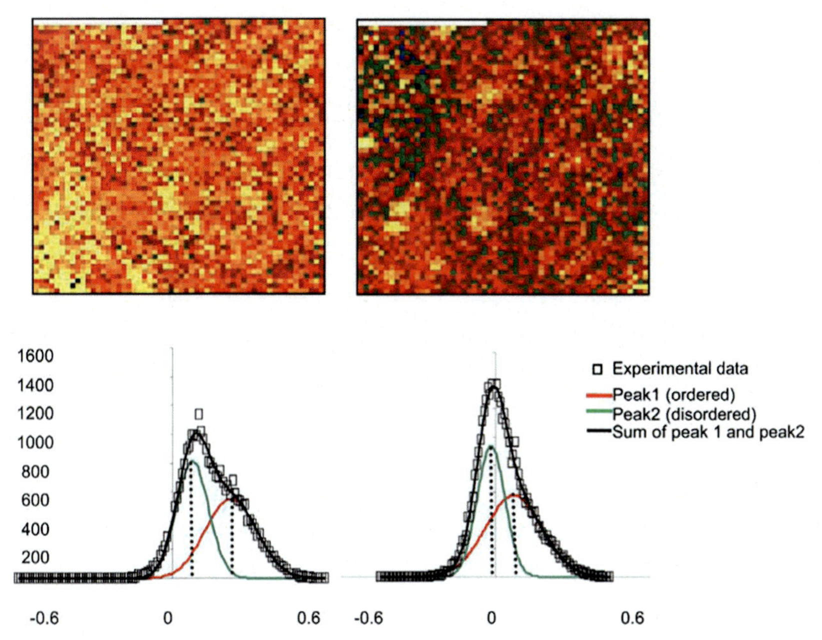

Fig. 6 Fluidization of membrane domains by 7ketocholesterol. Top panels: typical GP images for a control and 7ketocholesterol-treated cells. Botttom panels: GP histograms. *Adapted from Shentu, T. P., Titushkin, I., Singh, D. K., Gooch, K. J., Subbaiah, P. V., Cho, M., et al. (2010). oxLDL-induced decrease in lipid order of membrane domains is inversely correlated with endothelial stiffness and network formation.* American Journal of Physiology Cell Physiology *299(2), C218–229.*

showed that due to the different tilt in the orientation of cholesterol and 7ketocholesterol within the phospholipid bilayer, the presence of cholesterol increases the order parameter of all bonds in the lipid tails of the three major phospholipids, POPPC (1-palmitoyl-2-oleoyl-glycero-3-phosphocholine), DPPC (1,2-dipalmitoyl-sn-glycero-3-phosphocholine) and SM (sphingomyelin). In contrast, the presence of 7ketocholesterol decreased the order parameter of the lipid tails of POPC, DPPC and SM lipid species.

7.2 Oxidized phospholipids

We also found that similarly to oxysterols, incorporation of bio-active oxidized phospholipids, specifically, oxPAPC (1-palmitoyl-2-arachidonoyl-sn-glycero-phosphocholine), a component of the minimally oxidized LDL, a highly reactive oxLDL species, also disrupts lipid packing in endothelial membranes, as detected by Laurdan microscopy (Ayee et al., 2017).

The effect was observed for two oxPAPC species, POPC (5-oxovaleroyl)-sn-glycero-3-phosphocholine and PGPC (1-palmitoyl-2-glutaroyl-sn-glycero-3-phosphocholine), both with truncated tails generating a tilt in their orientations within the bilayer.

Thus, Laurdan microscopy provides a unique tool to clearly discriminate between the effects of different lipid species on lipid order/packing of biological membranes.

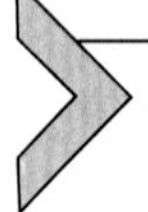

8. Challenges and new directions

8.1 Sensitivity to fading and development of new probes

The first limitation that discussed above is the rapid photo-bleaching of Laurdan dye, which makes it unsuitable for standard confocal and total internal reflection fluorescence (TIRF) systems, limiting its current use to multi-photon microscopy. To resolve this limitation, several new membrane probes that work similarly to Laurdan but are more stable have been developed. Specifically, Di-4-ANEPPDHQ, a voltage-sensitive dye that was first developed to analyze neural networks (Obaid, Loew, Wuskell, & Salzberg, 2004), was demonstrated to be sensitive to lipid packing and capable to identifying cholesterol-rich domains in model membranes (Jin, Millard, Wuskell, Clark, & Loew, 2005). Similarly to Laurdan, Di-4-ANEPPDHQ, has a spectral shift when incorporated in ordered vs. disordered domains (Jin et al., 2006). Di-4-ANEPPDHQ is more photostable than Laurdan and was successfully used in confocal and TIRF microscopy (Owen, Rentero, Magenau, Abu-Siniyeh, & Gaus, 2012). Interestingly, it was also shown that the GP values generated using Laurdan are more influenced by the temperature that those generated using Di-4-ANEPPDHQ, which were shown to be more sensitive to the cholesterol content of the membrane (Amaro, Reina, Hof, Eggeling, & Sezgin, 2017). Another lipid-packing sensitive dye that was shown greater photostability than original Laurdan dye is M-Laurdan a close derivative of Laurdan but with less sensitivity to photobleaching (Mazeres, Joly, Lopez, & Tardin, 2014). Moreover, an interesting modification to Laurdan dye was the addition of a membrane anchor that stabilizes it in the outer leaflet of the plasma membrane and prevents its internalization and labeling of the internal membrane (Danylchuk, Sezgin, Chabert, & Klymchenko, 2020). Development of these new dyes also opens the door to develop super-resolution applications to overcome the spatial resolution limitations. Indeed, the resolution of multi-photon or confocal microscopy is not sufficient for imaging of individual rafts unless the rafts coalesce. To this end, the attention is turned to super-resolution fluorescence microcopy that increases the lateral resolution from 200 nm diffraction limit of conventional microscopy to 5–30 nm range, as reviewed by (Owen and Gaus 2013). It still needs to be established whether Laurdan or its derivatives can be suitable for the super-resolution imaging. It is also important to note, however, that these environmentally-sensitive dyes were also shown to have significant effects on membrane properties, particularly at higher concentrations (Suhaj, Gowland, Bonini, Owen, & Lorenz, 2020).

8.2 Laurdan imaging in vivo

Another important advance is applications of Laurdan or its derivatives not only to individual cells but also to tissues, organs or even organisms to study changes in lipid packing in intact tissues. Laurdan imaging was successfully implemented to visual lipid packing in model organisms, such as zebra fish and in living vertebrate embryos (Owen, Magenau, Majumdar, & Gaus, 2010; Owen, Rentero, Magenau, Abu-Siniyeh, & Gaus, 2012). Development of another Laurdan derivative, C-Laurdan advances this possibility by increased sensitivity to membrane polarity and yielding brighter images in multi-photon microscopy (Kim et al., 2007). This latter approach was used to image ordered domains in brain slices at the depth of 100–250 μm in live tissue (Kim et al., 2008). Developing further applications for imaging lipid order of membrane in intact tissues will provide major insights into the impact of multiple pathological conditions on cell function.

Acknowledgments

Many thanks to Mr. Gregory Kowalsky for preparing the figures for the manuscript. I also thank Dr. Elizabeth Le Master, Mr Amit Paul, Mr. Victor Aguilar and Ms. Dana Lazarko for fruitful discussions and critical reading of the manuscript. The work is supported by NIH grants R01HL083298 and R01HL141120.

References

Aguilar, L. F., Pino, J. A., Soto-Arriaza, M. A., Cuevas, F. J., Sanchez, S., & Sotomayor, C. P. (2012). Differential dynamic and structural behavior of lipid-cholesterol domains in model membranes. *PLoS One*, 7(6), e40254.

Amaro, M., Reina, F., Hof, M., Eggeling, C., & Sezgin, E. (2017). Laurdan and Di-4-ANEPPDHQ probe different properties of the membrane. *Journal of Physics D: Applied Physics*, *50*(13), 134004.

Ayee, M. A., LeMaster, E., Shentu, T. P., Singh, D. K., Barbera, N., Soni, D., et al. (2017). Molecular-scale biophysical modulation of an endothelial membrane by oxidized phospholipids. *Biophysical Journal*, *112*(2), 325–338.

Ayee, M. A. A., Levitan I., (2021). Lipoprotein-induced increases in cholesterol and 7-ketocholesterol result in opposite molecular-scale biophysical effects on membrane

structure. Frontiers in Cardiovascular Medicine, 8, 715932. https://doi.org/10.3389/fcvm.2021.715932. eCollection 2021.
Bagatolli, L. A., Sanchez, S. A., Hazlett, T., & Gratton, E. (2003). Giant vesicles, Laurdan, and two-photon fluorescence microscopy: Evidence of lipid lateral separation in bilayers. *Methods Enzymology*, *360*, 481–500.
Baigent, C., Blackwell, L., Emberson, J., Holland, L. E., Reith, C., Bhala, N., et al. (2010). Efficacy and safety of more intensive lowering of LDL cholesterol: A meta-analysis of data from 170,000 participants in 26 randomised trials. *Lancet*, *376*(9753), 1670–1681.
Bogachkov, Y. Y., Chen, L., Le Master, E., Fancher, I. S., Zhao, Y., Aguilar, V., et al. (2020). LDL induces cholesterol loading and inhibits endothelial proliferation and angiogenesis in Matrigels: Correlation with impaired angiogenesis during wound healing. *American Journal of Physiology. Cell Physiology*, *318*(4), C762–C776.
Bonaventura, G., Barcellona, M. L., Golfetto, O., Nourse, J. L., Flanagan, L. A., & Gratton, E. (2014). Laurdan monitors different lipids content in eukaryotic membrane during embryonic neural development. *Cell Biochemistry and Biophysics*, *70*(2), 785–794.
Byfield, F. J., Tikku, S., Rothblat, G. H., Gooch, K. J., & Levitan, I. (2006). OxLDL increases endothelial stiffness, force generation and network formation. *Journal of Lipid Research*, *47*, 715–723.
Castelli, W. P., Anderson, K., Wilson, P. W., & Levy, D. (1992). Lipids and risk of coronary heart disease. The Framingham Study. *Annals of Epidemiology*, *2*(1-2), 23–28.
Cooper, R. A. (1977). Abnormalities of cell-membrane fluidity in the pathogenesis of disease. *The New England Journal of Medicine*, *297*(7), 371–377.
Danylchuk, D. I., Sezgin, E., Chabert, P., & Klymchenko, A. S. (2020). Redesigning solvatochromic probe laurdan for imaging lipid order selectively in cell plasma membranes. *Analytical Chemistry*, *92*(21), 14798–14805.
Dietrich, C., Bagatolli, L. A., Volovyk, Z. N., Thompson, N. L., Levi, M., Jacobson, K., et al. (2001). Lipid rafts reconstituted in model membranes. *Biophysical Journal*, *80*, 1417–1428.
Frank, P. G., Woodman, S. E., Park, D. S., & Lisanti, M. P. (2003). Caveolin, caveolae, and endothelial cell function. *Arteriosclerosis, Thrombosis, and Vascular Biology*, *23*(7), 1161–1168. (Epub ahead of print).
Garcia-Cruset, S., Carpenter, K. L., Guardiola, F., Stein, B. K., & Mitchinson, M. J. (2001). Oxysterol profiles of normal human arteries, fatty streaks and advanced lesions. *Free Radical Research*, *35*(1), 31–41.
Gaus, K., Gratton, E., Kable, E. P., Jones, A. S., Gelissen, I., Kritharides, L., et al. (2003). Visualizing lipid structure and raft domains in living cells with two-photon microscopy. *Proceedings of the National Academy of Sciences of the United States of America*, *100*(26), 15554–15559.
Gaus, K., Le Lay, S., Balasubramanian, N., & Schwartz, M. A. (2006). Integrin-mediated adhesion regulates membrane order. *The Journal of Cell Biology*, *174*(5), 725–734.
Gaus, K., Zech, T., & Harder, T. (2006). Visualizing membrane microdomains by Laurdan 2-photon microscopy. *Molecular Membrane Biology*, *23*(1), 41–48.
Golfetto, O., Hinde, E., & Gratton, E. (2013). Laurdan fluorescence lifetime discriminates cholesterol content from changes in fluidity in living cell membranes. *Biophysical Journal*, *104*(6), 1238–1247.
Gunther, G., Malacrida, L., Jameson, D. M., Gratton, E., & Sanchez, S. A. (2021). LAURDAN since weber: The quest for visualizing membrane heterogeneity. *Accounts of Chemical Research*, *54*(4), 976–987.
Hicks, J. B., Lai, Y., Sheng, W., Yang, X., Zhu, D., Sun, G. Y., et al. (2008). Amyloid-beta peptide induces temporal membrane biphasic changes in astrocytes through cytosolic phospholipase A2. *Biochimica et Biophysica Acta*, *1778*(11), 2512–2519.

Jay, A. G., & Hamilton, J. A. (2017). Disorder amidst membrane order: standardizing laurdan generalized polarization and membrane fluidity terms. *Journal of Fluorescence*, *27*(1), 243–249.

Jin, L., Millard, A. C., Wuskell, J. P., Clark, H. A., & Loew, L. M. (2005). Cholesterol-enriched lipid domains can be visualized by di-4-ANEPPDHQ with linear and nonlinear optics. *Biophysical Journal*, *89*(1), L04–L06.

Jin, L., Millard, A. C., Wuskell, J. P., Dong, X., Wu, D., Clark, H. A., et al. (2006). Characterization and application of a new optical probe for membrane lipid domains. *Biophysical Journal*, *90*(7), 2563–2575.

Kim, H. M., Choo, H.-J., Jung, S.-Y., Ko, Y.-G., Park, W.-H., Jeon, S.-J., et al. (2007). A two-photon fluorescent probe for lipid raft imaging: C-laurdan. *ChemBioChem*, *8*(5), 553–559.

Kim, H. M., Jeong, B. H., Hyon, J.-Y., An, M. J., Seo, M. S., Hong, J. H., et al. (2008). Two-photon fluorescent turn-on probe for lipid rafts in live cell and tissue. *Journal of the American Chemical Society*, *130*(13), 4246–4247.

Levitan, I., & Shentu, T. P. (2011). Impact of oxLDL on cholesterol-rich membrane rafts. *Journal of Lipids*, *2011*, 730209.

Levitan, I., Volkov, S., & Subbaiah, P. V. (2010). Oxidized LDL: Diversity, patterns of recognition and pathophysiology. *Antioxidants & Redox Signaling*, *13*, 39–75.

Lingwood, D., & Simons, K. (2010). Lipid rafts as a membrane-organizing principle. *Science*, *327*(5961), 46–50.

Ma, Y., Benda, A., Kwiatek, J., Owen, D. M., & Gaus, K. (2018). Time-resolved laurdan fluorescence reveals insights into membrane viscosity and hydration levels. *Biophysical Journal*, *115*(8), 1498–1508.

Mazeres, S., Joly, E., Lopez, A., & Tardin, C. (2014). Characterization of M-laurdan, a versatile probe to explore order in lipid membranes. *F1000Research*, *3*, 172.

Morandi, M. I., Kluzek, M., Wolff, J., Schroder, A., Thalmann, F., & Marques, C. M. (2021). Accumulation of styrene oligomers alters lipid membrane phase order and miscibility. *Proceedings of the National Academy of Sciences of the United States of America*, *118*(4), e2016037118.

Obaid, A. L., Loew, L. M., Wuskell, J. P., & Salzberg, B. M. (2004). Novel naphthylstyryl-pyridinium potentiometric dyes offer advantages for neural network analysis. *Journal of Neuroscience Methods*, *134*(2), 179–190.

Oh, M.-J., Zhang, C., LeMaster, E., Adamos, C., Berdyshev, E., Bogachkov, Y., et al. (2016). Oxidized LDL signals through Rho-GTPase to induce endothelial cell stiffening and promote capillary formation. *Journal of Lipid Research*, *57*(5), 791–808.

Owen, D. M., & Gaus, K. (2013). Imaging lipid domains in cell membranes: The advent of super-resolution fluorescence microscopy. *Frontiers in Plant Science*, *4*, 503.

Owen, D. M., Magenau, A., Majumdar, A., & Gaus, K. (2010). Imaging membrane lipid order in whole, living vertebrate organisms. *Biophysical Journal*, *99*(1), L7–L9.

Owen, D. M., Rentero, C., Magenau, A., Abu-Siniyeh, A., & Gaus, K. (2011). Quantitative imaging of membrane lipid order in cells and organisms. *Nature Protocols*, 7(1), 24–35.

Parasassi, T., De Stasio, G., d'Ubaldo, A., & Gratton, E. (1990). Phase fluctuation in phospholipid membranes revealed by Laurdan fluorescence. *Biophysical Journal*, *57*(6), 1179–1186.

Parasassi, T., De Stasio, G., Ravagnan, G., Rusch, R. M., & Gratton, E. (1991). Quantitation of lipid phases in phospholipid vesicles by the generalized polarization of Laurdan fluorescence. *Biophysical Journal*, *60*(1), 179–189.

Parasassi, T., Di Stefano, M., Loiero, M., Ravagnan, G., & Gratton, E. (1994). Influence of cholesterol on phospholipid bilayers phase domains as detected by Laurdan fluorescence. *Biophysical Journal*, *66*(1), 120–132.

Parasassi, T., & Gratton, E. (1995). Membrane lipid domains and dynamics as detected by Laurdan fluorescence. *Journal of Fluorescence*, *5*(1), 59–69.

Parasassi, T., Gratton, E., Yu, W. M., Wilson, P., & Levi, M. (1997). Two-photon fluorescence microscopy of laurdan generalized polarization domains in model and natural membranes. *Biophysical Journal*, *72*(6), 2413–2429.

Pike, L. J. (2006). Rafts defined: A report on the keystone symposium on lipid rafts and cell function. *Journal of Lipid Research*, *47*(7), 1597–1598.

Salvador-Castell, M., Brooks, N. J., Winter, R., Peters, J., & Oger, P. M. (2021). Non-polar lipids as regulators of membrane properties in archaeal lipid bilayer mimics. *International Journal of Molecular Sciences*, *22*(11).

Sanchez, S. A., Gunther, G., Tricerri, M. A., & Gratton, E. (2011). Methyl-beta-cyclodextrins preferentially remove cholesterol from the liquid disordered phase in giant unilamellar vesicles. *The Journal of Membrane Biology*, *241*(1), 1–10.

Sanchez, S. A., Tricerri, M. A., & Gratton, E. (2007). Interaction of high density lipoprotein particles with membranes containing cholesterol. *Journal of Lipid Research*, *48*(8), 1689–1700.

Shentu, T. P., Titushkin, I., Singh, D. K., Gooch, K. J., Subbaiah, P. V., Cho, M., et al. (2010). oxLDL-induced decrease in lipid order of membrane domains is inversely correlated with endothelial stiffness and network formation. *American Journal of Physiology Cell Physiology*, *299*(2), C218–C229.

Sonnino, S., & Prinetti, A. (2013). Membrane domains and the "lipid raft" concept. *Current Medicinal Chemistry*, *20*(1), 4–21.

Steinberg, D. (2002). Atherogenesis in perspective: Hypercholesterolemia and inflammation as partners in crime. *Nature Medicine*, *8*, 1211–1217.

Suhaj, A., Gowland, D., Bonini, N., Owen, D. M., & Lorenz, C. D. (2020). Laurdan and Di-4-ANEPPDHQ influence the properties of lipid membranes: A classical molecular dynamics and fluorescence study. *The Journal of Physical Chemistry B*, *124*(50), 11419–11430.

Wang, J., Megha, & London, E. (2004). Relationship between sterol/steroid structure and participation in ordered lipid domains (lipid rafts): Implications for lipid raft structure and function. *Biochemistry*, *43*, 1010–1018.

Weber, G., & Farris, F. J. (1979). Synthesis and spectral properties of a hydrophobic fluorescent probe: 6-Propionyl-2-(dimethylamino)naphthalene. *Biochemistry*, *18*(14), 3075–3078.

Xu, X., & London, E. (2000). The effect of sterol structure on membrane lipid domains reveals how cholesterol can induce lipid domain formation. *Biochemistry*, *39*, 843–849.

Yeagle, P. L. (1991). Modulation of membrane function by cholesterol. *Biochimie*, *73*, 1303.

Zidovetzki, R., & Levitan, I. (2007). Use of cyclodextrins to manipulate plasma membrane cholesterol content: Evidence, misconceptions and control strategies. *Biochimica et Biophysica Acta (BBA) - Biomembranes*, *1768*(6), 1311.

Fluorescence sensors for imaging membrane lipid domains and cholesterol

Francisco J. Barrantes*
Biomedical Research Institute (BIOMED), Catholic University of Argentina (UCA)–National Scientific and Technical Research Council (CONICET), Buenos Aires, Argentina
*Corresponding author: e-mail address: francisco_barrantes@uca.edu.ar

Contents

Current Topics in Membranes, Volume 88
ISSN 1063-5823
https://doi.org/10.1016/bs.ctm.2021.09.004

Abstract

Lipid membrane domains are supramolecular lateral heterogeneities of biological membranes. Of nanoscopic dimensions, they constitute specialized hubs used by the cell as transient signaling platforms for a great variety of biologically important mechanisms. Their property to form and dissolve in the bulk lipid bilayer endow them with the ability to engage in highly dynamic processes, and temporarily recruit subpopulations of membrane proteins in reduced nanometric compartments that can coalesce to form larger mesoscale assemblies. Cholesterol is an essential component of these lipid domains; its unique molecular structure is suitable for interacting intricately with crevices and cavities of transmembrane protein surfaces through its rough β face while "talking" to fatty acid acyl chains of glycerophospholipids and sphingolipids via its smooth α face.

Progress in the field of membrane domains has been closely associated with innovative improvements in fluorescence microscopy and new fluorescence sensors. These advances enabled the exploration of the biophysical properties of lipids and their supramolecular platforms. Here I review the rationale behind the use of biosensors over the last few decades and their contributions towards elucidation of the in-plane and transbilayer topography of cholesterol-enriched lipid domains and their molecular constituents. The challenges introduced by super-resolution optical microscopy are discussed, as well as possible scenarios for future developments in the field, including virtual ("no staining") staining.

1. Introduction

For many years, the unavailability of appropriate sensors precluded the unambiguous determination of the physical state of biological membranes and the possibility of unraveling the contribution of individual lipid species to the suspected heterogeneous distribution of lipids across the bilayer. Parallel developments of cell biological, biochemical and biophysical approaches in this area of research have produced steady advances in the membrane biology field, narrowing down the number of unknowns while posing fascinating new questions about the organization and dynamics of membranes.

From a historical perspective, interest in understanding the properties of lipids in membranes and micelles gained momentum in the early 1970s when various physical approaches converged to characterize lipid phases and polymorphisms using small-angle X-ray techniques pioneered by Vittorio Luzzati and coworkers in Gif-sur-Yvette (Luzzati, Gulik-Krzywicki, Rivas, Reiss-Husson, & Rand, 1968; Luzzati & Husson, 1962), nuclear magnetic resonance by Dennis Chapman and coworkers (Cater, Chapman, Hawes, & Saville, 1974) and electron spin resonance by Marsh and his group (Marsh, 1974). Initial attempts to use fluorescent probes to study membrane fragments from mitochondria employed the classical fluorochrome 8-anilino-1-naphthalene-sulfonic acid (ANS) introduced by Gregorio Weber in studies on protein conformation (Weber, 1952; Weber & Laurence, 1954) to gain information on the energy state of this organelle (Azzi, Chance, Radda, & Lee, 1969). Early applications of fluorescence techniques to the study of biomembranes can be traced back to pioneer studies in Gregorio Weber's laboratory. These studies were an extension of Weber's more general interest in fluorescence polarization theory and applications. The fluorescent probes perylene and 2-methylantracene were incorporated into synthetic micelles to apply Perrin's polarization equations in an early investigation of microviscosity and order in the hydrocarbon region of micelles (Shinitzky, Dianoux, Gitler, & Weber, 1971). Subsequent studies applied these techniques to study thermotropic phase transitions and the effect of cholesterol on the microviscosity and order of phospholipid-containing synthetic membranes (Cogan, Shinitzky, Weber, & Nishida, 1973; Jacobson & Wobschall, 1974). See chapter "Membrane tension" by Chao and Sachs for a thorough analysis of methods to investigate liquid-liquid phase transitions.

The topic of lipid domains in membranes is intimately associated with the physicochemical properties of cholesterol, owing to the crucial involvement of this neutral lipid in the formation of such specialized domains together with phosphoglycerolipids and sphingolipids with saturated acyl chains. As will become apparent throughout this review, efforts devoted to exploring the multiple properties of cholesterol in a lipid bilayer have also brought new insights in the area of lipid domains and, reciprocally, the tools developed for investigating domains have helped to unravel new properties of cholesterol.

Physicochemical studies on artificial membrane-mimetic systems consisting of phospholipid and cholesterol have documented the generation of separated and coexisting liquid-liquid phases, one rich in cholesterol and saturated lipids, the liquid-ordered phase (Lo), and the more "fluid" liquid-disordered (Ld) phase, enriched in unsaturated acyl chain lipids. These physicochemical properties of membrane lipids gave rise to the concept of lipid "rafts," a hypothesis of great gravitation in membrane biology for the last four decades (Ahmed, Brown, & London, 1997; Jacobson, Mouritsen, & Anderson, 2007; Simons & Ikonen, 1997). According to this concept, principles similar to those operating in the above-mentioned complex lipid mixtures in vitro are responsible for the self-organization of separate domains in natural cell membranes, generating lateral heterogeneities, i.e. laterally segregated regions or domains, in the lipid bilayer. In this scenario, the "raft" domains display the physicochemical characteristics of liquid-ordered (Lo) gels (Samsonov, Mihalyov, & Cohen, 2001) whereas the remainder of the lipids are purportedly in a liquid-disordered (Ld) phase. Because of their postulated highly dynamic nature and small lateral dimensions (10–200 nm) (Mayor & Rao, 2004; Sezgin, Levental, Mayor, & Eggeling, 2017) raft domains have been elusive to characterize. The interest in directly visualizing these domains has increasingly attracted membrane biologists because of the important functional properties attributed to these distinct regions of the membrane, especially in the realm of signal transduction, constituting hubs where certain membrane proteins are sorted out and platforms for a variety of signaling mechanisms.

Cholesterol remains the quintessential component of these liquid-ordered lipid domains. Unesterified cholesterol synthesized de novo in the endoplasmic reticulum (Maxfield & Tabas, 2005; Yeagle, 1989) or derived from external sources via low-density lipoprotein (LDL) receptor-mediated internalization (Brown & Goldstein, 1984) resides mostly (~90%) in the plasma membrane, where it can account for up to half of the total lipid content, depending on the cell type (Lange, Swaisgood, Ramos, & Steck, 1989; Lange, Ye, & Steck, 2004). However, only a small fraction of cholesterol (~15% of plasmalemmal lipids) termed "chemically active" cholesterol (Das, Brown, Anderson, Goldstein, & Radhakrishnan, 2014) is accessible for transport; the great majority of cell-surface cholesterol is in the form of complexes with phospholipids and SMs (also known as "chemically inactive" cholesterol) (Lange et al., 2004; Radhakrishnan, Anderson, & McConnell, 2000; Radhakrishnan & McConnell, 1999, 2000) and as such inaccessible for intracellular transport. Studies comparing the transport of newly synthesized cholesterol with that of viruses upon temperature arrest led to the conclusion that cholesterol is transported from the endoplasmic reticulum to the plasmalemma primarily via a non-vesicular process, independently of the classical exocytic (ER ->Golgi ->plasma membrane) trafficking route (Heino et al., 2000; Urbani & Simoni, 1990).

Early electron microscope studies using free-fracture cytochemistry (reviewed in Severs & Robenek, 1983) were probably among the first experimental data to trigger new ideas on the topography of membranes and the concept of cholesterol-rich Lo raft domains that has been so influential over the last 3 decades. The need to experimentally test these theories initially led to the use of biochemical methods combining detergent extraction with differential centrifugation. This indirect approach was and still is open to criticism due to its artifact-prone nature. New approaches have been introduced in recent years to avoid these artifacts. For instance, the observation that bilayer nanodisks formed by phospholipids and the amphipathic styrene maleic anhydride (SMA) copolymer preserve the functional and structural integrity of α-helical and β-barrel transmembrane proteins (Knowles et al., 2009) has made it possible to directly extract discrete patches of nanoscale dimensions ($\sim$10 nm) containing proteins with their surrounding lipid environment from the membrane without the use of detergents (Jamshad et al., 2011). Despite initial concerns similar to those raised against the Triton X-100/centrifugation assays (e.g. whether lipids exchanged with the rest of the bilayer and simply co-purified with the separated proteins (Cuevas Arenas et al., 2017)), these nanotechnology-based detergent-free methods for the extraction of membrane proteins with their vicinal lipid milieu via SMA-lipid particles (SMALPs) have more recently been validated in prokaryotic (Teo et al., 2019) and eukaryotic membranes like the retinal rod disk membranes (Sander et al., 2021).

The difficulty of using biochemical methods to test the occurrence of lipid domains in cells gave impetus to the development of fluorescent membrane sensors and in particular cholesterol probes aimed at identifying and characterizing these structures in situ, i.e. without destroying the integrity of the membrane and introducing the minimum possible structural modifications. As early as 1972 X-ray diffraction studies were conducted on phospholipid bilayers with incorporated ANS, 12-(9-anthroyl)-stearic acid and *N*-octadecylnaphthyl-2-amine 6-sulfonic acid to assess the location of these fluorescent molecules and the degree of perturbation they might introduce into the system (Lesslauer, Cain, & Blasie, 1972).

As discussed in this review, the design and synthesis of several organic compounds with appropriate fluorescence properties, the introduction of intrinsically fluorescent sterols like dehydroergosterol and cholestanetriol in combination with extracellular quenchers, and the discovery of fungal or bacterial toxins like filipin or perfringolysin O provided the tools to explore several hitherto unknown properties of cholesterol and learn about lipid domains in membranes, the subject of this review. I address visualization of lipid domains in situ, sensing transbilayer asymmetry and sensing membrane polarity. Finally, I discuss the new requirements imposed by the introduction of superresolution optical microscopy approaches, ending by describing possible future scenarios and developments in this field, including virtual ("no staining") staining.

2. Fluorescent sensors for imaging lipid domains in situ

The application of fluorescence microscopy in the field of membrane biology has a long and successful tradition, which evolved hand in hand with the development of appropriate sensors (for reviews see Demchenko, Mely, Duportail, & Klymchenko, 2009; Epand, Kraayenhof, Sterk, Sang, & Epand, 1996). Synthetic lipid membranes have provided useful model platforms to investigate several physicochemical properties of their counterpart natural biomembranes. One of the most extensively studied properties is the phase behavior exhibited when liquid–liquid phases coexist in the bilayer. This phenomenon is observed when phospholipids with saturated acyl chains and sterols condense to form a liquid-ordered (Lo) phase, which separates from a liquid-disordered (Ld) phase composed predominantly of lipids with unsaturated acyl chains (Bagatolli, 2006; Dietrich et al., 2001; Kahya, Scherfeld, Bacia, Poolman, & Schwille, 2003; Kahya, Scherfeld, & Schwille, 2005; Korlach, Schwille, Webb, & Feigenson, 1999; Veatch & Keller, 2002, 2003a, 2003b, 2005). Attempts to correlate the phase behavior of lipids in model systems at equilibrium with the behavior of lipids in natural membranes has often made use of fluorescence techniques (see e.g. Bacia, Scherfeld, Kahya, & Schwille, 2004; Baumgart et al., 2007). Giant vesicles derived from plasma membranes by chemically-induced blebbing of the plasma membrane of cultured cells have also been employed as a model to study phase segregation in membrane bilayers (Baumgart et al., 2007).

Sterol structure influences liquid-ordered domain formation. The behavior of different sterols was found to be similar in symmetric (mixtures of sphingomyelin (SM), 1,2-dioleoyl-sn-glycero-3-phosphocholine (DOPC) and cholesterol), and asymmetric vesicles (in which SM was introduced into the outer leaflet). Cholesterol and 7-dehydrocholesterol strongly stabilized ordered domains in symmetric model membranes, while lanosterol, epicholesterol and desmosterol produced a moderate level of stabilization and 4-cholesten-3-one did not stabilize Lo domains at all (St. Clair & London, 2019). The ability to support Lo domains decreased in the order 7-DHC >cholesterol > desmosterol > lanosterol > epicholesterol > 4-cholesten-3-one. Endocytosis levels and bacterial uptake are even more closely correlated with the ability of sterols to form ordered domains than previously thought, and do not necessarily require sterols to have a 3β-OH group (St. Clair & London, 2019).

The ideal condition for imaging lipid domains in cells or membrane model systems with fluorescence microscopy is that the fluorescent sensor exhibit a preferential partitioning for one type of domain only, or that it exhibit some spectroscopic property that differentiates between the two domains, such as a marked shift in the fluorescence emission. A systematic search for such sensors shows that most of the probes used for this purpose partition preferentially into the liquid-disordered phase and few do so in cholesterol-rich Lo domains.

2.1 Carbocyanines

Carbocyanines are voltage-sensitive compounds introduced in the early 1980s, originally for application in the Neurosciences (optical recordings of membrane potential) (see review in Honig & Hume, 1989) and gradually used more widely as neuronal tracers and for the study of membrane fluidity in a variety of cells and tissues. Carbocyanine dyes share two conjugated planar ring structures substituted at a single position with isopropyl, oxygen or sulfur, giving rise to the DiI (1,1',dioctadecyl-3,3,3'3'-tetramethylindocarbocyanine perchlorate), diO (3,3'-dioctadecyloxacarbocyanine) and diS families of fluorescent probes. One of the most used of these probes is diI-C18 (1,1'-dioctadecyl-3,3,3',3'-tetramethylindo-carbocyanine perchlorate). The length of the two acyl chains (e.g. 18 C atoms each in the case of DiI-C18) determines the affinity of the compound for the hydrophobic region of the membrane. Carbocyanines with short alkyl chains partition between the water phase and the membrane in a membrane potential-dependent manner, whereas the long-chain carbocyanine dyes are readily inserted into membrane bilayers.

In the field of physicochemistry of membranes, carbocyanines have been assayed in various model systems. The two carbocyanine dyes DiI16 and DiI18 (1,1'-dioctadecyl-3,3,3',3'-tetramethylindocarbocyanine perchlorate) have found application as sensors of membrane domains. For instance, giant unilamellar vesicles (GUVs) composed of DOPC/BSM/cholesterol (2:2:1) display separation into a liquid-disordered and a liquid-ordered phase at room temperature, and incorporation of the probes Fast DiO, Fast DiI and DiD-C18 make apparent the heterogeneous fluorescence distribution of the two coexisting phases (Baumgarten, Makielski, & Fozzard, 1991; Sezgin et al., 2012). In some cases the bulkiness of the fluorophore moiety of the probes results in mutual interferences. For instance, the partitioning of DiIC18 interferes with that of the phospholipid analogue BODIPY-PC [2-(4,4-difluoro-5,7-dimethyl-4-bora-3a,4a-diaza-s-indacene-3-pentanoyl)-1-hexa-decanoyl-sn-glycero-3-phosphocholine].

The long-chain carbocyanine probes of the DiI family are highly lipophilic, and as such are excellent markers of membranes for cell biology experiments requiring good long-term viability of the cells, strong emission, homogeneous labeling, and compatibility with aldehyde fixation, requirements fulfilled by these sensors. These properties have been successfully exploited for the cytochemical identification of neurons and retrograde labelling upon injection of the dye into embryonic preganglionic neurons (Godement, Vanselow, Thanos, & Bonhoeffer, 1987).

2.2 DPH diphenylhexatriene (DPH)

1,6-Diphenyl-1,3,5-hexatriene (DPH) is a linear molecule with two phenyl groups at the extremes of its linear hydrocarbon chain. It is barely soluble in aqueous media and highly fluorescent in organic solvents or in membranes. Among the earliest applications of the probe was the measurement of steady-state fluorescence anisotropy of DPH in lymphocyte and lymphoma cells (Shinitzky & Inbar, 1976) in conjunction with that of hydroxycoumarin and trans-parinaric acid to study the effects of cholesterol and cholest-4-en-3-one on the thermotropic behavior of dipalmitoylphosphatidylcholine bilayers. The authors demonstrated the broadening of the gel-to-liquid phase transition of the phospholipid by the sterols and one of the first indications of cholesterol-rich domains in bilayers (Ben-Yashar & Barenholz, 1989). The degree of polarization of DPH increases with increasing cholesterol content (Wharton, De Martinez, & Green, 1980). The position and relative orientation of a series of DPH derivatives in the membrane was established using nitroxide spin-labeled lipids as quenchers and parallax analysis (Kaiser & London, 1998). The cationic derivative of DPH, TMA-DPH (1-(4-trimethylammonium-phenyl)-6-phenyl-1,3,5-hexatriene), is more soluble than its parental compound and its polar moiety positions it in a shallower (0.3–0.4 nm) location in membrane bilayers, establishing electrostatic interactions with electronegative lipid atoms (do Canto et al., 2016). TMA-DPH has been used in studies on endocytic processes (Illinger, Duportail, Poirel-Morales, Gerard, & Kuhry, 1995; Illinger, Italiano, Beck, Waltzinger, & Kuhry, 1993; Illinger, Poindron, & Kuhry, 1991). Fluorescence lifetime measurement of TMA-DPH has mostly been used as a sensor of the exofacial membrane leaflet using either fluorescence anisotropy or fluorescence lifetime measurements (Chazotte, 2011). A phosphatidylcholine derivative of DPH senses slightly different regions of the plasma membrane than do DPH or TMA-DPH, but the probe has not had extensive use in membrane studies (Ferretti, Tangorra, Zolese, & Curatola, 1993; Tangorra, Ferretti, Zolese, & Curatola, 1994).

Steroids exert influence on Lo domain formation and stabilization, and on the extension of the ordered assemblies. Combining two independent biophysical methods (steady-state anisotropy measurements of DPH with temperature-dependent fluorescence quenching by a nitroxide spin-labeled phosphatidylcholine (PC) (12-SLPC)) we studied the effect of steroid structure on the above properties of liquid-ordered domains. Lo domain-promoting steroids were found to exhibit a small polar group at position C3, an isooctyl side chain bond at C17, absence of carbons attached to C23 (i.e., C24–C27), and absence of polar groups in the fused rings, with the exception of substitutions at position C3 in the A ring (Wenz & Barrantes, 2003).

2.3 Parinaric acid

cis-Parinaric acid (cis-trans-trans-cis-9,11,13, 15-octadecatetraenoic acid or 9Z,11E,13E,15Z-octadecatetraenoic acid), discovered by Tsujimoto and Koyanagi in 1933, is a naturally occurring conjugated polyunsaturated fatty acid found in the seeds of the Makita tree, indigenous to Fiji. It contains an unusual conjugated tetraene which confers fluorescence to the fatty acid. Upon excitation at 320 nm it emits at 432 nm. *cis*-parinaric acid has been employed for the measurement of lipid enzyme activities (phospholipase, lipase), and as an indicator of lipid peroxidation. The polarization of fluorescence and the lifetime of *cis*-parinaric acid have been exploited to learn about the physical status of biomembranes (Calafut, Dix, & Verkman, 1989; Ruggiero & Hudson, 1989a, 1989b). The ligand binding site for fatty acids and cholesterol carrier proteins (Schroeder, Myers-Payne,

Billheimer, & Wood, 1995; Stolowich et al., 1997) as well as its location in the membrane bilayer (Castanho, Prieto, & Acuña, 1996) and the role of cholesterol in liquid-ordered domains of caveolae (Gallegos, McIntosh, Atshaves, & Schroeder, 2004) have been investigated using this sensor.

2.4 NBD-derivatives

Adducts of the relatively small fluorophore 4-chloro-7-nitrobenzofurazan (NBD) have been used to label many different biological molecules, and lipids in particular, for instance fatty acids like the saturated octadecanoic acid derivative NBD-stearic acid [12-(*N*-methyl)-*N*-[(7-nitrobenz-2-oxa-1,3-diazol-4-yl)amino]-octadecanoic acid]. Phosphoglycerolipids such as PE have been tagged covalently with the fluorophore (NBD-DPPE [1,2-dipalmitoyl-sn-glycero-3-phosphoethanolamine-*N*-(7-nitro-2-1,3-benzoxadiazol-4-yl]), and this fluorescent lipid has been shown to be a sensor of the liquid-ordered phase together with the more bulky planar probe NAP (naphtho[2,3-a]pyrene) (Juhasz, Davis, & Sharom, 2010).

The so-called red-edge excitation shift or REES is observed with polar fluorophores in motionally restricted, viscous media, or in condensed phases where the dipolar relaxation time for the solvent shell around the fluorophore is comparable to or longer than its fluorescence lifetime (Chattopadhyay, 2003). NBD-tagged lipids show the REES effect, shifting their maximum fluorescence emission toward higher wavelengths, owing to the shift in the excitation wavelength toward the red edge of the absorption band. The REES effect was originally interpreted as stemming from the slow rates of solvent relaxation (reorientation) around the fluorophore in its excited state (Chattopadhyay, 2003), i.e. from mobility constraints of the solvent molecules surrounding the fluorescent sensor. Other authors considered that the REES effect is independent of temperature and cholesterol content, and therefore insensitive to water relaxation phenomena, attributing the phenomenon to the anomalous transverse location of the probe in the membrane (Amaro, Filipe, Prates Ramalho, Hof, & Loura, 2016). The peculiar orientation of the nitro group determines the fluorescence lifetime of NBD in the membrane (Filipe, Pokorná, Hof, Amaro, & Loura, 2019). The REES effect has been exploited to determine the topology of the membrane-bound colicin E1 channel in the bilayer, utilizing the emission of tryptophan residues (Tory & Merrill, 2002). Chattopadhyay and coworkers have made extensive use of REES of NBD-tagged phospholipids, which were reported to undergo a noticeable red-edge shift upon incorporation into artificial bilayers (Chattopadhyay & Mukherjee, 1993; Mukherjee & Chattopadhyay, 1995; Raghuraman & Chattopadhyay, 2004).

2.5 FITC-chitosan derivatives

Fluorescein-isothiocyanate (FITC)-labeled glycol chitosan molecules have been reported to label lipid-ordered domains in model or cell membranes (Jiang et al., 2016). Since the FITC-labeled glycol chitosan molecules do not completely insert into the lipid bilayer but interact presumably via attractive electrostatic interactions and/or hydrophobic interactions with the ordered domain lipids, it is considered that they do not disturb the membrane organization. This family of probes are non-permeant indirect reporter groups of cholesterol localization in lipid-ordered domains in living cells.

2.6 Polyethylene-glycol-derivatized cholesterol

Another membrane-impermeable sensor of cholesterol-rich ordered lipid domains is the fluorescent derivative of polyethylene oxide (poly (ethyleneglycol)cholesteryl ether (fPEG-cholesterol) introduced by Kobayashi and coworkers (Sato et al., 2004). The membrane impermeability of fPEG-Chol make it a sensor of liquid-ordered lipid microdomains located in the outer plasma membrane leaflet (Sato et al., 2004). fPEG-cholesterol has been used alone or in conjunction with the sphingomyelin-binding protein lysenin (Hullin-Matsuda & Kobayashi, 2007) to label cholesterol-rich lipid domains in living cells (Hullin-Matsuda & Kobayashi, 2007). fPEG-cholesterol has also been used to follow the endocytic internalization of the nicotinic receptor protein and cholesterol (Kamerbeek et al., 2013).

2.7 Intrinsically fluorescent cholesterol analogues dehydroergosterol and cholestanetriol

Two polyene sterols with a high degree of structural similarity to cholesterol have been extensively used in biophysical and cell biological studies of membranes addressing the location and trafficking of cholesterol in cells. These are dehydroergosterol ((DHE, $\Delta^{5,7,9(11),22}$-ergostatetraen-3β-ol) and cholestanetriol (5-β-cholestane-3-α,7-α, 12-α-triol, CTL) (Fig. 1).

DHE is a fluorescent derivative of ergosterol, a natural sterol found in yeasts such as *Saccharomyces cerevisiae*. CTL is the closest structural analogue of cholesterol, differing only in having two additional double bonds in the planar ring system while its aliphatic side chain is identical to that of

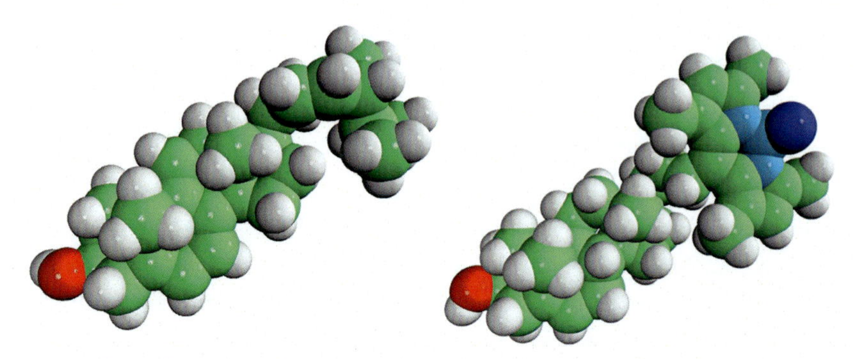

Fig. 1 Molecular models depicting one of the smallest cholesterol fluorescent sensors. Left: the naturally fluorescent cholesterol analogue dehydroergosterol (ergosta-5,7,9(11),22-tetraen-3β-ol, DHE), and right: the bulkier TopFluo-cholesterol (23-(dipyrrometheneboron difluoride)-24-norcholesterol). *Images provided by Avanti Polar Lipids.*

cholesterol. The three conjugated double bonds in the sterol backbone provide CTL with intrinsic fluorescence. Shared with DHE, this advantageous property is counterbalanced by the fact that absorption and emission occur in the UV region of the spectrum, thus reducing their usefulness for imaging live cells, particularly with conventional wide-field fluorescence microscopy, because of its damaging effects on cell viability (Schroeder et al., 2005). In addition, DHE has a high photobleaching rate. Nonetheless, DHE has been successfully combined with wide-field microscopy using quartz optics to study the cellular topography and traffic of this cholesterol analogue ((Hao et al., 2002; Hao, Mukherjee, & Maxfield, 2001; Mondal, Mesmin, Mukherjee, & Maxfield, 2009; Mukherjee, Zha, Tabas, & Maxfield, 1998; Wüstner, Herrmann, Hao, & Maxfield, 2002) and see review in (Maxfield & Wüstner, 2012)). The quartz optics can be bypassed by the use of multiphoton laser excitation (McIntosh et al., 2007; McIntosh et al., 2008).

Biophysical studies have shown that DHE induces the formation of liquid-ordered domains in bilayers and giant unilamellar vesicles (GUVs) containing phospholipids and cholesterol (Garvik, Benediktson, Simonsen, Ipsen, & Wüstner, 2009). Contrary to expectations, the probe does not exhibit red-edge excitation shift in model membranes, irrespective of the phase state of the membrane, an observation that has been interpreted as implying the probe's lack of environmental sensitivity, attributable to the very small change in its dipole moment upon excitation (Chattopadhyay, Biswas, Rukmini, Saha, & Samanta, 2021).

Unlike DHE and ergosterol, CTL exerts an ordering effect on the fatty acid acyl chains of the hydrophobic region of the bilayer(Solanko, Modzel, Solanko, & Wüstner, 2015). Both DHE and CTL induce ordering of phospholipid bilayers composed of palmitoyl-oleoyl-phosphatidylcholine or palmitoyl-oleoyl-phosphatidylcholine/cholesterol, although the degree of order is lower than that induced by cholesterol itself (Robalo, do Canto, Carvalho, Ramalho, & Loura, 2013).

DHE and CTL exhibit small differences in metabolism in comparison to cholesterol. For instance, DHE is unable to bind to several proteins involved in cholesterol metabolism in the ER and activate sterol regulatory element-binding protein 2 (Mesmin et al., 2011). Uptake and esterification of DHE by non-lipoprotein pathways is lower than that of cholesterol. DHE is also a less efficient substrate of acyl-CoA:cholesterol acyltransferase than cholesterol (Liu, Chang, Westover, Covey, & Chang, 2005). The applications of DHE or CTL in fluorescence microscopy of living cells are limited by their low quantum yield, high photobleaching rate, and their absorption and emission bands in the UV region of the spectrum, overlapping with the UV absorption of essentially all protein constituents of the cell, with inherent deleterious effects when irradiated in this spectral region using wide-field microscopy.

2.8 Sphingomyelin sensors

Sphingophosphorylcholine and a fatty acid coupled to a fluorophore like nitrobenzoxadiazole (NBD) or boron dipyrromethene difluoride (BODIPY) have been used as sphingomyelin sensors, and, indirectly, of ordered lipid domains in membranes. Short-chain fatty acid acyl moieties in these sensors favor partition of the probe into liquid-disordered bulk phase lipid domains, although natural sphingomyelins show preference for liquid-ordered domains (Kinoshita et al., 2017; Kishimoto, Ishitsuka, & Kobayashi, 2016). The water-soluble quencher dithionite selectively quenched NBD-labeled sphingomyelin in the outer leaflet of the membrane (McIntyre & Sleight, 1991). The pore-forming sphingomyelin-binding protein lysenin has been used to study the localization of the lipid in cells (Carquin et al., 2014) and discloses the occurrence of sphingomyelin-rich domains larger than 100 nm in diameter in the inner leaflet of the plasma membranes of human skin fibroblasts and neutrophils (Murate et al., 2015). A non-toxic mushroom protein termed Nakanori that specifically binds to a complex of sphingomyelin and cholesterol has been used to label cell-surface ordered domains and also lipid domains that colocalized with inner leaflet small GTPase H-Ras, but not K-Ras (Makino et al., 2017). Nakanori was also employed to study the altered distribution of cholesterol and sphingomyelin in Niemann-Pick type C fibroblasts (Makino et al., 2017).

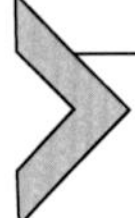

3. Sensing transbilayer topography of lipids, cholesterol, and lipid domains

The distribution of lipid components in the axis perpendicular to the plane of the membrane bilayer has interested biophysicists for decades. The transbilayer topography of lipids across the membrane bilayer, including that of cholesterol, is important for cell homeostasis and is hence under tight metabolic control. The cell maintains higher concentrations of PCs and SMs in the outer membrane leaflet, whereas phosphatidylethanolamine (PE), phosphatidylserine (PS) and phosphatidylinositol (PI) phosphoglycerolipids predominate in the inner leaflet (Kobayashi & Menon, 2018; van Meer, Voelker, & Feigenson, 2008). This varies between membranes: phosphatidylserine is mainly at the cytoplasmic leaflet in the plasmalemma, whereas it is in the lumenal leaflet in the endoplasmic reticulum (Kobayashi & Menon, 2018). Asymmetric distribution of lipids obviously obeys functional requirements, particularly in the plasma membrane, which separates two totally different universes. Maintaining asymmetrical topography requires active processes and energy expenditure: the free energy barrier associated with the spontaneous translocation of a charged lipid from one face of the membrane to the other is high.

The asymmetry does not pertain exclusively to the phospholipids; lipidomic studies have established that the compositional differences between the two membrane leaflets also includes the degree of unsaturation of the acyl chains and packing, as well as asymmetries in membrane protein shape and translational diffusion. Using enzymatic digestion combined with mass spectroscopy of red blood cells, the distribution of around 400 lipid species could be determined: the inner leaflet of the plasma membrane contains on average two-fold more unsaturated lipids than the exoplasmic leaflet, while the latter is more densely packed and hence less diffusive than the cytoplasmic-facing hemilayer (Lorent et al., 2020).

The distribution of cholesterol across the two leaflets of the plasma membrane is less clear than that of phospholipids; despite its relatively high concentration in plasma membranes the issue remains controversial, with some authors maintaining that cholesterol occurs mostly in the exofacial hemilayer and others sustaining the opposite view (Giang & Schick, 2016; Steck &

Lange, 2018; van Meer, 2011). Cholesterol transbilayer distribution depends on the ATP-dependent P4-ATPases, also termed "flippases," which actively translocate the aminophospholipids PS and PE to the cytosolic-facing hemilayer, whereas the ATP-binding cassette (ABC) ABCA1 and ABCG1 transporters translocate lipids in the opposite direction as exporters rather than "floppases" ((van Meer, 2011) and see review in (Liu, 2019)). Dysfunction of these transporters is known to occur in various diseased conditions like metabolic syndrome, atherosclerosis, cardiovascular disease, cystic fibrosis, retinal degeneration, and cholestasis (Liu, 2019).

A method based on tunable orthogonal cholesterol sensors has been developed to simultaneously quantify cholesterol in the two leaflets of the plasma membrane (Liu et al., 2017). The technique revealed marked transbilayer cholesterol asymmetry in several mammalian cells: cholesterol concentration in the inner leaflet was found to be ~12-fold lower than that in the outer leaflet. The asymmetry was maintained by active transport of cholesterol from the inner to the outer leaflet and by its retention at the exofacial leaflet. The increase in the cytoplasmic leaflet cholesterol level was triggered in a stimulus-specific manner -e.g. Wnt signaling activity- leading to the notion that cholesterol serves as a signaling function (Liu et al., 2017). Another bio-orthogonal cholesterol sensor has recently been developed and used in conjunction with super-resolution microscopy to image lipid domains of <50 nm diameter in the plasmalemma of live cells (Lorizate et al., 2021).

The transbilayer distribution of cholesterol remains controversial, partly because (i) the sterol does not occur in a single pool; (ii) cholesterol does not show an homogeneous in-plane distribution; (iii) it can flip-flop between the two leaflets of the bilayer (Ikonen, 2008); and (iv) the distribution of the cholesterol pool residing in liquid-ordered lipid domains ("rafts") is likely to differ from that of the "free" cholesterol pool in the membrane. In fact the original "raft" hypothesis favored the notion that Lo lipid domains occur in the outer, exofacial leaflet of the membrane (Simons & Ikonen, 1997). Ever since this seminal idea, the subject of cholesterol distribution across the bilayer has remained a contentious and methodologically difficult issue in membrane biology (see (Kobayashi & Menon, 2018; Murate et al., 2015; Murate & Kobayashi, 2016) for reviews on current techniques to study membrane asymmetry). Experimentally, the transbilayer topography of cholesterol has been studied in early work using the intrinsic fluorescent analogue DHE in combination with fluorescence quenching with trinitrobenzene sulfonic acid. DHE was preferentially incorporated into the inner, cytoplasmic-facing hemilayer of the plasma membrane of fibroblast (Hale & Schroeder, 1982) and other cell types (Kier, Sweet, Cowlen, & Schroeder, 1986).

Another aspect of membrane asymmetry that remains controversial is whether a cholesterol-rich ordered domain in one bilayer leaflet can induce ordered domain formation in the opposite membrane leaflet; in other words, whether there is a coupling between domains at opposite regions of the membrane hemilayers. Research on this topic has benefited from the use of artificial lipid vesicles (London, 2019). Using asymmetric and symmetric lipid vesicles composed of SM, cholesterol, and either unsaturated dioleoyl PC (DOPC) or 1-palmitoyl 2-oleoyl PC (POPC) the temperature dependence of Lo domain formation was studied using Förster resonance energy transfer (St. Clair, Kakuda, & London, 2020). In cholesterol-containing asymmetric SM + PC outside/PC inside vesicles, the PC-containing inner leaflet destabilized Lo domain formation in the outer leaflet; ordered domain formation was suppressed by asymmetry over the entire temperature range measured (St. Clair et al., 2020). SMs appear to play an important role in inter-leaflet coupling: SM in the outer leaflet appears to induce Lo domains in asymmetric vesicles (Lin & London, 2015) and long-chain SM depletes cholesterol from the cytoplasmic leaflet in asymmetric membranes (Karlsen, Bruhn, Pezeshkian, & Khandelia, 2021). The presence of SMs with long acyl chain (24 C atoms) in the outer leaflet suppresses Lo domains in the plasma membrane of HeLa cells, under which conditions cholesterol is preferentially found in the inner hemilayer (Courtney et al., 2018).

See chapter "Mass spectrometry-based lipid analysis and imaging" by Pathmasiri et al. for additional views on assessing membrane lipid asymmetry.

3.1 Intrinsically fluorescent cholesterol analogues, dehydroergosterol and cholestatrienol

In the context of transbilayer sidedness, DHE appears to be a reporter of asymmetry since it partitions in a non-homogeneous manner along the axis perpendicular to the plane of the lipid bilayer. Using trinitrobenzene sulfonic acid as quencher, DHE was shown to partition more favorably into the inner leaflet of the plasma membrane in fibroblast (Hale & Schroeder, 1982) and other cell types (Kier et al., 1986).

In terms of the lateral distribution of DHE in the plasma membrane, work by Wüstner showed that DHE colocalized with fluid membrane-preferring phospholipids in the plasmalemma and associated compartments like microvilli, filopodia, nanotubules and membrane blebs induced by

F-actin disruption, but not in cholesterol-rich domains (Wüstner, 2007). However, Wüstner's group later reported that DHE preferentially partitioned in the liquid-ordered membrane domains and induced the formation of such domains in supported bilayers and giant unilamellar vesicles (GUVs) made of dipalmitoyl-phosphatidylcholine, dioleoyl-phosphatidylcholine and Chol (Garvik et al., 2009). The ambiguity of these findings thus casts doubt on DHE's characterization as an "ideal" probe.

3.2 Filipin

Filipin, together with nystatin and amphotericin B, are antibiotics with primarily antifungal activity and a lower antibacterial activity, exerted by promoting membrane leakage. Filipin is a complex mixture of at least 4 polyene amphipathic macrolides isolated from the mycelium and culture filtrates of *Streptomyces filipinensis*. This organism has found increasing applications in antifungal biotechnology and as a model organism for understanding how butyrolactones control secondary metabolism and antibiotic production (Barreales, Payero, de Pedro, & Aparicio, 2018). The name filipin derives from its biological origin and this, in turn, from its geographical stem. Fig. 1 shows the structure of filipin.

Filipin was used initially to localize cholesterol in histochemical studies (Börnig & Geyer, 1974; Geyer & Börnig, 1975), and in electron microscope studies of freeze-fractured cells to study the relationship between cholesterol and coated pits/endocytic vesicles (Montesano, Vassalli, & Orci, 1981) as well as the topology of cholesterol in the Golgi complex (Orci et al., 1981). The advantages and disadvantages of filipin as a sensor for the localization of cholesterol in membranes, and its use and abuse, were recognized early (Miller, 1984).

Filipin binds free, unesterified cholesterol (Maxfield & Wüstner, 2012). The fact that it does not bind to esterified sterols provides a technical advantage since the lack of esterified cholesterol in lipid droplets improves the signal-to-noise ratio in cellular studies.

The mechanism through which filipin-cholesterol interactions are established in not totally clear. The three main hypotheses summarized by Prieto and coworkers (Castanho, Coutinho, & Prieto, 1992) are: (i) filipin:sterol form planar 1:1 complexes between the two hemilayers of the membrane; (ii) the complex is established at a 1:1 stoichiometry but at the membrane surface, with concomitant disordering of the bilayer, or (iii) the interaction takes place at the exofacial leaflet of the membrane and the deformation of the bilayer occurs due to the increased surface pressure. van Deenen and coworkers were among the first to study filipin with electron microscopy based on the ability of the antibiotic to induce the formation of pits in erythrocyte membranes, and the presumption that these occurred via filipin-cholesterol interactions (Kinsky, Luse, Zopf, Van Deenen, & Haxby, 1967). The fluorescence properties of filipin were exploited some years later to establish the location of cholesterol in sarcoplasmic reticulum membrane vesicles (Drabikowski, Lagwińska, & Sarzala, 1973). These authors measured excitation and emission spectra of filipin in dimethylformamide (exc.: 360 nm; emiss.: 480 nm) and determined the existence of energy transfer between tryptophan residues and filipin. The photophysical properties and topography of filipin in model membrane systems has been studied using fluorescence quenching with nitroxide spin labels (Castanho & Prieto, 1995).

One drawback of filipin is that it requires fixation of the cell, and hence it cannot be employed in dynamic studies with living cells. There are also drawbacks related to the low specificity of filipin, the first being that filipin does not enable the experimentalist to discern whether the identified cholesterol molecules are those originated by de novo synthesis or whether they are incorporated from the blood stream via lipoprotein-mediated mechanisms. The second, more important downside, is that filipin has been reported to recognize ganglioside GM1 and cholesterol with similar affinities (Arthur, Heinecke, & Seyfried, 2011). Despite these drawbacks filipin is still a useful sensor of cholesterol topography and has played an important role in our understanding of the pathophysiology of Niemann-Pick disease type C, a cholesterol lysosomal storage disease, over the course of more than three decades of research (Pentchev et al., 1985; Saha et al., 2020).

3.3 Perfringolysin O

Perfringolysin O, also called theta-toxin, is a cholesterol-binding cytolysin produced by the anaerobic bacterium *Clostridium perfringens*. Cholesterol-dependent cytolysins belong to a family of pore-forming toxins secreted by Gram-positive bacteria. The C-terminus of the cytolysins is also termed domain 4 or simply D4, a region of the protein consisting of two four-stranded β-sheets. Another cytolysin, pneumolysin, is a thiol-activated cytolysin toxin from *Streptococcus pneumoniae* (Saunders, Mitchell, Walker, Andrew, & Boulnois, 1989). Pneumolysin and perfringolysin O share a high degree of homology in their membrane-associating regions (Savinov &

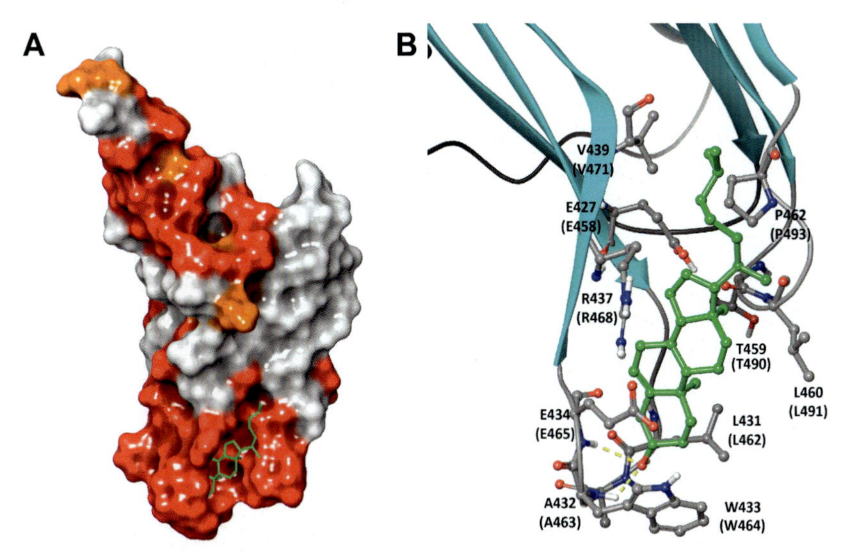

Fig. 2 (A) Induced-fitting docking model of the C-terminus (domain 4 or D4) of the cytolysin pneumolysin (surface rendering, where red are identical, orange conserved, and white non-conserved residues in comparison to perfringolysin O analogous residues) with a cholesterol molecule (green wire representation) docked on its surface. The D4 domain is responsible for the cholesterol-dependent membrane binding of cytolysins to cholesterol-containing membranes, thus explaining why this portion of the molecule has been used as a cholesterol sensor. (B) Detail of the cholesterol binding region in pneumolysin predicted by the induced-fitting model. The cholesterol molecule is shown in green ball-and-stick rendering and the pneumolysin undecapeptide as cyan arrows and grey ball-and-stick rendering. *Reproduced from Savinov, S. N., & Heuck, A. P. (2017). Interaction of cholesterol with perfringolysin O: What have we learned from functional analysis?* Toxins (Basel), *9(12). doi:10.3390/toxins9120381 under Creative Common CC BY license.*

Heuck, 2017). Fig. 2 shows the structure of pneumolysin with a cholesterol molecule docked on its surface. The mechanism of action of perfringolysin is relatively well known. It is initiated by the insertion of a large portion of its β-barrel structure into a cholesterol-containing region of the target membrane; the presence of cholesterol is an absolute requirement for this to occur.

Cytolysin binding to the exofacial hemilayer of cell membranes is the first in a series of steps that may lead to cell death upon infection with *C. perfringens* or *Streptococcus pneumoniae*. Binding is followed by insertion of the toxin, provided cholesterol is present at the cell surface at a certain threshold (40–60%) concentration (Maekawa & Fairn, 2015). Cholesterol is necessary and sufficient to trigger perfringolysin O binding to cells, and addition of SM inhibits binding (Flanagan, Tweten, Johnson, & Heuck, 2009). Insertion is followed by dimerization, multimerization, pre-pore formation, and pore formation. At low, sub-lytic concentrations, toxin binding to cells, e.g. mast cells, results in activation, degranulation, and transcriptional activation that leads to cytokine production. At higher toxin concentrations, large pores may be formed by high order multimers of the toxin as shown in Fig. 3, an oligomerization product that may lead to necroptotic cell death. Upon infection with *C. perfringens*, the toxin activates caspase-1 and the inflammasome cascade in the target cell which can progress to gas gangrene (Yamamura et al., 2019).

Epicholesterol does not interact with perfringolysin O, and its addition to the cell reduces toxin binding (Flanagan et al., 2009). A study found that the toxin did not discriminate between cholesterol-containing liquid-ordered domains and coprostanol-containing disordered lipid domains (Lin & London, 2013). These authors also confirmed that the toxin does not bind to epicholesterol, a raft-promoting 3α-OH sterol. Using a Förster resonance energy transfer assay in model membrane vesicles containing coexisting ordered and disordered lipid domains, perfringolysin O preferentially partitioned into ordered domains in vesicles containing high

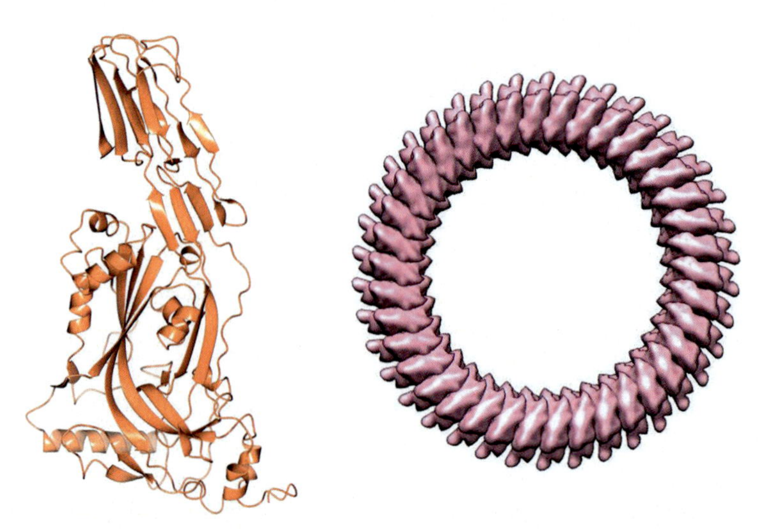

Fig. 3 Left: Crystal structure of a monomer of perfringolysin O from *C. perfringens* obtained by X-ray crystallography at 2.9 Å resolution (PDB entry: 1MJI). Right: pore structure of pneumolysin (PDB entry: 2BK1, EMDataResource EMD-1107) obtained by fitting the α-carbon trace of perfringolysin O onto a cryo-EM map of pneumolysin (Marshall et al., 2015). *Image created using CCP4mg molecular graphics (McNicholas, Potterton, Wilson, & Noble, 2011).*

and low-transition temperature lipids plus cholesterol or 1:1 (mol/mol) cholesterol/epicholesterol, whereas they preferred disordered domains in vesicles containing high-Tm and low-Tm lipids plus 1:1 (mol/mol) coprostanol/epicholesterol. The authors concluded that the association of proteins with ordered lipid domains is dependent upon both the association of the protein-bound sterol with ordered domains and hydrophobic match between the transmembrane segments and the ordered domains (Lin & London, 2013). One of the first cytochemical applications of perfringolysin using fluorescence microscopy revealed the sequestration of cholesterol in caveolae (Fujimoto, Hayashi, Iwamoto, & Ohno-Iwashita, 1997). The proteolytically-modified and biotinylated perfringolysin O has been used to image cholesterol-rich membranes by electron microscopy (Möbius et al., 2002).

An important biotechnological step towards the obtention of perfringolysin O in pure form and sufficient amounts was the cloning and expression of the gene coding for this cytolysin (Tweten, 1988). This enabled the characterization of peptide cleavage products of the toxin, such as the enzymatically cleaved C-terminal domain 4 (D4 fragment) that interacts with sterols (Nelson, Johnson, & London, 2008) and derivatization with fluorophores. The toxin's capacity to recognize cholesterol above a certain concentration in the exofacial leaflet of the plasma membrane has been exploited to produce various cholesterol biosensors, mostly based on fluorescent derivatives of the protein or its peptide fragments. A biosensor based on the D4 from *Clostridium perfringens* perfringolysin O toxin was found to recognize cholesterol in the inner leaflet of the plasma membrane (Maekawa & Fairn, 2015).

The genetically-encoded cholesterol biosensor based on the fourth domain of *C. perfringens* theta-toxin, D4H, was shown to recognize cholesterol in the cytoplasmic-facing leaflet of the plasma membrane and organelles (Maekawa & Fairn, 2015). D4H disassociates from the plasma membrane upon cholesterol extraction and after perturbations of cellular cholesterol trafficking. When used in combination with a recombinant version of the biosensor, these authors found that phosphatidylserine was essential for retaining cholesterol in the cytosolic leaflet of the plasma membrane. The authors indicated that the sensor is suitable for super-resolution microscopy applications.

The D4 peptide coupled to the green fluorescent protein (D4-GFP) and the biotinylated derivative (BCtheta) were reported to label specifically cholesterol-rich lipid domains (Ohno-Iwashita et al., 2004; Waheed et al., 2001) following biochemical criteria (detergent extraction with sucrose gradient centrifugation and enrichment in detergent-resistant membranes). The probes were also employed to determine whether the inner leaflet of the plasma membrane contained cholesterol-rich domains isolated using biochemical procedures. According to this criterion, the cytoplasmic-facing hemilayer was found to contain cholesterol-enriched domains (Hayashi, Shimada, Inomata, & Ohno-Iwashita, 2006). D4-GFP was also used to measure cholesterol content in the plasma membrane by flow cytometry (Wilhelm, Voilquin, Kobayashi, Tomasetto, & Alpy, 2019). Another fluorescent adduct (with enhanced GFP), EGFP-D4, was used to validate the inner-leaflet preferential partition of cholesterol in the bilayer. Controlled expression levels of wild-type EGFP-D4 and several mutants thereof showed that wild-type EGFP-D4 did not localize in the plasmalemma, whereas the D434A mutant did (Buwaneka, Ralko, Liu, & Cho, 2021) The specificity of perfringolysin O and its derivatives for cholesterol-rich Lo lipid domains contrasts with that of filipin, which binds in a non-specific manner to cholesterol.

Cholesterol is known to activate mTORC1 kinase, considered to be the master growth regulator, by recruiting the enzyme to the lysosomal surface. This activation requires the delivery of cholesterol across endoplasmic reticulum-lysosomal contacts via cholesterol carrier proteins like the oxysterol binding protein (OSBP). In cells lacking OSBP, the recruitment of mTORC1 is inhibited due to impaired cholesterol transport to lysosomes (Lim et al., 2019). These authors used the red fluorescent probe mCherry to label D4H*, an improved version of the original D4H in which Y415A and A463W mutations were introduced, and which binds to membranes when cholesterol exceeds 10% molar content. In non-permeabilized cells, the new biosensor mCherry-D4H* labeled only the outer leaflet of the plasmalemma. The probe was delivered to the cell interior via a liquid nitrogen pulse that permeabilized the plasma membrane but left intact the lysosomal membrane. mCherry-DH4* did not bind to lysosomal membranes, indicating cholesterol content below 10%. In contrast, the sensor bound to lysosomal membrane from patients carrying the pathogenic NPC1 mutant characteristic of Niemann-Pick disease type C (Lim et al., 2019).

3.4 Fluorescent theonellamides

Theonellamides are bicyclic dodecapeptides from marine sponges of the genus *Theonella*. They have cytotoxic and antifungal properties, inhibiting yeast cell growth by interfering with the endogenous sterol, ergosterol. Fluorescently-labeled derivatives of theonellamides (e.g. aminomethyl-coumarin or BODIPY-FL-labelled theonellamides) have been used as sterol sensors, recognizing 3β-hydroxy-sterols like cholesterol and ergosterol. In mammalian cultured cells the probes display a patchy distribution in the plasma membrane, similar to that exhibited by filipin. Although theonellamides exhibit low membrane permeability, they also label intracellular organelles (Nishimura et al., 2013).

3.5 Sterol-transfer protein, Osh4

The oxysterol-binding protein-related proteins (Orp) are conserved from yeast to humans and are implicated in the regulation of sterol homeostasis and in signal transduction pathways. Osh/Orp proteins transport sterols between organelles and are involved in phosphoinositide metabolism. Osh4 is one such yeast cytosolic sterol transfer protein known to bind cholesterol and 25-hydroxy-cholesterol and transport these sterols to the ER; in exchange it transports phosphatidylinositol-4-phosphate (PI(4)P) back (de Saint-Jean et al., 2011; Im, Raychaudhuri, Prinz, & Hurley, 2005). Osh4p has been shown to extract and transport dehydroergosterol, two processes inhibited by (PI(4)P) (de Saint-Jean et al., 2011). Recently eOsh4 was engineered from Osh4 to remove the sterol transfer activity and the affinity for 25-hydroxy-cholesterol and PI(4)P. eOsh4 acts as a ratiometric sensor WCR-eOsh4, which tightly binds vesicles containing low concentrations of cholesterol ($\leq$5 mol%). WCR-eOsh4 has high affinity and specificity for cholesterol-containing membranes and can be used to detect low-abundance cholesterol in the membrane, independent of its lipid environment (Buwaneka et al., 2021)

3.6 Correlative studies using fluorescent cholesterol sensors and other physical techniques

Cyclodextrins have been widely employed to manipulate the cholesterol content of biological membranes. In addition, cyclodextrins have been used to compare their ability to extract different fluorescent cholesterol analogues. Differences were observed in the kinetics of cyclodextrin-mediated extraction of NBD-cholesterol and two BODIPY-labeled cholesterol analogues from the membrane, which were followed by measuring the Förster resonance energy transfer between a rhodamine-labeled phosphatidylethanolamine and the cholesterol analogues (Milles et al., 2013). A multiplicity of factors determines the differential ability of cyclodextrin extraction, including the probes' hydrophobicities and their orientation within the bilayer, affinity with cyclodextrin and the impact of the latter on lipid order, among other variables.

Ganglioside and SM probes tagged with BODIPY, as well as NBD cholesterol derivatives PC and SM, were imaged using fluorescence microscopy and atomic force microscopy (AFM) to test the ability of the probes to sense lipid domains (Shaw et al., 2006). The authors observed that some of the probes purported to be reporters of liquid-ordered (raft) lipid domains partitioned into non-raft regions of the membrane, warranting caution in the use of fluorescence microscopy as the sole criterion for lipid domain identification.

Photolabeling techniques can afford a high degree of specificity to identify the amino acid residues in the target protein. A powerful combination is obtained by adding a fluorescent tag to the photolabeling compound. This has recently been accomplished with a series of cholesterol analogue photolabeling reagents to which fluorescent tags (e.g. the red-emitting TAMRAazide (5-carboxytetramethylrhodamine-azide)) were attached via click chemistry on their alkyne moiety. These cholesterol-mimetic compounds retain cholesterol's ability to suppress HMG-CoA synthase, the key enzyme for the de novo cholesterol synthesis, and may prove useful not only for identifying cholesterol sites on transport proteins but also for identifying the subcellular localization of the incumbent carrier proteins in dynamic studies of cholesterol trafficking.

3.7 Probes for specific phospholipids

Recent years have witnessed the design and production of sensors exploiting the known affinity of proteins for certain lipids and enzymes involved in lipid metabolism. These developments were aided by the discovery of fluorescent proteins and the biotechnology of these proteins. Fluorescent proteins currently cover a wide palette of the visible spectrum and new variants are increasingly being produced in the far-red/near-infrared and infrared regions, partly in response to requirements in the field of super-resolution microscopy. In this section, the use of fluorescent proteins is exemplified in their application to the study of some phospholipid membrane classes.

3.8 Polyphosphoinositide

Despite their relatively low amounts in membranes, polyphosphoinositides (poly-PIs) are an important class of lipids involved in the vesicular transport of proteins and lipids between the different compartments of eukaryotic cells, budding, membrane fusion, and cytoskeleton dynamics. Dysregulation of the enzymes involved in their metabolism is increasingly being related to oncogenesis, myopathies, and neuropathies (see review in (De Craene, Bertazzi, Bär, & Friant, 2017)). Tamas Balla and coworkers have for years studied the cell biology of this key class of lipids and the involvement of plekstrin homology (PH) domains. Early studies characterized the role phospholipase C δ, which specifically binds PtdIns(4,5)P2. One of the probes developed towards this end was the green fluorescent protein variant GFP-PH-PLC, the GFP-tagged PH domain of the phospholipase C δ (Varnai & Balla, 1998).

Other poly-PI sensors were subsequently developed, like the one employing the PtdIns4P binding domain of SidM (P4M) of the secreted effector protein SidM from the bacterial pathogen *Legionella pneumophila*. A green fluorescent protein adduct, GFP-P4Mx2 (GFP-conjugated tandem fusion of P4M domain consisting of amino acids 546–647 of *Legionella pneumophila* SidM) was produced to investigate the distribution of phosphatidylinositol 4-phosphate (PtdIns4P). Pools of this poly-PI were found in Golgi membranes, the plasma membrane, and Rab7-positive late endosomes/lysosomes (Hammond, Machner, & Balla, 2014), reflecting the wide distribution of this lipid class in the cell.

3.9 Phosphatidyserine

Using quick-freeze and freeze-fracture replica labeling electron microscopy it was recently possible to establish the transbilayer distribution of PS in the ER membrane (Tsuji et al., 2019). PS was found to reside predominantly in the exofacial, cytoplasmic-facing leaflet of the membrane. PS is maintained in the cytosolic face of the ER membrane by a mechanism similar to that operating in the plasmalemma that depends on TMEM16K family scramblases (Tsuji et al., 2019). The probe used by Tsuji and coworkers to identify PS was based on the protein evectin-2 (GST-2xPH) and its mutant GST-2xPH(K20E). Evectin-2 is a recycling endosome-resident protein (Kay & Fairn, 2019) essential for the retrograde transport from recycling endosomes to the trans-Golgi network (Uchida et al., 2011). Evectin-2 contains an N-terminal pleckstrin homology (PH) domain and a C-terminal hydrophobic region. It is the PH domain of evectin-2 that binds PS specifically. Green fluorescent protein (GFP) derivatives of two evectin-2 molecules in tandem (GFP-Evt2PH) (Uchida et al., 2011) were recently used by Li and coworkers (Li et al., 2021) as a live-cell sensor of PS together with Lact-C2-GFP, another genetically encoded fluorescent

PS biosensor based on the C2 domain of lactadherin (Lact-C2) (Yeung et al., 2008). Li and collaborators contrast live-cell sensors (GFP-Evt-2PH and Lact-C2-GFP) with "purified" sensors like GST-2xPH, the evectin fluorescent protein used by Tsuji and coworkers in their electron microscopy work (Tsuji et al., 2019).

The members of the evolutionarily conserved GRAMD1 family of ER-resident transport proteins (GRAMD1a/1b/1c) are known for their involvement in cholesterol homeostasis through regulated transport of the "accessible" cholesterol pool from the plasma membrane to the ER (see review in Naito and Saheki, 2021). The participation of the GRAMD1 proteins in lipid homeostasis goes beyond this important function: they harbor separate but structurally neighboring sites for cholesterol and anionic lipids like PS in their GRAM domain. This allows GRAMD1 proteins to function as a coincidence detector for cholesterol and PS (Ercan, Naito, Koh, Dharmawan, & Saheki, 2021). Some domains, including PH domains, contain synergistic binding sites that each exhibit low affinity for the two lipids. The simultaneous binding of two lipids to an individual domain occurs synergistically when both lipids are present in the target membrane (Lemmon, 2008). Ercan and coworkers stably expressed a fluorescence derivative of GRAMD1b (GFP-GRAM1b WT) and its mutant G187L (GFP-conjugated GRAM domain of GRAMD1b with a single G187L mutation in TKO cells) to follow the recruitment of the GRAMD1b protein to the ER and plasma membrane and the influence of the G187L mutation on GRAMD1-dependent cholesterol transport.

3.10 Phosphatidic acid

Structurally, phosphatidic acid (PA) is the simplest glycerophospholipid and a minority component in membranes. Despite its low abundance, it is a structural determinant of membrane shape depending on its location in the bilayer and displays various other roles, including second messenger functions in important cellular mechanisms, vesicular trafficking, cytoskeletal organization, secretion and cell proliferation (Liu, Su, & Wang, 2013). Proteins that possess PA binding domains like the yeast proteins Spo20p, Opi1p and Raf1 constitute possible targets for fluorescence derivatization. Spo20p has been found in the plasma membrane of mammalian cells and Opi1 p shuttles between ER and nuclear membranes. Fragments of these proteins containing the PA-binding domain were obtained as fusion proteins in *E. coli* by Vitale and coworkers (Kassas et al., 2017). Phagocytic cells like the macrophages require phospholipase-generated PA to remodel their plasma membrane and augment their surface area during phagocytosis. The GFP-fused PA-binding protein probes and GFP-PDE4A1, a GFP-tagged single exon in the unique N-terminal region of the cAMP-specific phosphodiesterase, were used to study the affinity of the binding domain for PA, the intracellular sites of PA synthesis in the course of phagocytosis, and the recruitment of PA to the plasma membrane (Kassas et al., 2017).

3.11 Glycerol-3-phosphate acyltransferase

SEIPIN is a protein implicated in both adipogenesis and lipid droplet expansion that interacts with microsomal isoforms of glycerol-3-phosphate acyltransferase (GPAT). Abnormally elevated in SEIPIN-deficient cells, this enzyme is associated with the block in adipogenesis and abnormal lipid droplet morphology observed in loss-of-function mutations in SEIPIN related with the Berardinelli-Seip congenital lipodystrophy 2 syndrome (Pagac et al., 2016). These authors designed a series of sensors (mCherry-tagged GPAT4, HA-tagged seipin, GFP-tagged seipin, mCherry-tagged seipin, and a combined HA/GFP/mCherry-fused human seipin) to study the relationship between the enzyme, SEIPIN, and the lipodystrophy.

4. Sensing membrane polarity

The issue of membrane polarity -often termed "membrane fluidity"- has been approached by a variety of biophysical methods such as electron spin resonance, nuclear magnetic resonance, and fluorescence spectroscopy. The three methodologies require appropriate sensors. In the fluorescence spectroscopy approach, fluidity of biomembranes is sensed indirectly through measurement of the polarity of the probe's microenvironment, usually reflected in characteristic spectral signatures of the probe calibrated in membrane-mimetic systems (Klymchenko, 2017).

4.1 Laurdan

Laurdan (6-dodecanoyl-2-dimethylamino naphthalene) was conceived by Gregorio Weber, together with other similar fluorescent probes like Prodan (6-propionyl-2-dimethylaminonaphthalene) as an environmental sensor (Weber & Farris, 1979). Laurdan was initially applied to characterize the polarity of protein regions in bovine serum albumin and myoglobin, and introduced in membrane biophysics only years later (Parasassi, Conti, &

Gratton, 1986). The groups of Parasassi, Gratton, Bagatolli and Gauss have contributed a variety of methodological developments covering both static and dynamic imaging modalities that contributed to establishing Laurdan not only as the gold standard in solvatochromic membrane polarity sensing but also as a reference multi-sensor of membrane-related phenomena (Bagatolli, Gratton, & Fidelio, 1998; Bagatolli, Maggio, Aguilar, Sotomayor, & Fidelio, 1997; Fiorini, Curatola, Kantar, Giorgi, & Gratton, 1993; Gaus et al., 2003; Gaus, Zech, & Harder, 2006; Levi, Wilson, Cooper, & Gratton, 1993; Parasassi, De Stasio, d'Ubaldo, & Gratton, 1990; Parasassi, De Stasio, Ravagnan, Rusch, & Gratton, 1991; Parasassi, Di Stefano, Ravagnan, Sapora, & Gratton, 1992; Parasassi & Gratton, 1992; Parasassi, Loiero, Raimondi, Ravagnan, & Gratton, 1993a, 1993b; Parasassi, Ravagnan, Rusch, & Gratton, 1993). Since it first began to be used forty years ago, Laurdan has become a well-established tool and continues to provide information in both cuvette fluorimetry and fluorescence microscopy studies.

Laurdan exhibits high sensitivity to the polarity of its environment and to the molecular dynamics of the surrounding dipoles. Dipolar relaxation processes are reflected as relatively large spectral shifts of its emission (Weber & Farris, 1979). Measurements of the so-called general polarization (GP) of Laurdan exploit its advantageous spectral properties, as initially measured by time-resolved fluorescence emission spectra in cuvette studies (Parasassi et al., 1986). Laurdan's molecular structure dictates the positioning of its lauric acid acyl chain in the bilayer hydrophobic core and its naphthalene aromatic ring at the hydrophilic/hydrophobic interface region of the membrane bilayer. In this region, the overwhelming "solvent" surrounding Laurdan naphthalene moiety is made up of dipole-carrying water molecules. Laurdan senses the relaxation rates of these water dipoles, with rotational times in the range of its fluorescence lifetime (Parasassi, Krasnowska, Bagatolli, & Gratton, 1998) and hence it reports on the vicinal dielectric milieu: when no relaxation occurs Laurdan fluorescence emission is blue-shifted and GP values are high, indicating low water content (and mobility), coincident with a tight packing of the lipid acyl chains such as observed in the liquid-ordered lipid phase. Laurdan does not show preference for a single lipid phase, partitioning ubiquitously into both Ld and Lo phases. The liquid-disordered phase allows more water molecules to populate the interface region; Laurdan senses the dipolar relaxation of this larger number of water molecules, affecting its own excited state dipole moment (Weber & Farris, 1979) and resulting in red-shifted emission and lower GP values. The position of the fluorescence emission maxima ($\sim$440 in the gel phase and 490 nm in the liquid-disordered phase) is maintained over a wide range of temperatures (Antollini et al., 1996; Bacalum, Zorilă, & Radu, 2013).

These properties of Laurdan fluorescence can be translated into the fluorescence microscopy realm to provide topographical information. Early measurements undertaken in membrane-mimetic synthetic lipid systems and isolated native membranes using two-photon microscopy led to the visualization of lateral heterogeneities in membranes rendered by Laurdan GP (Parasassi, Gratton, Yu, Wilson, & Levi, 1997). This paved the way for characterization of the domain separation and coexistence in GUVs (Bagatolli & Gratton, 1999; Bagatolli, Parasassi, & Gratton, 2000; Montes, Alonso, Goni, & Bagatolli, 2007) and planar bilayers (Brewer, de la Serna, Wagner, & Bagatolli, 2010). Laurdan GP in combination with fluorescence correlation spectroscopy (FCS) was subsequently employed to probe cell membrane heterogeneity and detect nanoscopic mobile structures, presumably lipid domains, containing tightly packed molecules (Sanchez, Tricerri, & Gratton, 2012).

Application of the phasor geometric plot method (Jameson, Gratton, & Hall, 1984) to fluorescence lifetime imaging (FLIM) of Laurdan in the fluorescence microscope opened up new possibilities to visualize membrane heterogeneities -lipid domains- with enhanced precision. Recent reviews discuss the contributions of the phasor approach and its applications to hyperspectral analysis and lifetime imaging (Gunther, Malacrida, Jameson, Gratton, & Sánchez, 2021; Ranjit, Malacrida, Jameson, & Gratton, 2018). The FLIM-phasor combination using Laurdan fluorescence lifetime was first applied to study the influence of cholesterol content and changes in membrane fluidity and phospholipid order in live cell measurements ((Golfetto, Hinde, & Gratton, 2013); reviewed in (Golfetto, Hinde, & Gratton, 2015)). Dual-channel recording of Laurdan FLIM with the phasor method facilitates the dissection (unmixing) of up to three components of the image (Fereidouni, Bader, & Gerritsen, 2012). In the case of Laurdan, the interesting distinction is between environmental polarity information vs. dipolar relaxation phenomena informing on water content in the membrane.

We have used Laurdan to explore the lipid microenvironment of the paradigm fast ligand-gated ion channel, the neurotransmitter receptor for acetylcholine. To this end we introduced a hitherto unexploited property of the probe, i.e. its ability to act as a Förster resonance energy transfer

(FRET) acceptor of tryptophan emission, using the transmembrane tryptophan residues of the nicotinic receptor as donors. We were thus able to characterize the physical state of environmental lipids within Förster distances from these membrane-embedded tryptophan residues in a native membrane environment (Antollini et al., 1996; Antollini & Barrantes, 1998; Antollini & Barrantes, 2002) (reviewed in (Antollini & Barrantes, 2007)) and see chapter "Methods for assessment of membrane protrusion dynamics" by Fauser et al. on the evaluation of membrane structure by Laurdan imaging.

4.2 di-n-ANEPPDHQ

In the early 90s Loew and coworkers introduced di-4-ANEPPS, the naphthyl analogue of the aminostyryl class of fluorescent dyes, as a fast potentiometric sensor for membrane studies (Loew et al., 1992). The compound exhibits an increase in fluorescence emission together with a red shift of its excitation spectrum upon hyperpolarization of the membrane. Di-4-ANEPPDHQ is one in a series of naphthylstyryl-pyridinium (di-n-ANEPPDHQ) fluorescent compounds introduced subsequently by Loew and his group (Obaid, Loew, Wuskell, & Salzberg, 2004). These probes have the same chromophore as the potentiometric sensor di-8-ANEPPS and the quaternary ammonium headgroup (DHQ) of RH795, thus resulting in two positive charges vs the neutral propylsulfonate-ring nitrogen combination (Obaid et al., 2004). Di-4-ANEPPDHQ can discriminate between liquid-ordered phases and liquid-disordered phases coexisting in model membranes using either linear or nonlinear microscopies (Jin, Millard, Wuskell, Clark, & Loew, 2005). The fluorescence emission spectrum of di-4-ANEPPDHQ is blue-shifted ~60 nm in the liquid-ordered phase (emission maximum 610 nm) compared with the liquid-disordered phase (emission maximum 560 nm) and exhibits strong second harmonic generation in the liquid-disordered phase compared with the liquid-ordered phase. These spectral changes can be conveniently followed by the simple GP approach based on the linear combination of the emission maxima. Tested in large unilamellar lipid vesicles, di-4-ANEPPDHQ showed a cumulative 60-nm difference in emission spectra in cholesterol-containing LUVs in the liquid-ordered state versus cholesterol-free vesicles in the liquid-disordered phase (Jin et al., 2006).

Using GP measurements, a comparative study of Laurdan vs di-4-ANEPPDHQ found the former to be more sensitive to temperature, and the naphtylstyryl probe more sensitive to cholesterol content (Amaro, Reina, Hof, Eggeling, & Sezgin, 2017). As shown in Fig. 4, the two probes exhibit an environmentally-sensitive shift in their emission spectra and can be used in a complementary manner to image simultaneously coexisting phases in the same specimen.

We have employed di-4-ANEPPDHQ imaging in an attempt to observe the distribution of ordered and disordered lipid regions and their topographical relationship to the location of the nicotinic acetylcholine receptor at the plasma membrane (Kamerbeek et al., 2013). We found that the receptor protein, imaged with fluorescent antibodies, was roughly equally distributed between diffraction-limited liquid-ordered and disordered domains. The proportion of receptor aggregates associated with liquid-ordered domains diminished upon cholesterol depletion, and this was accompanied by a decrease in di-4-ANEPDHQ GP values.

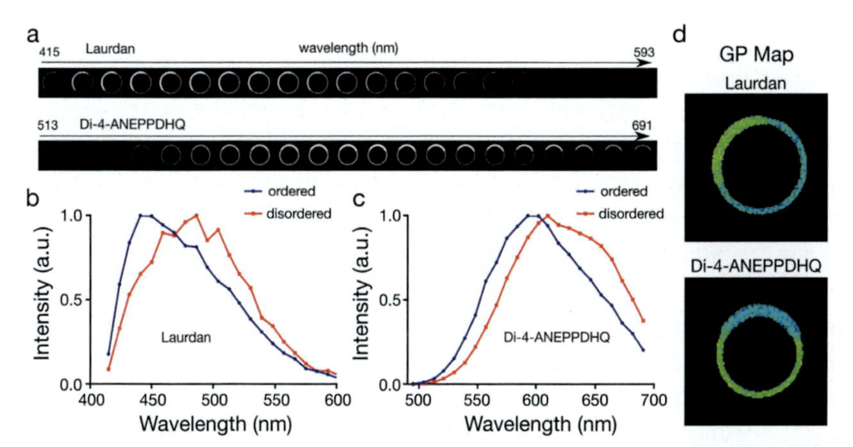

Fig. 4 Generalized polarization (GP) imaging of Laurdan and di-4-ANEPPDHQ in giant plasma membrane-derived vesicles with coexisting phases. (A) Spectral images of the vesicles. (B, C) fluorescence emission spectra of Laurdan and di-4-ANEPPDHQ in liquid-ordered (red) and disordered (blue) lipid domains coexisting in the vesicles. (D) Fluorescence microscopy images mapping the coexisting distribution of the two probes on the perimeter of the giant vesicles. *Reproduced from Amaro, M., Reina, F., Hof, M., Eggeling, C., & Sezgin, E. (2017). Laurdan and Di-4-ANEPPDHQ probe different properties of the membrane.* Journal of Physics D: Applied Physics, 50*(13), 134004. doi:10.1088/1361-6463/aa5dbc under the terms of the Creative Commons Attribution 3.0 license.*

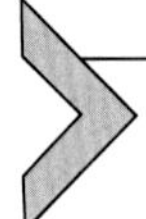

5. Super-resolution microscopy triggers the development of new sensors

Super-resolution optical microscopy (nanoscopy) revolutionized and revived light microscopy, establishing new standards in fluorescence microscopy (Hell, 2015). It has currently acquired the status of a mature technique and the applications cover a wide spectrum in many disciplines of biology, physics, materials science, microbiology, biotechnology, and chemistry (Betzig, 2015; Moerner, 2015).

The revolutionary facet of nanoscopy is related to its ability to circumvent the century-old dogma, the diffraction-barrier of light (Abbe, 1873): the degree of detail that can be resolved by a conventional light microscope is fundamentally limited by diffraction. Imaged with conventional lenses, an infinitely small point source produces a spot of finite volume, the point spread function (PSF), and two such small point sources that are closer together than the half-width of the PSF half-width will overlap and be observed as a single object. The resolution in the focal plane can be approximated as $0.5\ \lambda/\text{N.A.}$, with λ being the wavelength of light and N.A. the numerical aperture of the objective lens. Using visible light ($\lambda \sim 550$ nm) and a high-N.A. objective lens (N.A. ~ 1.4) the attainable resolution is ~ 200 nm. Nanoscopy has gone beyond this diffraction barrier, and this achievement was accompanied by corresponding advances in the design and chemical synthesis of organic compounds that fulfill the requirements of a given super-resolution microscopy modality.

Resolution can be improved by restriction of the fluorescence emission to an area much smaller than the PSF. Stimulated emission depletion (STED) nanoscopy is a direct scanning modality of super-resolution optical microscopy whose essential principle is the de-activation of fluorophores in the immediate perimetric volume surrounding the interrogating beam, thereby spatially confining the emission volume (Hell & Wichmann, 1994).

Under conditions used for single-molecule detection, photobleaching follows a two-step mechanism (Eggeling, Widengren, Rigler, & Seidel, 1998). The collision of $^3O_2$ molecular oxygen with the long-lived triplet state of fluorophores depletes the latter via triplet–triplet energy transfer annihilation, quenching fluorescence emission and returning the fluorophore to its ground singlet-state. In addition, reactive oxygen species such as singlet oxygen ($^1O_2$), superoxide, and hydroxy radical are produced (Helmerich, Beliu, Matikonda, Schnermann, & Sauer, 2021). The reactive oxygen species is doubly damaging in fluorescence microscopy since it hampers cell viability, reduces the fluorescence count rate in STED and induces blinking of single-molecule localizations in the other family of super-resolution microscopy modalities, the single-molecule localization microscopies (SMLM). In the latter approach, fluorophores are stochastically switched on and off ("blink"), and the localizations are estimated off-line by computing the centroids of their point-spread functions. The accuracy of the localizations depends on the density of the fluorophores and the number of photons collected per localization. Nanoscopy has therefore required special attention to the buffer conditions for fixed and live-cell imaging containing antioxidants such as *n*-propyl gallate, *p*-phenylenediamine or ascorbic acid to mitigate the effects of reactive oxygen species (Henriques, Griffiths, Hesper Rego, & Mhlanga, 2011) and optimization of fluorescent probes for STED (Connor, Byrne, Berselli, Long, & Keyes, 2019; Man et al., 2021; Nizamov et al., 2016; Xu et al., 2020) and of switchable probes for live-cell and SMLM (Grimm et al., 2015; Grimm et al., 2016; van de Linde, Heilemann, & Sauer, 2012) imaging modalities.

In the STED modality of nanoscopy, the off-switching of the fluorescence emitters by the doughnut-shaped depletion volume allows only the central emitters in the "on" emission state to contribute their photons during scanning (Wildanger, Rittweger, Kastrup, & Hell, 2008). The resolution increases as the depletion beam increases and restricts emission to smaller volumes. The number of possible excitation cycles that the fluorescent probe can withstand is an important requisite for the success of the STED technique, and this depends on the photophysical stability of the fluorophore (Gould, Hess, & Bewersdorf, 2012; van der Velde, Smit, Punter, & Cordes, 2018).

Photobleaching, the irreversible diminution of the fluorescence emission, or its transient counterpart, the conversion of the fluorophore into non-emissive species ("blinking") (Ha & Tinnefeld, 2012) can occur via intersystem crossing, chemical reactions, or redox processes that are enhanced in the case of fluorophores lying inside the cell, where the presence of e.g. glutathione or nicotinamide adenine dinucleotide compromise fluorescence emission intensity. The high (de)excitation illumination conditions imposed by nanoscopy have posed special requirements in terms of photostability of the probes. A major effort has been devoted to the design of new strategies for the organic synthesis of suitably photostable and photoconversion-resistant fluorophores, or probes for spectral unmixing, spectrally-resolved lifetime imaging and other nanoscopic modalities.

Rhodamines (Bossi, Belov, Polyakova, & Hell, 2006; Boyarskiy et al., 2008), caged NN rhodamines (Belov, Wurm, Boyarskiy, Jakobs, & Hell, 2010) and cell-penetrant rhodamine and fluorogenic carbopyronines with absorption maxima in the 500–630 nm region have been synthesized for live-cell STED microscopy (Butkevich et al., 2016). These probes are amenable to emission depletion via orange-red (618 nm) and near-infrared (775 nm) lasers. Triarylmethane fluorophores with increased stability and in particular improved resistance to oxidative "photobluing" via photoconversion, also aimed at STED microscopy of living cells (Butkevich, Bossi, Lukinavičius, & Hell, 2019) have been recently incorporated into this repertoire of new fluorescent probes.

Blue-shifted fluorescence excitation and emission maxima (hypochromic shift) in the range of 660-735 nm have been reported to occur with the photoproducts of far-red absorbing cyanine dyes Cy5 and Alexa-Fluor647. With the latter probe, anomalous emission due to a photoconverted blue-shifted fluorescent product of the cyanine core was observed when imaging HIV virions at the cell surface, and attributed to the high irradiation at 561 nm or 488 nm in addition to the 644 nm absorption maximum of the fluorophore (Dirix et al., 2018). The opposite, bathochromic shift ("photoredding") has also been reported for some fluorophores. Fluorophore photobleaching and blinking are both accompanied by diminution of the fluorescence intensity, whereas photobluing of photoredding lead to spectral inconsistency and misidentification of spectral signals, particularly important in multicolor fluorescence microscopy.

The now well-established HaloTag method is used in most cases to incorporate the fusion proteins into living cells. The HaloTag technique relies on the use of a protein sensor genetically fused to an engineered enzyme (modified *Rhodococcus rhodochrous* dehalogenase), which covalently links the protein of interest to its target. Only the 6'-carboxy isomers of the above rhodamine and fluorogenic carbopyronines have been successfully employed to tag proteins.

Photoactivatable fluorescent probes have so far had little application in STED microscopy due to their low solubility in aqueous media and because they require two-photon activation via the STED beam. Hell and coworkers have recently reported the synthesis of ONB-2SiR, a fluorophore that can be both photoactivated in the UV region and specifically de-excited by STED at 775 nm (Weber et al., 2021).

One field of fluorescence microscopy that has benefited enormously from the combined efforts of biology and organic chemistry is that of fluorescent proteins. Developments in this area have provided cell biologists the opportunity to exploit various photophysical properties of fluorescent proteins, i.e. photo-transformations like photo-switching (as used in STORM nanoscopy (Bates, Huang, Dempsey, & Zhuang, 2007; Bates, Huang, & Zhuang, 2008)), photoactivation (as exploited in PALM (Hess, Girirajan, & Mason, 2006; Manley et al., 2008)) or photoconversion (as utilized in PALM microscopy with mEos or Dendra2 fluorescent proteins (Adam et al., 2011)). The reader is referred to several reviews on the applications of fluorescent proteins (Fernández-Luna, Coto, & Costa, 2018; Kremers, Gilbert, Cranfill, Davidson, & Piston, 2011; Truong & Ferré-D'Amaré, 2019). Fluorescent proteins have been successfully used to trace proteins in cells, study the localization of membrane proteins or follow their lateral motion in living cells (Manley et al., 2008), but they have still not been applied in lipid domain studies. Due to their much lower background and phototoxicity compared to proteins with fluorescence emission in the visible region of the spectrum (Tosheva, Yuan, Matos Pereira, Culley, & Henriques, 2020), infrared fluorescent proteins offer interesting possibilities for development and application in super-resolution microscopy. This is a needy area and an opportunity for the bioengineering field, given the scarcity of near-infrared and infrared-emitting proteins with great potential for live-cell nanoscopy imaging (Hense et al., 2015; Matlashov et al., 2020; Shcherbakova et al., 2016; Yu et al., 2014).

The availability of new organic fluorophores suitable for super-resolution imaging has impacted on the field of lipid membrane domains, given the nanometric dimensions of these lateral heterogeneities and the increased resolution afforded by nanoscopy. GUVs and their analogous giant plasma membrane vesicles obtained from cells are convenient objects to test fluorescent sensor molecules in parallel under identical experimental conditions. That is precisely the strategy followed by Petra Schwille and coworkers in order to compare the behavior of several fluorescent lipid domain sensors. Label size, polarity, charge, and position of the probes on the one hand, and lipid headgroup and membrane composition on the other determined the complex partitioning of the sensors (Sezgin et al., 2012). "Raft" lipid analogues partitioned preferentially into the giant plasma membrane vesicles, whereas in synthetic GUVs they partitioned into liquid-disordered lipid domains.

A fluorescent glycerophospholipid analogue consisting of a saturated 18-carbon acyl chain PE and a hydrophilic far-red emitting fluorophore (KK114), tethered to the head group of PE by a long polyethylene glycol

(PEG) linker, was synthesized to probe diffusion characteristics of lipids in membranes (Honigmann, Mueller, Hell, & Eggeling, 2013). The lateral diffusion of the probe was studied with fluorescence correlation spectroscopy in the STED mode (STED-FCS), a technique which interrogates molecular motions in the femtoliter volumes interrogated by this technique. On a mica support, the lipid analogue diffuses freely in both the Ld and Lo phases, with diffusion coefficients of 1.8 $\mu m^2 s^{-1}$ and 0.7 $\mu m^2 s^{-1}$, respectively. Unlike most far-red emitting fluorescent lipid analogues, KK114-PEG-PE partitions predominantly into Lo domains in phase-separated ternary bilayers, but the tightly-packed lipid acyl chains in the Lo phase only slow down diffusion without producing anomalous sub-diffusion. In contrast, STED-FCS measurements on mica-supported membranes showed anomalous sub-diffusion, which the authors attributed to the transient partitioning of the lipid analogue into lipid domains of nanoscopic dimensions, where diffusion is slowed down.

A pyrene-based fluorescent ceramide conjugate, PyLa- C17Cer, has been synthesized and applied to living cells to identify lipid droplets using STED microscopy (Connor et al., 2019). The parent compound, PyLa, is a pyrene carboxyl core appended with 3,4-dimethylaminophenyl moiety. The probe permeates inside the cell and exhibits high fluorescent quantum yield, mega-Stokes shift, high photochemical stability, and low cytotoxicity. Since it does not show triplet emission at low temperatures, the authors suggest its use for fluorescence correlation spectroscopy (FCS), a highly suitable spectroscopic technique to match STED for the study of lipid dynamics in membranes (Eggeling, 2015).

The solvatochromic probe di-4-ANEPPDHQ introduced by Loew and colleagues has been employed as a sensor of membrane physical status. We employed this probe to depict the partitioning between ordered and disordered regions of the plasmalemma with diffraction-limited resolution in wide-field microscopy (Kamerbeek et al., 2013). In super-resolution microscopy, the probe has recently been used to map the lipid environment with sub-diffraction resolution on length scales below 300 nm (Nieves & Owen, 2020). A drawback of solvatochromic dyes is that they report on the membrane order of lipid domains but fail to provide accurate information about their lipid and protein composition (Lorizate et al., 2021).

The ability to measure the fluorescence spectra of individual fluorescent probes in a standard wide-field microscope added a new dimension to the fingerprinting of molecules via fluorescence microscopy. The rationale behind this approach is that averaging all photons stemming from multiple fluorescence emissions from a given fluorophore species into a single detector device sacrifices valuable information contained in the individual spectrum of each molecule. Using a grating or prism in the optical path of the microscope and an imaging detector to disperse the emission spectrum, however, diffracted images of fluorescent nanospheres with emission maxima separated by <20 nm and of single fluorescent probe molecules with 30 nm separation could be obtained (Heider, Barhoum, Peterson, Schaefer, & Harris, 2010). Based on this principle Xe and coworkers employed super-resolution microscopy to spectrally resolve peroxisomes, vimentin filaments, microtubules, and the outer mitochondrial membrane in fixed PtK2 cells. The method was coined spectrally-resolved STORM (Zhang, Kenny, Hauser, Li, & Xu, 2015). A gain in spatial resolution of ~10 nm was already apparent when adding the spectroscopic signature of the photon-emitting molecule, i.e. a four-fold improvement over photon localization without spectral discrimination (Dong et al., 2016). The group of Samuel Hess developed similar approaches using PALM single-molecule localization microscopy. In the spectrally-resolved modality, PALM imaging of fluorescent proteins having partially overlapping emission spectra like Dendra2-hemmaglutinin, PAmCherry-cofilin or PAmKate-transferrin receptor could be resolved at the single-molecule level (Mlodzianoski, Curthoys, Gunewardene, Carter, & Hess, 2016).

Nile Red (Nile Blue A oxazone), a low molecular weight organic chemical used for decades as a dye to label intracellular lipid storages in classical histological and histopathological studies, has found new applications in super-resolution imaging using the PAINT technique to localize single-molecules and extract spectral information; the acronym sPAINT (spectral PAINT) was coined in this case (Bongiovanni et al., 2016). This superresolution modality was applied to calibrate Nile Red properties on synthetic 10 nm unilamellar vesicles of known lipid composition and to measure the hydrophobicity of the plasmalemma of HEK-293 cells depleted of or enriched in cholesterol via methyl-β-cyclodextrin. A red-shift in the mean position and a reduction of the distribution-width of hydrophobic localizations was observed for the membranes of cholesterol-depleted cells compared with control cells (Bongiovanni et al., 2016).

A very similar study was conducted by the group of Zhang using the STORM technique and Nile Red as a sensor of membrane polarity on fixed cells. They could image distinct low-polarity regions of ~100 nm in diameter in the plasma membrane upon addition of cholesterol or treatment with cholera toxin in fixed cells but failed to find a similar pattern in native cells

(Moon et al., 2017). In agreement with the results of Bongiovanni et al. (Bongiovanni et al., 2016), Zhang and coworkers observed that cholesterol depletion with methyl-β-cyclodextrin led to a strong red-shift of the Nile Red single-molecule spectrum at the plasmalemma but not in internal organelle membranes. Addition of cholesterol led to blue-shifted emission puncta dispersed across the membrane. These observations led the authors to infer that the low-polarity regions corresponded to ordered lipid domains observable only in cholesterol-treated cells, suggesting that such domains may not be present in native cells (Moon et al., 2017). Exactly what Nile Red senses is still a matter of debate. According to Lorizate and coworkers (Lorizate et al., 2021) Nile-Red-based sensors used in recent studies (Danylchuk, Moon, Xu, & Klymchenko, 2019; Moon et al., 2017) do not report the occurrence of lipid heterogeneity in the plasma membrane of living cells but only a higher degree of order in outward and inward protrusions of the membrane, indicating a direct association of curvature with order. To overcome these constraints, a new biorthogonal cholesterol biosensor (chol-N_3) was developed and used to characterize lipid domains in the plasma membrane of live cells interrogated with STED nanoscopy, with good resolution along the axis perpendicular to the membrane and in 3D imaging of thick samples like brain slices (Lorizate et al., 2021). The size distribution of the presumptive lipid-domain STED puncta showed peaks at ~50 and 150 nm lateral width. Other recent studies using FCS-STED nanoscopy corroborate the usefulness of Nile Red-based sensors in studies on lipid domain packing and dynamics with long acquisition times (Carravilla et al., 2021). See Chapter 10 in this Volume for a comprehensive treatment of orthogonal lipid sensors.

6. Conclusions and future directions

Imaging and physical characterization of lipid domains, transbilayer and lateral (in-plane) asymmetries and other biophysical properties of cell membranes have experienced major and exciting advances in the last decades. Due to their higher photostability and brightness, fluorophore-tagged (extrinsic) lipid analogues have provided a wider range of possibilities than the intrinsically fluorescent cholesterol analogues dehydroergosterol or cholestanetriol; nevertheless, both types of probes still find specific applications in membrane research, as discussed in previous sections. The number of sensors for identifying molecular components has grown to include not only a sizeable number of lipid probes but also sensors to identify different enzymes and specific membrane proteins that indirectly report on lipid domains. This repertoire of biosensors has enabled the characterization of cell-surface binding partners, endocytic trafficking, lysosomal degradation, and disease conditions affecting such physiological phenomena. In addition, new biophysical techniques have become available, allowing the study of the physical properties underlying the structural and mechanistic bases of these phenomena.

The field of artificial intelligence (AI) and object recognition has had a major explosion in recent years, catalyzed by the inception of the convolutional, multilayer (and thus "deep") neural network termed AlexNet by Krizhevsky and coworkers (Krizhevsky, Sutskever, & Hinton, 2012) (see review in (Brent & Boucheron, 2018)). The branch of AI known as deep learning (DL) has penetrated many spheres of science. It is currently used extensively in several subdisciplines of biology, particularly in structural biology. In cryo-electron microscopy, for instance, it has become a valuable tool to automatize the "pruning" of macromolecular structures and eliminate false positives from thousands of single-molecule identifications collected via particle-picking algorithms (Sanchez-Garcia, Segura, Maluenda, Carazo, & Sorzano, 2018). DL has also found application in identifying "good" regions in electron microscope grids (Yokoyama et al., 2020), an otherwise time-consuming and tedious step, or estimating resolution in density maps (Avramov et al., 2019).

In fluorescence microscopy, DL methods are experiencing explosive developments that address many procedural steps of the different techniques and expand the current horizon of theoretical approaches to physical modeling of cell functioning. In combination with state-of-the-art molecular biology methods like genome-wide CRISPR (clusters of interspaced short palindromic repeats) screening technology, DL has recently found application in the classification of different cell phenotypes of interest upon photoactivation and isolation via flow cytometry; the approach, coined AI-photoswitchable screening (Kanfer et al., 2021).

One of the most exciting applications of DL in fluorescence microscopy is undoubtedly the training of neural networks to *predict* structures in microscope images. In this type of application, machine learning is used to train neural networks with information gained in gray-scale label-free micrographs e.g. structures in scanning or transmission electron micrographs into virtual "stained" micrographs of the same specimens (Christiansen et al., 2018; Helgadottir et al., 2021). Upon proper training, the networks can predict a given structure and color-code it for automatic identification of new samples. Working on an unstained specimen in a conventional bright-field or phase-contrast optical microscope, the DL approach replaces both the staining and the fluorescence microscopy steps with an in silico neural network solution that generates virtual fluorescence-stained images.

There are still some caveats. The cell is a three-dimensional object and understanding its complex organization and functioning requires tools to interrogate these two aspects concurrently, with the requisite spatial- and time-resolution imposed by the cell's workings. The limited contrast of the cellular structures still calls for imaging approaches that tag different components with distinct identifying sensors. This is still a major limitation of imaging techniques and fluorescence microscopy: despite the wide spectrum of sensors available, they do not suffice to differential label and individualize the thousands of molecular species present in the cell. When we speak of multi-color detection, we are still limited to a few digits of wavelengths that we can simultaneously explore. However, this is one of the problems tackled by the super-resolution microscopy method termed DNA-PAINT (DNA-based point accumulation for imaging in nanoscale topography), a nanoscopy approach that exploits programmable transient hybridization between short oligonucleotide strands (Jungmann, 2014; Schueder et al., 2017; van Wee, Filius, & Joo, 2021). One of the inherent advantages of DNA-PAINT is its ability to employ a much wider range of multiplexing than conventional fluorescence microscopy, which relies essentially in the use of multiple probes with different spectral properties, conventionally at most 3–4 (Bates et al., 2007; Dempsey, Vaughan, Chen, Bates, & Zhuang, 2011). Recently, Jungmann, Schwille and coworkers introduced a new approach for multiplexing imaging in stochastic super-resolution, a concept that should facilitate transcriptomic, proteomic and lipidomic studies of hundreds of molecular constituents. In their proof-of-concept work, they managed 124-color imaging within minutes by engineering DNA-PAINT blinking kinetics (Wade et al., 2019).

In summary, although the panoply of optical fluorescence techniques available for imaging lipid domains in cells is still limited, and suffers from various artifacts, we should be confident that the ingenuity of scientific progress will generate new approaches, instrumentation, and techniques to tackle the simultaneous interrogation of several biophysical properties of lipid domains with enhanced temporal and spatial resolution.

Acknowledgments

This work was written within the framework of the grant PICT 2015-2654 from the Ministry of Science, Technology and Innovative Production of Argentina.

Author contributions

I conceived and designed the study, searched the literature, interpreted the data and wrote the manuscript. I conceived the illustrations and had technical help to produce them.

Declaration of interests

The author declares no competing interests.

References

Abbe, E. (1873). Beitrage zur Theorie des Mikroskops und der mikroskopischen Wahrnehmung. *Archiv für Mikroskopische Anatomie*, *9*, 413–418. Retrieved from https://doi.org/10.1007/BF02956173.

Adam, V., Moeyaert, B., David, C. C., Mizuno, H., Lelimousin, M., Dedecker, P., et al. (2011). Rational design of photoconvertible and biphotochromic fluorescent proteins for advanced microscopy applications. *Chemistry & Biology*, *18*(10), 1241–1251. https://doi.org/10.1016/j.chembiol.2011.08.007.

Ahmed, S. N., Brown, D. A., & London, E. (1997). On the origin of sphingolipid/cholesterol-rich detergent-insoluble cell membranes: Physiological concentrations of cholesterol and sphingolipid induce formation of a detergent-insoluble, liquid-orderer lipid phase in model membranes. *The Biochemist*, *36*(36), 10944–10953.

Amaro, M., Filipe, H. A., Prates Ramalho, J. P., Hof, M., & Loura, L. M. (2016). Fluorescence of nitrobenzoxadiazole (NBD)-labeled lipids in model membranes is connected not to lipid mobility but to probe location. *Physical Chemistry Chemical Physics*, *18*(10), 7042–7054. https://doi.org/10.1039/c5cp05238f.

Amaro, M., Reina, F., Hof, M., Eggeling, C., & Sezgin, E. (2017). Laurdan and Di-4-ANEPPDHQ probe different properties of the membrane. *Journal of Physics D: Applied Physics*, *50*(13), 134004. https://doi.org/10.1088/1361-6463/aa5dbc.

Antollini, S. S., & Barrantes, F. J. (1998). Disclosure of discrete sites for phospholipid and sterols at the protein-lipid interface in native acetylcholine receptor-rich membrane. *Biochemistry*, *37*(47), 16653–16662. https://doi.org/10.1021/bi9808215.

Antollini, S. S., & Barrantes, F. J. (2002). Unique effects of different fatty acid species on the physical properties of the torpedo acetylcholine receptor membrane. *Journal of Biological Chemistry*, *277*(2), 1249–1254. Retrieved from http://www.ncbi.nlm.nih.gov/pubmed/11682474.

Antollini, S. S., & Barrantes, F. J. (2007). Laurdan studies of membrane lipid-nicotinic acetylcholine receptor protein interactions. *Methods in Molecular Biology*, *400*, 531–542. https://doi.org/10.1007/978-1-59745-519-0_36.

Antollini, S. S., Soto, M. A., Bonini de Romanelli, I. C., Gutierrez-Merino, C., Sotomayor, P., & Barrantes, F. J. (1996). Physical state of bulk and protein-associated lipid in nicotinic acetylcholine receptor-rich membrane studied by laurdan generalized polarization and fluorescence energy transfer. *Biophysical Journal*, *70*(3), 1275–1284. Retrieved from http://www.ncbi.nlm.nih.gov/pubmed/0008785283.

Arthur, J. R., Heinecke, K. A., & Seyfried, T. N. (2011). Filipin recognizes both GM1 and cholesterol in GM1 gangliosidosis mouse brain. *Journal of Lipid Research*, *52*(7), 1345–1351. https://doi.org/10.1194/jlr.M012633.

Avramov, T. K., Vyenielo, D., Gomez-Blanco, J., Adinarayanan, S., Vargas, J., & Si, D. (2019). Deep learning for validating and estimating resolution of cryo-electron microscopy density maps. *Molecules*, *24*(6), 1181. https://doi.org/10.3390/molecules24061181.

Azzi, A., Chance, B., Radda, G. K., & Lee, C. P. (1969). A fluorescence probe of energy-dependent structure changes in fragmented membranes. *Proceedings of the*

National Academy of Sciences of the United States of America, *62*(2), 612–619. https://doi.org/10.1073/pnas.62.2.612.
Bacalum, M., Zorilă, B., & Radu, M. (2013). Fluorescence spectra decomposition by asymmetric functions: Laurdan spectrum revisited. *Analytical Biochemistry*, *440*(2), 123–129. https://doi.org/10.1016/j.ab.2013.05.031.
Bacia, K., Scherfeld, D., Kahya, N., & Schwille, P. (2004). Fluorescence correlation spectroscopy relates rafts in model and native membranes. *Biophysical Journal*, *87*(2), 1034–1043. https://doi.org/10.1529/biophysj.104.040519.
Bagatolli, L. A. (2006). To see or not to see: Lateral organization of biological membranes and fluorescence microscopy. *Biochimica et Biophysica Acta*, *1758*(10), 1541–1556. Retrieved from http://www.ncbi.nlm.nih.gov/pubmed/16854370.
Bagatolli, L. A., & Gratton, E. (1999). Two-photon fluorescence microscopy observation of shape changes at the phase transition in phospholipid giant unilamellar vesicles. *Biophysical Journal*, 77(4), 2090–2101. Retrieved from http://www.ncbi.nlm.nih.gov/pubmed/10512829.
Bagatolli, L. A., Gratton, E., & Fidelio, G. D. (1998). Water dynamics in glycosphingolipid aggregates studied by LAURDAN fluorescence. *Biophysical Journal*, *75*(1), 331–341.
Bagatolli, L. A., Maggio, B., Aguilar, F., Sotomayor, C. P., & Fidelio, G. D. (1997). Laurdan properties in glycosphingolipid-phospholipid mixtures: A comparative fluorescence and calorimetric study. *Biochimica et Biophysica Acta*, *1325*(1), 80–90.
Bagatolli, L. A., Parasassi, T., & Gratton, E. (2000). Giant phospholipid vesicles: Comparison among the whole lipid sample characteristics using different preparation methods: A two photon fluorescence microscopy study. *Chemistry and Physics of Lipids*, *105*(2), 135–147. Retrieved from http://www.ncbi.nlm.nih.gov/pubmed/10823462.
Barreales, E. G., Payero, T. D., de Pedro, A., & Aparicio, J. F. (2018). Phosphate effect on filipin production and morphological differentiation in Streptomyces filipinensis and the role of the PhoP transcription factor. *PLoS One*, *13*(12), e0208278. https://doi.org/10.1371/journal.pone.0208278.
Bates, M., Huang, B., Dempsey, G. T., & Zhuang, X. (2007). Multicolor super-resolution imaging with photo-switchable fluorescent probes. *Science*, *317*, 1749–1753. Retrieved from https://doi.org/10.1126/science.1146598.
Bates, M., Huang, B., & Zhuang, X. (2008). Super-resolution microscopy by nanoscale localization of photo-switchable fluorescent probes. *Current Opinion in Chemical Biology*, *12*(5), 505–514. Retrieved from http://www.ncbi.nlm.nih.gov/pubmed/18809508.
Baumgart, T., Hammond, A. T., Sengupta, P., Hess, S. T., Holowka, D. A., Baird, B. A., et al. (2007). Large-scale fluid/fluid phase separation of proteins and lipids in giant plasma membrane vesicles. *Proceedings of the National Academy of Sciences of the United States of America*, *104*(9), 3165–3170. https://doi.org/10.1073/pnas.0611357104.
Baumgarten, C. M., Makielski, J. C., & Fozzard, H. A. (1991). External site for local anesthetic block of cardiac Na^+ channels. *Journal of Molecular and Cellular Cardiology*, *23*(Suppl. 1), 85–93.
Belov, V. N., Wurm, C. A., Boyarskiy, V. P., Jakobs, S., & Hell, S. W. (2010). Rhodamines NN: A novel class of caged fluorescent dyes. *Angewande Chemie International ed in English*. Retrieved from http://www.ncbi.nlm.nih.gov/pubmed/20391447.
Ben-Yashar, V., & Barenholz, Y. (1989). The interaction of cholesterol and cholest-4-en-3-one with dipalmitoylphosphatidylcholine. Comparison based on the use of three fluorophores. *Biochimica et Biophysica Acta*, *985*, 271–278.
Betzig, E. (2015). Single molecules, cells, and super-resolution optics (nobel lecture). *Angewandte Chemie (International Ed. in English)*. https://doi.org/10.1002/anie.201501003.
Bongiovanni, M. N., Godet, J., Horrocks, M. H., Tosatto, L., Carr, A. R., Wirthensohn, D. C., et al. (2016). Multi-dimensional super-resolution imaging enables

surface hydrophobicity mapping. *Nature Communications*, 7(1), 13544. https://doi.org/10.1038/ncomms13544.

Börnig, H., & Geyer, G. (1974). Staining of cholesterol with the fluorescent antibiotic "filipin". *Acta Histochemica*, *50*(1), 110–115.

Bossi, M., Belov, V., Polyakova, S., & Hell, S. W. (2006). Reversible red fluorescent molecular switches. *Angewandte Chemie (International Ed. in English)*, *45*(44), 7462–7465. Retrieved from http://www.ncbi.nlm.nih.gov/pubmed/17042053.

Boyarskiy, V. P., Belov, V. N., Medda, R., Hein, B., Bossi, M., & Hell, S. W. (2008). Photostable, amino reactive and water-soluble fluorescent labels based on sulfonated rhodamine with a rigidized xanthene fragment. *Chemistry*, *14*(6), 1784–1792. Retrieved from http://www.ncbi.nlm.nih.gov/pubmed/18058955.

Brent, R., & Boucheron, L. (2018). Deep learning to predict microscope images. *Nature Methods*, *15*(11), 868–870. https://doi.org/10.1038/s41592-018-0194-9.

Brewer, J., de la Serna, J. B., Wagner, K., & Bagatolli, L. A. (2010). Multiphoton excitation fluorescence microscopy in planar membrane systems. *Biochimica et Biophysica Acta*, *1798*(7), 1301–1308. Retrieved from http://www.ncbi.nlm.nih.gov/pubmed/20226161.

Brown, M. S., & Goldstein, J. L. (1984). How LDL receptors influence cholesterol and atherosclerosis. *Scientific American*, *251*, 52–60.

Butkevich, A. N., Bossi, M. L., Lukinavičius, G., & Hell, S. W. (2019). Triarylmethane fluorophores resistant to oxidative photobluing. *Journal of the American Chemical Society*, *141*(2), 981–989. https://doi.org/10.1021/jacs.8b11036.

Butkevich, A. N., Yu, G., Mitronova, S. C., Klocke, J. L., Kamin, D., Meineke, N. H., et al. (2016). Fluorescent rhodamines and fluorogenic carbopyronines for super-resolution STED microscopy in living cells. *Angewandte Chemie (International Ed. in English)*, *55*, 3290–3294. Retrieved from https://doi.org/10.1002/anie.201511018.

Buwaneka, P., Ralko, A., Liu, S.-L., & Cho, W. (2021). Evaluation of the available cholesterol concentration in the inner leaflet of the plasma membrane of mammalian cells. *Journal of Lipid Research*, *62*, 100084. https://doi.org/10.1016/j.jlr.2021.100084.

Calafut, T. M., Dix, J. A., & Verkman, A. S. (1989). Fluorescence depolarization of *cis*- and *trans*-parinaric acids in artificial and red cell membranes resolved by a double hindered rotational model. *Biochemistry*, *28*, 5051–5058.

Carquin, M., Pollet, H., Veiga-da-Cunha, M., Cominelli, A., Van Der Smissen, P., N'Kuli, F., et al. (2014). Endogenous sphingomyelin segregates into submicrometric domains in the living erythrocyte membrane. *Journal of Lipid Research*, *55*(7), 1331–1342. https://doi.org/10.1194/jlr.M048538.

Carravilla, P., Dasgupta, A., Zhurgenbayeva, G., Danylchuk, D. I., Klymchenko, A. S., Sezgin, E., et al. (2021). STED super-resolution imaging of membrane packing and dynamics by exchangeable polarity-sensitive dyes. *bioRxiv*. https://doi.org/10.1101/2021.06.05.446432.

Castanho, M. A. R. B., Coutinho, A., & Prieto, M. J. E. (1992). Absorption and fluorescence spectra of polyene antibiotics in the presence of cholesterol. *The Journal of Biological Chemistry*, *267*(1), 204–209.

Castanho, M., & Prieto, M. (1995). Filipin fluorescence quenching by spin-labeled probes: Studies in aqueous solution and in a membrane model system. *Biophysical Journal*, *69*(1), 155–168. https://doi.org/10.1016/S0006-3495(95)79886-1.

Castanho, M., Prieto, M., & Acuña, A. U. (1996). The transverse location of the flurescent probe *trans*-parinaric acid in lipid bilayers. *Biochimica Biophysica Acta Bio-Membranes*, *1279*, 164–168.

Cater, B. R., Chapman, D., Hawes, S. M., & Saville, J. (1974). Lipid phase transitions and drug interactions. *Biochimica et Biophysica Acta*, *363*(1), 54–69. https://doi.org/10.1016/0005-2736(74)90006-6.

Chattopadhyay, A. (2003). Exploring membrane organization and dynamics by the wavelength-selective fluorescence approach. *Chemistry and Physics of Lipids*, *122*(1-2), 3–17. Retrieved from http://www.ncbi.nlm.nih.gov/pubmed/12598034.

Chattopadhyay, A., Biswas, S. C., Rukmini, R., Saha, S., & Samanta, A. (2021). Lack of environmental sensitivity of a naturally occurring fluorescent analog of cholesterol. *Journal of Fluorescence*, *31*, 1401–1407. https://doi.org/10.1007/s10895-021-02767-4.

Chattopadhyay, A., & Mukherjee, S. (1993). Fluorophore environments in membrane-bound probes: A red edge excitation shift study. *Biochemistry*, *32*(14), 3804–3811. https://doi.org/10.1021/bi00065a037.

Chazotte, B. (2011). Labeling the plasma membrane with TMA-DPH. *Cold Spring Harbor Protocols*, *2011*(5). https://doi.org/10.1101/pdb.prot5622.

Christiansen, E. M., Yang, S. J., Ando, D. M., Javaherian, A., Skibinski, G., Lipnick, S., et al. (2018). In silico labeling: Predicting fluorescent labels in unlabeled images. *Cell*, *173*(3), 792–803.e719. https://doi.org/10.1016/j.cell.2018.03.040.

Cogan, U., Shinitzky, M., Weber, G., & Nishida, T. (1973). Microviscosity and order in the hydrocarbon region of phospholipid and phospholipid-cholesterol dispersions determined with fluorescent probes. *Biochemistry*, *12*(3), 521–528. https://doi.org/10.1021/bi00727a026.

Connor, O. D., Byrne, A., Berselli, G. B., Long, C., & Keyes, T. E. (2019). Mega-stokes pyrene ceramide conjugates for STED imaging of lipid droplets in live cells. *Analyst*, *144*(5), 1608–1621. https://doi.org/10.1039/c8an02260g.

Courtney, K. C., Pezeshkian, W., Raghupathy, R., Zhang, C., Darbyson, A., Ipsen, J. H., et al. (2018). C24 sphingolipids govern the transbilayer asymmetry of cholesterol and lateral organization of model and live-cell plasma membranes. *Cell Reports*, *24*(4), 1037–1049. https://doi.org/10.1016/j.celrep.2018.06.104.

Cuevas Arenas, R., Danielczak, B., Martel, A., Porcar, L., Breyton, C., Ebel, C., et al. (2017). Fast collisional lipid transfer among polymer-bounded nanodiscs. *Scientific Reports*, *7*, 45875. https://doi.org/10.1038/srep45875.

Danylchuk, D. I., Moon, S., Xu, K., & Klymchenko, A. S. (2019). Switchable solvatochromic probes for live-cell super-resolution imaging of plasma membrane organization. *Angewandte Chemie (International Ed. in English)*, *58*(42), 14920–14924. https://doi.org/10.1002/anie.201907690.

Das, A., Brown, M. S., Anderson, D. D., Goldstein, J. L., & Radhakrishnan, A. (2014). Three pools of plasma membrane cholesterol and their relation to cholesterol homeostasis. *eLife*, *3*, e02882. https://doi.org/10.7554/eLife.02882.

De Craene, J. O., Bertazzi, D. L., Bär, S., & Friant, S. (2017). Phosphoinositides, major actors in membrane trafficking and lipid signaling pathways. *International Journal of Molecular Sciences*, *18*(3). https://doi.org/10.3390/ijms18030634.

de Saint-Jean, M., Delfosse, V., Douguet, D., Chicanne, G., Payrastre, B., Bourguet, W., et al. (2011). Osh4p exchanges sterols for phosphatidylinositol 4-phosphate between lipid bilayers. *The Journal of Cell Biology*, *195*(6), 965–978. https://doi.org/10.1083/jcb.201104062.

Demchenko, A. P., Mely, Y., Duportail, G., & Klymchenko, A. S. (2009). Monitoring biophysical properties of lipid membranes by environment-sensitive fluorescent probes. *Biophysical Journal*, *96*(9), 3461–3470. Retrieved from http://www.ncbi.nlm.nih.gov/pubmed/19413953.

Dempsey, G. T., Vaughan, J. C., Chen, K. H., Bates, M., & Zhuang, X. (2011). Evaluation of fluorophores for optimal performance in localization-based super-resolution imaging. *Nature Methods*, *8*, 1027–1036. Retrieved from https://doi.org/10.1038/nmeth.1768.

Dietrich, C., Bagatolli, L. A., Volovyk, Z. N., Thompson, N. L., Levi, M., Jacobson, K., et al. (2001). Lipid rafts reconstituted in model membranes. *Biophysical Journal*, *80*(3), 1417–1428. Retrieved from http://www.ncbi.nlm.nih.gov/pubmed/11222302.

Dirix, L., Kennes, K., Fron, E., Debyser, Z., van der Auweraer, M., Hofkens, J., et al. (2018). Photoconversion of far-red organic dyes: Implications for multicolor super-resolution imaging. *ChemPhotoChem*, *2*(5), 433–441. https://doi.org/10.1002/cptc.201700216.

do Canto, A., Robalo, J. R., Santos, P. D., Carvalho, A. J. P., Ramalho, J. P. P., & Loura, L. M. S. (2016). Diphenylhexatriene membrane probes DPH and TMA-DPH: A comparative molecular dynamics simulation study. *Biochimica et Biophysica Acta*, *1858*(11), 2647–2661. https://doi.org/10.1016/j.bbamem.2016.07.013.

Dong, B., Almassalha, L., Urban, B. E., Nguyen, T.-Q., Khuon, S., Chew, T.-L., et al. (2016). Super-resolution spectroscopic microscopy via photon localization. *Nature Communications*, 7. https://doi.org/10.1038/ncomms12290.

Drabikowski, W., Lagwińska, E., & Sarzala, M. G. (1973). Filipin as a fluorescent probe for the location of cholesterol in the membranes of fragmented sarcoplasmic reticulum. *Biochimica et Biophysica Acta*, *291*(1), 61–70. https://doi.org/10.1016/0005-2736(73)90060-6.

Eggeling, C. (2015). Super-resolution optical microscopy of lipid plasma membrane dynamics. *Essays in Biochemistry*, *57*, 69–80. https://doi.org/10.1042/bse0570069.

Eggeling, C., Widengren, J., Rigler, R., & Seidel, C. A. M. (1998). Photobleaching of fluorescent dyes under conditions used for single-molecule detection: Evidence of two-step photolysis. *Analytical Chemistry*, *70*(13), 2651–2659.

Epand, R. F., Kraayenhof, R., Sterk, G. J., Sang, H. W. W. F., & Epand, R. M. (1996). Fluorescent probes of membrane surface properties. *Biochimica et Biophysica Acta. Bio-Membranes*, *1284*(2), 191–195.

Ercan, B., Naito, T., Koh, D. H. Z., Dharmawan, D., & Saheki, Y. (2021). Molecular basis of accessible plasma membrane cholesterol recognition by the GRAM domain of GRAMD1b. *The EMBO Journal*, *40*(6), e106524. https://doi.org/10.15252/embj.2020106524.

Fereidouni, F., Bader, A. N., & Gerritsen, H. C. (2012). Spectral phasor analysis allows rapid and reliable unmixing of fluorescence microscopy spectral images. *Optics Express*, *20*(12), 12729–12741. https://doi.org/10.1364/oe.20.012729.

Fernández-Luna, V., Coto, P. B., & Costa, R. D. (2018). When fluorescent proteins meet white light-emitting diodes. *Angewandte Chemie (International Ed. in English)*, *57*(29), 8826–8836. https://doi.org/10.1002/anie.201711433.

Ferretti, G., Tangorra, A., Zolese, G., & Curatola, G. (1993). Properties of a phosphatidylcholine derivative of diphenyl hexatriene (DPH-PC) in lymphocyte membranes. A comparison with DPH and the cationic derivative TMA-DPH using static and dynamic fluorescence. *Membrane Biochemistry*, *10*(1), 17–27. https://doi.org/10.3109/09687689309150249.

Filipe, H. A. L., Pokorná, Š., Hof, M., Amaro, M., & Loura, L. M. S. (2019). Orientation of nitro-group governs the fluorescence lifetime of nitrobenzoxadiazole (NBD)-labeled lipids in lipid bilayers. *Physical Chemistry Chemical Physics*, *21*(4), 1682–1688. https://doi.org/10.1039/c8cp06064a.

Fiorini, R., Curatola, G., Kantar, A., Giorgi, P. L., & Gratton, E. (1993). Use of Laurdan fluorescence in studying plasma membrane organization of polymorphonuclear leukocytes during the respiratory burst. *Photochemistry and Photobiology*, *57*, 438–441.

Flanagan, J. J., Tweten, R. K., Johnson, A. E., & Heuck, A. P. (2009). Cholesterol exposure at the membrane surface is necessary and sufficient to trigger perfringolysin O binding. *Biochemistry*, *48*(18), 3977–3987. https://doi.org/10.1021/bi9002309.

Fujimoto, T., Hayashi, M., Iwamoto, M., & Ohno-Iwashita, Y. (1997). Crosslinked plasmalemmal cholesterol is sequestered to caveolae: Analysis with a new cytochemical probe. *The Journal of Histochemistry and Cytochemistry*, *45*(9), 1197–1205.

Gallegos, A. M., McIntosh, A. L., Atshaves, B. P., & Schroeder, F. (2004). Structure and cholesterol domain dynamics of an enriched caveolae/raft isolate. *The Biochemical Journal*, *382*(Pt. 2), 451–461. https://doi.org/10.1042/bj20031562.

Garvik, O., Benediktson, P., Simonsen, A. C., Ipsen, J. H., & Wüstner, D. (2009). The fluorescent cholesterol analog dehydroergosterol induces liquid-ordered domains in model membranes. *Chemistry and Physics of Lipids*, *159*(2), 114–118. https://doi.org/10.1016/j.chemphyslip.2009.03.002.

Gaus, K., Gratton, E., Kable, E. P., Jones, A. S., Gelissen, I., Kritharides, L., et al. (2003). Visualizing lipid structure and raft domains in living cells with two-photon microscopy. *Proceedings of the National Academy of Sciences of the United States of America*, *100*(26), 15554–15559. Retrieved from http://www.ncbi.nlm.nih.gov/pubmed/14673117.

Gaus, K., Zech, T., & Harder, T. (2006). Visualizing membrane microdomains by Laurdan 2-photon microscopy. *Molecular Membrane Biology*, *23*(1), 41–48. Retrieved from http://www.ncbi.nlm.nih.gov/pubmed/16611579.

Geyer, G., & Börnig, H. (1975). "Filipin"—A histochemical fluorochrome for cholesterol. *Acta Histochemica. Supplementband*, *15*, 207–212.

Giang, H., & Schick, M. (2016). On the puzzling distribution of cholesterol in the plasma membrane. *Chemistry and Physics of Lipids*, *199*, 35–38. https://doi.org/10.1016/j.chemphyslip.2015.12.002.

Godement, P., Vanselow, J., Thanos, S., & Bonhoeffer, F. (1987). A study in developing visual systems with a new method of staining neurones and their processes in fixed tissue. *Development*, *101*(4), 697–713.

Golfetto, O., Hinde, E., & Gratton, E. (2013). Laurdan fluorescence lifetime discriminates cholesterol content from changes in fluidity in living cell membranes. *Biophysical Journal*, *104*(6), 1238–1247. https://doi.org/10.1016/j.bpj.2012.12.057.

Golfetto, O., Hinde, E., & Gratton, E. (2015). The Laurdan spectral phasor method to explore membrane micro-heterogeneity and lipid domains in live cells. *Methods in Molecular Biology*, *1232*, 273–290. https://doi.org/10.1007/978-1-4939-1752-5_19.

Gould, T. J., Hess, S. T., & Bewersdorf, J. (2012). Optical nanoscopy: From acquisition to analysis. *Annual Review of Biomedical Engineering*, *14*, 231–254. https://doi.org/10.1146/annurev-bioeng-071811-150025.

Grimm, J. B., English, B. P., Chen, J., Slaughter, J. P., Zhang, Z., Revyakin, A., et al. (2015). A general method to improve fluorophores for live-cell and single-molecule microscopy. *Nature Methods*, *12*(3), 244–250. https://doi.org/10.1038/nmeth.3256. http://www.nature.com/nmeth/journal/v12/n3/abs/nmeth.3256.html#supplementary-information.

Grimm, J. B., English, B. P., Choi, H., Muthusamy, A. K., Mehl, B. P., Dong, P., et al. (2016). Bright photoactivatable fluorophores for single-molecule imaging. *Nature Methods*, *13*(12), 985–988. https://doi.org/10.1038/nmeth.4034. http://www.nature.com/nmeth/journal/v13/n12/abs/nmeth.4034.html#supplementary-information.

Gunther, G., Malacrida, L., Jameson, D. M., Gratton, E., & Sánchez, S. A. (2021). LAURDAN since Weber: The quest for visualizing membrane heterogeneity. *Accounts of Chemical Research*, *54*(4), 976–987. https://doi.org/10.1021/acs.accounts.0c00687.

Ha, T., & Tinnefeld, P. (2012). Photophysics of fluorescent probes for single-molecule biophysics and super-resolution imaging. *Annual Review of Physical Chemistry*, *63*, 595–617. https://doi.org/10.1146/annurev-physchem-032210-103340.

Hale, J. E., & Schroeder, F. (1982). Asymmetric transbilayer distribution of sterol across plasma membranes determined by fluorescence quenching of dehydroergosterol. *European Journal of Biochemistry*, *122*, 649–661.

Hammond, G. R., Machner, M. P., & Balla, T. (2014). A novel probe for phosphatidylinositol 4-phosphate reveals multiple pools beyond the Golgi. *The Journal of Cell Biology*, *205*(1), 113–126. https://doi.org/10.1083/jcb.201312072.

Hao, M., Lin, S. X., Karylowski, O. J., Wustner, D., McGraw, T. E., & Maxfield, F. R. (2002). Vesicular and non-vesicular sterol transport in living cells. The endocytic recycling compartment is a major sterol storage organelle. *The Journal of Biological Chemistry*, *277*(1), 609–617. Retrieved from http://www.ncbi.nlm.nih.gov/pubmed/11682487.

Hao, M., Mukherjee, S., & Maxfield, F. R. (2001). Cholesterol depletion induces large scale domain segregation in living cell membranes. *Proceedings of the National Academy of Sciences of the United States of America*, *98*(23), 13072–13077. Retrieved from http://www.ncbi.nlm.nih.gov/pubmed/11698680.

Hayashi, M., Shimada, Y., Inomata, M., & Ohno-Iwashita, Y. (2006). Detection of cholesterol-rich microdomains in the inner leaflet of the plasma membrane. *Biochemical and Biophysical Research Communications*, *351*(3), 713–718. https://doi.org/10.1016/j.bbrc.2006.10.088.

Heider, E. C., Barhoum, M., Peterson, E. M., Schaefer, J., & Harris, J. M. (2010). Identification of single fluorescent labels using spectroscopic microscopy. *Applied Spectroscopy*, *64*(1), 37–45. https://doi.org/10.1366/000370210790572034.

Heino, S., Lusa, S., Somerharju, P., Ehnholm, C., Olkkonen, V. M., & Ikonen, E. (2000). Dissecting the role of the golgi complex and lipid rafts in biosynthetic transport of cholesterol to the cell surface. *Proceedings of the National Academy of Sciences of the United States of America*, *97*(15), 8375–8380. Retrieved from http://www.ncbi.nlm.nih.gov/pubmed/10890900.

Helgadottir, S., Midtvedt, B., Pineda, J., Sabirsh, A., Adiels, C. B., Romeo, S., et al. (2021). Extracting quantitative biological information from bright-field cell images using deep learning. *Biophysics Reviews*, *2*(3), 031401. https://doi.org/10.1063/5.0044782.

Hell, S. W. (2015). Nanoscopy with focused light (nobel lecture). *Angewandte Chemie (International Ed. in English)*. https://doi.org/10.1002/anie.201504181.

Hell, S. W., & Wichmann, J. (1994). Breaking the diffraction resolution limit by stimulated emission: Stimulated-emission-depletion fluorescence microscopy. *Optics Letters*, *19*, 780–782. Retrieved from https://doi.org/10.1364/OL.19.000780.

Helmerich, D. A., Beliu, G., Matikonda, S. S., Schnermann, M. J., & Sauer, M. (2021). Photoblueing of organic dyes can cause artifacts in super-resolution microscopy. *Nature Methods*. https://doi.org/10.1038/s41592-021-01061-2.

Henriques, R., Griffiths, C., Hesper Rego, E., & Mhlanga, M. M. (2011). PALM and STORM: Unlocking live-cell super-resolution. *Biopolymers*, *95*(5), 322–331. https://doi.org/10.1002/bip.21586.

Hense, A., Prunsche, B., Gao, P., Ishitsuka, Y., Nienhaus, K., & Nienhaus, G. U. (2015). Monomeric Garnet, a far-red fluorescent protein for live-cell STED imaging. *Scientific Reports*, *5*, 18006. https://doi.org/10.1038/srep18006.

Hess, S. T., Girirajan, T. P., & Mason, M. D. (2006). Ultra-high resolution imaging by fluorescence photoactivation localization microscopy. *Biophysics Journal*, *91*(11), 4258–4272. Retrieved from http://www.ncbi.nlm.nih.gov/pubmed/16980368.

Honig, M. G., & Hume, R. I. (1989). Dil and diO: Versatile fluorescent dyes for neuronal labelling and pathway tracing. *Trends in Neurosciences*, *12*(9), 333–335. 340-331.

Honigmann, A., Mueller, V., Hell, S. W., & Eggeling, C. (2013). STED microscopy detects and quantifies liquid phase separation in lipid membranes using a new far-red emitting fluorescent phosphoglycerolipid analogue. *Faraday Discussions*, *161*, 77–89. discussion 113–150. https://doi.org/10.1039/c2fd20107k.

Hullin-Matsuda, F., & Kobayashi, T. (2007). Monitoring the distribution and dynamics of signaling microdomains in living cells with lipid-specific probes. *Cellular and Molecular Life Sciences*, *64*(19-20), 2492–2504. https://doi.org/10.1007/s00018-007-7281-x.

Ikonen, E. (2008). Cellular cholesterol trafficking and compartmentalization. *Nature Reviews Molecular Cell Biology*, *9*(2), 125–138. Retrieved from http://www.ncbi.nlm.nih.gov/pubmed/18216769.

Illinger, D., Duportail, G., Poirel-Morales, N., Gerard, D., & Kuhry, J.-G. (1995). A comparison of the fluorescence properties of TMA-DPH as a probe fro plasma membrane and for endocytic membrane. *Biochimica et Biophysica Acta, 1239*, 58–66.

Illinger, D., Italiano, L., Beck, J.-P., Waltzinger, C., & Kuhry, J.-G. (1993). Comparative evolution of endocytosis levels and of the cell surface area during the L929 cell cycle: A fluorescence study with TMA-DPH. *Biology of the Cell, 79*, 265–268.

Illinger, D., Poindron, P., & Kuhry, J.-G. (1991). Fluid phase endocytosis investigated by fluorescence with trimethylamino-diphenylhexatriene in L929 cells; the influence of temperature and of cytoskeleton depolymerizing drugs. *Biology of the Cell, 73*, 131–138.

Im, Y. J., Raychaudhuri, S., Prinz, W. A., & Hurley, J. H. (2005). Structural mechanism for sterol sensing and transport by OSBP-related proteins. *Nature, 437*(7055), 154–158. https://doi.org/10.1038/nature03923.

Jacobson, K., Mouritsen, O. G., & Anderson, R. G. (2007). Lipid rafts: At a crossroad between cell biology and physics. *Nature Cell Biology, 9*(1), 7–14. https://doi.org/10.1038/ncb0107-7.

Jacobson, K., & Wobschall, D. (1974). Rotation of fluorescent probes localized within lipid bilayer membranes. *Chemistry and Physics of Lipids, 12*(2), 117–131. https://doi.org/10.1016/0009-3084(74)90049-8.

Jameson, D. M., Gratton, E., & Hall, R. D. (1984). The measurement and analysis of heterogeneous emissions by multifrequency phase and modulation fluorometry. *Applied Spectroscopy Reviews, 20*, 55–106.

Jamshad, M., Lin, Y. P., Knowles, T. J., Parslow, R. A., Harris, C., Wheatley, M., et al. (2011). Surfactant-free purification of membrane proteins with intact native membrane environment. *Biochemical Society Transactions, 39*(3), 813–818. https://doi.org/10.1042/bst0390813.

Jiang, Y. W., Guo, H. Y., Chen, Z., Yu, Z. W., Wang, Z., & Wu, F. G. (2016). In situ visualization of lipid raft domains by fluorescent glycol chitosan derivatives. *Langmuir, 32*(26), 6739–6745. https://doi.org/10.1021/acs.langmuir.6b00193.

Jin, L., Millard, A. C., Wuskell, J. P., Clark, H. A., & Loew, L. M. (2005). Cholesterol-enriched lipid domains can be visualized by di-4-ANEPPDHQ with linear and nonlinear optics. *Biophysical Journal, 89*(1), L04–L06. Retrieved from http://www.ncbi.nlm.nih.gov/pubmed/15879475.

Jin, L., Millard, A. C., Wuskell, J. P., Dong, X., Wu, D., Clark, H. A., et al. (2006). Characterization and application of a new optical probe for membrane lipid domains. *Biophysical Journal, 90*(7), 2563–2575. Retrieved from http://www.ncbi.nlm.nih.gov/pubmed/16415047.

Juhasz, J., Davis, J. H., & Sharom, F. J. (2010). Fluorescent probe partitioning in giant unilamellar vesicles of 'lipid raft' mixtures. *The Biochemical Journal, 430*(3), 415–423. https://doi.org/10.1042/bj20100516.

Jungmann, R. (2014). Multiplexed 3D cellular super-resolution imaging with DNA-PAINT and exchange-PAINT. *Nature Methods, 11*, 313–318. Retrieved from https://doi.org/10.1038/nmeth.2835.

Kahya, N., Scherfeld, D., Bacia, K., Poolman, B., & Schwille, P. (2003). Probing lipid mobility of raft-exhibiting model membranes by fluorescence correlation spectroscopy. *The Journal of Biological Chemistry, 278*(30), 28109–28115. Retrieved from http://www.ncbi.nlm.nih.gov/pubmed/12736276.

Kahya, N., Scherfeld, D., & Schwille, P. (2005). Differential lipid packing abilities and dynamics in giant unilamellar vesicles composed of short-chain saturated glycerol-phospholipids, sphingomyelin and cholesterol. *Chemistry and Physics of Lipids, 135*(2), 169–180. Retrieved from http://www.ncbi.nlm.nih.gov/pubmed/15869751.

Kaiser, R. D., & London, E. (1998). Location of diphenylhexatriene (DPH) and its derivatives within membranes: Comparison of different fluorescence quenching analyses of membrane depth. *Biochemistry, 37*(22), 8180–8190. https://doi.org/10.1021/bi980064a.

Kamerbeek, C. B., Borroni, V., Pediconi, M. F., Sato, S. B., Kobayashi, T., & Barrantes, F. J. (2013). Antibody-induced acetylcholine receptor clusters inhabit liquid-ordered and liquid-disordered domains. *Biophysical Journal, 105*(7), 1601–1611. https://doi.org/10.1016/j.bpj.2013.08.039.

Kanfer, G., Sarraf, S. A., Maman, Y., Baldwin, H., Dominguez-Martin, E., Johnson, K. R., et al. (2021). Image-based pooled whole-genome CRISPRi screening for subcellular phenotypes. *The Journal of Cell Biology, 220*(2). https://doi.org/10.1083/jcb.202006180.

Karlsen, M. L., Bruhn, D. S., Pezeshkian, W., & Khandelia, H. (2021). Long chain sphingomyelin depletes cholesterol from the cytoplasmic leaflet in asymmetric lipid membranes. *RSC Advances, 11*(37), 22677–22682. https://doi.org/10.1039/D1RA01464A.

Kassas, N., Tanguy, E., Thahouly, T., Fouillen, L., Heintz, D., Chasserot-Golaz, S., et al. (2017). Comparative characterization of phosphatidic acid sensors and their localization during frustrated phagocytosis. *The Journal of Biological Chemistry, 292*(10), 4266–4279. https://doi.org/10.1074/jbc.M116.742346.

Kay, J. G., & Fairn, G. D. (2019). Distribution, dynamics and functional roles of phosphatidylserine within the cell. *Cell Communication and Signaling: CCS, 17*(1), 126. https://doi.org/10.1186/s12964-019-0438-z.

Kier, A. B., Sweet, W. D., Cowlen, M. S., & Schroeder, F. (1986). Regulation of transbilayer distribution of a fluorescent sterol in tumor cell plasma membranes. *Biochimica et Biophysica Acta, 861*(2), 287–301.

Kinoshita, M., Suzuki, K. G., Matsumori, N., Takada, M., Ano, H., Morigaki, K., et al. (2017). Raft-based sphingomyelin interactions revealed by new fluorescent sphingomyelin analogs. *The Journal of Cell Biology, 216*(4), 1183–1204. https://doi.org/10.1083/jcb.201607086.

Kinsky, S. C., Luse, S. A., Zopf, D., Van Deenen, L. L. M., & Haxby, J. (1967). Interaction of filipin and derivatives with erythrocyte membranes and lipid dispersions: Electron microscopic observations. *Biochimica et Biophysica Acta (BBA) - Biomembranes, 135*(5), 844–861. https://doi.org/10.1016/0005-2736(67)90055-7.

Kishimoto, T., Ishitsuka, R., & Kobayashi, T. (2016). Detectors for evaluating the cellular landscape of sphingomyelin- and cholesterol-rich membrane domains. *Biochimica et Biophysica Acta (BBA) - Molecular and Cell Biology of Lipids, 1861*(8, Part B), 812–829. https://doi.org/10.1016/j.bbalip.2016.03.013.

Klymchenko, A. S. (2017). Solvatochromic and fluorogenic dyes as environment-sensitive probes: Design and biological applications. *Accounts of Chemical Research, 50*(2), 366–375. https://doi.org/10.1021/acs.accounts.6b00517.

Knowles, T. J., Finka, R., Smith, C., Lin, Y. P., Dafforn, T., & Overduin, M. (2009). Membrane proteins solubilized intact in lipid containing nanoparticles bounded by styrene maleic acid copolymer. *Journal of the American Chemical Society, 131*(22), 7484–7485. https://doi.org/10.1021/ja810046q.

Kobayashi, T., & Menon, A. K. (2018). Transbilayer lipid asymmetry. *Current Biology, 28*(8), R386–R391. https://doi.org/10.1016/j.cub.2018.01.007.

Korlach, J., Schwille, P., Webb, W. W., & Feigenson, G. W. (1999). Characterization of lipid bilayer phases by confocal microscopy and fluorescence correlation spectroscopy. *Proceedings of the National Academy of Sciences of the United States of America, 96*(15), 8461–8466. Retrieved from http://www.ncbi.nlm.nih.gov/pubmed/10411897.

Kremers, G. J., Gilbert, S. G., Cranfill, P. J., Davidson, M. W., & Piston, D. W. (2011). Fluorescent proteins at a glance. *Journal of Cell Science, 124*(Pt. 2), 157–160. https://doi.org/10.1242/jcs.072744.

Krizhevsky, A., Sutskever, I., & Hinton, G. E. (2012). *NIPS'12 Proc. 25th Int. Conf. Neural Inf. Process. Syst.*

Lange, Y., Swaisgood, M. H., Ramos, B. V., & Steck, T. L. (1989). Plasma membranes contain half the phospholipid and 90% of the cholesterol and sphingomyelin in cultured human fibroblasts. *The Journal of Biological Chemistry, 264*, 3786–3793.

Lange, Y., Ye, J., & Steck, T. L. (2004). How cholesterol homeostasis is regulated by plasma membrane cholesterol in excess of phospholipids. *Proceedings of the National Academy of Sciences of the United States of the America*, *101*(32), 11664–11667. Retrieved from http://www.ncbi.nlm.nih.gov/pubmed/15289597.

Lemmon, M. A. (2008). Membrane recognition by phospholipid-binding domains. *Nature Reviews Molecular Cell Biology*, *9*(2), 99–111. Retrieved from http://www.ncbi.nlm.nih.gov/pubmed/18216767.

Lesslauer, W., Cain, J. E., & Blasie, J. K. (1972). X-ray diffraction studies of lecithin bimolecular leaflets with incorporated fluorescent probes. *Proceedings of the National Academy of Sciences of the United States of America*, *69*(6), 1499–1503. https://doi.org/10.1073/pnas.69.6.1499.

Levi, M., Wilson, P. V., Cooper, O. J., & Gratton, E. (1993). Lipid phases in renal brush border membranes revealed by Laurdan fluorescence. *Photochemistry and Photobiology*, *57*, 420–425.

Li, Y. E., Wang, Y., Du, X., Zhang, T., Mak, H. Y., Hancock, S. E., et al. (2021). TMEM41B and VMP1 are scramblases and regulate the distribution of cholesterol and phosphatidylserine. *Journal of Cell Biology*, *220*(6). https://doi.org/10.1083/jcb.202103105.

Lim, C. Y., Davis, O. B., Shin, H. R., Zhang, J., Berdan, C. A., Jiang, X., et al. (2019). ER-lysosome contacts enable cholesterol sensing by mTORC1 and drive aberrant growth signalling in Niemann-Pick type C. *Nature Cell Biology*, *21*(10), 1206–1218. https://doi.org/10.1038/s41556-019-0391-5.

Lin, Q., & London, E. (2013). Transmembrane protein (perfringolysin o) association with ordered membrane domains (rafts) depends upon the raft-associating properties of protein-bound sterol. *Biophysical Journal*, *105*(12), 2733–2742. https://doi.org/10.1016/j.bpj.2013.11.002.

Lin, Q., & London, E. (2015). Ordered raft domains induced by outer leaflet sphingomyelin in cholesterol-rich asymmetric vesicles. *Biophysical Journal*, *108*(9), 2212–2222. https://doi.org/10.1016/j.bpj.2015.03.056.

Liu, X. (2019). ABC family transporters. *Advances in Experimental Medicine and Biology*, *1141*, 13–100. https://doi.org/10.1007/978-981-13-7647-4_2.

Liu, J., Chang, C. C., Westover, E. J., Covey, D. F., & Chang, T. Y. (2005). Investigating the allosterism of acyl-CoA:cholesterol acyltransferase (ACAT) by using various sterols: In vitro and intact cell studies. *The Biochemical Journal*, *391*(Pt 2), 389–397. https://doi.org/10.1042/bj20050428.

Liu, S. L., Sheng, R., Jung, J. H., Wang, L., Stec, E., O'Connor, M. J., et al. (2017). Orthogonal lipid sensors identify transbilayer asymmetry of plasma membrane cholesterol. *Nature Chemical Biology*, *13*(3), 268–274. https://doi.org/10.1038/nchembio.2268.

Liu, Y., Su, Y., & Wang, X. (2013). Phosphatidic acid-mediated signaling. *Advances in Experimental Medicine and Biology*, *991*, 159–176. https://doi.org/10.1007/978-94-007-6331-9_9.

Loew, L. M., Cohen, L. B., Dix, J., Fluhler, E. N., Montana, V., Salama, G., et al. (1992). A naphthyl analog of the aminostyryl pyridinium class of potentiometric membrane dyes shows consistent sensitivity in a variety of tissue, cell, and model membrane preparations. *The Journal of Membrane Biology*, *130*, 1–10.

London, E. (2019). Membrane structure–function insights from asymmetric lipid vesicles. *Accounts of Chemical Research*. https://doi.org/10.1021/acs.accounts.9b00300.

Lorent, J. H., Levental, K. R., Ganesan, L., Rivera-Longsworth, G., Sezgin, E., Doktorova, M., et al. (2020). Plasma membranes are asymmetric in lipid unsaturation, packing and protein shape. *Nature Chemical Biology*, *16*(6), 644–652. https://doi.org/10.1038/s41589-020-0529-6.

Lorizate, M., Terrones, O., Nieto-Garai, J. A., Rojo-Bartolomé, I., Ciceri, D., Morana, O., et al. (2021). Super-resolution microscopy using a bioorthogonal-based cholesterol probe provides unprecedented capabilities for imaging nanoscale lipid heterogeneity in living cells. *Small Methods*, 2100430. https://doi.org/10.1002/smtd.202100430.

Luzzati, V., Gulik-Krzywicki, T., Rivas, E., Reiss-Husson, F., & Rand, R. P. (1968). X-ray study of model systems: Structure of the lipid-water phases in correlation with the chemical composition of the lipids. *The Journal of General Physiology*, *51*(5), 37–43.

Luzzati, V., & Husson, F. (1962). The structure of the liquid-crystalline phasis of lipid-water systems. *The Journal of Cell Biology*, *12*(2), 207–219. https://doi.org/10.1083/jcb.12.2.207.

Maekawa, M., & Fairn, G. D. (2015). Complementary probes reveal that phosphatidylserine is required for the proper transbilayer distribution of cholesterol. *Journal of Cell Science*, *128*(7), 1422–1433. https://doi.org/10.1242/jcs.164715 %J Journal of Cell Science.

Makino, A., Abe, M., Ishitsuka, R., Murate, M., Kishimoto, T., Sakai, S., et al. (2017). A novel sphingomyelin/cholesterol domain-specific probe reveals the dynamics of the membrane domains during virus release and in Niemann-Pick type C. *FASEB Journal : Official Publication of the Federation of American Societies for Experimental Biology*, *31*(4), 1301–1322. https://doi.org/10.1096/fj.201500075R.

Man, Z., Cui, H., Lv, Z., Xu, Z., Wu, Z., Wu, Y., et al. (2021). Organic nanoparticles-assisted low-power STED nanoscopy. *Nano Letters*. https://doi.org/10.1021/acs.nanolett.1c00161.

Manley, S., Gillette, J. M., Patterson, G. H., Shroff, H., Hess, H. F., Betzig, E., et al. (2008). High-density mapping of single-molecule trajectories with photoactivated localization microscopy. *Nature Methods*, *5*(2), 155–157. Retrieved from http://www.ncbi.nlm.nih.gov/pubmed/18193054.

Marsh, D. (1974). An interacting spin label study of lateral expansion in dipalmitoyllecithin-cholesterol bilayers. *Biochimica et Biophysica Acta*, *363*(3), 373–386. https://doi.org/10.1016/0005-2736(74)90076-5.

Marshall, J. E., Faraj, B. H., Gingras, A. R., Lonnen, R., Sheikh, M. A., El-Mezgueldi, M., et al. (2015). The crystal structure of pneumolysin at 2.0 Å resolution reveals the molecular packing of the pre-pore complex. *Scientific Reports*, *5*, 13293. https://doi.org/10.1038/srep13293.

Matlashov, M. E., Shcherbakova, D. M., Alvelid, J., Baloban, M., Pennacchietti, F., Shemetov, A. A., et al. (2020). A set of monomeric near-infrared fluorescent proteins for multicolor imaging across scales. *Nature Communications*, *11*(1), 239. https://doi.org/10.1038/s41467-019-13897-6.

Maxfield, F. R., & Tabas, I. (2005). Role of cholesterol and lipid organization in disease. *Nature*, *438*(7068), 612–621. Retrieved from http://www.ncbi.nlm.nih.gov/pubmed/16319881.

Maxfield, F. R., & Wüstner, D. (2012). Analysis of cholesterol trafficking with fluorescent probes. *Methods in Cell Biology*, *108*. https://doi.org/10.1016/b978-0-12-386487-1.00017-1.

Mayor, S., & Rao, M. (2004). Rafts: Scale-dependent, active lipid organization at the cell surface. *Traffic*, *5*(4), 231–240. Retrieved from http://www.ncbi.nlm.nih.gov/pubmed/15030564.

McIntosh, A. L., Atshaves, B. P., Huang, H., Gallegos, A. M., Kier, A. B., & Schroeder, F. (2008). Fluorescence techniques using dehydroergosterol to study cholesterol trafficking. *Lipids*, *43*(12), 1185–1208. https://doi.org/10.1007/s11745-008-3194-1.

McIntosh, A. L., Atshaves, B. P., Huang, H., Gallegos, A. M., Kier, A. B., Schroeder, F., et al. (2007). Multiphoton laser-scanning microscopy and spatial analysis of dehydroergosterol distributions on plasma membrane of living cells. In T. J. McIntosh (Ed.), *Lipid Rafts* (pp. 85–105). Totowa, NJ: Humana Press.

McIntyre, J. C., & Sleight, R. G. (1991). Fluorescence assay for phospholipid membrane asymmetry. *Biochemistry, 30,* 11819–11824.

McNicholas, S., Potterton, E., Wilson, K. S., & Noble, M. E. (2011). Presenting your structures: The CCP4mg molecular-graphics software. *Acta Crystallographica. Section D, Biological Crystallography, 67*(Pt. 4), 386–394. https://doi.org/10.1107/s0907444911007281.

Mesmin, B., Pipalia, N. H., Lund, F. W., Ramlall, T. F., Sokolov, A., Eliezer, D., et al. (2011). STARD4 abundance regulates sterol transport and sensing. *Molecular Biology of the Cell, 22*(21), 4004–4015. https://doi.org/10.1091/mbc.E11-04-0372.

Miller, R. G. (1984). The use and abuse of filipin to localize cholesterol in membranes. *Cell Biology International Reports, 8*(7), 519–535. https://doi.org/10.1016/0309-1651(84)90050-x.

Milles, S., Meyer, T., Scheidt, H. A., Schwarzer, R., Thomas, L., Marek, M., et al. (2013). Organization of fluorescent cholesterol analogs in lipid bilayers—Lessons from cyclodextrin extraction. *Biochimica et Biophysica Acta, 1828*(8), 1822–1828. https://doi.org/10.1016/j.bbamem.2013.04.002.

Mlodzianoski, M. J., Curthoys, N. M., Gunewardene, M. S., Carter, S., & Hess, S. T. (2016). Super-resolution imaging of molecular emission spectra and single molecule spectral fluctuations. *PLoS One, 11*(3), e0147506. https://doi.org/10.1371/journal.pone.0147506.

Möbius, W., Ohno-Iwashita, Y., van Donselaar, E. G., Oorschot, V. M., Shimada, Y., Fujimoto, T., et al. (2002). Immunoelectron microscopic localization of cholesterol using biotinylated and non-cytolytic perfringolysin O. *The Journal of Histochemistry and Cytochemistry, 50*(1), 43–55. https://doi.org/10.1177/002215540205000105.

Moerner, W. E. (2015). Single-molecule spectroscopy, imaging, and photocontrol: Foundations for super-resolution microscopy (nobel lecture). *Angewandte Chemie (International Ed. in English).* https://doi.org/10.1002/anie.201501949.

Mondal, M., Mesmin, B., Mukherjee, S., & Maxfield, F. R. (2009). Sterols are mainly in the cytoplasmic leaflet of the plasma membrane and the endocytic recycling compartment in CHO cells. *Molecular Biology of the Cell, 20*(2), 581–588. Retrieved from http://www.ncbi.nlm.nih.gov/pubmed/19019985.

Montes, L. R., Alonso, A., Goni, F. M., & Bagatolli, L. A. (2007). Giant unilamellar vesicles electroformed from native membranes and organic lipid mixtures under physiological conditions. *Biophysics Journal, 93*(10), 3548–3554. Retrieved from http://www.ncbi.nlm.nih.gov/pubmed/17704162.

Montesano, R., Vassalli, P., & Orci, L. (1981). Structural heterogeneity of endocytic membranes in macrophages as revealed by the cholesterol probe, filipin. *Journal of Cell Science, 51,* 95–107.

Moon, S., Yan, R., Kenny, S. J., Shyu, Y., Xiang, L., Li, W., et al. (2017). Spectrally resolved, functional super-resolution microscopy reveals nanoscale compositional heterogeneity in live-cell membranes. *Journal of the American Chemical Society, 139*(32), 10944–10947. https://doi.org/10.1021/jacs.7b03846.

Mukherjee, S., & Chattopadhyay, A. (1995). Wavelength-selective fluorescence as a novel tool to study organization and dynamics in complex biological systems. *Journal of Fluorescence, 5,* 237–245.

Mukherjee, S., Zha, X. H., Tabas, I., & Maxfield, F. R. (1998). Cholesterol distribution in living cells: Fluorescence imaging using dehydroergosterol as a fluorescent cholesterol analog. *Biophysical Journal, 75*(4), 1915–1925.

Murate, M., Abe, M., Kasahara, K., Iwabuchi, K., Umeda, M., & Kobayashi, T. (2015). Transbilayer distribution of lipids at nano scale. *Journal of Cell Science, 128*(8), 1627–1638. https://doi.org/10.1242/jcs.163105.

Murate, M., & Kobayashi, T. (2016). Revisiting transbilayer distribution of lipids in the plasma membrane. *Chemistry and Physics of Lipids*, *194*, 58–71. https://doi.org/10.1016/j.chemphyslip.2015.08.009.

Naito, T., & Saheki, Y. (2021). GRAMD1-mediated accessible cholesterol sensing and transport. *Biochimica et Biophysica Acta - Molecular and Cell Biology of Lipids*, *1866*(8), 158957. https://doi.org/10.1016/j.bbalip.2021.158957.

Nelson, L. D., Johnson, A. E., & London, E. (2008). How interaction of perfringolysin O with membranes is controlled by sterol structure, lipid structure, and physiological low pH: Insights into the origin of perfringolysin O-lipid raft interaction. *The Journal of Biological Chemistry*, *283*(8), 4632–4642. https://doi.org/10.1074/jbc.M709483200.

Nieves, D. J., & Owen, D. M. (2020). Quantitative mapping of membrane nanoenvironments through single-molecule imaging of solvatochromic probes. *bioRxiv*. https://doi.org/10.1101/2020.07.19.209908 %J.

Nishimura, S., Ishii, K., Iwamoto, K., Arita, Y., Matsunaga, S., Ohno-Iwashita, Y., et al. (2013). Visualization of sterol-rich membrane domains with fluorescently-labeled theonellamides. *PLoS One*, *8*(12), e83716. https://doi.org/10.1371/journal.pone.0083716.

Nizamov, S., Sednev, M. V., Bossi, M. L., Hebisch, E., Frauendorf, H., Lehnart, S. E., et al. (2016). "Reduced" coumarin dyes with an O-phosphorylated 2,2-dimethyl-4-(hydroxymethyl)-1,2,3,4-tetrahydroquinoline fragment: Synthesis, spectra, and STED microscopy. *Chemistry – A European Journal*, *22*(33), 11631–11642. https://doi.org/10.1002/chem.201601252.

Obaid, A. L., Loew, L. M., Wuskell, J. P., & Salzberg, B. M. (2004). Novel naphthylstyryl-pyridium potentiometric dyes offer advantages for neural network analysis. *Journal of Neuroscience Methods*, *134*(2), 179–190. Retrieved from http://www.ncbi.nlm.nih.gov/pubmed/15003384.

Ohno-Iwashita, Y., Shimada, Y., Waheed, A. A., Hayashi, M., Inomata, M., Nakamura, M., et al. (2004). Perfringolysin O, a cholesterol-binding cytolysin, as a probe for lipid rafts. *Anaerobe*, *10*(2), 125–134. https://doi.org/10.1016/j.anaerobe.2003.09.003.

Orci, L., Montesano, R., Meda, P., Malaisse-Lagae, F., Brown, D., Perrelet, A., et al. (1981). Heterogeneous distribution of filipin-cholesterol complexes across the cisternae of the Golgi apparatus. *Proceedings of the National Academy of Sciences of the United States of America*, *78*, 293–297.

Pagac, M., Cooper, D. E., Qi, Y., Lukmantara, I. E., Mak, H. Y., Wu, Z., et al. (2016). SEIPIN regulates lipid droplet expansion and adipocyte development by modulating the activity of glycerol-3-phosphate acyltransferase. *Cell Reports*, *17*(6), 1546–1559. https://doi.org/10.1016/j.celrep.2016.10.037.

Parasassi, T., Conti, F., & Gratton, E. (1986). Time-resolved fluorescence emission spectra of Laurdan in phospholipid vesicles by multifrequency phase and modulation fluorometry. *Cellular and Molecular Biology*, *32*(1), 103–108.

Parasassi, T., De Stasio, G., d'Ubaldo, A., & Gratton, E. (1990). Phase fluctuation in phospholipid membranes revealed by Laurdan fluorescence. *Biophysical Journal*, *57*, 1179–1186.

Parasassi, T., De Stasio, G., Ravagnan, G., Rusch, R. M., & Gratton, E. (1991). Quantitation of lipid phases in phospholipid vesicles by the generalized polarization of Laurdan fluorescence. *Biophysical Journal*, *60*, 179–189.

Parasassi, T., Di Stefano, M., Ravagnan, G., Sapora, O., & Gratton, E. (1992). Membrane aging during cell growth ascertained by laurdan generalized polarization. *Experimental Cell Research*, *202*, 432–439.

Parasassi, T., & Gratton, E. (1992). Packing of phospholipid vesicles studied by oxygen quenching of Laurdan fluorescence. *Journal of Fluorescence*, *2*, 167–174.

Parasassi, T., Gratton, E., Yu, W. M., Wilson, P., & Levi, M. (1997). Two-photon fluorescence microscopy of Laurdan generalized polarization domains in model and natural membranes. *Biophysical Journal*, *72*(6), 2413–2429.

Parasassi, T., Krasnowska, E. K., Bagatolli, L., & Gratton, E. (1998). Laurdan and prodan as polarity-sensitive fluorescent membrane probes. *Journal of Fluorescence*, *8*(4), 365–373. https://doi.org/10.1023/A:1020528716621.

Parasassi, T., Loiero, M., Raimondi, M., Ravagnan, G., & Gratton, E. (1993a). Absence of lipid gel-phase domains in seven mammalian cell lines and in four primary cell types. *Biochimica Biophysica Acta Bio-Membrane*, *1153*, 143–154.

Parasassi, T., Loiero, M., Raimondi, M., Ravagnan, G., & Gratton, E. (1993b). Effect of cholesterol on phospholipid phase domains as detected by laurdan generalized polarization. *Biophysical Journal*, *64*, A72.

Parasassi, T., Ravagnan, G., Rusch, R. M., & Gratton, E. (1993). Modulation and dynamics of phase properties in phospholipid mixtures detected by Laurdan fluorescence. *Photochemistry and Photobiology*, *57*, 403–410.

Pentchev, P. G., Comly, M. E., Kruth, H. S., Vanier, M. T., Wenger, D. A., Patel, S., et al. (1985). A defect in cholesterol esterification in Niemann-Pick disease (type C) patients. *Proceedings of the National Academy of Sciences of the United States of America*, *82*(23), 8247–8251. https://doi.org/10.1073/pnas.82.23.8247.

Radhakrishnan, A., Anderson, T. G., & McConnell, H. M. (2000). Condensed complexes, rafts, and the chemical activity of cholesterol in membranes. *Proceedings of the National Academy of Sciences of the United States of America*, *97*(23), 12422–12427. Retrieved from http://www.ncbi.nlm.nih.gov/pubmed/11050164.

Radhakrishnan, A., & McConnell, H. M. (1999). Cholesterol-phospholipid complexes in membranes. *Journal of the American Chemical Society*, *121*(2), 486–487.

Radhakrishnan, A., & McConnell, H. M. (2000). Chemical activity of cholesterol in membranes. *Biochemistry*, *39*(28), 8119–8124. Retrieved from http://www.ncbi.nlm.nih.gov/pubmed/0010889017.

Raghuraman, H., & Chattopadhyay, A. (2004). Influence of lipid chain unsaturation on membrane-bound melittin: A fluorescence approach. *Biochimica et Biophysica Acta*, *1665*(1-2), 29–39. Retrieved from http://www.ncbi.nlm.nih.gov/pubmed/15471568.

Ranjit, S., Malacrida, L., Jameson, D. M., & Gratton, E. (2018). Fit-free analysis of fluorescence lifetime imaging data using the phasor approach. *Nature Protocols*, *13*(9), 1979–2004. https://doi.org/10.1038/s41596-018-0026-5.

Robalo, J. R., do Canto, A. M., Carvalho, A. J., Ramalho, J. P., & Loura, L. M. (2013). Behavior of fluorescent cholesterol analogues dehydroergosterol and cholestatrienol in lipid bilayers: A molecular dynamics study. *The Journal of Physical Chemistry. B*, *117*(19), 5806–5819. https://doi.org/10.1021/jp312026u.

Ruggiero, A., & Hudson, B. (1989a). Analysis of the anisotropy decay of trans-parinaric acid in lipid bilayers. *Biophysical Journal*, *55*, 1125–1135.

Ruggiero, A., & Hudson, B. (1989b). Critical density fluctuations in lipid bilayers detected by fluorescence lifetime heterogeneity. *Biophysical Journal*, *55*, 1111–1124.

Saha, P., Shumate, J. L., Caldwell, J. G., Elghobashi-Meinhardt, N., Lu, A., Zhang, L., et al. (2020). Inter-domain dynamics drive cholesterol transport by NPC1 and NPC1L1 proteins. *eLife*, *9*. https://doi.org/10.7554/eLife.57089.

Samsonov, A. V., Mihalyov, I., & Cohen, F. S. (2001). Characterization of cholesterol-sphingomyelin domains and their dynamics in bilayer membranes. *Biophysics Journal*, *81*(3), 1486–1500. Retrieved from http://www.ncbi.nlm.nih.gov/pubmed/11509362.

Sanchez, S. A., Tricerri, M. A., & Gratton, E. (2012). Laurdan generalized polarization fluctuations measures membrane packing micro-heterogeneity in vivo. *Proceedings of the National Academy of Sciences of the United States of America*, *109*(19), 7314–7319. https://doi.org/10.1073/pnas.1118288109.

Sanchez-Garcia, R., Segura, J., Maluenda, D., Carazo, J. M., & Sorzano, C. O. S. (2018). Deep Consensus, a deep learning-based approach for particle pruning in cryo-electron microscopy. *IUCrJ*, *5*(Pt. 6), 854–865. https://doi.org/10.1107/S2052252518014392.

Sander, C. L., Sears, A. E., Pinto, A. F. M., Choi, E. H., Kahremany, S., Gao, F., et al. (2021). Nano-scale resolution of native retinal rod disk membranes reveals differences in lipid composition. *Journal of Cell Biology*, *220*(8). https://doi.org/10.1083/jcb.202101063.

Sato, S. B., Ishii, K., Makino, A., Iwabuchi, K., Yamaji-Hasegawa, A., Senoh, Y., et al. (2004). Distribution and transport of cholesterol-rich membrane domains monitored by a membrane-impermeant fluorescent polyethylene glycol-derivatized cholesterol. *The Journal of Biological Chemistry*, *279*(22), 23790–23796. Retrieved from http://www.ncbi.nlm.nih.gov/pubmed/15026415.

Saunders, F. K., Mitchell, T. J., Walker, J. A., Andrew, P. W., & Boulnois, G. J. (1989). Pneumolysin, the thiol-activated toxin of Streptococcus pneumoniae, does not require a thiol group for in vitro activity. *Infection and Immunity*, *57*(8), 2547–2552. https://doi.org/10.1128/iai.57.8.2547-2552.1989.

Savinov, S. N., & Heuck, A. P. (2017). Interaction of cholesterol with perfringolysin O: What have we learned from functional analysis? *Toxins (Basel)*, *9*(12). https://doi.org/10.3390/toxins9120381.

Schroeder, F., Atshaves, B. P., Gallegos, A. M., McIntosh, A. L., Liu, J. C. S., Kier, A. B., et al. (2005). Chapter 1 lipid rafts and caveolae organization. *Advances in Molecular and Cell Biology*, *36*, 1–36. Elsevier.

Schroeder, F., Myers-Payne, S. C., Billheimer, J. T., & Wood, W. G. (1995). Probing the ligand binding sites of fatty acid and sterol carrier proteins: Effects of ethanol. *Biochemistry*, *34*(37), 11919–11927. Retrieved from http://www.ncbi.nlm.nih.gov/pubmed/7547928.

Schueder, F., Strauss, M. T., Hoerl, D., Schnitzbauer, J., Schlichthaerle, T., Strauss, S., et al. (2017). Universal super-resolution multiplexing by DNA exchange. *Angewandte Chemie (International Ed. in English)*, *56*(14), 4052–4055. https://doi.org/10.1002/anie.201611729.

Severs, N. J., & Robenek, H. (1983). Detection of microdomains in biomembranes. An appraisal of recent developments in freeze-fracture cytochemistry. *Biochimica et Biophysica Acta*, *737*(3-4), 373–408. https://doi.org/10.1016/0304-4157(83)90007-2.

Sezgin, E., Levental, I., Grzybek, M., Schwarzmann, G., Mueller, V., Honigmann, A., et al. (2012). Partitioning, diffusion, and ligand binding of raft lipid analogs in model and cellular plasma membranes. *Biochimica et Biophysica Acta*, *1818*(7), 1777–1784. https://doi.org/10.1016/j.bbamem.2012.03.007.

Sezgin, E., Levental, I., Mayor, S., & Eggeling, C. (2017). The mystery of membrane organization: Composition, regulation and roles of lipid rafts. *Nature Reviews. Molecular Cell Biology*. https://doi.org/10.1038/nrm.2017.16.

Shaw, J. E., Epand, R. F., Epand, R. M., Li, Z., Bittman, R., & Yip, C. M. (2006). Correlated fluorescence-atomic force microscopy of membrane domains: Structure of fluorescence probes determines lipid localization. *Biophysical Journal*, *90*(6), 2170–2178. Retrieved from http://www.ncbi.nlm.nih.gov/pubmed/16361347.

Shcherbakova, D. M., Baloban, M., Emelyanov, A. V., Brenowitz, M., Guo, P., & Verkhusha, V. V. (2016). Bright monomeric near-infrared fluorescent proteins as tags and biosensors for multiscale imaging. *Nature Communications*, 7, 12405. https://doi.org/10.1038/ncomms12405.

Shinitzky, M., Dianoux, A. C., Gitler, C., & Weber, G. (1971). Microviscosity and order in the hydrocarbon region of micelles and membranes determined with fluorescent probes. I. Synthetic micelles. *Biochemistry*, *10*(11), 2106–2113. https://doi.org/10.1021/bi00787a023.

Shinitzky, M., & Inbar, M. (1976). Microviscosity parameters and protein mobility in biological membranes. *Biochimica et Biophysica Acta*, *433*, 133–149.

Simons, K., & Ikonen, E. (1997). Functional rafts in cell membranes. *Nature*, *387*(6633), 569–572. Retrieved from http://www.ncbi.nlm.nih.gov/pubmed/9177342.

Solanko, K. A., Modzel, M., Solanko, L. M., & Wüstner, D. (2015). Fluorescent sterols and cholesteryl esters as probes for intracellular cholesterol transport. *Lipid Insights*, *8*(Suppl. 1), 95–114. https://doi.org/10.4137/lpi.S31617.

St. Clair, J. W., Kakuda, S., & London, E. (2020). Induction of ordered lipid raft domain formation by loss of lipid asymmetry. *Biophysical Journal*, *119*(3), 483–492. https://doi.org/10.1016/j.bpj.2020.06.030.

St. Clair, J. W., & London, E. (2019). Effect of sterol structure on ordered membrane domain (raft) stability in symmetric and asymmetric vesicles. *Biochimica et Biophysica Acta - Biomembranes*, *1861*(6), 1112–1122. https://doi.org/10.1016/j.bbamem.2019.03.012.

Steck, T. L., & Lange, Y. (2018). Transverse distribution of plasma membrane bilayer cholesterol: Picking sides. *Traffic*, *19*(10), 750–760. https://doi.org/10.1111/tra.12586.

Stolowich, N. J., Frolov, A., Atshaves, B., Murphy, E. J., Jolly, C. A., Billheimer, J. T., et al. (1997). The sterol carrier protein-2 fatty acid binding site: An NMR, circcular dichroic, and fluorescence spectroscopic determination. *The Biochemist*, *36*(7), 1719–1729.

Tangorra, A., Ferretti, G., Zolese, G., & Curatola, G. (1994). Study of plasma membrane heterogeneity using a phosphatidylcholine derivative of 1,6-diphenyl-1,3,5-hexatriene [2-(3-(diphenylhexatriene)propanoyl)-3-palmitoyl-L-α-phosphatidylcholine]. *Journal of Fluorescence*, *4*(4), 357–360. https://doi.org/10.1007/bf01881456.

Teo, A. C. K., Lee, S. C., Pollock, N. L., Stroud, Z., Hall, S., Thakker, A., et al. (2019). Analysis of SMALP co-extracted phospholipids shows distinct membrane environments for three classes of bacterial membrane protein. *Scientific Reports*, *9*(1), 1813. https://doi.org/10.1038/s41598-018-37962-0.

Tory, M. C., & Merrill, A. R. (2002). Determination of membrane protein topology by red-edge excitation shift analysis: Application to the membrane-bound colicin E1 channel peptide. *Biochimica et Biophysica Acta*, *1564*(2), 435–448. https://doi.org/10.1016/s0005-2736(02)00493-5.

Tosheva, K. L., Yuan, Y., Matos Pereira, P., Culley, S., & Henriques, R. (2020). Between life and death: Strategies to reduce phototoxicity in super-resolution microscopy. *Journal of Physics D: Applied Physics*, *53*(16), 163001. https://doi.org/10.1088/1361-6463/ab6b95.

Truong, L., & Ferré-D'Amaré, A. R. (2019). From fluorescent proteins to fluorogenic RNAs: Tools for imaging cellular macromolecules. *Protein Science*, *28*(8), 1374–1386. https://doi.org/10.1002/pro.3632.

Tsuji, T., Cheng, J., Tatematsu, T., Ebata, A., Kamikawa, H., Fujita, A., et al. (2019). Predominant localization of phosphatidylserine at the cytoplasmic leaflet of the ER, and its TMEM16K-dependent redistribution. *Proceedings of the National Academy of Sciences of the United States of America*, *116*(27), 13368–13373. https://doi.org/10.1073/pnas.1822025116.

Tweten, R. K. (1988). Cloning and expression in Escherichia coli of the perfringolysin O (theta-toxin) gene from Clostridium perfringens and characterization of the gene product. *Infection and Immunity*, *56*(12), 3228–3234. https://doi.org/10.1128/iai.56.12.3228-3234.1988.

Uchida, Y., Hasegawa, J., Chinnapen, D., Inoue, T., Okazaki, S., Kato, R., et al. (2011). Intracellular phosphatidylserine is essential for retrograde membrane traffic through endosomes. *Proceedings of the National Academy of Sciences of the United States of America*, *108*(38), 15846–15851. https://doi.org/10.1073/pnas.1109101108.

Urbani, L., & Simoni, R. D. (1990). Cholesterol and vesicular stomatitis virus G protein take separate routes from the endoplasmic reticulum to the plasma membrane. *The Journal of Biological Chemistry*, *265*(4), 1919–1923.

van de Linde, S., Heilemann, M., & Sauer, M. (2012). Live-cell super-resolution imaging with synthetic fluorophores. *Annual Review of Physical Chemistry*, *63*, 519–540. https://doi.org/10.1146/annurev-physchem-032811-112012.

van der Velde, J., Smit, J., Punter, M., & Cordes, T. (2018). Self-healing dyes for super-resolution microscopy. *bioRxiv*. https://doi.org/10.1101/373852.

van Meer, G. (2011). Dynamic transbilayer lipid asymmetry. *Cold Spring Harbor Perspectives in Biology*, *3*(5). https://doi.org/10.1101/cshperspect.a004671.

van Meer, G., Voelker, D. R., & Feigenson, G. W. (2008). Membrane lipids: Where they are and how they behave. *Nature Reviews. Molecular Cell Biology*, *9*(2), 112–124. Retrieved from http://www.ncbi.nlm.nih.gov/pubmed/18216768.

van Wee, R., Filius, M., & Joo, C. (2021). Completing the canvas: Advances and challenges for DNA-PAINT super-resolution imaging. *Trends in Biochemical Sciences*. https://doi.org/10.1016/j.tibs.2021.05.010.

Varnai, P., & Balla, T. (1998). Visualization of phosphoinositides that bind pleckstrin homology domains: Calcium- and agonist-induced dynamic changes and relationship to myo-[3H]inositol-labeled phosphoinositide pools. *The Journal of Cell Biology*, *143*(2), 501–510. Retrieved from http://www.ncbi.nlm.nih.gov/pubmed/9786958.

Veatch, S. L., & Keller, S. L. (2002). Organization in lipid membranes containing cholesterol. *Physical Review Letters*, *89*(26), 268101. Retrieved from http://www.ncbi.nlm.nih.gov/pubmed/12484857.

Veatch, S. L., & Keller, S. L. (2003a). A closer look at the canonical 'Raft Mixture' in model membrane studies. *Biophysics Journal*, *84*(1), 725–726. Retrieved from http://www.ncbi.nlm.nih.gov/pubmed/12524324.

Veatch, S. L., & Keller, S. L. (2003b). Separation of liquid phases in giant vesicles of ternary mixtures of phospholipids and cholesterol. *Biophysical Journal*, *85*(5), 3074–3083. Retrieved from http://www.ncbi.nlm.nih.gov/pubmed/14581208.

Veatch, S. L., & Keller, S. L. (2005). Seeing spots: Complex phase behavior in simple membranes. *Biochimica et Biophysica Acta*, *1746*(3), 172–185. Retrieved from http://www.ncbi.nlm.nih.gov/pubmed/16043244.

Wade, O. K., Woehrstein, J. B., Nickels, P. C., Strauss, S., Stehr, F., Stein, J., et al. (2019). 124-Color super-resolution imaging by engineering DNA-PAINT blinking kinetics. *Nano Letters*, *19*(4), 2641–2646. https://doi.org/10.1021/acs.nanolett.9b00508.

Waheed, A. A., Shimada, Y., Heijnen, H. F., Nakamura, M., Inomata, M., Hayashi, M., et al. (2001). Selective binding of perfringolysin O derivative to cholesterol-rich membrane microdomains (rafts). *Proceedings of the National Academy of Sciences of the United States of America*, *98*(9), 4926–4931. https://doi.org/10.1073/pnas.091090798.

Weber, G. (1952). Polarization of the fluorescence of macromolecules. II. Fluorescent conjugates of ovalbumin and bovine serum albumin. *Biochemical Journal*, *51*(2), 155–167. https://doi.org/10.1042/bj0510155.

Weber, G., & Farris, F. J. (1979). Synthesis and spectral properties of a hydrophobic fluorescent probe: 6-Propionyl-2-(dimethylamino)naphthalene. *Biochemistry*, *18*(14), 3075–3078.

Weber, M., Khan, T. A., Patalag, L. J., Bossi, M., Leutenegger, M., Belov, V. N., et al. (2021). Photoactivatable fluorophore for stimulated emission depletion (STED) microscopy and bioconjugation technique for hydrophobic labels. *Chemistry*, *27*(1), 451–458. https://doi.org/10.1002/chem.202004645.

Weber, G., & Laurence, D. J. (1954). Fluorescent indicators of adsorption in aqueous solution and on the solid phase. *The Biochemical Journal*, *56*(325th Meeting), xxxi.

Wenz, J. J., & Barrantes, F. J. (2003). Steroid structural requirements for stabilizing or disrupting lipid domains. *Biochemistry*, *42*(48), 14267–14276. https://doi.org/10.1021/bi035759c.

Wharton, S. A., De Martinez, S. G., & Green, C. (1980). Use of fluorescent probes in the study of phospholipid—sterol bilayers. *The Biochemical Journal, 191*(3), 785–790. https://doi.org/10.1042/bj1910785.
Wildanger, D., Rittweger, E., Kastrup, L., & Hell, S. W. (2008). STED microscopy with a supercontinuum laser source. *Optics Express, 16*(13), 9614–9621. Retrieved from http://www.ncbi.nlm.nih.gov/pubmed/18575529.
Wilhelm, L. P., Voilquin, L., Kobayashi, T., Tomasetto, C., & Alpy, F. (2019). Intracellular and plasma membrane cholesterol labeling and quantification using filipin and GFP-D4. *Methods in Molecular Biology, 1949*, 137–152. https://doi.org/10.1007/978-1-4939-9136-5_11.
Wüstner, D. (2007). Plasma membrane sterol distribution resembles the surface topography of living cells. *Molecular Biology of the Cell, 18*(1), 211–228. Retrieved from http://www.ncbi.nlm.nih.gov/pubmed/17065557.
Wüstner, D., Herrmann, A., Hao, M., & Maxfield, F. R. (2002). Rapid nonvesicular transport of sterol between the plasma membrane domains of polarized hepatic cells. *The Journal of Biological Chemistry, 277*(33), 30325–30336. Retrieved from http://www.ncbi.nlm.nih.gov/pubmed/12050151.
Xu, R., Xu, Y., Wang, Z., Zhou, Y., Dang, D., & Meng, L. (2020). Recent advances on organic fluorescent probes for stimulated emission depletion (STED) microscopy. *Combinatorial Chemistry & High Throughput Screening*. https://doi.org/10.2174/1386207323666200917104203.
Yamamura, K., Ashida, H., Okano, T., Kinoshita-Daitoku, R., Suzuki, S., Ohtani, K., et al. (2019). Inflammasome activation induced by perfringolysin O of Clostridium perfringens and its involvement in the progression of gas gangrene. *Frontiers in Microbiology, 10*, 2406. https://doi.org/10.3389/fmicb.2019.02406.
Yeagle, P. L. (1989). Lipid regulation of cell membrane structure and function. *The FASEB Journal, 3*, 1833–1842.
Yeung, T., Gilbert, G. E., Shi, J., Silvius, J., Kapus, A., & Grinstein, S. (2008). Membrane phosphatidylserine regulates surface charge and protein localization. *Science, 319*(5860), 210–213. https://doi.org/10.1126/science.1152066.
Yokoyama, Y., Terada, T., Shimizu, K., Nishikawa, K., Kozai, D., Shimada, A., et al. (2020). Development of a deep learning-based method to identify "good" regions of a cryo-electron microscopy grid. *Biophysical Reviews, 12*(2), 349–354. https://doi.org/10.1007/s12551-020-00669-6.
Yu, D., Gustafson, W. C., Han, C., Lafaye, C., Noirclerc-Savoye, M., Ge, W. P., et al. (2014). An improved monomeric infrared fluorescent protein for neuronal and tumour brain imaging. *Nature Communications, 5*, 3626. https://doi.org/10.1038/ncomms4626.
Zhang, Z., Kenny, S. J., Hauser, M., Li, W., & Xu, K. (2015). Ultrahigh-throughput single-molecule spectroscopy and spectrally resolved super-resolution microscopy. *Nature Methods, 12*(10), 935–938. https://doi.org/10.1038/nmeth.3528.

Mass spectrometry-based lipid analysis and imaging

Koralege C. Pathmasiri[a], Thu T.A. Nguyen[a], Nigina Khamidova[a], and Stephanie M. Cologna[a,b,*]

[a]Department of Chemistry, University of Illinois at Chicago, Chicago, IL, United States
[b]Laboratory of Integrated Neuroscience, University of Illinois at Chicago, Chicago, IL, United States
*Corresponding author: e-mail address: cologna@uic.edu

Contents

Abstract

Mass spectrometry imaging (MSI) is a powerful tool for in situ mapping of analytes across a sample. With growing interest in lipid biochemistry, the ability to perform such mapping without antibodies has opened many opportunities for MSI and lipid analysis. Herein, we discuss the basics of MSI with particular emphasis on MALDI mass spectrometry and lipid analysis. A discussion of critical advancements as well as protocol details are provided to the reader. In addition, strategies for improving the detection of lipids, as well as applications in biomedical research, are presented.

Current Topics in Membranes, Volume 88
ISSN 1063-5823
https://doi.org/10.1016/bs.ctm.2021.10.005

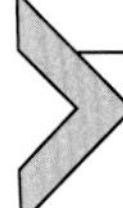

1. Introduction to matrix-assisted laser desorption/ionization mass spectrometry imaging (MALDI MSI)

Mass spectrometry has emerged as a powerful technique for the identification and quantification of analytes from an array of specimen types. The initial step in any mass spectrometry analysis of a molecule is ionization, which can be defined as the process in which a molecule or an atom becomes a positively or negatively charged ion after interacting with an energy source (Peacock, Zhang, & Trimpin, 2017). Depending on the nature of the sample and the ionization technique, the basic steps in ionization can vary. For example, if the sample is introduced into the mass spectrometer as a liquid, the first step will be vaporization followed by ionization. In mass spectrometry analysis, the analyte molecules and subsequent ions must be transferred into the gas phase to allow for detection.

Matrix-assisted laser desorption/ionization (MALDI) is an ionization technique that relies on two critical elements: a laser source and a matrix which facilitates ionization. A MALDI matrix is typically characterized as an organic molecule with π-electron conjugation, which allows this molecule to absorb laser energy. It is critical for the matrix molecules to sufficiently absorb the majority of the laser energy to prevent the fragmentation of the analytes. For this reason, it is important that the laser wavelength matches the solid-state absorption spectra of the matrix (Niehaus, Schnapp, Koch, Soltwisch, & Dreisewerd, 2017). Prior to spotting the sample onto the MALDI plate, the analyte is mixed with the matrix in solution, followed by solvent evaporation, causing the analyte and the matrix molecules to co-crystalize. The sample is then ready for laser irradiation. The central role of the matrix is to act as an energy buffer between the analyte and the laser (Chang et al., 2007; Karas & Krüger, 2003; Zenobi & Knochenmuss, 1998). In addition, heat generated upon the energy absorption by matrix helps the desorption of analyte molecules from the plate. Due to its strength in conserving the intact structure of molecules, MALDI is considered a soft ionization technique in mass spectrometry.

Fig. 1 is a depiction of the MALDI ionization process. Desorbed analytes can be ionized either as positively or negatively charged ions depending on the functional groups on the molecule. Some analytes can be ionized as both positive and negative ions, while others have higher ionization efficiency for only one ion type. Accordingly, the mass spectrometer instrument can be operated in either positive or negative mode for the detection of the

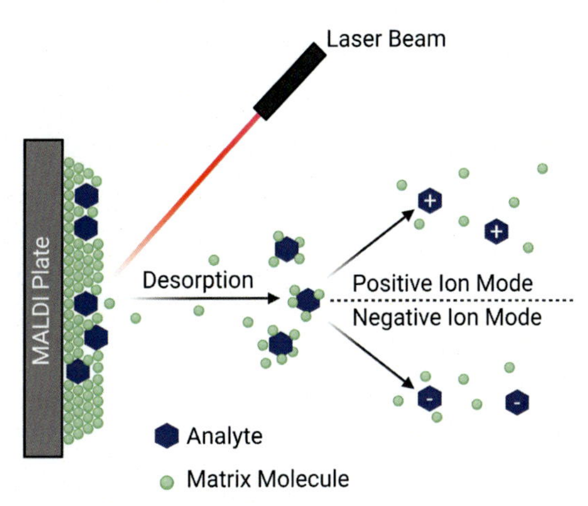

Fig. 1 Ionization process in MALDI. A laser is irradiated onto a surface and energy is absorbed by the matrix and analyte molecules to promote desorption. Analytes are ionized as positively or negatively charged ions before entering the mass analyzer.

respective ions. It is important to know how a particular analyte will ionize with high efficiency because the ability to detect such molecules will rely on this feature. Moreover, the ion suppression due to background and overlapping ions in each mode should be considered.

MALDI mass spectrometry imaging (MALDI MSI) uses the MALDI technique to map analytes of interest in samples such as animal tissue sections (Harris, Roseborough, Mor, Yeung, & Whitehead, 2020; Mallah et al., 2018), plant tissue slices (Boughton & Thinagaran, 2018; Korte, Yagnik, Feenstra, & Lee, 2015; Susniak, Krysa, Gieroba, Komaniecka, & Sroka-Bartnicka, 2020), microbial cultures (Debois et al., 2014; Vergeiner, Schafferer, Haas, & Müller, 2014), among others. MALDI can be regarded as the more commonly used and well-established ionization technique used in MSI compared to other ionization techniques (Castellanos et al., 2019; Pierson et al., 2020; Santoro et al., 2020; Vanbellingen et al., 2016). With the development of laser technology, MALDI MSI has improved in terms of spatial resolution and analysis speed (Balluff, Hopf, Porta Siegel, Grabsch, & Heeren, 2021; Bowman, Blakney, et al., 2020; Bowman, Bogie, et al., 2020; Prentice & Caprioli, 2016; Prentice, Chumbley, & Caprioli, 2015; RamalloGuevara, Paulssen, Popova, Hopf, & Levkin, 2021; Treu, Kokesch-Himmelreich, Walter, Hölscher, & Römpp, 2020). Among its advantages, mass spectrometry imaging allows multiplexed spatial mapping and quantification of a vast number of molecular classes without any labeling (Chumbley et al., 2016; Korte & Lee, 2013; Korte et al., 2015; Porta, Lesur, Varesio, & Hopfgartner, 2015; Prentice, Chumbley, Hachey, Norris, & Caprioli, 2016; Prentice, McMillen, & Caprioli, 2019; Rzagalinski & Volmer, 2017). Sample preparation for MALDI MSI includes slicing the tissue or other sample of interest into thin sections and applying a uniform matrix coating prior to loading into the mass spectrometer. Inside the mass spectrometer, the sample stage will move step-by-step in x and y directions while a stationary laser is used to desorb and ionize at each pixel to cover the region of interest. The mass spectrum for a given mass range is generated at each laser spot, including signals for multiple analyte molecules. This data will be used to regenerate heat maps for any detected molecule from the sample using data analysis software compatible with the data file. MALDI MSI workflow is described in detail later in this chapter under the topic "MALDI MSI lipid imaging workflow".

1.1 MALDI matrix

As noted above, a MALDI matrix is commonly an organic molecule capable of absorbing UV light (typically 337 or 355 nm) and can be easily co-crystallized with the analyte molecules of interest (Allwood, Dreyfus, Perera, & Dyer, 1997). In addition, a matrix molecule should be inert or have low reactivity to prevent being quickly oxidized by air or reacting with the analytes and should be stable under vacuum. There has been active research on developing matrices and derivatizing current matrices for numerous applications to overcome challenges. These challanges include, interference of matrix signal in the low molecular weight range, low ionization efficiency for some molecules, and low spot heterogeneity (Fan et al., 2020; Leopold, Popkova, Engel, & Schiller, 2018; Zhou, Fülöp, & Hopf, 2021). With the increased interest in MALDI MSI, new challenges with matrix application has surfaced. Since the amount of matrix significantly affects signal intensity, one must ensure a uniform matrix coating (Smolira & Wessely-Szponder, 2015). In addition, the formation of matrix crystals after application will also impact the signal, reproducibility, and image spatial resolution (Li et al., 2016). Therefore, the formation of smaller crystals is ideal for MALDI MSI. Another crucial factor is the vacuum stability of the matrix (Yang, Norris, & Caprioli, 2018). Since MSI experiment can take up to few hours per sample, the matrix should be stable from subliming out in the vacuum. The ability of the matrix to tolerate salts is also crucial in MALDI MSI analyses of biological tissues (Börnsen, 2000; Wang, Liu, & Wu, 2011).

The selection of a matrix is vital in a MALDI experiment since different matrices will show varying ionization efficiencies for an analyte of interest. The matrix used for an analysis of a specific molecule is typically selected using current knowledge of commonly used matrices, or by "trial and error" method using commonmatrices (Demeure, Quinton, Gabelica, & De Pauw, 2007; Smolira & Wessely-Szponder, 2015). The most popular matrices for MALDI MSI oflipids are 1,5-Diaminonapthalene (DAN), used in both positive and negative ionization modes, and 2,5-dihydroxybenzoic acid (DHB), which is more appropriate in positive mode ionization. Recent research presents hydralazine (a phthalazine derivative) as a universal matrix that can be used effectively in both positive and negative ion modes in lipid MALDI MSI analyses (Tang, Gordon, Wang, Chen, & Li, 2021). This matrix also allows a wider m/z range (50 Da–20,000 Da). Moreover, the authors demonstrated that hydralazine could also be used as a matrix for protein and small-molecule MSI experiments. Table 1 shows some of the most widely used matrices with their typical applications. Routinely used matrices can be categorized by their molecular structures into few following categories (Leopold et al., 2018).

1. *Benzoic acid derivatives*: 2,5-dihydroxybenzoic acid (DHB), 2, 5-dihydroxyterephthalic acid, 3,7-dihydroxy-2-naphthoic acid, 1, 4-dihydroxy-2-naphthoic acid
2. *Cinnamic acid derivatives*: α-cyano-4-hydroxycinnamic acid (CHCA), sinapinic acid, 4-chloro-a-cyanocinnamic acid (ClCCA)
3. *Heterocycles*: 9-aminoacridine (9-AA), 3-hydroxypicolinic acid, 4-hydroxycoumarin
4. *Other matrices*: 1,5-diaminonaphthalene (1,5-DAN), dithranol, 1,8-bis (dimethylamino)naphthalene (DMAN)

As shown in Table 1, CHCA is most commonly used for peptide analyses. Unlike peptides, matrix selection for lipids is more complicated due to different lipid classes with varied ionization efficiency using different matrices. The following section further discusses matrices that are used for different lipid classes.

Free fatty acids (FFA) are low molecular weight lipids that have been challenging to analyze using routinely used MALDI matrices. Despite the good ionization of FFA, lower molecular weight interference from the matrix can be a major issue in FFA MALDI analysis (Gladchuk et al., 2020). Matrices such as meso-tetrakis(pentafluorophenyl)porphyrin (MTPP) with higher molecular weight (974.57 g/mol) has been used to overcome the matrix ion overlap issue in FFA analysis (Ayorinde, Hambright, Porter, & Keith,

Table 1 Commonly used matrices and their applications.

Matrix name	**Application**
2,5-Dihydroxybenzoic acid (DHB)	More suitable in positive ion mode. Lipids, peptides and small molecule metabolites analysis (Janda et al., 2021; Strupat, Karas, & Hillenkamp, 1991; Wei et al., 2015)
1,5-Diaminonaphthalene (DAN)	Can be used in both positive and negative mode. Widely used in MALDI-MSI. Analysis of phospholipids, small metabolites and as a reductive matrix (Dong et al., 2013; Fukuyama, Iwamoto, & Tanaka, 2006; Korte & Lee, 2014)
9-Aminoacridine (9-AA)	Used in negative ion mode. Phospholipids, metabolites and Sulfatide analysis (Cerruti et al., 2012; Vermillion-Salsbury & Hercules, 2002)
α-Cyano-4-hydroxycinnamic acid (CHCA)	Most widely used matrix in protein and peptide analysis (Gobom et al., 2001; Watanabe et al., 2013; Zhu & Papayannopoulos, 2003)
Sinapinic acid	Used in positive ion mode. Protein and polypeptide analysis (Salum, Giudicessi, Schmidt De León, Camperi, & Erra-Balsells, 2017)
Trihydroxyacetophenone (THAP)	Neutral lipid, Glycosphingolipid, oligonucleotides and phosphorylated peptides (Kühn-Hölsken, Lenz, Sander, Lührmann, & Urlaub, 2005; Stübiger & Belgacem, 2007)
3-Hydroxypicolinic acid (HPA)	Oligonucleotide and sophorolipid analysis (Yatim, Wan Muhammad Zulkifli, Majid, Foster, & Hayes, 2020)

1999). The downside of this matrix is the low ionization efficiency for unsaturated FFA. There have been efforts to incorporate matrices such as 1,8-bis(dimethylamino)naphthalene (DMAN) (Staab & Saupe, 1988) and 9-aminoacridine (9-AA) (Shroff, Muck, & Svatoš, 2007) for FFA, albeit each comes with limitations.

Cholesterol and cholesteryl esters are molecules that attract high interest across a broad range of biomedical research and human health areas such as cardiovascular disease (Berger, Raman, Vishwanathan, Jacques, & Johnson, 2015), neurodegeneration (Dai et al., 2021) and genetic disorders. Cholesterol is known to have low ionization efficiency in mass spectrometry analysis and thus is also challenging in MALDI MSI. A unique feature of cholesterol analysis via mass spectrometry, compared to other lipids, is the neutral loss of water that occurs during the ionization process (Cologna, 2019; Johnson, ten Brink, & Jakobs, 2001). In MALDI, DHB is typically used for the analysis of cholesterol, and expected spectral features include the dehydrated molecule with both H^+ (denoted $[M+H]^+$) and Na^+ (denoted $[M+Na]^+$) adducts (Schiller et al., 2001). However, in the analysis of cholesteryl esters, the water loss is not seen because the free hydroxyl group is not present. Thus, MALDI and MALDI-MSI of cholesterol esters is more straightforward and commonly uses norharmane (NOR) (McMillen, Fincher, Klein, Spraggins, & Caprioli, 2020) matrix. An approach to improve the ionization of molecules such as cholesterol MALDI is to derivatize the molecule. Derivatization of a molecule is simply modifying the natural molecule using reactive chemical reagents. This can be done either in solution or on tissue. The purpose of derivatization of a molecule is to introduce a functional group with higher ionization efficiency and to increase the molecular weight of the analyte so that they appear in less crowded area of the mass spectrum. On-tissue enzyme-assisted derivatization has been developed recently by Roberto et al. to quantitively analyze cholesterol in mouse brain tissue (Angelini et al., 2021).

Diacylglycerols (DAG) and triacylglycerols (TAG) are neutral, energy storage lipids, mainly found in adipose tissues. Both DAG and TAG can be detected in the positive ion mode as alkali earth metal adduct ions (e.g., Na^+ and K^+) with the use of matrices like DHB and CHCA derivatives; however, the protonated molecule of TAG are considered less stable and can decompose before reaching the detector (Gidden, Liyanage, Durham, & Lay, 2007).

Phospholipids include several subclasses of lipids, including phosphatidylcholine (PC), phosphatidylethanolamine (PE), phosphatidylinositol (PI), phosphatidylglycerol (PG), and phosphatidylserine (PS). DAN, DHB and 9-AA are common matrices that can be used in phospholipid MALDI MSI of these types of lipids (Cerruti, Benabdellah, Laprévote, Touboul, & Brunelle, 2012; Strnad et al., 2019). Phospholipids such as PC are readily ionized in positive ion mode. Although PE also has a nitrogen atom, the ionization efficiency of PE is comparatively lower than PC. The negatively charged PI, PS, and PA lipids ionize in negative ion mode but with lower ionization efficiencies. 9-AA has been reported as a suitable matrix for the negative ion mode MALDI MSI analysis of very low abundant highly bioactive phosphatidylinositol phosphates, including PIP_2 (Pathmasiri et al., 2020; Vermillion-Salsbury & Hercules, 2002). 9-AA matrix has a comparatively lower background and finer crystal size than other matrices, making 9-AA a better candidate for MALDI MSI. In addition, 9-AA can be used in the positive ion mode and has a higher affinity towards Na^+ itself, thereby reducing the availability of Na^+ ions and promoting the formation of more H^+ adducts which avoid the dilution of signals.

1.2 Desorption electrospray ionization mass spectrometry imaging (DESI-MSI)

In addition to MALDI, there are a few other ionization techniques used for MSI. DESI is another soft ionization mode that does not use a matrix for analysis (Takáts, Wiseman, Gologan, & Cooks, 2004). In DESI, the analyte molecules are desorbed with the use of a spray under ambient conditions. As shown in Fig. 2, the spray is produced using a solvent flow with a nebulizing gas through a nozzle. Analytes are desorbed and ionized upon the impact of charged solvent droplets from the spray on the sample, which is then transported to the inlet of a mass spectrometer (Fig. 2). With the movement of the target sample reference to stationary spray, spectra across the sample are generated with corresponding coordinates, which are then saved for the regeneration of images. Unlike MALDI, DESI presents no additional background from the matrix, making DESI more reliable than MALDI for the analysis of lower mass analytes; however, some challenges are associated with the spatial resolution that can be achieved (Eberlin et al., 2011). A few recent

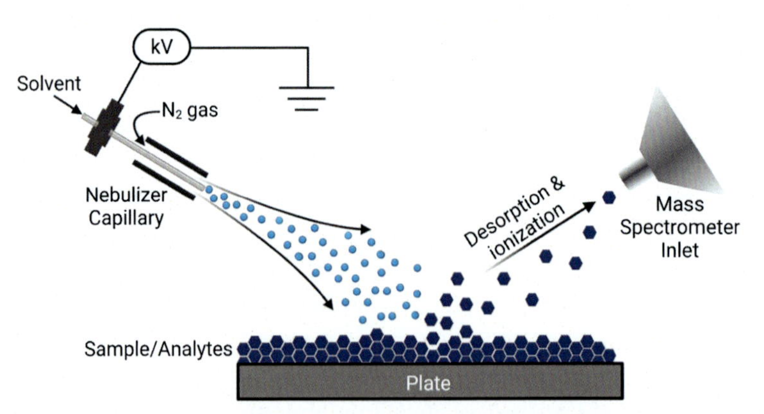

Fig. 2 Ionization process in DESI. Analyte molecules are desorbed and ionized upon the impact of charged solvent droplets created in the DESI spray. Desorbed and ionized analyte molecules then enter the mass spectrometer.

examples have proven that optimization of DESI-MSI can be useful for the analysis of larger molecules, including intact proteins (Ambrose et al., 2017; Honarvar & Venter, 2017; Shin, Drolet, Mayer, Dolence, & Basile, 2007). Parameters that can be optimized in DESI to improve ionization efficiency include solvent system, the flow rate of the solvent, capillary tube temperature, sheath gas pressure, reflection angle, incident angle, and ESI voltage. The spatial resolution of DESI-MSI is limited by the size of the spray, and nanoDESI has been developed to improve the spatial resolution in the imaging mode (Nguyen et al., 2019). DESI has been used in imaging of a wide variety of lipid classes, including cholesterol (Wu, Ifa, Manicke, & Cooks, 2009).

1.3 Secondary ion mass spectrometry (SIMS)

SIMS has been used widely for inorganic chemistry research and can be used to analyze both atoms and intact molecules (Eriksson, Masaki, Yao, Hayasaka, & Setou, 2013; Hiraoka, Asakawa, Fujimaki, Takamizawa, & Mori, 2006; Joo & Liang, 2013). Due to the popularity of MSI in biological tissues and the growing need to increase the spatial resolution in MSI, SIMS has gained significant attention. In SIMS, a primary ion beam is focused on the sample that is capable of desorbing analyte molecules creating secondary ions (analyte ions). The analyte ions are then entered into the mass analyzer. SIMS provides the best spatial resolution among the different mass spectrometry imaging techniques so far, ranging from micrometers to nanometers (Gamble & Anderton, 2016, Klitzing, Weber, & Kraft, 2013; Passarelli et al., 2017).

2. MALDI MSI lipid imaging workflow

The main steps in sample preparation for MALDI MSI experiments are provided in Fig. 3, which includes tissue sectioning, removal of interfering molecules by washing, application of the matrix coating, data acquisition, lipid identification, and data analysis. Each of these steps has been optimized for different applications by researchers (Bodzon-Kulakowska et al., 2020; Stanback et al., 2021; Vandenbosch et al., 2021; Visscher et al., 2019). In this section, the MALDI MSI workflow will be discussed in detail to provide a further understanding of each step and to explain recent advances in the field with a focus on tissue analysis.

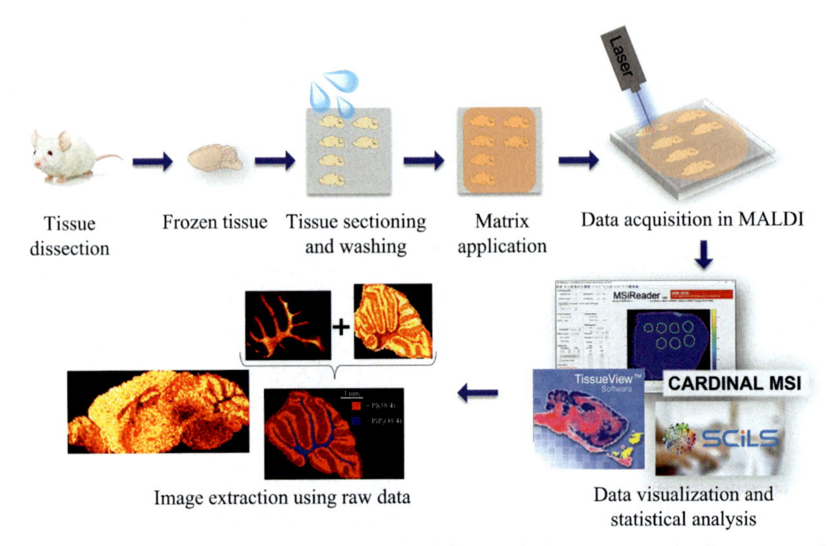

Fig. 3 MALDI-MSI workflow. MALDI-MSI workflow includes tissue sectioning, removal of interfering molecules by washing, application of the matrix coating, data acquisition, lipid identification and data analysis.

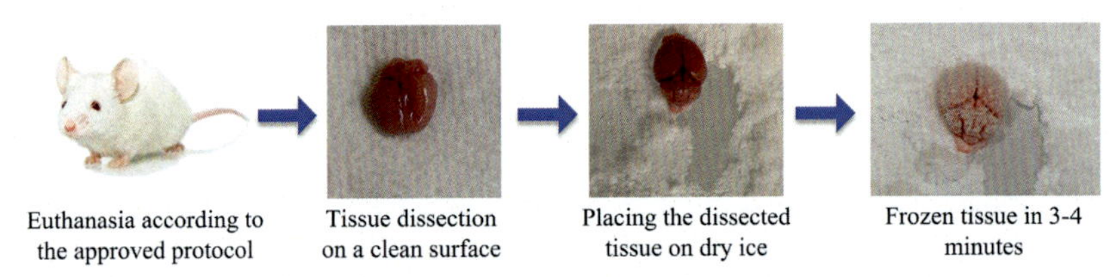

Fig. 4 Freezing dissected tissue. Freshly dissected tissues are flash frozen on a flat dry ice surface to allow preservation of tissue integrity.

2.1 Tissue sectioning for MALDI MSI

After dissecting the tissue of interest, tissues should be flash-frozen on dry ice, as shown in Fig. 4. Careful dissection and flash freezing help to protect the integrity of the tissue, which is crucial to obtaining quality images. Freezing the tissue on a nonsolid surface such as foil placed on dry ice could result in deformed tissue. Since the dry ice around the tissue evaporates readily, it exerts minimum pressure on tissue by dry ice. Flash freezing by immersing the tissue in chilled cryogenic solutions or liquid nitrogen can also be used. Care must be taken when immersing tissue in chilled solutions due to the possibility of tissue damage from sudden and extreme temperature changes. Dissected frozen tissue can be stored at $-80\,^{\circ}\mathrm{C}$ until sectioning.

The sectioning plane is selected depending on the area of interest in the tissue. Frozen tissues need to be sectioned into thin slices (~ 10 μm thick) and mounted onto the MALDI plate or ITO-coated slides. Careful and proper sectioning is essential since any damages such as scratches will impact the matrix deposition and the resulting image and data quality. A cryostat is used to section the tissues for MALDI MSI, which is operated at cryogenic temperatures (generally −10 °C to −30 °C depending on the tissue type). As an example, a mouse brain tissue is usually sectioned at −10 °C while mouse liver tissue is sectioned at −20 °C.

The following are some common problems that arise in tissue sectioning for MALDI MSI and possible solutions.

- Folding or curling problems: The use of an artistic paint brush to control the section is a way to control folding and curling. Curling can also happen if the cryostat blade is not positioned correctly or if not sharp enough. Most cryostats have an anti-roll plate that goes on the blade, leaving small spacing for the section to minimize rolling the section. Optimizing the sectioning temperature also helps smooth the section without rolling.
- Streaking problems: Streaks present on the sections as they are produced can also be due to sectioning at incorrect temperatures. Increasing the chamber temperature a few degrees and trying again will help determine if the temperature is an issue. A damaged blade, damaged anti-roll plate, or improper positioning of the anti-roll plate may also produce streaking.
- Thaw mounting problems: Moving the section onto the stainless steel MALDI plate can be tricky and must be done with extra care. Using an artistic paintbrush to gently slide the section onto the MALDI plate is a good way to move sections. After positioning the tissue on the MALDI plate, an artistic paintbrush can be used to press the section onto MALDI plate and brush gently to make sure the tissue will not roll before getting thaw mounted. Thaw mounting the section fast is always a better approach to minimize rolling.
- Problems in sectioning soft or small tissues: A common technique followed for the thin sectioning of the whole body and soft specimens is to embed the specimen in an embedding media. The most widely used embedding media are incompatible with MALDI-MSI since these media are polymer-based. Contamination of polymer can cause ion suppression in MALDI and interference with matrix deposition. Kimberly et al. investigated several embedding media compatible with mass spectrometry and found 5% carboxymethyl cellulose +10% gelatin-based media as

an optimal embedding medium for MALDI MSI cryosectioning (Nelson, Daniels, Fournie, & Hemmer, 2013). Embedded sectioning is also useful in sectioning smaller tissues, like mouse intestines, which can be challenging to section when mounted on OCT. Adhesive films can be used to assist the sectioning of whole-body specimens embedded in mass spectrometry-friendly media (Kawamoto, 2003). In this case, the adhesive film is fastened onto the cutting surface to assist the section to be intact after cryosection and can be mounted on indium tin oxide (ITO) coated glass slides. Both nonconductive and conductive adhesive films have been used. However, conductive films have shown improved signals compared to nonconductive films (Saigusa et al., 2019)

2.2 Sample storage for MALDI MSI

Tissue sections mounted on MALDI plates or ITO glass slides can be stored at cryogenic temperatures until analysis. Since lipids are relatively less stable molecules, easily delocalize and diffuse over time, the proper storage is crucial to preserve the integrity as well as spatial distribution. Specifically, phospholipids are prone to quick degradation and decrease in abundance over time, resulting in the increase oxidized phospholipids and lysophospholipids (Patterson, Thomas, & Chaurand, 2014). In addition, the degradation of phospholipids could give rise to free fatty acid signals and promote fatty acid dimerization in MSI (Dill, Eberlin, Costa, Ifa, & Cooks, 2011). Delocalization or degradation of lipid in a study comparing diseased and control tissues could lead to false biological interpretation and identification of incorrect biomarkers.

Lukowski et al. has investigated the effect of storage conditions on data quality and reproducibility in MALDI MSI using human kidney tissue as the test specimen (Lukowski et al., 2020). They further show the storage of tissue at cryogenic temperatures can minimize lipid oxidation. Furthermore, minimizing the exposure of the tissue sections to the atmosphere can lower lipid oxidization. This can be accomplished by coating the tissue sections with the matrix before freezing or storing the samples under vacuum at cryogenic temperatures (Lukowski et al., 2020). In addition to storage at cryogenic temperature, minimizing the freeze-thaw cycles is also essential to avoid lipid degradation. Freeze-thawing has a higher impact on phospholipid and triacylglycerol degradation resulting accumulation of free fatty acids (van der Vusse et al., 1989). Freeze-thaw cycles can be avoided by planning the experiments well, including tissue dissection and sectioning. If the

MSI study includes a comparison of two groups like diseased and control, it is crucial to maintain consistency in sample preparation and storage. If possible, samples from all the groups should be mounted on the same MALDI plate to avoid inconsistency.

Formalin-fixed paraffin-embedded (FFPE) tissue blocks or sections are commonly used to store biological specimen in most tissue libraries. But the use of FFPE tissue in MSI is not straightforward due to paraffin which can interfere highly with mass spectrometry analysis. When using FFPE tissues for MSI, paraffin has to be removed and can be done using optimized washing techniques (Angel, Mehta, Norris-Caneda, & Drake, 2018; Hermann et al., 2020). There have been concerns in the field about FFPE tissues that could have metabolite loss or modification due to the processing (Nilsson et al., 2015). Achim et al. has performed a comprehensive study to compare the analysis of the metabolites in fresh frozen (FF) tissues and FFPE tissues (Buck, Ly, et al., 2015). This study showed more than 70% overlap in identified metabolites between FF and FFPE tissues. The authors further demonstrate the conservation of biomolecular information at the metabolite level, which could be assigned to clinical and histology parameters.

2.3 Washing tissue sections for lipid imaging

After sectioning and mounting the tissues on to MALDI plate, tissues can be washed depending on the type of analysis. Washing techniques and the buffer composition must be selected carefully since the optimal washing conditions are different for various analytes. Since biological samples have an extremely complex molecular composition, the removal of salts and other highly abundant "noise" molecules can improve the detection of analytes of interest. For example, in the analysis of protein and small metabolites by MALDI MSI, hydrophobic washing buffers are used to wash away the lipids. Washing tissue sections with chloroform at −20 °C has shown to remove glycerophospholipids and glycerolipids while improving ionization of adenosine triphosphate (ATP) related metabolites, amino acid derivatives, glucose derivatives, and glycolysis metabolites (Yang et al., 2018). Using organic washes in combination with optimized aqueous-based buffers has shown increased sensitivity of proteins in MALDI MSI (Thomas, Patterson, Laveaux Charbonneau, & Chaurand, 2013). A recent review by Vu et al. also highlighted numerous methods of washing that improve the detection of neuropeptides (Vu, DeLaney, & Li, 2021). For the analysis of lipids, hydrophilic washing buffers are used to wash away blood and salt on the tissue sections, which will interfere with ionization in mass spectrometry. Aqueous washes contribute to diminished sodium and potassium adduct formation by washing away salts and thereby enhancing the H^+ adduct formation in positive ion mode. Commonly used aqueous washings buffers are ammonium formate-based and ammonium acetate-based buffers with a near-neutral pH value (Angel, Spraggins, Baldwin, & Caprioli, 2012).

2.4 Matrix application

Following the appropriate washing protocol, tissues should be dried to remove any remaining washing buffer and moisture. At this point in the protocol, the user can weigh the MALDI plate or ITO slide in order to calculate the matrix density after matrix deposition. Obtaining an optical image of the tissue section(s) before matrix application can also be useful in subsequent data analysis and data interpretation. Application of a selected matrix to obtain a homogeneous layer on the tissue slices is required to obtain reliable and data. Poor matrix deposition or the incorrect amount of matrix density leads to a lower signal-to-noise ratio and results in poor images have been observed in our laboratory. When developing new methods for MALDI MSI, it is important not only to select the appropriate matrix but also to optimize the applied matrix density for the analysis. Based on our experience, a higher matrix density is needed for the analysis of molecules over ~1000 Da, but this could vary depending on the molecular properties of the analyte. When conducting a large study using MALDI MSI, it is critical to record the matrix density and obtain consistent matrix density throughout the experiment. It is also important to consider the crystal size and homogeneity obtained using different matrix application techniques since these can affect the final resolution and quality of the images (Goodwin, 2012; Thomas, Charbonneau, Fournaise, & Chaurand, 2012). Different techniques available for matrix application are discussed below in detail.

2.4.1 Airbrush matrix application

Artistic airbrush has been widely used from the early stage of MALDI MSI for matrix application (DeKeyser, Kutz-Naber, Schmidt, Barrett-Wilt, & Li, 2007; Shimma et al., 2007). In this technique, the matrix is dissolved in an organic-based solvent and is sprayed using the airbrush directly on the MALDI plate. Spray velocity and the movement of the spray across the MALDI plate have to be controlled manually and are critical in reproducibility. The setup for matrix application using the airbrush technique is

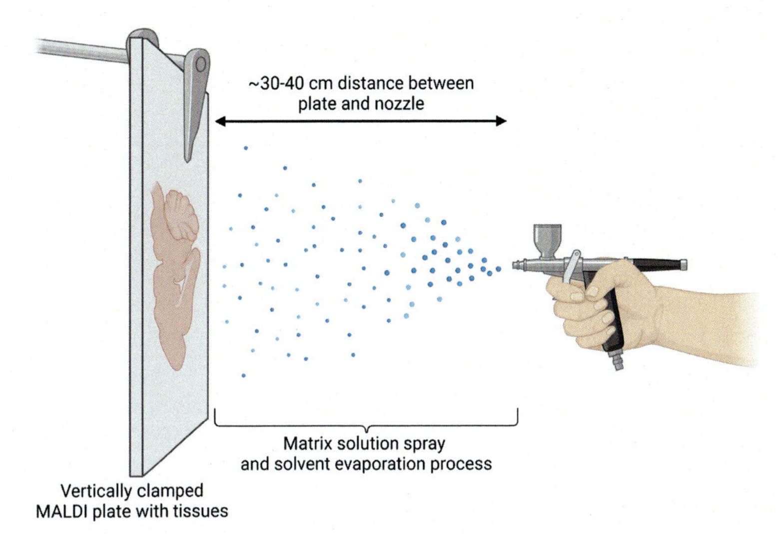

Fig. 5 Matrix application using artistic airbrush. Matrix is dissolved in an organic-based solvent that is sprayed uniformly on the vertically clamped MALDI plate maintaining an appropriate distance between plate and the airbrush nozzle.

simple and can be easily assembled. As shown in the Fig. 5, MALDI plate is clamped vertically, minimizing the area of the plate covered by the clamp and tissue facing outside. Setup should be used inside a hood since organic solvents are sprayed, and the vapor of the matrix could contaminate the atmosphere. An airbrush connected to a nitrogen gas tank should be washed well by spraying just the solvent a few times. This will help to clear any clogs and remove air from the nozzle. Fill the paint chamber with the matrix solution and test the airbrush for a stable spray. Trapped air bubbles and particulate in the nozzle could give rise to sputtering and results inconsistent matrix coating. The airbrush should be held straight and perpendicular to the target plate, and a consistent distance between the MALDI plate and the airbrush nozzle must be maintained throughout the matrix application. The distance between the plate and airbrush nozzle should be sufficient to allow the excess solvent to get vaporized in the spray and create matrix aerosols (typically 30–40 cm distance is maintained as shown in Fig. 5) (DeKeyser et al., 2007). It is necessary to maintain a constant spray velocity and be aware of the remaining level of matrix solution in the airbrush to avoid running out, which will result in creating large solvent droplets and sputtering. Unlike in sublimation matrix application technique, using airbrush technique, it is possible to even coat the tissues with mixtures of. A few drawbacks of using the airbrush technique include poor reproducibility, larger matrix crystals with varying sizes, the possibility of delocalization of analytes on the tissue, and difficulty of finding a robust solvent system (Gemperline, Rawson, & Li, 2014).

2.4.2 Matrix application by sublimation

Sublimation is the transition of a solid substance into the gaseous state without going through a liquid state. Most solid substances can be sublimed by manipulating the pressure and the temperature. This process is used to apply the matrix in MALDI MSI using a sublimation apparatus (Hankin, Barkley, & Murphy, 2007). As shown in Fig. 6, a vacuum pump is connected to the sublimation chamber through a cold trap, and the sublimation chamber is placed on a heating mantle or sand bath. Matrix of interest is placed in the bottom of sublimation chamber directly, or matrix can be first dissolved in a volatile organic solvent then transfer into the sublimation chamber followed by evaporating the solvent to create matrix layer on the bottom of the sublimation chamber. Washed and dried tissues on the MALDI plate are pre-weighed and attached to the bottom of the glass condenser using conductive copper tape, as shown in Fig. 6. Care should be taken to make sure there is proper contact of the MALDI plate with the glass

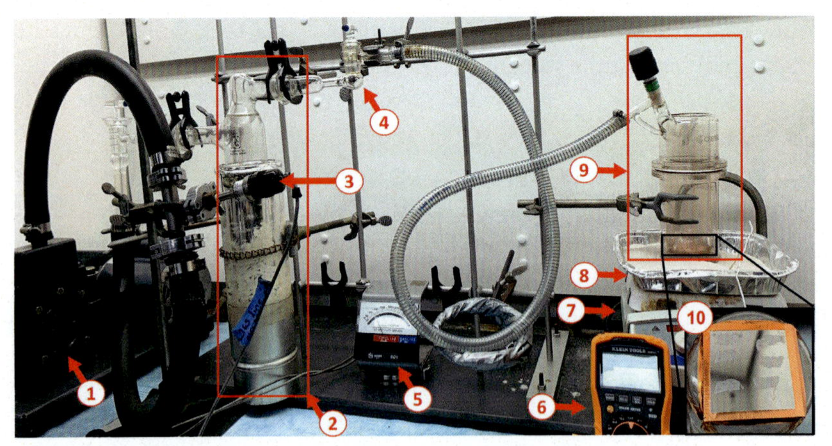

Fig. 6 Sublimation matrix application setup. 1 = Rough pump, 2 = Cold trap, 3 = Thermocouple vacuum gauge, 4 = Valve before sublimation chamber, 5 = Vacuum gauge readout, 6 = Multimeter with temperature monitoring probe (to monitor temperature on sand bath), 7 = Hot plate, 8 = Sand bath, 9 = Sublimation chamber (on bottom) and glass condenser (on top) setup, 10 = MALDI plate with tissues mounted onto the bottom of glass condenser.

condenser since the heat transfer between plate and condenser is critical for successful sublimation. The use of conductive copper tape facilitate efficient heat transfer and uniform matrix coating. It is also essential to make sure the tissues are dehydrated and free of moisture since the moisture on the tissue will interfere with sublimation. After preparing the MALDI plate and the matrix in the sublimation chamber, the hot plate is set to the desired temperature and let equilibrate at the set temperature. Change in the temperature or pressure during sublimation could result in incorrect matrix density. The only way to control the amount of matrix deposited on the MALDI plate is to optimize sublimation temperature, pressure, and time (Thomas et al., 2012). Reproducible matrix density can be obtained by maintaining the above parameters consistent in every experiment. Once the temperature is stable at the desired temperature, and the matrix chamber is sealed, an ice slush is added to cool the MALDI plate, which must be at a lower temperature than the matrix. Sublimation is initiated by opening the valve to the vacuum supply, which will bring the entire setup, including the sublimation chamber, to lower pressure (20–100 mTorr).

The main benefit of sublimation is the fine crystal size which is important to improve image resolution in MALDI MSI (Hankin et al., 2007). During the process of sublimation, impurities from the matrix are also reduced because of phase transitions that take place. Yang et al. incorporated hydration/recrystallization after matrix application by sublimation, which yielded matrix crystal between 0.5 and 3 μm in size, to further improve detection of proteins between 3 and 30 kDa (Yang & Caprioli, 2011). In contrast to airbrush spraying, sublimation results in a very uniform coating of the matrix. Fernández et al. developed a stainless steel sublimation apparatus to improve the reproducibility in matrix application (Fernández, Garate, Martín-Saiz, Galetich, & Fernández, 2019). This setup provides better control of the sample temperature during sublimation, vacuum inside the sublimation chamber, and sublimation temperature. Xie et al. developed a matrix sublimation setup with the ability to control the crystallization temperature (Xie, Wu, Hung, Chen, & Chan, 2021). The authors show the ability of the device to obtain reproducible and fine crystal size, which improved the image resolution as well as ion intensity. Interestingly, combining sublimation matrix deposition with sodium salt doping of tissue section has shown the ability to image phospholipid and neutral lipids simultaneously (Dufresne, Patterson, Norris, & Caprioli, 2019). This method shows the ability to image neutral lipids with low ionization efficiency, including cholesterol esters, cerebrosides, and triglycerides.

2.4.3 Automated matrix deposition

Automated matrix spray systems for MALDI MSI matrix application has been developed and are commercially available. In this system, a spray of the matrix is created with the aid of a heated sheath gas. Heated sheath gas aid for the rapid evaporation of solvent protecting the tissues from getting wet as well as the sheath gas could maintain a stable spray. In these systems, a user can control the flow rate of the matrix, the temperature of the sheath gas, distance from the nozzle to sample, and speed of the movement of the nozzle. These parameters should be optimized for different matrices since different matrices will behave differently in the deposition process. Automatic sprayers provide more controlled matrix density, crystal size, uniform coating, and reproducible matrix application, providing better consistency in data. In a comparison of airbrush matrix application, sublimation, and automated matrix application, authors have shown that automated matrix deposition improved the number of metabolites detected (Gemperline et al., 2014). This study also talks about the combined use of sublimation and automated spray technique to improve the detection of metabolites. Compared to matrix application using the airbrush method, both sublimation and automated spray provide higher reproducibility and lower analyte diffusion, while automated systems show the most reliable and reproducible matrix application. Automated matrix application takes place under more wet conditions compared to sublimation. This could allow better analyte extraction and co-crystallization leading to improved sensitivity.

Even though automated matrix deposition has many advantages, access to these systems is limited due to the high cost. Since sublimation and airbrush matrix application systems can be assembled by the user, the cost is lower than automated systems. A higher level of training and experience will be needed for both airbrush and sublimation techniques to get a uniform and reproducible matrix coating. Time consumption for matrix application is also higher in both airbrush and sublimation techniques.

2.5 On tissue chemical derivatization

Due to the highly diverse chemical structures of lipids, these molecules show a wide range of different chemical features. Diverse functional groups of lipids create high variability in mass spectrometry ionization efficiency for different lipid classes. This can create challenges in detecting low abundant lipids or lipids with lower ionization efficiency. To solve this issue, researchers have utilized the diverse functional groups of lipids to selectively modify lipid molecules for improved ionization (Jiang, Ory, & Han, 2007; Pan, Qin, & Han, 2021; Wang, Krull, Liu, & Orr, 2003). Not only does chemical derivation improve ionization, but it also can improve the resolution and ability to detect lipids isomers (Wang, Palavicini, Cseresznye, & Han, 2017; Zhang, Shang, Ouyang, & Xia, 2020) along with determining the unsaturation position fatty acids (Jeck, Korf, Vosse, & Hayen, 2019). With the increase in demand to image analytes using MALDI MSI, more chemical derivatization reactions have been developed. Derivatization of targeted functional groups and analytes isdiscussed below in detail.

2.5.1 Chemical derivatization of amines

Beyond being highly abundant in proteins, multiple lipid classes, including phosphatidylethanolamines (PE), phosphatidylserine (PS), and amine-containing sphingolipids. Derivatization of amine groups can be applied in a global lipidomic experiment to improve ionization and identification of all amine-containing lipids.

Trans-cinnamaldehyde/4-hydroxy-3-methoxycinnamaldehyde: Trans-cinnamaldehyde (CA) is a naturally occurring and commercially available aromatic compound. Primary amines can readily react with the aldehyde group in CA to make a stable Schiff base with significantly increased ionization efficiency (Guo et al., 2020; Manier et al., 2011). Manier et al. also demonstrated the possibility of pre-coating the MALDI plate with CA via spin coating technique for efficient sample preparation to perform MALDI MSI of isoniazid, an anti-tuberculosis drug. Guo et al. utilized CA derivatization for the imaging of small endogenous metabolites using MALDI (Guo et al., 2020). This reagent also has been utilized to improve ionization of polar amino neurotransmitters in brain tissues and successfully imaged using MALDI ionization where the distribution of 23 endogenous metabolites, such as glycine, alanine, valine, GABA, and dopamine, were reported (Esteve, Tolner, Shyti, van den Maagdenberg, & McDonnell, 2016).

2-nitrobenzaldehyde (NBA): Similarly, Li et al. developed a derivatization strategy for neuropeptides using NBA reagent, which promotes a nanosecond timescale photochemically promoted click reaction (Li et al., 2019). The photochemically active NBA reagent is irradiated with nanosecond solid-state Nd:YAG laser to create a reactive 2-nitrobenzoic anion that reacts with a primary amine. This derivatization shifts the mass of the analyte by $(n + 1)^{*}133$ Da and improves the ionization efficiency of neurotransmitters. Proof of concept experiments demonstrated that with the use of NBA, higher-quality images (high image contrast, less lateral diffusion) of analytes were observed for several neuropeptides, small proteins, and lipids such as monomethyl-phosphatidylethanolamine, phosphatidylcholine, and triacylglycerol. However, the authors also noted a downside for this technique: the possibility of diluting the signal intensity of peptides, caused by the difference in labeling efficiency of NBA on several lysine residues of a peptide, increasing the complexity of the mass spectra and signal overlapping between several peptides. This can cause issues with accurate protein and peptide identification and quantification (Li et al., 2019).

p-N,N,N-trimethylammonioanilyl N-hydroxysuccinimidyl carbamate iodide (TAHS): TAHS was originally developed by Shimbo et al. for the derivatization of amino acids in liquid chromatography-mass spectrometry (LC-MS) (Shimbo, Yahashi, Hirayama, Nakazawa, & Miyano, 2009). Toue and coworkers used this derivatization strategy to image and quantify amino acids in human colon cancer specimens by MSI (Toue et al., 2014).

2,4-diphenyl-pyranylium tetrafluoroborate (DPP-TFB): DPP-TFB is a pyrylium salt that specifically reacts with primary amines to yield charged quaternary amino groups under mild conditions (Shariatgorji et al., 2014). A significant increase in ionization efficiency has enabled to successful image of low abundant neurotransmitters in tissues of animal models, mouse, rat, and primate, for Parkinson's disease. In addition, when comparing the use of multiple pyrylium salts in MSI of primary amine-containing metabolites, Shariatgorji and colleagues were able to use used 2,4-diphenyl-pyranylium (DPP) ion as a reactive matrix, rather than just derivatization agent, due to its ability to absorb energy from a laser with a wavelength of 355 nm (Shariatgorji et al., 2015). This further reduces the number of steps needed to prepare samples for MSI experiments. They successfully detected and quantified the amount of β-*N*-methylamino-L-alanine, an environmental toxin that typically has poor ionizability, in rat brain tissues and highlighted DPP's ability to form relatively small and homogeneous crystals, allowing for higher spatial resolution (20 μm) (Shariatgorji et al., 2015). DPP-TFB was also used in a recent DESI-MSI study on the brain tissue of a murine model of alkaptonuria, a tyrosine metabolic disorder (Davison et al., 2019) where the level changes in abundance of several neurotransmitters such as tyrosine, tyramine, and noradrenaline with nitisinone treatment. Detection of other molecules like serotonin, tryptophan, and glutamate was also reported in this study even though they showed no changes in diseased and treated animals compared to controls. In addition to derivatization, reactive matrices have been used to increase the ionization efficiency of primary and secondary amines, including neurotransmitters for MALDI MSI (Shariatgorji et al., 2019).

2.5.2 Chemical derivatization of carbonyl functional group

Carbonyl functionality is mainly comprised of ketones and aldehydes that can be derivatized using hydrazine-based reagents. Hydrazine-based reagents have also been used as reactive matrices in MALDI MSI, which facilitate in absorbing UV light and derivatization of carbonyl groups (Flinders, Morrell, Marshall, Ranshaw, & Clench, 2015). Reagents can be sprayed on the tissue sections, and a longer incubation time is needed for this reaction. Similarly, carboxylic acids are widely found in biomolecules, including lipids and proteins. Derivatization of carboxylic acid enables improved ionization and detection in the positive ion mode by masking the acidity of the group. Due to the acidic nature introduced by carboxylic acid, FFA cannot be ionized in positive ion mode without derivatization (Ren et al., 2019; Wu, Comi, Li, Rubakhin, & Sweedler, 2016).

2-picolylamine (PA): 2-picolylamine is a widely used chemical derivatization reagent that has been used to derivatize compounds with carboxylic acid functional groups in clinical and pharmaceutical applications (Kanemitsu et al., 2019). Secondary and aromatic amine functionalities in 2-picolylamine derivatives introduce readily ionizing features to the molecule in positive ion mode (Nagatomo, Okada, Ichimura, Tsuneyama, & Inoue, 2018). Wu et al. utilized this reagent to derivatize free fatty acids on tissue and successfully imaged using MALDI MSI (Wu et al., 2016). Derivatization reagent mixture is sprayed on the dried tissue slices using electrospray or airbrush technique. Wu et al. demonstrated the efficient derivatization and detection of free fatty acids using this approach, and electrospray deposition of the derivatization reagent compared to airbrush deposition showed higher sensitivity, higher spatial resolution with lower analyte delocalization.

N,N-dimethylpiperazine iodide (DMPI): Another commercially available derivatizing agent for carboxylic acids is DMPI which generates a stable amide bond and a quaternary amine with high ionization efficiency in the positive ion mode (Wang, Wang, Zhang, Sun, & Guo, 2019). Wang et al. utilized the DMPI derivatization method in imaging FFA and phospholipids simultaneously in thyroid cancer tissues (Wang et al., 2019). In this approach, tissue slices were dried in a desiccator, and derivatization reagents were applied to the tissue using an electrospray device. They optimized the reaction conditions, including parameters for the electrospray device, to obtain a highly efficient derivatization reaction on tissue. Moreover, Wang and coworkers compared the limit of detection for free fatty acids derivatized with 2-picolylamine and DMPI in solution and as well as on tissue. They found that the DMPI derivatized FFA have higher ionization efficiency compared to 2-picolylamine derivatized fatty acids in solution and on tissue for imaging applications.

2.5.3 Chemical derivatization of double bonds

Beyond challenges associated with ionization in MALDI MSI experiments, another area that has seen growth is in the ability to determine the position of unsaturation along the lipid fatty acid chains. Due to the nature of the mass spectrometry measurement, the mass of the molecule is not indicative of the double bond position. While fragmentation strategies (e.g., tandem MS or MS/MS) have been employed, there still lends the need for alternative strategies. One such is the role of derivatization of double bonds to enable position determination by mass spectrometry (Esch & Heiles, 2020; Yang, Dilthey, & Gross, 2013). Several examples are provided below.

Paternò–Büchi Reaction: The Paternò–Büchi reaction is a widely used photoactive organic reaction that uses high energy UV light or visible light with a catalyst to excite a carbonyl group for [2 + 2] cycloaddition resulting in oxetane (D'Auria, 2019; Rykaczewski & Schindler, 2020). This reaction has been used to identify and quantify lipid double bond location isomers using mass spectrometry lipidomics (Ma et al., 2016). Moreover, Ma et al. used acetone as the derivatization reagent and employed Paternò–Büchi as an online photochemical reaction in liquid chromatography where derivatized lipids are observed with a + 58 Da mass shift. When coupled with tandem MS, fragmentation of derivatized lipids at the original double bond location can be determined. Bednaric et al. optimized a Paternò–Büchi reaction for MALDI MSI with a few modifications to be able to determine the double bond position of lipids in the MSI experiment (Bednařík, Bölsker, Soltwisch, & Dreisewerd, 2018). Even though acetone is the well-established derivatization reagent in the common Paternò–Büchi reaction, it is not ideal for MALDI MSI owing to the tendency to delocalize analytes by condensed acetone droplets. As a replacement for acetone, research groups have investigated benzaldehyde as a derivatization reagent. UV light-activated Paternò–Büchi reaction using benzaldehyde results in a four-membered oxetane ring system that readily fragments in tandem MS experiments yielding distinct fragment ions. In this work, they were able to localize carbon-carbon double-bond positions in phospho- and glycolipids and report for the first-time phosphatidylserines double bond isomers show the different distribution in the gray and white matter of mice cerebella.

Meta-chloroperoxybenzoic acid (mCPBA) epoxidation: Kuo and coworkers used Meta-chloroperoxybenzoic acid (mCPBA), a commonly used epoxidation reagent in organic chemistry, to localize double bond position in lipids (Kuo et al., 2019). Epoxide results from the above reaction yield diagnostic fragment ions under collision-induced dissociation enabling identification of unsaturation position. mCPBA reagent in methanol is sprayed on the tissue using a commercial airbrush. This approach has been able to use not only to identify double bond location but also to quantify fatty acid isomers using diagnostic fragment ions. In this highlighted paper, the authors demonstrated the use of mCPBA in accurately assigning and quantifying several C=C isomers of several FAs, including FA 18:3 ω-3 and ω-6 isomers, FA 20:3 ω-6 and ω-9 isomers. Moreover, combined with LC-MS analyses, the authors performed pilot studies on C=C lipid isomers in cancerous human sera, 3T3-L1 adipocytes, and mouse kidney tissues. Notably, using DESI-MSI, an increase in the level of Δ11 isomer of FA 18:1 and a decrease in Δ9 isomer was observed in the kidney medulla. Interestingly, the authors observed differential distribution between C=C isomers in distinct tissue regions as the Δ11 isomer of PG 16:0_18:1 was significantly elevated in the tumor cell region.

2.5.4 Cholesterol derivatization for analysis in MSI

Cholesterol is one of the main structures as well as regulatory lipids in mammalian cells. Cholesterol in the cell membrane plays a vital role in maintaining membrane stability (Yamaguchi & Ishimatu, 2020), among other functions. Additionally, cholesterol serves as the precursor for steroid hormones, including sex hormones, vitamin D, and *gluco* and mineralo corticosteroids (Cortes et al., 2014). Although cholesterol is a highly abundant lipid, detection of cholesterol in mass spectrometry has been challenging due to its very low ionization efficiency, as noted earlier. The low ionization efficiency of cholesterol makes it even more challenging to analyze in MSI. Several chemical derivatization strategies have been employed to improve ionization and image cholesterol in animal tissue sections for MSI studies (Angelini et al., 2021; Mielczarek, Slowik, Kotlinska, Suder, & Bodzon-Kulakowska, 2021).

Wu et al. derivatized cholesterol using betaine aldehyde, which selectively reacts with alcohols to form a hemiacetal salt with a permanent positive charge (Wu et al., 2009). This study used DESI to image cholesterol after derivatization. Mielczarek adapted this approach in MALDI MSI and successfully imaged derivatized cholesterol in different mouse tissues

(Mielczarek et al., 2021). Since betaine is reactive alcohol, solvent selection for sample preparation and reagent dilution has to be done carefully. In this study, 100% acetonitrile and acetonitrile-water mixture has been tested to prepare the reagent, and the results suggest the use of acetonitrile: water mixture yields higher signal. In addition, the number of betaine layers applied on the tissue for the cholesterol derivatization has an impact on the derivatized cholesterol signal. Mielczarek et al. presented optimization of the reaction conditions and MALDI matrices for successful on-tissue derivatization and imaging using MALDI MSI for cholesterol.

Angelini and coworkers developed a method for quantitative and qualitative imaging of cholesterol using another on-tissue derivatization approach (Angelini et al., 2021). In this study, an enzyme-assisted on-tissue derivatization was used. Cholesterol oxidase is used to oxidize the 3β-hydroxy group to yield a carbonyl group that can be reacted with Girard-P hydrazine to introduce a bulky group with a stable positive charge. The resulting derivatized cholesterol is readily ionized in the positive ion mode and fragment to yields diagnostic fragment ions. This study showed proof of concept mapping of cholesterol across different regions in the mouse brain and also shows reduction of cholesterol in $Npc1^{-/-}$ mouse brain. Girand-T hydrosine has been used to derivatize other steroid molecules with the carbonyl group and detected in the MSI (Barré et al., 2016; Cobice et al., 2013).

2.6 Quantification in MALDI MSI

Quantification in mass spectrometry can be divided into two main types being: relative and absolute quantification. Quantification of lipids in mass spectrometry is more challenging than proteins due to several reasons, including high molecular diversity of lipids resulting in varied ionization efficiency and difficulty in finding the most suitable internal standard (Wang, Wang, & Han, 2017; Yang & Han, 2011). Developments in relative and absolute quantification using MALDI MSI are discussed below.

Relative quantification is the comparison of the abundance level of one or more analytes among different sample groups (Bantscheff, Schirle, Sweetman, Rick, & Kuster, 2007). Performing relative quantification in MALDI MSI include using the signal intensity of analytes that are normalized among samples using different approaches such as normalization to a reference peak or total ion chromatogram (TIC) (Fonville et al., 2012; Rzagalinski & Volmer, 2017; Thomas et al., 2013). Proper normalization is required in MALDI MSI quantification due to the heterogeneity of biological tissues, potential artifacts caused in sample preparation, and batch effects (Boskamp et al., 2021; Pirman, Kiss, Heeren, & Yost, 2013). Isotopically labeled internal standards used for normalization can be uniformly applied underneath or on top of tissue using similar techniques discussed in the matrix application section (Dewez, De Pauw, Heeren, & Balluff, 2021; Pirman & Yost, 2011). A common color scale is used across all sample groups for each analyte to represent the normalized abundance.

Absolute quantification is done with the use of calibration curves and has been a growing interest in the MALDI MSI field (Ellis, Bruinen, & Heeren, 2014; Tobias & Hummon, 2020). Important considerations in using calibration curves for absolute quantification and ways to improve accuracy have been described by Zabell et al. and are applicable in MALDI MSI (Zabell, Lytle, & Julian, 2016). Calibration standards are spotted on the MALDI plate close to the tissue, and the normalized signal intensity per area for the ROI of each calibrant spot is typically used to plot the calibration curve (Porta et al., 2015). Another method that uses a mimetic tissue model spiked with calibration standards is developed to improve the reliability of the use of calibration curve in MALDI MSI absolute quantification (Barry, Groseclose, & Castellino, 2019; Groseclose & Castellino, 2013). This approach mimics the biological tissue conditions in terms of ion suppression and sample complexity. Another strategy to incorporate calibration standards in MALDI MSI is the use of a reference tissue to spot the standards (Buck, Halbritter et al., 2015). It is recommended to image the calibration spots in the same imaging run to minimize errors in quantification (Tobias & Hummon, 2020).

3. MALDI MSI data analysis

In a MALDI MSI experiment, the mass spectrum for a selected mass range is saved at every pixel. This data is used to map the distribution of the relative abundance of a particular ion in the imaged area. A color scale is used to represent the relative abundance of the detected ion throughout the tissue creating a two-dimensional ion density map or MSI image. With the increase of the MSI applications, there has been a number of instrumental developments and methodology developments to improve the data quality and molecular coverage in MSI experiments. Data analysis and interpretation tools also have been developed to improve the capabilities of data representation along with developing the use of statistical analysis in MSI. Several widely used MSI data analysis tools are discussed briefly below.

3.1 MSiReader

MSiReader is a free, open-source, and user-friendly software for mass spectrometry imaging data analysis based on the MATLAB platform (Bokhart, Nazari, Garrard, & Muddiman, 2018; Robichaud, Garrard, Barry, & Muddiman, 2013). MSiReader supports multiple input-file formats, namely, imzML, mzXML, IMG, and ASCII files. MSiReader provides useful features, including tools to plot the average mass spectrum for a region of interest, normalize the signal in each pixel across the tissue, peak finding using the parabolic fitting, saving and reloading MATLAB sessions, colocalization plots using multiple channels, and batch image generation. MSiReader offers a free standalone version that can be used without MATLAB installation. A detailed description of available tools is given in the MSiReader user guide (https://msireader.wordpress.ncsu.edu/).

3.2 SCiLS

SCiLS is commercial software that couples with Bruker mass spectrometers. This software can load multiple data sets at the same time. SCiLS is one of the few software programs that can align multiple data files to construct three-dimensional MSI images. Spatial segmentation is a well-developed feature in SCiLS which allows the identification of different regions in the imaged specimen based on the MSI data. For spatial segmentation, noise reduction is done with the use of an algorithm to look at peak shape and intensity (Alexandrov, 2012, 2010), followed by hierarchical clustering. This software was initially introduced to perform the analysis on compressed binary data that results in a smaller file size and has a fast-processing time. Currently, SCiLS can be used for all vendor files making it accessible for more users.

3.3 Cardinal

Cardinal is an open-source R package written for statistical analysis and visualization of MSI data in biological samples, like animal tissues (Bemis et al., 2015). Cardinal supports imzML and Analyze7.5 data formats as input files. This package includes data processing tools including total ion current normalization, baseline correction using median interpolation, peak detection, and peak alignment using mean spectrum. Similar to SCiLS, Cardinal also offers a spatial segmentation feature based on principle component analysis and Spatially-Aware and Spatially Aware Structurally Adaptive (SASA) segmentation (Alexandrov & Kobarg, 2011). Cardinal has the ability to automate the data analysis and uses R scripts by the user to generate reports.

3.4 The consensus algorithm for replicated data analysis

Tobias et al. presented an R-based algorithm to generate consensus mass spectra using biological and technical replicate analysis (Tobias, Olson, & Cologna, 2018). The goal of many MSI experiments on biological tissues is to identify spatial and abundance differences in one set of samples compared to another sample set. As an example, to determine the alteration of lipids in the brain tissue of a diseased mouse model compared to a healthy control mouse model. To address such biological questions, it is crucial to incorporate biological replicates in the study. In addition, since mass spectrometry imaging involves sample preparation steps that can significantly affect the results, it is important to include technical replicated in the MSI study. The consensus algorithm includes the initial peak filtering step based on intensity frequency distribution to yield the most significant ion in the spectrum. The QT clustering algorithm is used to normalize the spectra to a unit vector, and cluster the peaks in multiple replicate data files. These normalized and peak clustered data are finally used in making consensus spectrum using dot product analysis. With the use of this algorithm, multiple replicates from two groups of samples can be analyzed to find the lipid alteration. This increases the reliability of data by including the biological variability into consideration.

4. Lipid identification in MSI

High confidence identification of detected lipids in an MSI analysis is crucial and challenging due to the structural diversity of lipids and the presence of isomers. The specificity of lipid identification is increased with the amount of structural information uncovered (for example: fatty acid identification, fatty acid position determination, double bond position identification, and stereochemistry) (Bonney & Prentice, 2021). As an initial step, identification of ions from the background is needed to narrow down lipid candidates. Signal/noise ratio and ability to see physical features of the tissue in the extracted image for a particular m/z value could be helpful to decide if the signal is from the tissue or a matrix ion or background ion. Accurate mass matching is considered and basic level of lipid annotation. The use of mass spectrometers with high mass resolution and maintaining higher mass accuracy can improve the confidence of lipid identification by accurate mass (Bowman, Blakney, et al., 2020; Bowman, Bogie, et al., 2020; Cornett, Frappier, & Caprioli, 2008). On tissue tandem, mass spectrometry can be used to generate a characteristic fragmentation pattern to identify the lipids with higher confidence.

Lipid isomer identification in MSI is challenging due to the absence of chromatographic separation and has been a major area that has been improved in recent years. Pham et al. introduced radical directed dissociation (RDD) to generate radicals under UV irradiation, resulting in improved fragmentation of lipids (Pham, Ly, Trevitt, Mitchell, & Blanksby, 2012). Recently, ultraviolet photodissociation (UVPD) has been used in double bond position identification and positional isomer identification (Fang, Rustam, Palmieri, Sieber, & Reid, 2020; Williams, Klein, Greer, & Brodbelt, 2017). Gas-phase ion/ion reaction, which improves the fragmentation to yield more molecular information, has been employed in MSI for lipid identification (Randolph, Shenault, Blanksby, & McLuckey, 2020; Specker, Van Orden, Ridgeway, & Prentice, 2020). Ion molecular reaction, which allows the reaction of ions with molecules like ozone in the gas phase, also yields modifications to ions to improve fragmentation (Paine et al., 2018; Thomas et al., 2008). Electron induced dissociation (EID), which is another type of ion activation method, has been used in MSI to generate more informative fragmentation for identification (Baba, Campbell, Le Blanc, & Baker, 2017; Born & Prentice, 2020; Campbell & Baba, 2015; Jones, Thompson, Carter, & Kane, 2015).

5. Applications of MALDI MSI in Niemann-Pick type C

MALDI MSI is a suitable technique to identify alteration of biomolecules in abundance as well as in distribution in animal tissue. This technique is widely used in lipid imaging given the challenges associated with antibodies for lipids. Performing MSI instead of analyzing a tissue lysate by mass spectrometry adds highly valuable spatial information that is lost in analyzing a tissue lysate. Fig. 7 shows two examples of the importance of having spatial information. Our laboratory studies Niemann-Pick Type C, a fatal neurodegenerative lipid storage disorder. Using a mouse model that is null for the Npc1 gene, Fig. 7, shows the accumulation of phosphatidylserine (36:2) in 7-week-old *Npc1*$^{-/-}$ mouse cerebellar gray matter compared to age-matched control (Fig. 7A and B). Interestingly accumulation of phosphatidic acid (36:2) is observed in the mutant mouse cerebellum and is quite pronounced in cerebellar lobule X compared to the age-matched control (Fig. 7C and D). This observation suggests both above lipids are more affected in the *Npc1*$^{-/-}$ mouse cerebellum than the other brain regions,

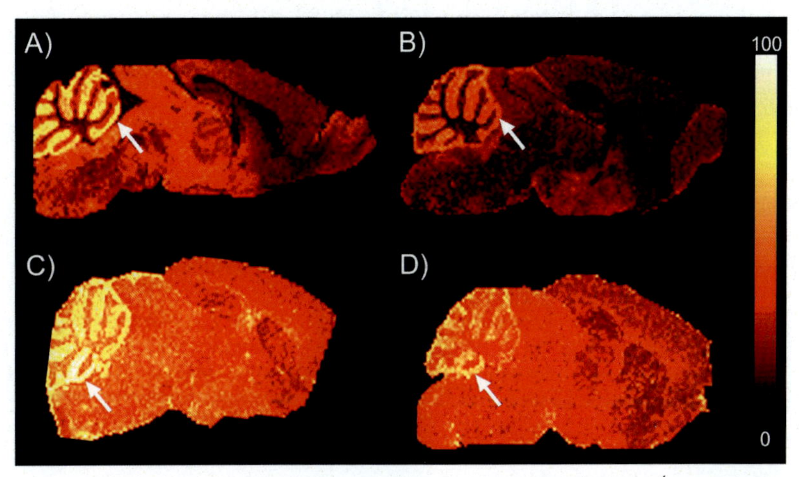

Fig. 7 MALDI-MSI shows alterations of lipids in 7-week-old *Npc1*$^{-/-}$ mouse brain. (A) phosphatidylserine [PS(36:2)] is increased in the gray matter of the *Npc1*$^{-/-}$ mouse cerebellum. (B) Normal distribution of PS(36:2) in a wild-type (*Npc1*$^{+/+}$) mouse cerebellum. (C) Phosphatidic acid [PA(36:2)] is increased in lobule X of the *Npc1*$^{-/-}$ mouse cerebellum. (D) Distribution of PA(36:2) in wild-type (*Npc1*$^{+/+}$) mouse. Lipid assignments are determined by accurate mass matching. PS(36:2) $m/z = 786.53$ is detected as [M+H] and identified with a mass error of 3 ppm. PA(36:2) $m/z = 699.50$ is detected as [M-H] and identified with a mass error of 4 ppm.

and these lipids are highly abundant in gray matter compared to lipid-rich white matter. If these lipids are analyzed by a brain lysate, the above information is lost. Apart from studying diseases, MALDI MSI has also been used in pharmaceutical research to obtain information about drug distribution (Giordano et al., 2016; Liu et al., 2013; Tang et al., 2019), drug delivery (Handler et al., 2021; Sugihara, Watanabe, & Végvári, 2016; Xue et al., 2018), pharmacokinetics (Nishidate et al., 2019) and drug metabolism (Prideaux, Staab, & Stoeckli, 2010; Schulz, Becker, Groseclose, Schadt, & Hopf, 2019).

6. Current perspective

MALDI MSI has developed rapidly in the past few years and is being used heavily in biological and biochemical applications. The development of new instruments and methods for MALDI MSI has enabled researchers to image more molecules with higher spatial resolution in a shorter time. Currently, MALDI MSI is rising in pharmaceutical research due to its ability to monitor drug metabolism, delivery, and distribution. Areas that are currently being developed in the field include improvement of spatial resolution in order to be able to image at the cellular and subcellular level and improvement in molecular identification.

References

Alexandrov, T. (2012). MALDI imaging mass spectrometry: Statistical data analysis and current computational challenges. *BMC Bioinformatics*, *13*(Suppl 16), S11. https://doi.org/10.1186/1471-2105-13-S16-S11.

Alexandrov, T., Becker, M., Deininger, S.-O., Ernst, G., Wehder, L., Grasmair, M., et al. (2010). Spatial segmentation of imaging mass spectrometry data with edge-preserving image Denoising and clustering. *Journal of Proteome Research*, *9*(12), 6535–6546. https://doi.org/10.1021/pr100734z.

Alexandrov, T., & Kobarg, J. H. (2011). Efficient spatial segmentation of large imaging mass spectrometry datasets with spatially aware clustering. *Bioinformatics (Oxford, England)*, *27*(13), i230–i238. https://doi.org/10.1093/bioinformatics/btr246.

Allwood, D. A., Dreyfus, R. W., Perera, I. K., & Dyer, P. E. (1997). Optical absorption of matrix compounds for laser-induced desorption and ionization (MALDI). *Applied Surface Science*, *109-110*, 154–157. https://doi.org/10.1016/S0169-4332(96)00652-6.

Ambrose, S., Housden, N. G., Gupta, K., Fan, J., White, P., Yen, H. Y., et al. (2017). Native desorption electrospray ionization liberates soluble and membrane protein complexes from surfaces. *Angewandte Chemie (International Ed. in English)*, *56*(46), 14463–14468. https://doi.org/10.1002/anie.201704849.

Angel, P. M., Mehta, A., Norris-Caneda, K., & Drake, R. R. (2018). MALDI imaging mass spectrometry of N-glycans and tryptic peptides from the same formalin-fixed, paraffin-embedded tissue section. *Methods in Molecular Biology*, *1788*, 225–241. https://doi.org/10.1007/7651_2017_81.

Angel, P. M., Spraggins, J. M., Baldwin, H. S., & Caprioli, R. (2012). Enhanced sensitivity for high spatial resolution lipid analysis by negative ion mode matrix assisted laser desorption ionization imaging mass spectrometry. *Analytical Chemistry*, *84*(3), 1557–1564. https://doi.org/10.1021/ac202383m.

Angelini, R., Yutuc, E., Wyatt, M. F., Newton, J., Yusuf, F. A., Griffiths, L., et al. (2021). Visualizing cholesterol in the brain by on-tissue derivatization and quantitative mass spectrometry imaging. *Analytical Chemistry*, *93*(11), 4932–4943. https://doi.org/10.1021/acs.analchem.0c05399.

Ayorinde, F. O., Hambright, P., Porter, T. N., & Keith, Q. L., Jr. (1999). Use of meso-tetrakis(pentafluorophenyl)porphyrin as a matrix for low molecular weight alkylphenol ethoxylates in laser desorption/ ionization time-of-flight mass spectrometry. *Rapid Communications in Mass Spectrometry*, *13*(24), 2474–2479. https://doi.org/10.1002/(sici)1097-0231(19991230)13:24<2474::Aid-rcm814>3.0.Co;2-0.

Baba, T., Campbell, J. L., Le Blanc, J. C. Y., & Baker, P. R. S. (2017). Distinguishing cis and trans isomers in intact complex lipids using electron impact excitation of ions from organics mass spectrometry. *Analytical Chemistry*, *89*(14), 7307–7315. https://doi.org/10.1021/acs.analchem.6b04734.

Balluff, B., Hopf, C., Porta Siegel, T., Grabsch, H. I., & Heeren, R. M. A. (2021). Batch effects in MALDI mass spectrometry imaging. *Journal of the American Society for Mass Spectrometry*, *32*(3), 628–635. https://doi.org/10.1021/jasms.0c00393.

Bantscheff, M., Schirle, M., Sweetman, G., Rick, J., & Kuster, B. (2007). Quantitative mass spectrometry in proteomics: A critical review. *Analytical and Bioanalytical Chemistry*, *389*(4), 1017–1031. https://doi.org/10.1007/s00216-007-1486-6.

Barré, F. P. Y., Flinders, B., Garcia, J. P., Jansen, I., Huizing, L. R. S., Porta, T., et al. (2016). Derivatization strategies for the detection of triamcinolone Acetonide in cartilage by using matrix-assisted laser desorption/ionization mass spectrometry imaging. *Analytical Chemistry*, *88*(24), 12051–12059. https://doi.org/10.1021/acs.analchem.6b02491.

Barry, J. A., Groseclose, M. R., & Castellino, S. (2019). Quantification and assessment of detection capability in imaging mass spectrometry using a revised mimetic tissue model. *Bioanalysis*, *11*(11), 1099–1116. https://doi.org/10.4155/bio-2019-0035.

Bednařík, A., Bölsker, S., Soltwisch, J., & Dreisewerd, K. (2018). An on-tissue Paternò-Büchi reaction for localization of carbon-carbon double bonds in phospholipids and glycolipids by matrix-assisted laser-desorption-ionization mass-spectrometry imaging. *Angewandte Chemie (International Ed. in English)*, *57*(37), 12092–12096. https://doi.org/10.1002/anie.201806635.

Bemis, K. D., Harry, A., Eberlin, L. S., Ferreira, C., van de Ven, S. M., Mallick, P., et al. (2015). Cardinal: An R package for statistical analysis of mass spectrometry-based imaging experiments. *Bioinformatics (Oxford, England)*, *31*(14), 2418–2420. https://doi.org/10.1093/bioinformatics/btv146.

Berger, S., Raman, G., Vishwanathan, R., Jacques, P. F., & Johnson, E. J. (2015). Dietary cholesterol and cardiovascular disease: A systematic review and meta-analysis. *The American Journal of Clinical Nutrition*, *102*(2), 276–294. https://doi.org/10.3945/ajcn.114.100305.

Bodzon-Kulakowska, A., Arena, R., Mielczarek, P., Hartman, K., Kozol, P., Gibula-Tarlowska, E., et al. (2020). Mouse single oocyte imaging by MALDI-TOF MS for lipidomics. *Cytotechnology*, *72*(3), 455–468. https://doi.org/10.1007/s10616-020-00393-9.

Bokhart, M. T., Nazari, M., Garrard, K. P., & Muddiman, D. C. (2018). MSiReader v1.0: Evolving open-source mass spectrometry imaging software for targeted and untargeted analyses. *Journal of the American Society for Mass Spectrometry*, *29*(1), 8–16. https://doi.org/10.1007/s13361-017-1809-6.

Bonney, J. R., & Prentice, B. M. (2021). Perspective on emerging mass spectrometry Technologies for Comprehensive Lipid Structural Elucidation. *Analytical Chemistry*, *93*(16), 6311–6322. https://doi.org/10.1021/acs.analchem.1c00061.

Born, M.-E. N., & Prentice, B. M. (2020). Structural elucidation of phosphatidylcholines from tissue using electron induced dissociation. *International Journal of Mass Spectrometry*, *452*, 116338. https://doi.org/10.1016/j.ijms.2020.116338.

Börnsen, K. O. (2000). Influence of salts, buffers, detergents, solvents, and matrices on MALDI-MS protein analysis in complex mixtures. *Methods in Molecular Biology*, *146*, 387–404. https://doi.org/10.1385/1-59259-045-4:387.

Boskamp, T., Casadonte, R., Hauberg-Lotte, L., Deininger, S., Kriegsmann, J., & Maass, P. (2021). Cross-normalization of MALDI mass spectrometry imaging data improves site-to-site reproducibility. *Analytical Chemistry*, *93*(30), 10584–10592. https://doi.org/10.1021/acs.analchem.1c01792.

Boughton, B. A., & Thinagaran, D. (2018). Mass spectrometry imaging (MSI) for plant metabolomics. *Methods in Molecular Biology*, *1778*, 241–252. https://doi.org/10.1007/978-1-4939-7819-9_17.

Bowman, A. P., Blakney, G. T., Hendrickson, C. L., Ellis, S. R., Heeren, R. M. A., & Smith, D. F. (2020). Ultra-high mass resolving power, mass accuracy, and dynamic range MALDI mass spectrometry imaging by 21-T FT-ICR MS. *Analytical Chemistry*, *92*(4), 3133–3142. https://doi.org/10.1021/acs.analchem.9b04768.

Bowman, A. P., Bogie, J. F. J., Hendriks, J. J. A., Haidar, M., Belov, M., Heeren, R. M. A., et al. (2020). Evaluation of lipid coverage and high spatial resolution MALDI-imaging capabilities of oversampling combined with laser post-ionisation. *Analytical and Bioanalytical Chemistry*, *412*(10), 2277–2289. https://doi.org/10.1007/s00216-019-02290-3.

Buck, A., Halbritter, S., Späth, C., Feuchtinger, A., Aichler, M., Zitzelsberger, H., et al. (2015). Distribution and quantification of irinotecan and its active metabolite SN-38 in colon cancer murine model systems using MALDI MSI. *Analytical and Bioanalytical Chemistry*, *407*(8), 2107–2116. https://doi.org/10.1007/s00216-014-8237-2.

Buck, A., Ly, A., Balluff, B., Sun, N., Gorzolka, K., Feuchtinger, A., et al. (2015). High-resolution MALDI-FT-ICR MS imaging for the analysis of metabolites from formalin-fixed, paraffin-embedded clinical tissue samples. *The Journal of Pathology*, *237*(1), 123–132. https://doi.org/10.1002/path.4560.

Campbell, J. L., & Baba, T. (2015). Near-complete structural characterization of phosphatidylcholines using electron impact excitation of ions from organics. *Analytical Chemistry*, *87*(11), 5837–5845. https://doi.org/10.1021/acs.analchem.5b01460.

Castellanos, A., Ramirez, C. E., Michalkova, V., Nouzova, M., Noriega, F. G., & Francisco, F. L. (2019). Three dimensional secondary ion mass spectrometry imaging (3D-SIMS) of Aedes aegypti ovarian follicles. *Journal of Analytical Atomic Spectrometry*, *34*(5), 874–883. https://doi.org/10.1039/c8ja00425k.

Cerruti, C. D., Benabdellah, F., Laprévote, O., Touboul, D., & Brunelle, A. (2012). MALDI imaging and structural analysis of rat brain lipid negative ions with 9-Aminoacridine matrix. *Analytical Chemistry*, *84*(5), 2164–2171. https://doi.org/10.1021/ac2025317.

Chang, W. C., Huang, L. C. L., Wang, Y.-S., Peng, W.-P., Chang, H. C., Hsu, N. Y., et al. (2007). Matrix-assisted laser desorption/ionization (MALDI) mechanism revisited. *Analytica Chimica Acta*, *582*(1), 1–9. https://doi.org/10.1016/j.aca.2006.08.062.

Chumbley, C. W., Reyzer, M. L., Allen, J. L., Marriner, G. A., Via, L. E., Barry, C. E., 3rd, et al. (2016). Absolute quantitative MALDI imaging mass spectrometry: A case of rifampicin in liver tissues. *Analytical Chemistry*, *88*(4), 2392–2398. https://doi.org/10.1021/acs.analchem.5b04409.

Cobice, D. F., Mackay, C. L., Goodwin, R. J. A., McBride, A., Langridge-Smith, P. R., Webster, S. P., et al. (2013). Mass spectrometry imaging for dissecting steroid Intracrinology within target tissues. *Analytical Chemistry*, *85*(23), 11576–11584. https://doi.org/10.1021/ac402777k.

Cologna, S. M. (2019). Mass spectrometry imaging of cholesterol. In A. Rosenhouse-Dantsker, & A. N. Bukiya (Eds.), *Cholesterol modulation of protein function: Sterol specificity and indirect mechanisms* (pp. 155–166). Cham: Springer International Publishing.

Cornett, D. S., Frappier, S. L., & Caprioli, R. M. (2008). MALDI-FTICR imaging mass spectrometry of drugs and metabolites in tissue. *Analytical Chemistry*, *80*(14), 5648–5653. https://doi.org/10.1021/ac800617s.

Cortes, V. A., Busso, D., Maiz, A., Arteaga, A., Nervi, F., & Rigotti, A. (2014). Physiological and pathological implications of cholesterol. *Frontiers in Bioscience (Landmark Edition)*, *19*, 416–428. https://doi.org/10.2741/4216.

Dai, L., Zou, L., Meng, L., Qiang, G., Yan, M., & Zhang, Z. (2021). Cholesterol metabolism in neurodegenerative diseases: Molecular mechanisms and therapeutic targets. *Molecular Neurobiology*, *58*(5), 2183–2201. https://doi.org/10.1007/s12035-020-02232-6.

D'Auria, M. (2019). The Paternò–Büchi reaction – A comprehensive review. *Photochemical & Photobiological Sciences*, *18*(10), 2297–2362. https://doi.org/10.1039/C9PP00148D.

Davison, A. S., Strittmatter, N., Sutherland, H., Hughes, A. T., Hughes, J., Bou-Gharios, G., et al. (2019). Assessing the effect of nitisinone induced hypertyrosinaemia on monoamine neurotransmitters in brain tissue from a murine model of alkaptonuria using mass spectrometry imaging. *Metabolomics*, *15*(5), 68. https://doi.org/10.1007/s11306-019-1531-4.

Debois, D., Jourdan, E., Smargiasso, N., Thonart, P., De Pauw, E., & Ongena, M. (2014). Spatiotemporal monitoring of the antibiome secreted by bacillus biofilms on plant roots using MALDI mass spectrometry imaging. *Analytical Chemistry*, *86*(9), 4431–4438. https://doi.org/10.1021/ac500290s.

DeKeyser, S. S., Kutz-Naber, K. K., Schmidt, J. J., Barrett-Wilt, G. A., & Li, L. (2007). Imaging mass spectrometry of neuropeptides in decapod crustacean neuronal tissues. *Journal of Proteome Research*, *6*(5), 1782–1791. https://doi.org/10.1021/pr060603v.

Demeure, K., Quinton, L., Gabelica, V., & De Pauw, E. (2007). Rational selection of the optimum MALDI matrix for top-down proteomics by in-source decay. *Analytical Chemistry*, *79*(22), 8678–8685. https://doi.org/10.1021/ac070849z.

Dewez, F., De Pauw, E., Heeren, R. M. A., & Balluff, B. (2021). Multilabel per-pixel quantitation in mass spectrometry imaging. *Analytical Chemistry*, *93*(3), 1393–1400. https://doi.org/10.1021/acs.analchem.0c03186.

Dill, A. L., Eberlin, L. S., Costa, A. B., Ifa, D. R., & Cooks, R. G. (2011). Data quality in tissue analysis using desorption electrospray ionization. *Analytical and Bioanalytical Chemistry*, *401*(6), 1949. https://doi.org/10.1007/s00216-011-5249-z.

Dong, W., Shen, Q., Baibado, J. T., Liang, Y., Wang, P., Huang, Y., et al. (2013). Phospholipid analyses by MALDI-TOF/TOF mass spectrometry using 1,5-diaminonaphthalene as matrix. *International Journal of Mass Spectrometry*, *343-344*, 15–22. https://doi.org/10.1016/j.ijms.2013.04.004.

Dufresne, M., Patterson, N. H., Norris, J. L., & Caprioli, R. M. (2019). Combining salt doping and matrix sublimation for high spatial resolution MALDI imaging mass spectrometry of neutral lipids. *Analytical Chemistry*, *91*(20), 12928–12934. https://doi.org/10.1021/acs.analchem.9b02974.

Eberlin, L. S., Liu, X., Ferreira, C. R., Santagata, S., Agar, N. Y. R., & Cooks, R. G. (2011). Desorption electrospray ionization then MALDI mass spectrometry imaging of lipid and protein distributions in single tissue sections. *Analytical Chemistry*, *83*(22), 8366–8371. https://doi.org/10.1021/ac202016x.

Ellis, S. R., Bruinen, A. L., & Heeren, R. M. A. (2014). A critical evaluation of the current state-of-the-art in quantitative imaging mass spectrometry. *Analytical and Bioanalytical Chemistry*, *406*(5), 1275–1289. https://doi.org/10.1007/s00216-013-7478-9.

Eriksson, C., Masaki, N., Yao, I., Hayasaka, T., & Setou, M. (2013). MALDI imaging mass spectrometry-a mini review of methods and recent developments. *Mass Spectrometry (Tokyo, Japan)*, *2*(Spec Iss), S0022-S0022. https://doi.org/10.5702/massspectrometry.S0022.

Esch, P., & Heiles, S. (2020). Investigating C[double bond, length as m-dash]C positions and hydroxylation sites in lipids using Paternò-Büchi functionalization mass spectrometry. *Analyst*, *145*(6), 2256–2266. https://doi.org/10.1039/c9an02260k.

Esteve, C., Tolner, E. A., Shyti, R., van den Maagdenberg, A. M. J. M., & McDonnell, L. A. (2016). Mass spectrometry imaging of amino neurotransmitters: A comparison of derivatization methods and application in mouse brain tissue. *Metabolomics*, *12*, 30. https://doi.org/10.1007/s11306-015-0926-0.

Fan, B., Zhou, H., Wang, Y., Zhao, Z., Ren, S., Xu, L., et al. (2020). Surface siloxane-modified silica materials combined with metal–organic frameworks as novel MALDI matrixes for the detection of low-MW compounds. *ACS Applied Materials & Interfaces*, *12*(33), 37793–37803. https://doi.org/10.1021/acsami.0c11404.

Fang, M., Rustam, Y., Palmieri, M., Sieber, O. M., & Reid, G. E. (2020). Evaluation of ultraviolet photodissociation tandem mass spectrometry for the structural assignment of unsaturated fatty acid double bond positional isomers. *Analytical and Bioanalytical Chemistry*, *412*(10), 2339–2351. https://doi.org/10.1007/s00216-020-02446-6.

Fernández, R., Garate, J., Martín-Saiz, L., Galetich, I., & Fernández, J. A. (2019). Matrix sublimation device for MALDI mass spectrometry imaging. *Analytical Chemistry*, *91*(1), 803–807. https://doi.org/10.1021/acs.analchem.8b04765.

Flinders, B., Morrell, J., Marshall, P. S., Ranshaw, L. E., & Clench, M. R. (2015). The use of hydrazine-based derivatization reagents for improved sensitivity and detection of carbonyl containing compounds using MALDI-MSI. *Analytical and Bioanalytical Chemistry*, *407*(8), 2085–2094. https://doi.org/10.1007/s00216-014-8223-8.

Fonville, J. M., Carter, C., Cloarec, O., Nicholson, J. K., Lindon, J. C., Bunch, J., et al. (2012). Robust data processing and normalization strategy for MALDI mass spectrometric imaging. *Analytical Chemistry*, *84*(3), 1310–1319. https://doi.org/10.1021/ac201767g.

Fukuyama, Y., Iwamoto, S., & Tanaka, K. (2006). Rapid sequencing and disulfide mapping of peptides containing disulfide bonds by using 1,5-diaminonaphthalene as a reductive matrix. *Journal of Mass Spectrometry*, *41*(2), 191–201. https://doi.org/10.1002/jms.977.

Gamble, L. J., & Anderton, C. R. (2016). Secondary ion mass spectrometry imaging of tissues, cells, and microbial systems. *Microscopy Today*, *24*(2), 24–31. https://doi.org/10.1017/S1551929516000018.

Gemperline, E., Rawson, S., & Li, L. (2014). Optimization and comparison of multiple MALDI matrix application methods for small molecule mass spectrometric imaging. *Analytical Chemistry*, *86*(20), 10030–10035. https://doi.org/10.1021/ac5028534.

Gidden, J., Liyanage, R., Durham, B., & Lay, J. O., Jr. (2007). Reducing fragmentation observed in the matrix-assisted laser desorption/ionization time-of-flight mass spectrometric analysis of triacylglycerols in vegetable oils. *Rapid Communications in Mass Spectrometry*, *21*(13), 1951–1957. https://doi.org/10.1002/rcm.3041.

Giordano, S., Morosi, L., Veglianese, P., Licandro, S. A., Frapolli, R., Zucchetti, M., et al. (2016). 3D mass spectrometry imaging reveals a very heterogeneous drug distribution in tumors. *Scientific Reports*, *6*(1), 37027. https://doi.org/10.1038/srep37027.

Gladchuk, A., Shumilina, J., Kusnetsova, A., Bureiko, K., Billig, S., Tsarev, A., et al. (2020). High-throughput fingerprinting of Rhizobial free fatty acids by chemical thin-film deposition and matrix-assisted laser desorption/ionization mass spectrometry. *Methods and Protocols*, *3*(2). https://doi.org/10.3390/mps3020036.

Gobom, J., Schuerenberg, M., Mueller, M., Theiss, D., Lehrach, H., & Nordhoff, E. (2001). α-Cyano-4-hydroxycinnamic acid affinity sample preparation. A protocol for MALDI-MS peptide analysis in proteomics. *Analytical Chemistry*, *73*(3), 434–438. https://doi.org/10.1021/ac001241s.

Goodwin, R. J. A. (2012). Sample preparation for mass spectrometry imaging: Small mistakes can lead to big consequences. *Journal of Proteomics*, *75*(16), 4893–4911. https://doi.org/10.1016/j.jprot.2012.04.012.

Groseclose, M. R., & Castellino, S. (2013). A mimetic tissue model for the quantification of drug distributions by MALDI imaging mass spectrometry. *Analytical Chemistry*, *85*(21), 10099–10106. https://doi.org/10.1021/ac400892z.

Guo, S., Tang, W., Hu, Y., Chen, Y., Gordon, A., Li, B., et al. (2020). Enhancement of on-tissue chemical derivatization by laser-assisted tissue transfer for MALDI MS imaging. *Analytical Chemistry*, *92*(1), 1431–1438. https://doi.org/10.1021/acs.analchem.9b04618.

Handler, A. M., Pommergaard Pedersen, G., Troensegaard Nielsen, K., Janfelt, C., Just Pedersen, A., & Clench, M. R. (2021). Quantitative MALDI mass spectrometry imaging for exploring cutaneous drug delivery of tofacitinib in human skin. *European Journal of Pharmaceutics and Biopharmaceutics*, *159*, 1–10. https://doi.org/10.1016/j.ejpb.2020.12.008.

Hankin, J. A., Barkley, R. M., & Murphy, R. C. (2007). Sublimation as a method of matrix application for mass spectrometric imaging. *Journal of the American Society for Mass Spectrometry*, *18*(9), 1646–1652. https://doi.org/10.1016/j.jasms.2007.06.010.

Harris, A., Roseborough, A., Mor, R., Yeung, K. K., & Whitehead, S. N. (2020). Ganglioside detection from formalin-fixed human brain tissue utilizing MALDI imaging mass spectrometry. *Journal of the American Society for Mass Spectrometry*, *31*(3), 479–487. https://doi.org/10.1021/jasms.9b00110.

Hermann, J., Noels, H., Theelen, W., Lellig, M., Orth-Alampour, S., Boor, P., et al. (2020). Sample preparation of formalin-fixed paraffin-embedded tissue sections for MALDI-mass spectrometry imaging. *Analytical and Bioanalytical Chemistry*, *412*(6), 1263–1275. https://doi.org/10.1007/s00216-019-02296-x.

Hiraoka, K., Asakawa, D., Fujimaki, S., Takamizawa, A., & Mori, K. (2006). Electrosprayed droplet impact/secondary ion mass spectrometry. *The European Physical Journal D - Atomic, Molecular, Optical and Plasma Physics*, *38*(1), 225–229. https://doi.org/10.1140/epjd/e2005-00282-6.

Honarvar, E., & Venter, A. R. (2017). Ammonium bicarbonate addition improves the detection of proteins by desorption electrospray ionization mass spectrometry. *Journal of the American Society for Mass Spectrometry*, *28*(6), 1109–1117. https://doi.org/10.1007/s13361-017-1628-9.

Janda, M., Seah, B. K. B., Jakob, D., Beckmann, J., Geier, B., & Liebeke, M. (2021). Determination of abundant metabolite matrix adducts illuminates the dark metabolome of MALDI-mass spectrometry imaging datasets. *Analytical Chemistry*, *93*(24), 8399–8407. https://doi.org/10.1021/acs.analchem.0c04720.

Jeck, V., Korf, A., Vosse, C., & Hayen, H. (2019). Localization of double-bond positions in lipids by tandem mass spectrometry succeeding high-performance liquid chromatography with post-column derivatization. *Rapid Communications in Mass Spectrometry*, *33*(Suppl 1), 86–94. https://doi.org/10.1002/rcm.8262.

Jiang, X., Ory, D. S., & Han, X. (2007). Characterization of oxysterols by electrospray ionization tandem mass spectrometry after one-step derivatization with dimethylglycine. *Rapid Communications in Mass Spectrometry*, *21*(2), 141–152. https://doi.org/10.1002/rcm.2820.

Johnson, D. W., ten Brink, H. J., & Jakobs, C. (2001). A rapid screening procedure for cholesterol and dehydrocholesterol by electrospray ionization tandem mass spectrometry. *Journal of Lipid Research*, *42*(10), 1699–1705.

Jones, J. W., Thompson, C. J., Carter, C. L., & Kane, M. A. (2015). Electron-induced dissociation (EID) for structure characterization of glycerophosphatidylcholine: Determination of double-bond positions and localization of acyl chains. *Journal of Mass Spectrometry*, *50*(12), 1327–1339. https://doi.org/10.1002/jms.3698.

Joo, S., & Liang, H. (2013). Secondary ion mass spectroscopy (SIMS). In Q. J. Wang, & Y.-W. Chung (Eds.), *Encyclopedia of tribology* (pp. 2989–2994). Boston, MA: Springer US.

Kanemitsu, Y., Mishima, E., Maekawa, M., Matsumoto, Y., Saigusa, D., Yamaguchi, H., et al. (2019). Comprehensive and semi-quantitative analysis of carboxyl-containing metabolites related to gut microbiota on chronic kidney disease using 2-picolylamine isotopic labeling LC-MS/MS. *Scientific Reports*, *9*(1), 19075. https://doi.org/10.1038/s41598-019-55600-1.

Karas, M., & Krüger, R. (2003). Ion formation in MALDI: The cluster ionization mechanism. *Chemical Reviews*, *103*(2), 427–440. https://doi.org/10.1021/cr010376a.

Kawamoto, T. (2003). Use of a new adhesive film for the preparation of multi-purpose fresh-frozen sections from hard tissues, whole-animals, insects and plants. *Archives of Histology and Cytology*, *66*(2), 123–143. https://doi.org/10.1679/aohc.66.123.

Klitzing, H. A., Weber, P. K., & Kraft, M. L. (2013). Secondary ion mass spectrometry imaging of biological membranes at high spatial resolution. In A. A. Sousa, & M. J. Kruhlak (Eds.), *Nanoimaging: Methods and protocols* (pp. 483–501). Totowa, NJ: Humana Press.

Korte, A. R., & Lee, Y. J. (2013). Multiplex mass spectrometric imaging with polarity switching for concurrent acquisition of positive and negative ion images. *Journal of the American Society for Mass Spectrometry*, *24*(6), 949–955. https://doi.org/10.1007/s13361-013-0613-1.

Korte, A. R., & Lee, Y. J. (2014). MALDI-MS analysis and imaging of small molecule metabolites with 1,5-diaminonaphthalene (DAN). *Journal of Mass Spectrometry*, *49*(8), 737–741. https://doi.org/10.1002/jms.3400.

Korte, A. R., Yagnik, G. B., Feenstra, A. D., & Lee, Y. J. (2015). Multiplex MALDI-MS imaging of plant metabolites using a hybrid MS system. *Methods in Molecular Biology*, *1203*, 49–62. https://doi.org/10.1007/978-1-4939-1357-2_6.

Kühn-Hölsken, E., Lenz, C., Sander, B., Lührmann, R., & Urlaub, H. (2005). Complete MALDI-ToF MS analysis of cross-linked peptide-RNA oligonucleotides derived from nonlabeled UV-irradiated ribonucleoprotein particles. *RNA*, *11*(12), 1915–1930. https://doi.org/10.1261/rna.2176605.

Kuo, T. H., Chung, H. H., Chang, H. Y., Lin, C. W., Wang, M. Y., Shen, T. L., et al. (2019). Deep Lipidomics and molecular imaging of unsaturated lipid isomers: A universal strategy initiated by mCPBA epoxidation. *Analytical Chemistry*, *91*(18), 11905–11915. https://doi.org/10.1021/acs.analchem.9b02667.

Leopold, J., Popkova, Y., Engel, K. M., & Schiller, J. (2018). Recent developments of useful MALDI matrices for the mass spectrometric characterization of lipids. *Biomolecules*, *8*(4), 173. https://doi.org/10.3390/biom8040173.

Li, G., Ma, F., Cao, Q., Zheng, Z., DeLaney, K., Liu, R., et al. (2019). Nanosecond photochemically promoted click chemistry for enhanced neuropeptide visualization and rapid protein labeling. *Nature Communications*, *10*(1), 4697. https://doi.org/10.1038/s41467-019-12548-0.

Li, S., Zhang, Y., Liu, J.a., Han, J., Guan, M., Yang, H., et al. (2016). Electrospray deposition device used to precisely control the matrix crystal to improve the performance of MALDI MSI. *Scientific Reports*, *6*(1), 37903. https://doi.org/10.1038/srep37903.

Liu, X., Ide, J. L., Norton, I., Marchionni, M. A., Ebling, M. C., Wang, L. Y., et al. (2013). Molecular imaging of drug transit through the blood-brain barrier with MALDI mass spectrometry imaging. *Scientific Reports*, *3*(1), 2859. https://doi.org/10.1038/srep02859.

Lukowski, J. K., Pamreddy, A., Velickovic, D., Zhang, G., Pasa-Tolic, L., Alexandrov, T., et al. (2020). Storage conditions of human Kidney tissue sections affect spatial Lipidomics analysis reproducibility. *Journal of the American Society for Mass Spectrometry*, *31*(12), 2538–2546. https://doi.org/10.1021/jasms.0c00256.

Ma, X., Chong, L., Tian, R., Shi, R., Hu, T. Y., Ouyang, Z., et al. (2016). Identification and quantitation of lipid C=C location isomers: A shotgun lipidomics approach enabled by photochemical reaction. *Proceedings of the National Academy of Sciences*, *113*(10), 2573–2578. https://doi.org/10.1073/pnas.1523356113.

Mallah, K., Quanico, J., Trede, D., Kobeissy, F., Zibara, K., Salzet, M., et al. (2018). Lipid changes associated with traumatic brain injury revealed by 3D MALDI-MSI. *Analytical Chemistry*, *90*(17), 10568–10576. https://doi.org/10.1021/acs.analchem.8b02682.

Manier, M. L., Reyzer, M. L., Goh, A., Dartois, V., Via, L. E., Barry, C. E., 3rd, et al. (2011). Reagent precoated targets for rapid in-tissue derivatization of the anti-tuberculosis drug isoniazid followed by MALDI imaging mass spectrometry. *Journal of the American Society for Mass Spectrometry*, *22*(8), 1409–1419. https://doi.org/10.1007/s13361-011-0150-8.

McMillen, J. C., Fincher, J. A., Klein, D. R., Spraggins, J. M., & Caprioli, R. M. (2020). Effect of MALDI matrices on lipid analyses of biological tissues using MALDI-2 postionization mass spectrometry. *Journal of Mass Spectrometry*, *55*(12), e4663. https://doi.org/10.1002/jms.4663.

Mielczarek, P., Slowik, T., Kotlinska, J. H., Suder, P., & Bodzon-Kulakowska, A. (2021). The study of derivatization prior MALDI MSI analysis-charge tagging based on the cholesterol and betaine aldehyde. *Molecules (Basel, Switzerland)*, *26*(9), 2737. https://doi.org/10.3390/molecules26092737.

Nagatomo, R., Okada, Y., Ichimura, M., Tsuneyama, K., & Inoue, K. (2018). Application of 2-Picolylamine Derivatized ultra-high performance liquid chromatography tandem mass spectrometry for the determination of short-chain fatty acids in feces samples. *Analytical Sciences*, *34*(9), 1031–1036. https://doi.org/10.2116/analsci.18SCP10.

Nelson, K. A., Daniels, G. J., Fournie, J. W., & Hemmer, M. J. (2013). Optimization of whole-body zebrafish sectioning methods for mass spectrometry imaging. *Journal of Biomolecular Techniques*, *24*(3), 119–127. https://doi.org/10.7171/jbt.13-2403-002.

Nguyen, S. N., Kyle, J. E., Dautel, S. E., Sontag, R., Luders, T., Corley, R., et al. (2019). Lipid coverage in Nanospray desorption electrospray ionization mass spectrometry imaging of mouse lung tissues. *Analytical Chemistry*, *91*(18), 11629–11635. https://doi.org/10.1021/acs.analchem.9b02045.

Niehaus, M., Schnapp, A., Koch, A., Soltwisch, J., & Dreisewerd, K. (2017). New insights into the wavelength dependence of MALDI mass spectrometry. *Analytical Chemistry*, *89*(14), 7734–7741. https://doi.org/10.1021/acs.analchem.7b01744.

Nilsson, A., Goodwin, R. J., Shariatgorji, M., Vallianatou, T., Webborn, P. J., & Andrén, P. E. (2015). Mass spectrometry imaging in drug development. *Analytical Chemistry*, *87*(3), 1437–1455. https://doi.org/10.1021/ac504734s.

Nishidate, M., Hayashi, M., Aikawa, H., Tanaka, K., Nakada, N., Miura, S. I., et al. (2019). Applications of MALDI mass spectrometry imaging for pharmacokinetic studies during drug development. *Drug Metabolism and Pharmacokinetics*, *34*(4), 209–216. https://doi.org/10.1016/j.dmpk.2019.04.006.

Paine, M. R. L., Poad, B. L. J., Eijkel, G. B., Marshall, D. L., Blanksby, S. J., Heeren, R. M. A., et al. (2018). Mass spectrometry imaging with isomeric resolution enabled by ozone-induced dissociation. *Angewandte Chemie (International Ed. in English)*, *57*(33), 10530–10534. https://doi.org/10.1002/anie.201802937.

Pan, M., Qin, C., & Han, X. (2021). Quantitative analysis of Polyphosphoinositide, Bis(monoacylglycero)phosphate, and Phosphatidylglycerol species by shotgun Lipidomics after methylation. In F.-F. Hsu (Ed.), *Mass spectrometry-based Lipidomics: Methods and protocols* (pp. 77–91). New York, NY: Springer US.

Passarelli, M. K., Pirkl, A., Moellers, R., Grinfeld, D., Kollmer, F., Havelund, R., et al. (2017). The 3D OrbiSIMS—Label-free metabolic imaging with subcellular lateral resolution and high mass-resolving power. *Nature Methods*, *14*(12), 1175–1183. https://doi.org/10.1038/nmeth.4504.

Pathmasiri, K. C., Pergande, M. R., Tobias, F., Rebiai, R., Rosenhouse-Dantsker, A., Bongarzone, E. R., et al. (2020). Mass spectrometry imaging and LC/MS reveal decreased cerebellar phosphoinositides in Niemann-pick type C1-null mice. *Journal of Lipid Research*, *61*(7), 1004–1013. https://doi.org/10.1194/jlr.RA119000606.

Patterson, N. H., Thomas, A., & Chaurand, P. (2014). Monitoring time-dependent degradation of phospholipids in sectioned tissues by MALDI imaging mass spectrometry. *Journal of Mass Spectrometry*, *49*(7), 622–627. https://doi.org/10.1002/jms.3382.

Peacock, P. M., Zhang, W.-J., & Trimpin, S. (2017). Advances in ionization for mass spectrometry. *Analytical Chemistry*, *89*(1), 372–388. https://doi.org/10.1021/acs.analchem.6b04348.

Pham, H. T., Ly, T., Trevitt, A. J., Mitchell, T. W., & Blanksby, S. J. (2012). Differentiation of complex lipid isomers by radical-directed dissociation mass spectrometry. *Analytical Chemistry*, *84*(17), 7525–7532. https://doi.org/10.1021/ac301652a.

Pierson, E. E., Midey, A. J., Forrest, W. P., Shah, V., Olivos, H. J., Shrestha, B., et al. (2020). Direct drug analysis in polymeric implants using desorption electrospray ionization - mass spectrometry imaging (DESI-MSI). *Pharmaceutical Research*, *37*(6), 107. https://doi.org/10.1007/s11095-020-02823-x.

Pirman, D. A., Kiss, A., Heeren, R. M. A., & Yost, R. A. (2013). Identifying tissue-specific signal variation in MALDI mass spectrometric imaging by use of an internal standard. *Analytical Chemistry*, *85*(2), 1090–1096. https://doi.org/10.1021/ac3029618.

Pirman, D. A., & Yost, R. A. (2011). Quantitative tandem mass spectrometric imaging of endogenous acetyl-l-carnitine from piglet brain tissue using an internal standard. *Analytical Chemistry*, *83*(22), 8575–8581. https://doi.org/10.1021/ac201949b.

Porta, T., Lesur, A., Varesio, E., & Hopfgartner, G. (2015). Quantification in MALDI-MS imaging: What can we learn from MALDI-selected reaction monitoring and what can we expect for imaging? *Analytical and Bioanalytical Chemistry*, *407*(8), 2177–2187. https://doi.org/10.1007/s00216-014-8315-5.

Prentice, B. M., & Caprioli, R. M. (2016). The need for speed in matrix-assisted laser desorption/ionization imaging mass spectrometry. *Postdoc Journal: A Journal of Postdoctoral Research and Postdoctoral Affairs*, *4*(3), 3–13.

Prentice, B. M., Chumbley, C. W., & Caprioli, R. M. (2015). High-speed MALDI MS/MS imaging mass spectrometry using continuous raster sampling. *Journal of Mass Spectrometry*, *50*(4), 703–710. https://doi.org/10.1002/jms.3579.

Prentice, B. M., Chumbley, C. W., Hachey, B. C., Norris, J. L., & Caprioli, R. M. (2016). Multiple time-of-flight/time-of-flight events in a single laser shot for improved matrix-assisted laser desorption/ionization tandem mass spectrometry quantification. *Analytical Chemistry*, *88*(19), 9780–9788. https://doi.org/10.1021/acs.analchem.6b02821.

Prentice, B. M., McMillen, J. C., & Caprioli, R. M. (2019). Multiple TOF/TOF events in a single laser shot for multiplexed lipid identifications in MALDI imaging mass spectrometry. *International Journal of Mass Spectrometry*, *437*, 30–37. https://doi.org/10.1016/j.ijms.2018.06.006.

Prideaux, B., Staab, D., & Stoeckli, M. (2010). Applications of MALDI-MSI to pharmaceutical research. *Methods in Molecular Biology*, *656*, 405–413. https://doi.org/10.1007/978-1-60761-746-4_23.

RamalloGuevara, C., Paulssen, D., Popova, A. A., Hopf, C., & Levkin, P. A. (2021). Fast Nanoliter-scale cell assays using droplet microarray-mass spectrometry imaging. *Advanced Biology*, *5*(3), e2000279. https://doi.org/10.1002/adbi.202000279.

Randolph, C. E., Shenault, D. S. M., Blanksby, S. J., & McLuckey, S. A. (2020). Structural elucidation of ether Glycerophospholipids using gas-phase ion/ion charge inversion chemistry. *Journal of the American Society for Mass Spectrometry*, *31*(5), 1093–1103. https://doi.org/10.1021/jasms.0c00025.

Ren, H., Chen, W., Wang, H., Kang, Y., Zhu, X., Li, J., et al. (2019). Quantitative analysis of free fatty acids in gout by disposable paper-array plate based MALDI MS. *Analytical Biochemistry*, *579*, 38–43. https://doi.org/10.1016/j.ab.2019.05.013.

Robichaud, G., Garrard, K. P., Barry, J. A., & Muddiman, D. C. (2013). MSiReader: An open-source interface to view and analyze high resolving power MS imaging files on Matlab platform. *Journal of the American Society for Mass Spectrometry*, *24*(5), 718–721. https://doi.org/10.1007/s13361-013-0607-z.

Rykaczewski, K. A., & Schindler, C. S. (2020). Visible-light-enabled Paternò–Büchi reaction via triplet energy transfer for the synthesis of Oxetanes. *Organic Letters*, *22*(16), 6516–6519. https://doi.org/10.1021/acs.orglett.0c02316.

Rzagalinski, I., & Volmer, D. A. (2017). Quantification of low molecular weight compounds by MALDI imaging mass spectrometry—A tutorial review. *Biochimica et Biophysica Acta, Proteins and Proteomics*, *1865*(7), 726–739. https://doi.org/10.1016/j.bbapap.2016.12.011.

Saigusa, D., Saito, R., Kawamoto, K., Uruno, A., Kano, K., Aoki, J., et al. (2019). Conductive adhesive film expands the utility of matrix-assisted laser desorption/ionization mass spectrometry imaging. *Analytical Chemistry*, *91*(14), 8979–8986. https://doi.org/10.1021/acs.analchem.9b01159.

Salum, M. L., Giudicessi, S. L., Schmidt De León, T., Camperi, S. A., & Erra-Balsells, R. (2017). Application of Z-sinapinic matrix in peptide MALDI-MS analysis. *Journal of Mass Spectrometry*, *52*(3), 182–186. https://doi.org/10.1002/jms.3908.

Santoro, A. L., Drummond, R. D., Silva, I. T., Ferreira, S. S., Juliano, L., Vendramini, P. H., et al. (2020). In situ DESI-MSI Lipidomic profiles of breast cancer molecular subtypes and precursor lesions. *Cancer Research*, *80*(6), 1246–1257. https://doi.org/10.1158/0008-5472.Can-18-3574.

Schiller, J., Zschörnig, O., Petkovic, M., Müller, M., Arnhold, J., & Arnold, K. (2001). Lipid analysis of human HDL and LDL by MALDI-TOF mass spectrometry and 31P-NMR. *Journal of Lipid Research*, *42*(9), 1501–1508. https://doi.org/10.1016/S0022-2275(20)34196-1.

Schulz, S., Becker, M., Groseclose, M. R., Schadt, S., & Hopf, C. (2019). Advanced MALDI mass spectrometry imaging in pharmaceutical research and drug development. *Current Opinion in Biotechnology*, *55*, 51–59. https://doi.org/10.1016/j.copbio.2018.08.003.

Shariatgorji, M., Nilsson, A., Fridjonsdottir, E., Vallianatou, T., Källback, P., Katan, L., et al. (2019). Comprehensive mapping of neurotransmitter networks by MALDI–MS imaging. *Nature Methods*, *16*(10), 1021–1028. https://doi.org/10.1038/s41592-019-0551-3.

Shariatgorji, M., Nilsson, A., Goodwin, R. J. A., Källback, P., Schintu, N., Zhang, X., et al. (2014). Direct targeted quantitative molecular imaging of neurotransmitters in brain tissue sections. *Neuron*, *84*(4), 697–707. https://doi.org/10.1016/j.neuron.2014.10.011.

Shariatgorji, M., Nilsson, A., Källback, P., Karlsson, O., Zhang, X., Svenningsson, P., et al. (2015). Pyrylium salts as reactive matrices for MALDI-MS imaging of biologically active primary amines. *Journal of the American Society for Mass Spectrometry*, *26*(6), 934–939. https://doi.org/10.1007/s13361-015-1119-9.

Shimbo, K., Yahashi, A., Hirayama, K., Nakazawa, M., & Miyano, H. (2009). Multifunctional and highly sensitive Precolumn reagents for amino acids in liquid chromatography/tandem mass spectrometry. *Analytical Chemistry*, *81*(13), 5172–5179. https://doi.org/10.1021/ac900470w.

Shimma, S., Sugiura, Y., Hayasaka, T., Hoshikawa, Y., Noda, T., & Setou, M. (2007). MALDI-based imaging mass spectrometry revealed abnormal distribution of phospholipids in colon cancer liver metastasis. *Journal of Chromatography. B, Analytical Technologies in the Biomedical and Life Sciences*, *855*(1), 98–103. https://doi.org/10.1016/j.jchromb.2007.02.037.

Shin, Y.-S., Drolet, B., Mayer, R., Dolence, K., & Basile, F. (2007). Desorption electrospray ionization-mass spectrometry of proteins. *Analytical Chemistry*, *79*(9), 3514–3518. https://doi.org/10.1021/ac062451t.

Shroff, R., Muck, A., & Svatoš, A. (2007). Analysis of low molecular weight acids by negative mode matrix-assisted laser desorption/ionization time-of-flight mass spectrometry. *Rapid Communications in Mass Spectrometry*, *21*(20), 3295–3300. https://doi.org/10.1002/rcm.3216.

Smolira, A., & Wessely-Szponder, J. (2015). Importance of the matrix and the matrix/sample ratio in MALDI-TOF-MS analysis of cathelicidins obtained from porcine neutrophils. *Applied Biochemistry and Biotechnology*, *175*(4), 2050–2065. https://doi.org/10.1007/s12010-014-1405-1.

Specker, J. T., Van Orden, S. L., Ridgeway, M. E., & Prentice, B. M. (2020). Identification of phosphatidylcholine isomers in imaging mass spectrometry using gas-phase charge inversion ion/ion reactions. *Analytical Chemistry*, *92*(19), 13192–13201. https://doi.org/10.1021/acs.analchem.0c02350.

Staab, H. A., & Saupe, T. (1988). "Proton Sponges" and the geometry of hydrogen bonds: Aromatic nitrogen bases with exceptional Basicities. *Angewandte Chemie International Edition*, *27*(7), 865–879. https://doi.org/10.1002/anie.198808653.

Stanback, A. E., Conroy, L. R., Young, L. E. A., Hawkinson, T. R., Markussen, K. H., Clarke, H. A., et al. (2021). Regional N-glycan and lipid analysis from tissues using MALDI-mass spectrometry imaging. *STAR Protocols*, *2*(1), 100304. https://doi.org/10.1016/j.xpro.2021.100304.

Strnad, Š., Pražienková, V., Sýkora, D., Cvačka, J., Maletínská, L., Popelová, A., et al. (2019). The use of 1,5-diaminonaphthalene for matrix-assisted laser desorption/ionization mass spectrometry imaging of brain in neurodegenerative disorders. *Talanta*, *201*, 364–372. https://doi.org/10.1016/j.talanta.2019.03.117.

Strupat, K., Karas, M., & Hillenkamp, F. (1991). 2,5-Dihydroxybenzoic acid: A new matrix for laser desorption—Ionization mass spectrometry. *International Journal of Mass Spectrometry and Ion Processes*, *111*, 89–102. https://doi.org/10.1016/0168-1176(91)85050-V.

Stübiger, G., & Belgacem, O. (2007). Analysis of lipids using 2,4,6-trihydroxyacetophenone as a matrix for MALDI mass spectrometry. *Analytical Chemistry*, *79*(8), 3206–3213. https://doi.org/10.1021/ac062236c.

Sugihara, Y., Watanabe, K.-I., & Végvári, Á. (2016). Novel insights in drug metabolism by MS imaging. *Bioanalysis*, *8*(6), 575–588. https://doi.org/10.4155/bio-2015-0020.

Susniak, K., Krysa, M., Gieroba, B., Komaniecka, I., & Sroka-Bartnicka, A. (2020). Recent developments of MALDI MSI application in plant tissues analysis. *Acta Biochimica Polonica*, *3*(67), 277–281. https://doi.org/10.18388/abp.2020_5394.

Takáts, Z., Wiseman, J. M., Gologan, B., & Cooks, R. G. (2004). Mass spectrometry sampling under ambient conditions with desorption electrospray ionization. *Science*, *306*(5695), 471–473. https://doi.org/10.1126/science.1104404.

Tang, W., Chen, J., Zhou, J., Ge, J., Zhang, Y., Li, P., et al. (2019). Quantitative MALDI imaging of spatial distributions and dynamic changes of Tetrandrine in multiple organs of rats. *Theranostics*, *9*(4), 932–944. https://doi.org/10.7150/thno.30408.

Tang, W., Gordon, A., Wang, F., Chen, Y., & Li, B. (2021). Hydralazine as a versatile and universal matrix for high-molecular coverage and dual-polarity matrix-assisted laser desorption/ionization mass spectrometry imaging. *Analytical Chemistry*, *93*, 9083–9093. https://doi.org/10.1021/acs.analchem.1c00498.

Thomas, A., Charbonneau, J. L., Fournaise, E., & Chaurand, P. (2012). Sublimation of new matrix candidates for high spatial resolution imaging mass spectrometry of lipids: Enhanced information in both positive and negative polarities after 1,5-Diaminonapthalene deposition. *Analytical Chemistry*, *84*(4), 2048–2054. https://doi.org/10.1021/ac2033547.

Thomas, M. C., Mitchell, T. W., Harman, D. G., Deeley, J. M., Nealon, J. R., & Blanksby, S. J. (2008). Ozone-induced dissociation: Elucidation of double bond position within mass-selected lipid ions. *Analytical Chemistry*, *80*(1), 303–311. https://doi.org/10.1021/ac7017684.

Thomas, A., Patterson, N. H., Laveaux Charbonneau, J., & Chaurand, P. (2013). Orthogonal organic and aqueous-based washes of tissue sections to enhance protein sensitivity by MALDI imaging mass spectrometry. *Journal of Mass Spectrometry*, *48*(1), 42–48. https://doi.org/10.1002/jms.3114.

Thomas, A., Patterson, N. H., Marcinkiewicz, M. M., Lazaris, A., Metrakos, P., & Chaurand, P. (2013). Histology-driven data Mining of Lipid Signatures from multiple imaging mass spectrometry analyses: Application to human colorectal cancer liver metastasis biopsies. *Analytical Chemistry*, *85*(5), 2860–2866. https://doi.org/10.1021/ac3034294.

Tobias, F., & Hummon, A. B. (2020). Considerations for MALDI-based quantitative mass spectrometry imaging studies. *Journal of Proteome Research*, *19*(9), 3620–3630. https://doi.org/10.1021/acs.jproteome.0c00443.

Tobias, F., Olson, M. T., & Cologna, S. M. (2018). Mass spectrometry imaging of lipids: Untargeted consensus spectra reveal spatial distributions in Niemann-pick disease type C1. *Journal of Lipid Research*, *59*(12), 2446–2455. https://doi.org/10.1194/jlr.D086090.

Toue, S., Sugiura, Y., Kubo, A., Ohmura, M., Karakawa, S., Mizukoshi, T., et al. (2014). Microscopic imaging mass spectrometry assisted by on-tissue chemical derivatization for visualizing multiple amino acids in human colon cancer xenografts. *Proteomics*, *14*(7–8), 810–819. https://doi.org/10.1002/pmic.201300041.

Treu, A., Kokesch-Himmelreich, J., Walter, K., Hölscher, C., & Römpp, A. (2020). Integrating high-resolution MALDI imaging into the development pipeline of anti-tuberculosis drugs. *Journal of the American Society for Mass Spectrometry*, *31*(11), 2277–2286. https://doi.org/10.1021/jasms.0c00235.

van der Vusse, G. J., de Groot, M. J. M., Willemsen, P. H. M., van Bilsen, M., Schrijers, A. H. G. J., & Reneman, R. S. (1989). Degradation of phospholipids and triacylglycerol, and accumulation of fatty acids in anoxic myocardial tissue, disrupted by freeze-thawing. In G. J. Van Der Vusse (Ed.), *Lipid metabolism in normoxic and ischemic heart* (pp. 83–90). Boston, MA: Springer US.

Vanbellingen, Q. P., Castellanos, A., Rodriguez-Silva, M., Paudel, I., Chambers, J. W., & Fernandez-Lima, F. A. (2016). Analysis of chemotherapeutic drug delivery at the single cell level using 3D-MSI-TOF-SIMS. *Journal of the American Society for Mass Spectrometry*, *27*(12), 2033–2040. https://doi.org/10.1007/s13361-016-1485-y.

Vandenbosch, M., Nauta, S. P., Svirkova, A., Poeze, M., Heeren, R. M. A., Siegel, T. P., et al. (2021). Sample preparation of bone tissue for MALDI-MSI for forensic and (pre) clinical applications. *Analytical and Bioanalytical Chemistry*, *413*(10), 2683–2694. https://doi.org/10.1007/s00216-020-02920-1.

Vergeiner, S., Schafferer, L., Haas, H., & Müller, T. (2014). Improved MALDI-TOF microbial mass spectrometry imaging by application of a dispersed solid matrix. *Journal of the American Society for Mass Spectrometry*, *25*(8), 1498–1501. https://doi.org/10.1007/s13361-014-0923-y.

Vermillion-Salsbury, R. L., & Hercules, D. M. (2002). 9-Aminoacridine as a matrix for negative mode matrix-assisted laser desorption/ionization. *Rapid Communications in Mass Spectrometry*, *16*(16), 1575–1581. https://doi.org/10.1002/rcm.750.

Visscher, M., Moerman, A. M., Burgers, P. C., Van Beusekom, H. M. M., Luider, T. M., Verhagen, H. J. M., et al. (2019). Data processing pipeline for lipid profiling of carotid atherosclerotic plaque with mass spectrometry imaging. *Journal of the American Society for Mass Spectrometry*, *30*(9), 1790–1800. https://doi.org/10.1007/s13361-019-02254-y.

Vu, N. Q., DeLaney, K., & Li, L. (2021). Neuropeptidomics: Improvements in mass spectrometry imaging analysis and recent advancements. *Current Protein & Peptide Science*, *22*(2), 158–169. https://doi.org/10.2174/1389203721666201116115708.

Wang, Y., Krull, I. S., Liu, C., & Orr, J. D. (2003). Derivatization of phospholipids. *Journal of Chromatography. B, Analytical Technologies in the Biomedical and Life Sciences*, *793*(1), 3–14. https://doi.org/10.1016/s1570-0232(03)00359-3.

Wang, H.-Y. J., Liu, C. B., & Wu, H.-W. (2011). A simple desalting method for direct MALDI mass spectrometry profiling of tissue lipids. *Journal of Lipid Research*, *52*(4), 840–849. https://doi.org/10.1194/jlr.D013060.

Wang, M., Palavicini, J. P., Cseresznye, A., & Han, X. (2017). Strategy for quantitative analysis of isomeric Bis(monoacylglycero)phosphate and Phosphatidylglycerol species by shotgun Lipidomics after one-step methylation. *Analytical Chemistry*, *89*(16), 8490–8495. https://doi.org/10.1021/acs.analchem.7b02058.

Wang, M., Wang, C., & Han, X. (2017). Selection of internal standards for accurate quantification of complex lipid species in biological extracts by electrospray ionization mass spectrometry—What, how and why? *Mass Spectrometry Reviews*, *36*(6), 693–714. https://doi.org/10.1002/mas.21492.

Wang, S.-S., Wang, Y.-J., Zhang, J., Sun, T.-Q., & Guo, Y.-L. (2019). Derivatization strategy for simultaneous molecular imaging of phospholipids and low-abundance free fatty acids in thyroid cancer tissue sections. *Analytical Chemistry*, *91*(6), 4070–4076. https://doi.org/10.1021/acs.analchem.8b05680.

Watanabe, M., Terasawa, K., Kaneshiro, K., Uchimura, H., Yamamoto, R., Fukuyama, Y., et al. (2013). Improvement of mass spectrometry analysis of glycoproteins by MALDI-MS using 3-aminoquinoline/α-cyano-4-hydroxycinnamic acid. *Analytical and Bioanalytical Chemistry*, *405*(12), 4289–4293. https://doi.org/10.1007/s00216-013-6771-y.

Wei, Y., Zhang, Y., Lin, Y., Li, L., Liu, J., Wang, Z., et al. (2015). A uniform 2,5-dihydroxybenzoic acid layer as a matrix for MALDI-FTICR MS-based lipidomics. *Analyst*, *140*(4), 1298–1305. https://doi.org/10.1039/c4an01964d.

Williams, P. E., Klein, D. R., Greer, S. M., & Brodbelt, J. S. (2017). Pinpointing double bond and sn-positions in Glycerophospholipids via hybrid 193 nm ultraviolet Photodissociation (UVPD) mass spectrometry. *Journal of the American Chemical Society, 139*(44), 15681–15690. https://doi.org/10.1021/jacs.7b06416.

Wu, Q., Comi, T. J., Li, B., Rubakhin, S. S., & Sweedler, J. V. (2016). On-tissue derivatization via electrospray deposition for matrix-assisted laser desorption/ionization mass spectrometry imaging of endogenous fatty acids in rat brain tissues. *Analytical Chemistry, 88*(11), 5988–5995. https://doi.org/10.1021/acs.analchem.6b01021.

Wu, C., Ifa, D. R., Manicke, N. E., & Cooks, R. G. (2009). Rapid, direct analysis of cholesterol by charge labeling in reactive desorption electrospray ionization. *Analytical Chemistry, 81*(18), 7618–7624. https://doi.org/10.1021/ac901003u.

Xie, H., Wu, R., Hung, Y. L. W., Chen, X., & Chan, T. D. (2021). Development of a matrix sublimation device with controllable crystallization temperature for MALDI mass spectrometry imaging. *Analytical Chemistry, 93*(16), 6342–6347. https://doi.org/10.1021/acs.analchem.1c00260.

Xue, J., Liu, H., Chen, S., Xiong, C., Zhan, L., Sun, J., et al. (2018). Mass spectrometry imaging of the in situ drug release from nanocarriers. *Science Advances, 4*(10), eaat9039. https://doi.org/10.1126/sciadv.aat9039.

Yamaguchi, T., & Ishimatu, T. (2020). Effects of cholesterol on membrane stability of human erythrocytes. *Biological & Pharmaceutical Bulletin, 43*(10), 1604–1608. https://doi.org/10.1248/bpb.b20-00435.

Yang, J., & Caprioli, R. M. (2011). Matrix sublimation/recrystallization for imaging proteins by mass spectrometry at high spatial resolution. *Analytical Chemistry, 83*(14), 5728–5734. https://doi.org/10.1021/ac200998a.

Yang, K., Dilthey, B. G., & Gross, R. W. (2013). Identification and quantitation of fatty acid double bond positional isomers: A shotgun lipidomics approach using charge-switch derivatization. *Analytical Chemistry, 85*(20), 9742–9750. https://doi.org/10.1021/ac402104u.

Yang, K., & Han, X. (2011). Accurate quantification of lipid species by electrospray ionization mass spectrometry - meet a key challenge in lipidomics. *Metabolites, 1*(1), 21–40. https://doi.org/10.3390/metabo1010021.

Yang, H., Ji, W., Guan, M., Li, S., Zhang, Y., Zhao, Z., et al. (2018). Organic washes of tissue sections for comprehensive analysis of small molecule metabolites by MALDI MS imaging of rat brain following status epilepticus. *Metabolomics, 14*(4), 50. https://doi.org/10.1007/s11306-018-1348-6.

Yang, J., Norris, J. L., & Caprioli, R. (2018). Novel vacuum stable ketone-based matrices for high spatial resolution. *MALDI Imaging Mass Spectrometry, 53*(10), 1005–1012. https://doi.org/10.1002/jms.4277.

Yatim, A. R. M., Wan Muhammad Zulkifli, W. N. F., Majid, A. M. S., Foster, J. L., & Hayes, D. G. (2020). 3-Hydroxypicolinic acid as an effective matrix for Sophorolipid structural elucidation using matrix-assisted laser desorption ionization time-of-flight mass spectrometry. *Journal of Surfactants and Detergents, 23*(3), 565–571. https://doi.org/10.1002/jsde.12394.

Zabell, A. P. R., Lytle, F. E., & Julian, R. K. (2016). A proposal to improve calibration and outlier detection in high-throughput mass spectrometry. *Clinical Mass Spectrometry, 2*, 25–33. https://doi.org/10.1016/j.clinms.2016.12.003.

Zenobi, R., & Knochenmuss, R. (1998). Ion formation in. *MALDI Mass Spectrometry., 17*(5), 337–366. https://doi.org/10.1002/(SICI)1098-2787(1998)17:5<337::AID-MAS2>3.0.CO;2-S.

Zhang, W., Shang, B., Ouyang, Z., & Xia, Y. (2020). Enhanced phospholipid isomer analysis by online photochemical derivatization and RPLC-MS. *Analytical Chemistry, 92*(9), 6719–6726. https://doi.org/10.1021/acs.analchem.0c00690.

Zhou, Q., Fülöp, A., & Hopf, C. (2021). Recent developments of novel matrices and on-tissue chemical derivatization reagents for MALDI-MSI. *Analytical and Bioanalytical Chemistry*, *413*(10), 2599–2617. https://doi.org/10.1007/s00216-020-03023-7.

Zhu, X., & Papayannopoulos, I. A. (2003). Improvement in the detection of low concentration protein digests on a MALDI TOF/TOF workstation by reducing alpha-cyano-4-hydroxycinnamic acid adduct ions. *Journal of Biomolecular Techniques: JBT*, *14*(4), 298–307.

Deciphering lipid transfer between and within membranes with time-resolved small-angle neutron scattering

Ursula Perez-Salas[a,*], Sumit Garg[a], Yuri Gerelli[b], and Lionel Porcar[c]
[a]Physics Department, University of Illinois at Chicago, Chicago, IL, United States
[b]Department of Life and Environmental Sciences, Universita' Politecnica delle Marche, Ancona, Italy
[c]Institut Laue Langevin, Grenoble Cedex 9, France
*Corresponding author: e-mail address: ursulaps@uic.edu

Contents

Abstract

This review focuses on time-resolved neutron scattering, particularly time-resolved small angle neutron scattering (TR-SANS), as a powerful in situ noninvasive technique to investigate intra- and intermembrane transport and distribution of lipids and sterols in lipid membranes. In contrast to using molecular analogues with potentially large

Current Topics in Membranes, Volume 88
ISSN 1063-5823
https://doi.org/10.1016/bs.ctm.2021.10.004

chemical tags that can significantly alter transport properties, small angle neutron scattering relies on the relative amounts of the two most abundant isotope forms of hydrogen: protium and deuterium to detect complex membrane architectures and transport processes unambiguously. This review discusses advances in our understanding of the mechanisms that sustain lipid asymmetry in membranes—a key feature of the plasma membrane of cells—as well as the transport of lipids between membranes, which is an essential metabolic process.

Abbreviation

AMP antimicrobial peptide
CM contrast matched
DMPC 1,2-dimyristoyl-*sn*-glycero-3-phosphocholine
DPPC 1,2-dipalmitoyl-*sn*-glycero-3-phosphocholine
HDL high density lipoproteins
LDL low density lipoproteins
MD molecular dynamics
NMR nuclear magnetic resonance
NR neutron reflectivity
POPA 1-palmitoyl-2-oleoyl-glycero-3-phosphatidic acid
POPC 1-palmitoyl-2-oleoyl-glycero-3-phosphocholine
POPG 1-palmitoyl-2-oleoyl-sn-glycero-3-phosphoglycerol
SANS small-angle neutron scattering
SAXS small-angle X-ray scattering
SLD scattering length density
TR-NR time-resolved neutron reflectivity
TR-SANS time-resolved small-angle neutron scattering

1. Introduction

Lipids are essential components of cellular membranes (Yeagle, 1993). The structure of membranes consists of a continuous double layer (bilayer) of lipid molecules in which membrane proteins are embedded. Cells use this structure as the boundary that separates their interior from the environment that surrounds them (the extracellular space). Eukaryotic cells also bound organelles, having specialized functions, with membranes that require unique protein and lipid compositions (van Meer, Voelker, & Feigenson, 2008). Further, some of these membranes, like the cell's boundary, the plasma membrane (PM), are known to have a strict asymmetric distribution of lipids between the cytosolic facing leaflet and the extracellular facing leaflet and this distribution is responsible for the physiological fate of cells (Bretscher, 1972; Kobayashi & Menon, 2018; Opdenkamp, 1979; van Meer et al., 2008). For example, serine lipids, normally in the inner cytosolic facing leaflet, when present in the outer exocellular leaflet, signal for phagocytosis and blood coagulation (Fadok et al., 1992).

Indeed, transbilayer flip-flop rates and energetics can influence inter-organelle lipid transport by rearranging lipids from inner to outer leaflets or vice versa; this directly affects membrane curvature, for example, and consequently vesicle budding and fission and vesicle fusion (Lev, 2006; Sprong, van der Sluijs, & van Meer, 2001). In addition to an asymmetric distribution of lipids across membranes, the distribution of molecular components in membranes can be laterally heterogeneous (Devaux & Morris, 2004; Gupta, Korte, Herrmann, & Wohland, 2020; Simons & Ikonen, 1997) and implicated in signaling processes through the PM (Carbone et al., 2017; Stone, Shelby, Nunez, Wisser, & Veatch, 2017; Xiao, McAtee, & Su, 2021).

As a result, the quest to understand lipid trafficking as it relates to lipid homeostasis and metabolism, and how this lipid organization leads to proper membrane function, has been the focus of numerous studies for half a century (Holthuis & Levine, 2005; Nicolson, 2014). These studies have lead to significant breakthroughs, including those that lead to the 2013 Nobel Award in Medicine and Physiology for the finding that vesicular transport plays a major part in protein and lipid translocation along several energy-dependent pathways (Mellman & Emr, 2013). Further, it was discovered that nonvesicular transport mechanisms, including the spontaneous movement of lipids, also play critical roles in lipid homeostasis (Lev, 2010) as demonstrated by the existence of such transport, even under conditions in which vesicular transport is blocked (Kaplan & Simoni, 1985; Vance, Aasman, & Szarka, 1991).

One way to gauge the energetic toll of lipid transport is to quantify the passive movement of lipids between and within membranes. Unfortunately, the field has moved slowly due to the wide variation in the rates of lipid transfer between and particularly within membranes. Even in studies of model membrane systems, where lipid composition is controlled, the reported transfer rates have been inconsistent. For example, the reported half-life for cholesterol's transmembrane flipping varies by five to six orders of magnitude, ranging from several hours (Brasaemle, Robertson, & Attie, 1988; Poznansky & Lange, 1978; Rodrigueza, Wheeler, Klimuk, Kitson, & Hope, 1995) to a few minutes or seconds (Backer & Dawidowicz, 1981; John, Kubelt, Muller, Wustner, & Herrmann, 2002;

Leventis & Silvius, 2001; Schroeder et al., 1996; Steck, Ye, & Lange, 2002), and to even a few milliseconds (Baral, Levental, & Lyman, 2020; Bruckner, Mansy, Ricardo, Mahadevan, & Szostak, 2009) to tens of nanoseconds (Bennett, MacCallum, Hinner, Marrink, & Tieleman, 2009; Gu, Baoukina, & Tieleman, 2019). Noninvasive approaches like time-resolved small angle neutron scattering (TR-SANS) or sum-frequency generation vibrational spectroscopy (SFGVP) have shown that the movement of lipids is extremely sensitive to slight chemical structure differences, finding that the transfer rates of unaltered lipid molecules are dramatically different from their chemically tagged counterparts (for example, due to a fluorescent label) (Garg, Porcar, Woodka, Butler, & Perez-Salas, 2011; Liu & Conboy, 2005). Even studies using the same lipids and similar time resolved noninvasive approaches to investigate the movement of lipids across the bilayer (flip-flop) found drastically different results: in single flat supported membranes flip-flop of lipids are found to be several orders of magnitude faster (Anglin & Conboy, 2009; Gerelli, Porcar, Lombardi, & Fragneto, 2013) than in vesicles, which are reported to take hours (Liu, Kelley, Batchu, Porcar, & Perez-Salas, 2020; Marquardt et al., 2017; Nakano, Fukuda, Kudo, Endo, & Handa, 2007). As it turned out, the surface supporting the membranes produce membrane defects (Marquardt et al., 2017), as well as a broadening of the melting phase transition of lipids (Gerelli, 2019) and surface driven lipid packing constraints (Wah et al., 2017) that promote fast flip-flop.

The past decade and in particular the past 5 years have seen progress in the revision of protocols to remove possible artifacts and biases that may be responsible for these hugely varying reports. Indeed, it became clear that the use of chemical tags, extraneous compounds, and even a supporting surface affect or influence lipid transport. Hence the use of nonperturbing approaches has become a strict requirement for a detailed study of the behavior of lipids in membranes. As a result, neutron scattering and in particular TR-SANS coupled with contrast matching, has emerged as a powerful tool to study lipid transport which, in addition, can track the movement of lipids in situ, removing the need to do step-wise sampling of the kinetic process.

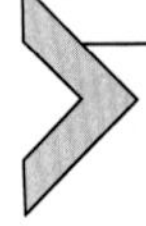

2. Small angle neutron scattering (SANS) in the study of membranes

2.1 SANS nuts and bolts

SANS is an ideal technique to obtain structural information of particles, such as lipid vesicles and here we will briefly review the basics of SANS to describe the strengths and advantages of this method (Mahieu & Gabel, 2018; Qian, Sharma, & Clifton, 2020). A careful and very detailed description of SANS, with an emphasis on biological systems, can be found in these references (Hamley, 2021; Sivia, 2011; Svergun, 2010; Svergun, Feĭgin, & Taylor, 1987).

In a small angle scattering experiment, a collimated incoming beam of neutrons—produced in nuclear reactors or in an accelerator-based spallation facility—impinge on a sample (typically in a 300 μL quartz or other high neutron transmission cuvette) and the scattered neutrons are detected on a 2 dimensional ^{3}He detector. Neutrons and other subatomic particles are characterized by a wavelength, λ, in the same way that X-rays, visible light, and other types of radiation are. For SANS, the neutron wavelength typically varies between 1 and 20 Å. The wave nature of the neutrons when scattered by nanoparticles ranging between 1 and 1000 nm results in an interference pattern that is then captured on a 2D detector. A schematic of the scattering process and the intensity pattern obtained from a solution of particles is shown in Fig. 1A.

The intensity pattern on the 2D detector for a random distribution of nanoparticles is radially symmetric and therefore can be radially averaged, as shown in the schematic in Fig. 1A. After being corrected by the empty sample container (quartz cell for example), sample transmission factor and incident neutron flux, an intensity versus Q curve in absolute scale (cm^{-1}) as shown in Fig. 1B is obtained. Q, as shown in Fig. 1A, corresponds to the magnitude of the neutrons' momentum direction change due to an elastic scattering event with the sample. In atomic units, Q is related to the scattering angle θ (see Fig. 1A) as follows:

$$Q = \frac{4\pi}{\lambda} sin\left(\frac{\theta}{2}\right) \tag{1}$$

Since Q is inversely proportional to λ, it is inverselely porporional to length. Consequently, the larger length scales of the system, such as size and shape and overall composition, are captured in the low Q part of the spectra while smaller length scales such as the bilayer's structure and leaflet composition are captured in the higher Q range of the spectra (see Fig. 1B). Direct evaluation of the scattertered intensity pattern provides information on the particles' size and their inner structure as well as information relating to correlated distances between particles, typically found in concentrated solutions. The scattering intensity pattern however has significantly more detailed information which is retrieved through the use of models whose parameters

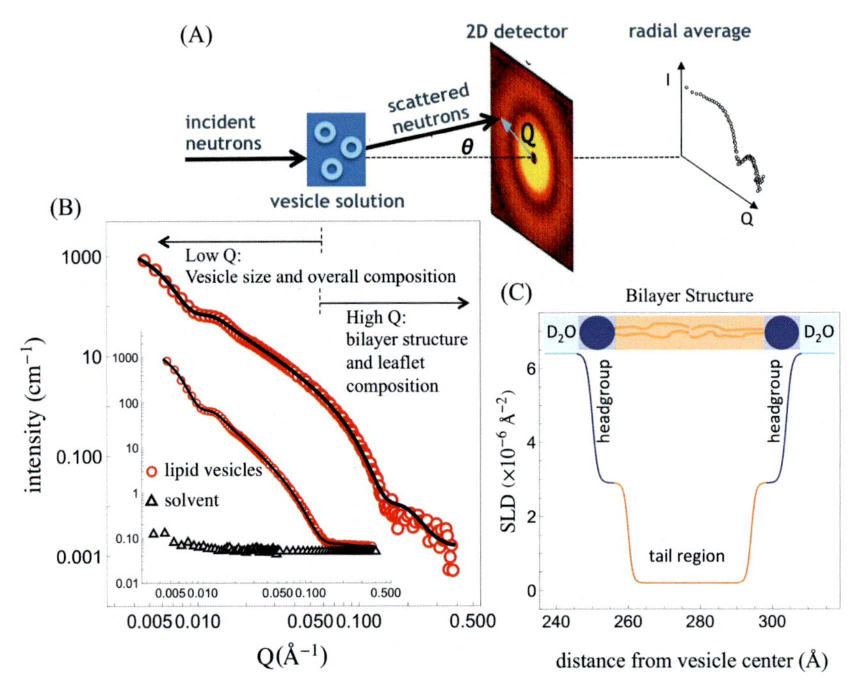

Fig. 1 (A) Neutron scattering schematic showing the interference pattern emerging from the scattering of neutrons by the sample, such as a solution of lipid vesicles, on a 2D detector which can be radially averaged if scattering is isotropic. (B) Scattering curve from unilamellar lipid vesicles, where the background has been removed and where the intensity, in the high Q (Q ~ 0.4/Å) has reached 10^{-3}/cm. Hence, in this case, the data is reliable up to Q ~ 0.4/Å. Inset: Data before background subtraction. The near Q-independent solvent scattering, in this case D_2O, corresponds to the background signal. The line through the data corresponds to a fit for vesicles having a mean diameter of 50 nm—provided by the low Q—and inner structure corresponding to the headgroup and tail regions with distinct SLDs—provided by the high Q. (C) Scattering length density membrane profile corresponding to the fit shown in (B).

are optimized by fitting algorithms. Based on the best-fit values of these parameters, the experimenter makes conclusions about the nature of the scatterer.

In the simple case of a dilute solution of homogeneous particles, the intensity is given by:

$$I(Q) - I_{background} = v\left(SLD_{particle} - SLD_{solvent}\right)^2 P(Q)_{particle} \tag{2}$$

Here $P(Q)$, the form factor, represents the model for the particles and contains the details of their shape and size. v corresponds to the volume fraction of particles. The scattering length density, or SLD, corresponds to a measure of the interaction of neutrons with the atomic and isotopic make-up of the particles and solvent. For example, at room temperature (~20 °C), the SLD of H_2O is -0.56×10^{-6} Å^{-2}, and of D_2O it is 6.37×10^{-6} Å^{-2}, corresponding to an order of magnitude difference in SLD values plus a sign reversal, while mixtures of H_2O and D_2O produce SLD values in between, given by a volume fraction relationship; for lipids (including lipid tails and headgroup), the SLD value is typically 0.2×10^{-6} Å^{-2}, but when hydrogens are replaced with deuteriums the SLD changes significantly, up to ~6.5×10^{-6} Å^{-2} when most hydrogens are replaced with deuteriums. Hence, the contrast term in Eq. (2), $SLD_{particle} - SLD_{solvent}$, can be exquisitely manipulated by the use of hydrogen to deuterium isotopes in the system, which makes the use of SANS so advantageous in the study of biological systems, where hydrogen (and therefore its substitution with deuterium) is abundant.

In Eq. (2) the particles are characterized by a single homogeneous SLD. In general, however, particles can certainly have inner structure, leading to regions with different SLDs. For example, lipid vesicles, depending on the contrast condition and on the experiment's Q-range and resolution, may reveal their inner structure: four onion-like layers corresponding to the two leaflets, each with a headgroup region and a tail region (as shown in Fig. 1C). In this case, Eq. (2) has to be modified to contain additional terms from these contributions.

In a system consisting of a solution of particles, such as vesicle dispersions in an aqueous solvent, the solvent background scattering, $I_{background}$, is nearly Q-independent as shown in the inset of Fig. 1B. When the $I_{background}$ is removed from the intensity, $I(Q)$, the scattering signal can reach values as low as 10^{-3}/cm as shown in Fig. 1B, corresponding to attainable Q values typically between ~0.3/Å and 0.4/Å. This Q range corresponds to a spatial resolution of ~10 Å. Although a scattering signal down to 10^{-4}/cm with reasonable statistics can be reached, it would take extremely long counting times and additional sample environment considerations which make it impractical in the typical beamtime awarded to use the instrument.

Although small angle scattering with X-rays (SAXS) has a higher flux than SANS and higher spatial resolution, the contrast variation tool-set available to SANS makes this the technique of choice in many contexts, particularly relating to direct measurement of membrane structures (e.g., membrane asymmetry (Liu et al., 2020; Nguyen et al., 2019) and/or domain formation (Heberle et al., 2013, 2016), or highlighting particular lipid species, like cholesterol (Garg et al., 2014, 2011). Still, SAXS and SANS are certainly excellent complementary techniques when studying membrane structures in model lipid vesicles (Eicher et al., 2017; Heberle & Pabst, 2017) or in the complexity of cells (Semeraro, Devos, Porcar, Forsyth, & Narayanan, 2017; Semeraro, Marx, Frewein, & Pabst, 2021; Semeraro, Marx, Mandl, et al., 2021).

2.2 Contrast and contrast matching

As mentioned above, the difference in the scattering between hydrogen and deuterium makes SANS particularly powerful because it can straightforwardly reveal a specific process of interest or it can highlight a specific feature within a biological complex by eliminating the contribution from any other feature or process that is not the one of interest. Eliminating a particular signal is done through contrast matching which is a technique that is implemented straightforwardly. For example, if the scattering from the particles described by Eq. (2) is to be eliminated, the procedure consists on obtaining the solvent condition (a mixture of D_2O and H_2O) that brings the contrast term to 0 and produces an intensity that is flat and indistinguishable from the background scattering. To do this, one has to measure the scattering of the particles in solvents having several D_2O/H_2O ratios, and then, from a linear fit to the square root of the average low Q intensity minus the background vs the D_2O/H_2O ratio, we can obtain the zero intensity condition, known as the contrast match-point. Fig. 2 shows a schematic of vesicles in three different solvents, including the contrast match point as well as the contrast scattering series for deuterated POPC (1-palmitoyl-2-oleoyl phosphocholine) where the palmitoyl tail is fully deuterated (designated dPOPC). The scattering curves shown in the figure include the measurement of these vesicles in their contrast-matched (CM) solvent (48.6% D_2O), where indeed the corresponding scattering is flat as the solvent's SLD now matches the SLD of the vesicles. Fig. 2 also shows the linear fit to the square root of the background-subtracted low Q average intensity, from which the contrast matched point was determined. Then, from this invisible scaffold, any third component will then be revealed. For example, when studying cholesterol transport or cholesterol solubility in membranes, it is advantageous to eliminate the contribution from the phospholipids (Garg et al., 2014, 2011); in the studies of membrane proteins, it is advantageous to eliminate the contribution of the lipid scaffolding (Heinrich, Kienzle, Hoogerheide, & Losche, 2020; Johansen, Pedersen, Porcar, Martel, & Arleth, 2018); in the study of protein complexes,

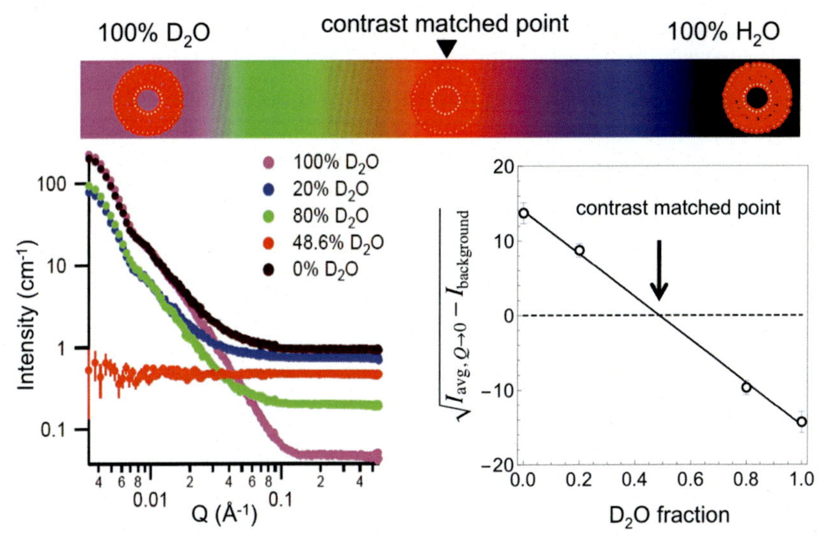

Fig. 2 Top: Schematic of vesicles in different solvent conditions. Lower Left: Scattering from dPOPC vesicles in five solvents: 100% D_2O, 80% D_2O, 48.6% D_2O (CM-point), 20% D_2O, 0% D_2O (100% H_2O). Lower Right: Square root of the average low Q intensity minus the background vs the D_2O/H_2O ratio. The contrast-matched (CM) point, corresponds to a 0.486 D_2O fraction from the linear fit, which indeed corresponds to fully CM vesicles as shown from the flat scattering in this solvent condition on the scattering plot shown on the left.

it is advantageous to obtain separate signals from distinct protein domains (Mahieu & Gabel, 2018; Sugiyama et al., 2014; Zaccai et al., 2016). This "highlighting" strategy has some parallel with ^{1}H NMR where the use of deuterium substitutions eliminate their peak contributions to the spectra. As has recently been highlighted, the diversity and complexity of the systems probed with neutrons, including membranes directly derived from cells, are dependent on our ability to deuterate selected parts of the system (Ashkar et al., 2018). This selective deuteration allows contrast conditions that highlight membranes in all their complexity, as shown recently by Nickels et al. in their study of nanoscale domains in the plasma membrane of gram-positive bacteria in vivo (Nickels et al., 2017). Although it is not necessarily trivial to perform high levels of deuterium substitutions, hydrogen is abundant in biological systems, and—to a lesser or greater extent, depending on the system's components—makes this approach possible. As such, in support of its user community, national neutron facilities have deuteration initiatives to deuterate materials that are usually not available elsewhere, such as in the production of deuterated unsaturated lipids (Chakraborty et al., 2020; Darwish et al., 2013) and deuterated cholesterol (Moulin et al., 2018; Nickels et al., 2015).

Although deuteration does not change the chemical identity of lipids, deuteration does slightly affect their density and melting transition and can modify hydrogen bonding (Bryant et al., 2019; Luchini et al., 2018). The use of D_2O also affects the solubility of biomoleculaes compared to H_2O and modifies the solvent's pH (Efimova, Haemers, Wierczinski, Norde, & van Well, 2007). Indeed, D_2O has a significant effect on living organisms (Thomson, 1960). In spite of these effects, we have found that the transport properties of lipids appear not affected, as will be shown later in this chapter.

In this chapter, we will present work with lipids with all their hydrogens (also referred to as *hydrogenated* lipids) as well as lipids where selected hydrogens have been replaced with deuteriums, which we refer to as *deuterated* lipids. To highlight these differences, we use the letter h or d next to the lipid's acronym; for example, POPC is either hydrogenated, hPOPC or with 31 substituted deuteriums in the palmitoyl tail, dPOPC. Contrast matched points for the lipids used in the data presented in this chapter are: 13% D_2O for hPOPC, 48.6% D_2O for dPOPC, 56% D_2O for dPOPS (1-palmitoyl (d31)-2-oleoyl phosphoserine) and 87% D_2O for dDMPC (di-myristoyl (d54) phosphocholine) and 92% dDPPC (di-palmitoyl (d62) phosphocholine).

2.3 Time-resolved small angle neutron scattering (TR-SANS)

Time-resolved small angle neutron scattering has been successfully applied to the study of the transfer of lipids and sterols between and within membranes using unilamellar vesicles (Breidigan, Krzyzanowski, Liu, Porcar, & Perez-Salas, 2017; Garg et al., 2011; Nakano et al., 2007, 2009) and lipid nanodiscs (Nakano et al., 2009; Xia et al., 2015) by tracking structure and composition changes as a function of time. TR-SANS has also been used in the study of lipid exchange between vesicles and lipoproteins (Maric et al., 2019), and the transfer of lipids in the presence of transport modulators, which does not require any additional experimental design (Maric et al., 2019; Nakao, Kimura, Sakai, Ikeda, & Nakano, 2021; Nguyen et al., 2019, 2021; Nielsen, Bjørnestad, Pipich, Jenssen, & Lund, 2021; Nielsen, Prévost, Jenssen, & Lund, 2020). More recently, TR-SANS has been used to exclusively study lipid flip-flop in vesicles with compositional asymmetry across membrane leaflets and changes in these kinetics due to protein interactions (Marx, Frewein, et al., 2021; Marx, Semeraro, et al., 2021; Nguyen, DiPasquale, Rickeard, Doktorova, et al., 2019). In addition to lipids and sterols, TR-SANS has been applied to other types of systems that exchange molecules such as the exchange of polymer chains between polymer micelles(Choi, Bates, & Lodge, 2011; Choi, Lodge, & Bates, 2010; Lund, Willner, Stellbrink, Lindner, & Richter, 2006).

The smallest temporal step with TR-SANS is in the subsecond range due to the relatively low number of neutrons produced at neutron facilities compared to the very large production of photons at X-ray facilities. More intense neutron sources are being planned or are under construction and they will certainly provide shorter time scales.

2.4 Measurement of the transfer of lipids and sterols between membranes using TR-SANS

To measure the transfer rate of a "probe" molecule of interest between membranes the approach is to have two vesicle populations, one enriched with the "probe" molecule, called donor vesicles, and the other devoid of the "probe" molecule, called acceptor vesicles. A schematic of possible contrast-matching schemes used to measure the transfer of one lipid species or a sterol between vesicles is shown in Fig. 3. At $t=0$, the two vesicle populations are mixed with "ideal" contrast conditions. For fast processes ($<\sim$s), their capture requires the use of a stopped-flow apparatus (Cuevas Arenas et al., 2017) and a multitude of repeated experiments to obtain good statistics while for slow exchange, a single manual mixing is enough. Upon mixing, the exchange of the "probe" molecule between vesicles follows and is continuously tracked with TR-SANS until no more changes are detected in the scattering, indicating that the "probe" molecule is evenly distributed between all vesicles. TR-SANS directly detects the transport of the "probe" molecule from donor-to-acceptor vesicles without the need to physically separate donor from acceptor vesicles as other approaches have required, such as having to use centrifugation (Doktorova et al., 2018; Liu et al., 2020; Wimley & Thompson, 1991), filtration (Yancey et al., 1996), column separation (Dawoud & Abdou, 2021; McLean & Phillips, 1981), as well as other techniques (Sahoo et al., 2021).

The contrast scheme of Fig. 3 is not unique and certainly others are applicable too (Nakano et al., 2007; Wah et al., 2017). The advantage of the scheme shown in Fig. 3 is that it tracks a single species by removing

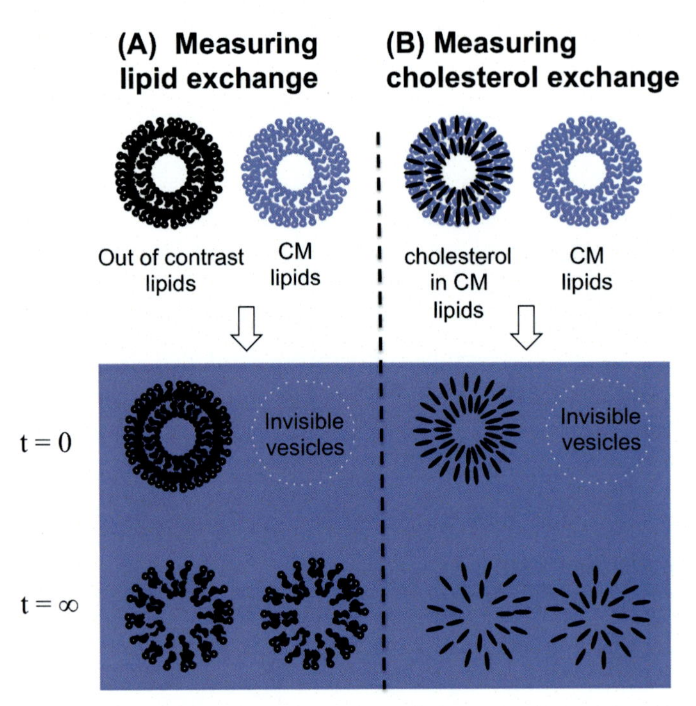

Fig. 3 Schematic of a scheme that focuses on highlighting one lipid species by contrast matching all other contributions. Therefore, the lipid species out of contrast is the only contribution to the scattering. (A) Schematic to highlight one lipid species while to other is CM. (B) Schematic to highlight cholesterol exchange in CM d-lipids.

the contribution of all other species in the system and therefore inarguably showing that changes in the scattering can only be due to the redistribution of that one species, such as was the case for cholesterol (Breidigan et al., 2017; Garg et al., 2011).

In the case of a dilute mixture of donor and acceptor vesicles, the scattering intensity has the contribution from both populations; following from Eq. (2) this is given by:

$$I(Q) - I_{incoh} = v_d(SLD_d - SLD_{solvent})^2 P(Q)_d + v_a(SLD_a - SLD_{solvent})^2 P(Q)_a. \quad (3)$$

Here, the subscripts d and a correspond to the donor and acceptor populations, respectively.

If the volume fraction of each population is the same and the vesicles are the same size, which is done by extruding the vesicles with the same filter size, the analysis of the experiment is simplified significantly. In this case, the changes in the scattering are only due to changes in composition in the vesicles, reflected in their SLDs. Hence Eq. (3) is simplified to:

$$I(Q,t) - I_{incoh} = vP(Q)\left((SLD_d(t) - SLD_{solvent})^2 + (SLD_a(t) - SLD_{solvent})^2\right) \quad (4)$$

The SLD of the vesicle can be obtained by averaging the SLD of the "probe" molecule (SLD_P) with that of the membrane ($SLD_{membrane}$) based on their corresponding volume fractions in the vesicle:

$$\mathrm{SLD_d(t)} = \phi_\mathrm{d}(\mathrm{t})\mathrm{SLD_P} + (1 - \phi_\mathrm{d}(\mathrm{t}))\mathrm{SLD_{membrane}} \quad (5a)$$

$$\mathrm{SLD_a(t)} = \phi_\mathrm{a}(\mathrm{t})\mathrm{SLD_P} + (1 - \phi_\mathrm{a}(\mathrm{t}))\mathrm{SLD_{membrane}} \quad (5b)$$

where $\phi_d(t)$ and $\phi_a(t)$ are the time-dependent volume fractions of the "probe" molecule in the donor and acceptor vesicles respectively, and where:

$$\phi_\mathrm{a}(\mathrm{t}) = \phi_\mathrm{d}(0) - \phi_\mathrm{d}(\mathrm{t}) \quad (6)$$

As mentioned earlier, the SLD of the solvent can be tuned by changing the D_2O/H_2O ratio and it can be set such that $SLD_{membrane} = SLD_{solvent}$ (as schematically shown in Fig. 3). In this case, the scattering intensity from the system is reduced to:

$$I(Q,t) - I_{incoh} = vP(Q)\left((\phi_d(t)\Delta SLD)^2 + ((\phi_d(0) - \phi_d(t))\Delta SLD)^2\right) \quad (7)$$

where $\phi_d(0)$ is the initial volume fraction of the "probe" molecule in the donor vesicles and $\Delta SLD = SLD_P - SLD_{membrane}$. Hence, the final expression for the intensity is given by:

$$I(Q,t) = \beta(Q)\,\tilde{I}(t.)$$

where $\beta(Q)$ is a time-independent prefactor corresponding to the scattering from the donor vesicles:

$$\beta(Q) = v\Delta\mathrm{SLD}^2\phi_d^2(0)P(Q) \quad (8)$$

while $\tilde{I}(t)$ correspond to the compositional changes in the vesicles as a result of the transfer of only the "probe" molecule:

$$\widetilde{I}(t) = \varphi_d^2(t) + (1 - \varphi_d(t))^2 \tag{9}$$

Here, $\varphi_d(t) = \phi_d(t)/\phi_d(0)$. At $t=0$, $\varphi_d=1$ and thus $\widetilde{I} = 1$, which reflects that, initially, the acceptor vesicles are invisible to neutrons. At $t \rightarrow \infty, \varphi_d = 1/2$ which in turn results in $\widetilde{I} = 1/2$, meaning that the overall scattered intensity drops by half at equilibrium when all vesicles have the same concentration of the "probe" molecule. Fig. 4A shows the initial and final scattering curves corresponding to the transfer of hPOPC in CM dPOPC vesicles as well as the transfer of dPOPC in CM hPOPC vesicles. The initial scattering curves change in intensity by half when they reach the equilibrium state since the donor-to-acceptor vesicles concentrations are the same. As shown in the plot, the acceptor vesicles were indeed CM since their scattering signal is flat. The signal from CM hPOPC vesicles is higher than the signal from CM dPOPC vesicles because of the higher content of hydrogen—coming mostly from H_2O. Hydrogen, in contrast to deuterium, produces a significantly higher background (for 48.6% D_2O it is 0.66/cm while for 13% D_2O it is ~1/cm). Also shown is a plot of the normalized intensity change due to the redistribution of hPOPC or dPOPC between the vesicles in the system. As expected, with a ratio of donors to acceptors of one, $\widetilde{I} = 1/2$ at equilibrium. Given that these curves overlap, we conclude that both hPOPC and dPOPC have the same transfer characteristics and that deuteration does not affect the transport mechanism. Fig. 4B shows the case of cholesterol transfer in CM dPOPC vesicles and the corresponding changes in the normalized intensity due to the redistribution of cholesterol between equal donor-to-acceptor vesicle populations. As with POPC transfer, at equilibrium $\widetilde{I} = 1/2$.

If the ratio of donor-to-acceptor vesicles is not the same then Eqs. (8) and (9) have to be modified as follows:

$$\beta(Q) = \nu_d \Delta \mathrm{SLD}^2 \phi_d^2(0) P(Q) \tag{10a}$$

$$\widetilde{I}(t) = \varphi_d^2(t) + \frac{\nu_a}{\nu_d}\varphi_a^2(t) = \varphi_d^2(t) + \frac{\nu_d}{\nu_a}(1 - \varphi_d(t))^2 \tag{10b}$$

where,

$$\varphi_a(t) = \phi_a(t)/\phi_d(0) = \frac{\nu_d}{\nu_a}(1 - \varphi_d(t)) \tag{10c}$$

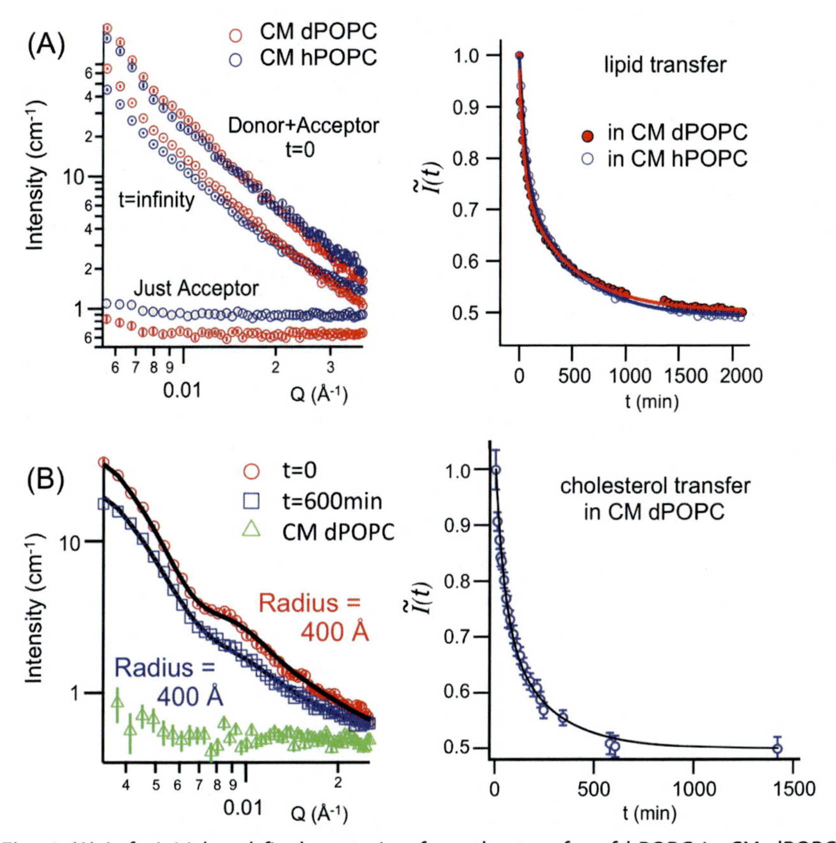

Fig. 4 (A) Left: Initial and final scattering from the transfer of hPOPC in CM dPOPC 100 nm in diameter vesicles as well as the transfer of dPOPC in CM hPOPC 100 nm in diameter vesicles at 75 °C. Scattering from CM acceptor vesicles is flat, with a higher overall signal from CM hPOPC vesicles due to the higher hydrogen content. TR-SANS data was acquired using a resolution that is not optimized to resolve structure but to obtain high flux and be able to track the changes over time with high statistics. Also, noteworthy, is that the compositional changes are captured in the low Q region of the spectra ($Q_{max} < 0.04$). Right: Normalized total intensity change due to the transfer of hPOPC between donor and acceptor vesicles as well as the transfer of dPOPC between donor and acceptor vesicles. The lines through the data are fits using first-order kinetic equations (Eqs. 12a–12d). Because these curves overlap, we conclude that deuteration does not affect the transport characteristics of POPC. The donor-to-acceptor vesicles concentrations, being the same, shows an equilibrium value for the normalized total intensity of 0.5. (B) Left: $t = 0$ and $t = 600$ min scattering curves from the transfer of cholesterol in CM dPOPC at 50 °C. t = 0, was measured without the acceptor vesicles and was used to normalize the total intensity as a function of time. The lines through the data correspond to fits using the vesicle form factor for a symmetric vesicle. Right: Normalized total intensity changes due to the transfer of cholesterol from donor-to-acceptor vesicles. The ratio of donor-to-acceptor vesicles is 1, and therefore, at equilibrium, the normalized intensity value is also 0.5. Lines through the data are a fit using Eqs. (12a)–(12d). *Panel (B) left figure is reproduced from Garg, S., Porcar, L., Woodka, A. C., Butler, P. D., & Perez-Salas, U. (2011). Noninvasive neutron scattering measurements reveal slower cholesterol transport in model lipid membranes. Biophysical Journal, 101(2), 370–377. doi:10.1016/j.bpj.2011.06.014.*

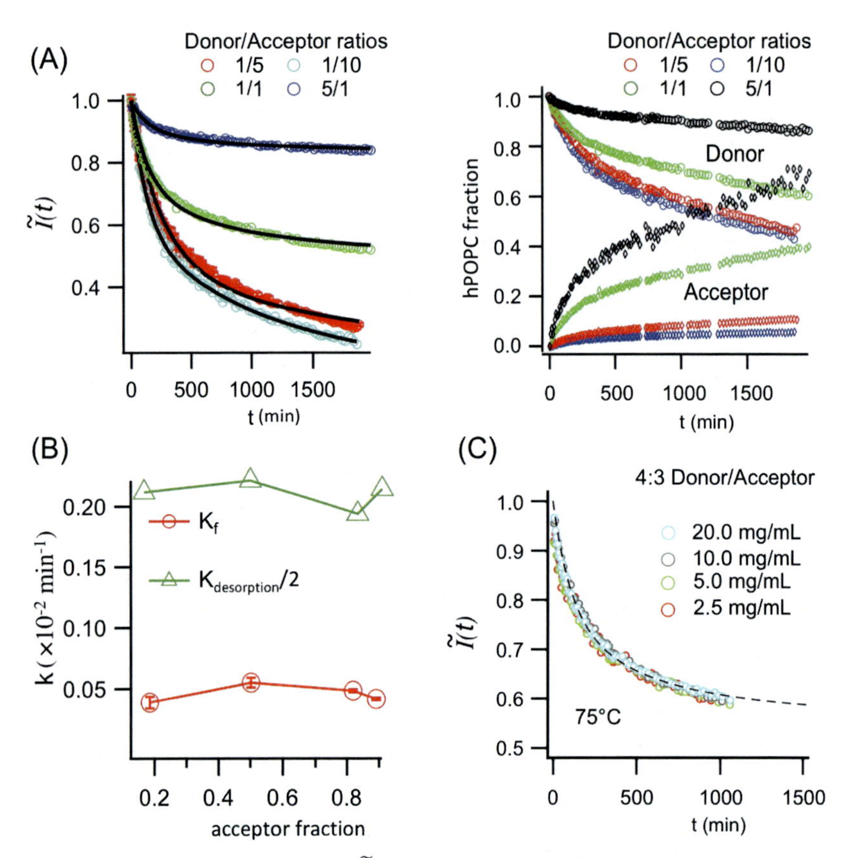

Fig. 5 (A) Left: Normalized intensity, $\widetilde{I}(t)$, for the transfer of hPOPC between donor and acceptor vesicles measured at constant total lipid concentration of 20 mg/mL, and where, initially, the acceptor vesicles are CM to the solvent. Shown are donor-to-acceptor ratios: 1:5, 1:1, 5:1 and 1:10. The kinetics were performed at 65 °C. The continuous lines correspond to fits to the data using Eqs. (12a)–(12d) and (13a)–(13c). Right: The fractions of hPOPC in donor and acceptor vesicles as a function of time derived from $\widetilde{I}(t)$ using Eqs. (10b) and (10c). (B) Respective rates for flip-flop and membrane desorption, k_f and $k_{desorption}$, obtained from the fits presented in (A). These rates are found to be independent of donor-to-acceptor ratio. (C) $\widetilde{I}(t)$ for 4:3 donor-to-acceptor ratio at 20, 10, 5 and 2.5 mg/mL highlighting free-diffusion of lipids rather than a transfer due to collisions (Jones & Thompson, 1989). The rates found from the fits (dashed line) are the same within the error bars: $k_f = 0.001 \pm 0.0001\ \text{min}^{-1}$ and $k_{desorption}/2 = 0.006 \pm 0.0002\ \text{min}^{-1}$. These data were taken at 75 °C.

Fig. 5A shows the normalized intensity, $\widetilde{I}(t)$, for the transfer of hPOPC between donor and acceptor vesicles having different population ratios. Because at equilibrium the "probe" is distributed evenly over all vesicles, with $\varphi_{d,a}(t=\infty) = \frac{\nu_d}{\nu_d + \nu_a}$, then, from Eq. (10b),

$$\widetilde{I}(t=\infty) = \frac{v_d^2}{(v_d + v_a)^2} + \frac{v_d v_a}{(v_d + v_a)^2} = \frac{v_d}{v_d + v_a}$$

Hence, when more acceptor vesicles are present than donor vesicles, the equilibrium intensity drops below ½, and when the number of donors dominates, the equilibrium intensity will be above ½, as shown in the figure. Indeed the equilibrium intensity is ½ when the ratio of donors to acceptors is 1:1.

As expressed by Eq. (10b), it is possible to obtain the overall composition changes in both the donor and acceptor vesicles directly from $\widetilde{I}(t)$. Fig. 5A also shows the corresponding compositional changes in donor and acceptor vesicles directly derived from $\widetilde{I}(t, v_d, v_a)$.

Although in this calculation we set $SLD_{membrane} = SLD_{solvent}$ following the scheme shown in Fig. 3, this is not a requirement and experiments can certainly be done using other contrast schemes (Nakano et al., 2007; Wah et al., 2017). For example, the use of the final equilibrium SLD of the vesicles may be the preferred solvent condition (Nakano et al., 2007; Nakao et al., 2021). The strength of the approach described in Fig. 3, however, is that the scattering comes only from the "probe" molecule with no other contribution, which is a procedure we used to track cholesterol (Breidigan et al., 2017; Garg et al., 2011) and shown in Fig. 4B. Indeed, if the simplifications used above are not valid then the new conditions have to be implemented starting with Eq. (3). Although the calculations may become more innvolved, the equation certainly still holds.

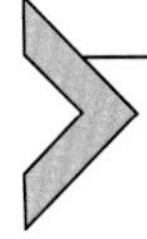

3. Kinetic and thermodynamic characteristics of the transport of lipids and sterols between and within membranes obtained from TR-SANS measurements

3.1 Obtaining transfer coefficients: Exchange and flip-flop of lipids and sterols in membranes

To extract the transfer coefficients from the scattering curves, we propose a simple first-order transfer between donor and acceptor vesicles as used previously (Garg et al., 2011). This model supposes that the transfer of the "probe" molecules originates from two types of pools: one pool resides in the outer leaflet of the membranes and is directly available to exchange with the outer leaflets of other membranes, while the other pool resides in the inner leaflet of the membrane which can only exchange with other vesicles after it flips to the outer leaflet. If the intramembrane flipping rate is slow and hinders the intermembrane exchange rate, then both transfer contributions can be accessed by TR-SANS; otherwise, only the exchange process is captured. But how slow is "slow"? An analysis to this question was performed by Wah et al. (Wah et al., 2017) and found that in order to distinguish these two processes the flip-flop rate's upper limit has to be about 1.5 times the exchange rate. Currently the literature reports flop-flop events that range from minutes to days; however, the current technical limit is in the subsecond to seconds range with the use of a stop-flow apparatus and multiple measurements as mentioned earlier.

Let, C_{in_d} and C_{out_d} be the concentration of the "probe" molecule in the inner and outer leaflet of the donor population such that:

$$\varphi_d = C_{in\_d} + C_{out\_d} \tag{11a}$$

and

$$1 - \varphi_d = C_{in\_a} + C_{out\_a} \tag{11b}$$

where C_{in_a} and C_{out_a} are the concentration of the "probe" molecule in the inner and outer leaflet of the acceptor population. Then the time-varying concentration of the "probe" molecule in the leaflets of donor and acceptor vesicles is described by Eqs. (12a)–(12d):

$$\frac{dC(t)_{in\_d}}{dt} = -k_f\left(C(t)_{in\_d} - C(t)_{out\_d}\right) \tag{12a}$$

$$\frac{dC(t)_{out\_d}}{dt} = k_f\left(C(t)_{in\_d} - C(t)_{out\_d}\right) - k_{ex}C(t)_{out\_d} + k'_{ex}C(t)_{out\_a} \tag{12b}$$

$$\frac{dC(t)_{out\_a}}{dt} = k_f\left(C(t)_{in\_a} - C(t)_{out\_a}\right) - k'_{ex}C(t)_{out\_a} + k_{ex}C(t)_{out\_d} \tag{12c}$$

$$\frac{dC(t)_{in\_a}}{dt} = -k_f\left(C(t)_{in\_a} - C(t)_{out\_a}\right) \tag{12d}$$

where k_f corresponds to the rate coefficient for intraleaflet flip-flop, and where, k_{ex} and k_{ex}' correspond to the rate coefficients for exchange between the donor population and acceptor population.

If we assume the transfer of the "donor" molecules between vesicles happens through the aqueous phase via desorption from the bilayer (and where we also assume that the concentration of the "donor" molecule in the solvent is saturated and not changing in time) we find, as obtained by

Jones et al. (Jones & Thompson, 1989), that the exchange rate, k_{ex}, is a function of the relative population of acceptors to donors and given by:

$$k_{ex} = \frac{\nu_a}{\nu_d + \nu_a} k_{desorption} \tag{13a}$$

where $k_{desorption}$ is the desorption rate from the bilayer into the aqueous phase. When $\nu_a = \nu_d$, the exchange rate is directly related the desorption rate:

$$k_{ex,\nu_d=\nu_a} = \frac{1}{2} k_{desorption} \tag{13b}$$

In addition, k_{ex} and k_{ex}' are related by:

$$k'_{ex} = \frac{v_d}{v_a} k_{ex} = \frac{v_d}{v_d + v_a} k_{desorption} \tag{13c}$$

In Fig. 5A, the continuous lines through the data correspond to fits to $\widetilde{I}(t)$ using the kinetic model described by Eqs. (12a)–(12d) and (13a)–(13c). From the fits we obtained the rates for lipid desorption, $k_{desorption}$, and for lipid flip-flop, k_f. As shown in Fig. 5B, these rates are independent of the donor-to-acceptor ratio, as expected.

When flip-flop is not rate-limiting to the exchange process, Eqs. (12a)–(12d) reduce to:

$$\frac{d\varphi_d(t)}{dt} = -\frac{k_{ex}}{2}\varphi_d(t) + \frac{k'_{ex}}{2}(1 - \varphi_d(t)) \tag{14}$$

where the exchange coefficients are given by Eqs. (13a)–(13c).

A comparison between the case where we consider flip-flop between inner and outer leaflets as rate-limiting to the exchange process (Eqs. 12a–12d) and the case where we only consider an exchange process (Eq. 14) is shown in Fig. 6. The plot on the left corresponds to the transfer of hPOPC in CM dPOPC vesicles, while the plot on the right corresponds to the transfer of cholesterol in CM dPOPC vesicles. This experimental scheme (schematically outlined in Fig. 3) assures that the intensity changes are only due to hPOPC or cholesterol transferring between donor and acceptor vesicles. As shown, we clearly find that the model where flipping is limiting the exchange process is the one that best describes both data sets (Garg, Porcar, Woodka, Butler, & Perez-Salas, 2012).

In the case where the transfer does not primarily happen through the aqueous phase, but through vesicle collisions, Jones et al. propose an effective concentration-dependent exchange rate (Jones & Thompson, 1989). Fig. 5C shows the normalized intensity changes for the transfer of

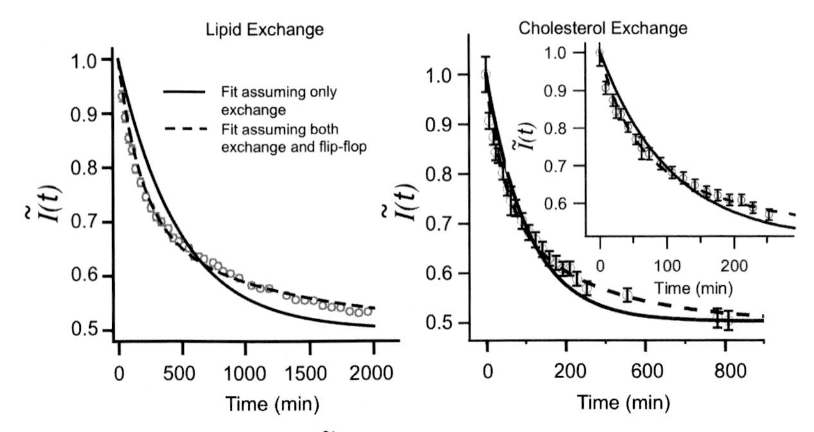

Fig. 6 Normalized total intensity, $\tilde{I}(t)$, for the transfer of hPOPC (left) and cholesterol (right) in CM 100 nm in diameter dPOPC vesicles having a 1:1 donor-to-acceptor ratio. The lines through the data compare the case where flip-flop between leaflets is rate limiting to the exchange process and the case when it is not (hence we only detect exchange). The model where flipping is limiting the exchange process (dashed line) is the one that best describes both data sets. Inset: a close-up on the kinetic intensity change at short times for the case of cholesterol exchange. *Figure on the right is reproduced from Garg, S., L. Porcar, A. C. Woodka, P. D. Butler, and U. Perez-Salas. (2012). Response to "how slow is the Transbilayer diffusion (Flip-flop) of cholesterol? Biophysical Journal, 102(4), 947–949.*

hPOPC between CM dPOPC vesicles having a 4:3 donor-to-acceptor population ratio at four concentrations, 2.5, 5, 10 and 20 mg/mL. The data, as well as the fits, show that there are no perceptible concentration-dependent effects.

If we take a closer look at the schematic in Fig. 3, there is a fundamental difference between the two experiments. In Fig. 3A, for the case of lipids, there is no mass exchange—the exchange is isotopic—while in Fig. 3B, which is the case of cholesterol, the transfer is driven by a redistribution of mass. In the case of mass transfer, such asymmetry in the composition of the vesicles could generate a driving force to equilibrate the chemical potential such that the exchange becomes faster than in the isotopic exchange case. Fig. 7 compares the two cases. In one case, we studied the isotopic exchange of 35 mol% of h and d DMPC in CM dDPPC vesicles while in the other case we studied the redistribution of 35 mol% hDMPC between CM dDPPC vesicles. We find that the isotope exchange is captured quite accurately by Eqs. (12a)–(12d). In the case of mass transfer, while the fit is not as good, the model captures the overall behavior of the data

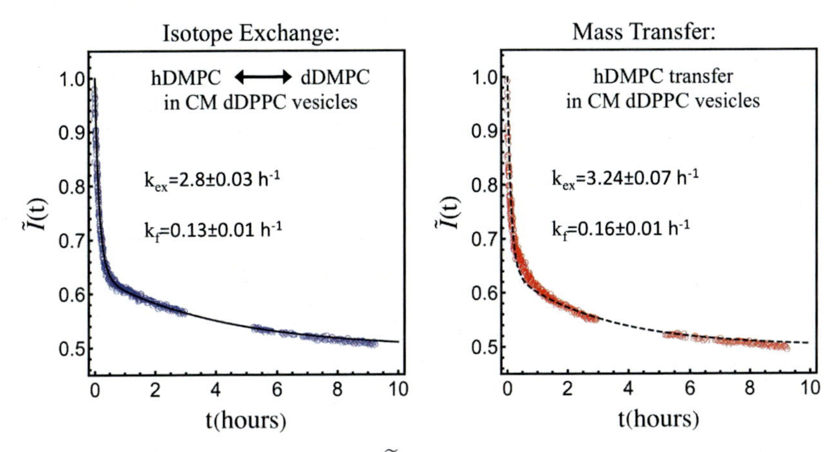

Fig. 7 Normalized total intensity, $\widetilde{I}(t)$, for, left, isotopic transfer: hDMPC + dDPPC ⇔ dDMPC + dDPPC and, right, mass transfer: hDMPC + dDPPC ⇔ dDPPC. The concentration of hDMPC in the donor vesicles was 35 mol%. The concentration of dDMPC in the acceptor vesicles, in the case of the isotope exchange experiment, was also 35 mol%. The temperature was set to 65 °C. The lines through the data correspond to fits using the exchange and flipping model (Eqs. 12a–12d).

and both fits produce rates that differ by only factors of order 1 as shown in the figure. Hence, in this case, mass transfer is having a negligible effect on the rates of transfer.

3.2 Obtaining the energetics of lipid and sterol transport

The rate constants measured are temperature dependent, increasing with increasing temperature and generally following an Arrhenius behavior. This behavior establishes a linear relation between the natural logarithm of the rates and the inverse of the absolute temperature. Fig. 8A shows the temperature dependent $\widetilde{I}(t, T)$ for the transfer of hPOPC in CM dPOPC membranes as well as the transfer of cholesterol in CM dPOPC membranes. Fig. 8B shows the corresponding Arrhenius plot for the transfer rate coefficients, flipping and exchange for both cholesterol (red) and hPOPC (blue). The activation energy, E_a, for flipping and for exchange is obtained from the slope in the Arrhenius plot. In addition to the activation energy, it is possible to extract thermodynamic parameters according to Eyring's transition state theory (Eyring, 1935; Laidler & King, 1983) through the implementation by Homan et al. (Homan & Pownall, 1988) where the activation entropy, $\Delta S^{\ddagger}$, and the activation enthalpy, $\Delta H^{\ddagger}$, are related as follows:

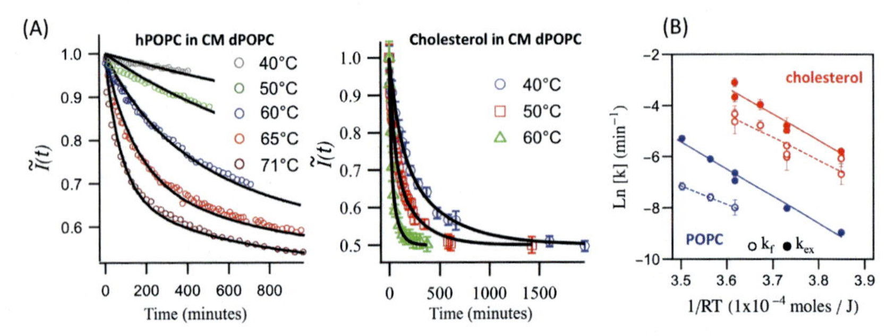

Fig. 8 (A) Normalized total intensity, $\widetilde{I}(t,T)$, for the transfer of hPOPC in CM dPOPC vesicles (left) and cholesterol transfer in CM dPOPC vesicles (right) as a function of temperature. Lines through the data correspond to fits using Eqs. (12a)–(12d). (B) Arrhenius plot for the rates of exchange and flip-flop for cholesterol (red) and hPOPC (blue) in dPOPC CM vesicles obtained from the fits shown in (A) as a function of temperature. Open symbols correspond to flip-flop rates and solid symbols correspond to exchange rates. *Cholesterol data reproduced from Garg, S., Porcar, L., Woodka, A. C., Butler, P. D., & Perez-Salas, U. (2011). Noninvasive neutron scattering measurements reveal slower cholesterol transport in model lipid membranes. Biophysical Journal, 101(2), 370–377. doi:10.1016/j.bpj.2011.06.014.*

$$e^{\Delta S^{\ddagger}/R} = \frac{N_A h}{RT} \kappa_{T^*} e^{-\Delta H^{\ddagger}/RT} \tag{15}$$

where N_A, h, R are Avogadro's number, Plank's constant and the gas constant, respectively. T is temperature in Kelvins and κ_{T*} corresponds to the rate extrapolated to 37 °C (in absolute temperature, 310 K). The activation enthalpy is related to the activation energy as follows$\Delta H^{\ddagger} = E_a - RT$and the difference between the activation enthalpy and the activation entropy term $T\Delta S^{\ddagger}$ is the activation free energy: $\Delta G^{\ddagger} = \Delta H^{\ddagger} - T\Delta S^{\ddagger}$.

Table 1 shows the corresponding thermodynamic values obtained through this analysis for hPOPC as well as for cholesterol transfer, flip-flop and exchange, in CM dPOPC vesicles. Although we will discuss in more detail these results in the section below, we can make the following observations. We find that the rates for exchange and flip-flop for POPC and the energetics for these processes are consistent with values found in the literature. However, for cholesterol, the results shown in Fig. 8 and Table 1 are surprising. Fig. 8B shows that the exchange and flip-flop rates of POPC are about 20 times slower than those found for cholesterol, but cholesterol is found to flip-flop surprisingly slow too, taking many hours at physiological temperatures. We also find that the energetics for cholesterol exchange is similar to those of POPC. This is perhaps not too surprising

Table 1 Thermodynamic parameters for cholesterol and POPC exchange and flipping at 37 °C.

	K (h^{-1})	T1/2 (h)	Ea (KJ)	$\Delta H^{\ddagger}$(KJ)	$T\Delta S^{\ddagger}$(KJ)	$\Delta G^{\ddagger}$(KJ)
	POPC					
Flipping	0.0029 ± 0.0001	234 ± 1	72 ± 2	69 ± 2	43 ± 2	112 ± 1
Exchange	0.0040 ± 0.0001	158 ± 1	106 ± 7	103 ± 7	8 ± 7	111 ± 5
	Cholesterol					
Flipping	0.05 ± 0.03	13 ± 3.7	90 ± 14	88 ± 14	17 ± 14	105 ± 10
Exchange	0.09 ± 0.02	8.5 ± 0.3	104 ± 5	101 ± 5	2 ± 5	104 ± 3

knowing the location of cholesterol in the membrane (Waldie et al., 2019). On the other hand, the energetics for flipping is slightly lower for cholesterol, which can be understood in terms of its smaller hydrophilic volume.

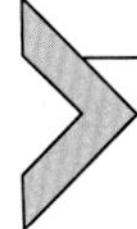

4. Transport behavior of lipids and sterols in membranes

4.1 Exchange and flip-flop behavior of lipids in model membranes

The passive exchange of lipids between membranes and flip-flop within membranes has been demonstrated to be slow. But it is also because lipids move slowly that lipid gradients between different membranes and within membranes can be established. The structural characteristics of lipids, imparting in them slow transport through an aqueous environment as well as between membrane leaflets, allows for some passive regulation mechanism to maintain composition gradients without having, potentially, a significant contribution from ATP-dependent mechanisms to maintain them.

Tail structure, tail length as well as lipid headgroup type determine the time-scale of this passive regulation. Using TR-SANS, we have started to quantify lipid transport characteristics and passive energetic landscapes. We found, that DMPC (dimyristoyl phosphocholine), which is a two tail 14 carbon (C14) long saturated phosphocholine lipid, exchanges and flip-flops in DMPC membranes faster than in DPPC (dipalmitoyl phosphocholine) membranes, where DPPC is a two tail 16 carbon (C16) long saturated phosphocholine lipid. Although this is an expected trend, it was interestingly to find that it is the membrane thickness (a change of ~8 Å (Kucerka, Nieh, & Katsaras, 2011)) that produces the largest effect.

Flipping and exchange transport rates for DMPC in DMPC membranes at 65 °C can be extracted from the thermodynamic parameters reported by Nakano et al. (Nakano et al., 2007), and giving a flip-flop rate of $0.63\,h^{-1}$ and an exchange rate of $4.2\,h^{-1}$. Comparing to the values found in DPPC membranes (Fig. 7) we see a slowdown by a factor of ~4 for flipping and ~1.3 for exchange. It is not too surprising to find roughly the same value for the exchange of DMPC between DMPC and DPPC (C14 vs C16). However, for flipping, it is clear that DMPC will necessitate more energy to flip inside a thicker bilayer. DPPC flip-flop in DPPC membranes at 65 °C, on the other hand, was found to be $0.03\,h^{-1}$ by Marquardt et al. (Marquardt et al., 2017) using ^{1}H NMR, which is roughly a factor of 5 slower than DMPC in DPPC membranes. NMR is an alternative technique that can be used to measure lipid flip flop in vesicles. We recently found that in the case of DPPC, flip flop rates measured by ^{1}H NMR and SANS were consistent (Liu et al., 2020). Hence, both the hydrophobic volume of the molecule that flips and the host membrane thickness are important determinants in the flip-flop rates.

Another important potential factor affecting the rates is the tail structure. DPPC and POPC membranes have a similar thickness (within ~1 Å (Kucerka et al., 2011)), however, POPC has one 18 carbon long monounsaturated tail and a second C16 saturated tail. From Table 1, we see that POPC flips in POPC membranes at a rate of $0.03\,h^{-1}$ at 65 °C, which is the same flip-flop rate for DPPC in DPPC at 65 °C. Perhaps surprising is that the differences in the tail order between these two lipids (Seelig & Seelig, 1977) is not showing a difference in the flip-flop rate in this case. Hence, the dominant effect for flipping is a correlation between membrane thickness and the tail length of the lipid "probe."

Although in this case we do not have the exchange rate of DPPC between DPPC membranes we anticipate them to be similar to those of POPC because they have similar tail lengths and have the same headgroup. Comparing the exchange rate of DMPC in both C14 and C16 and the exchange rate of POPC between POPC membranes at 65 °C we find that POPC exchanges about 30 times slower than DMPC.

In addition to the rates, we find that the thermodynamic parameters for lipid exchange and flip flop obtained from TR-SANS measurements for DMPC by Nakano et al. (Nakano et al., 2007) and for POPC (as shown above) show similar trends: the activation energy to flip is slightly lower than to exchange (by ~20 to 30 kJ/mol) while the corresponding free energies are slightly lower for DMPC (around 100 kJ/mol) than for POPC (around 110–115 kJ/mol).

These trends were also captured by MD simulations (Sapay, Bennett, & Tieleman, 2009). The simulations show that the energy barrier for lipids to desorb from the membrane into the aqueous medium and the energy barrier to flip across the bilayer center are similar and increase with bilayer thickness. Interestingly, Sapay et al. also find that these energy barriers in membranes that have similar bilayer thickness but different saturation (for example, DPPC vs POPC) are nearly identical. A quantitative comparison between the free energies obtained by the simulations and TR-SANS, however, show that the free energies in the simulations are lower than what is obtained in TR-SANS experiments. In the case of POPC, the difference is not large, with simulations predicting an energy barrier of ~95 kJ for both lipid desorption from the membrane and lipid flip-flop. In the case of DMPC the difference is a factor of more than 2, with simulations predicting a value of 40 kJ/mol.

Although membrane order (between POPC and DPPC) has no detectable effect on lipid flip-flop, as discussed above, Nakano et al. (Nakano, Fukuda, Kudo, Matsuzaki, et al., 2009) showed that membrane order induced by cholesterol has a significant effect. Using TR-SANS Nakano et al. showed that DMPC membranes with cholesterol can slow down the flip-flop rates of DMPC significantly; at 40 mol%, the highest concentration of cholesterol studied, the flip-flop rate of DMPC had decreased by a factor of at least ~20, while at 20 mol% the flip-flop rate had only slowed down by a factor of ~4. Yet the exchange of DMPC between membranes remained unaffected.

MD simulations studying the process of flip-flop of lipids in the presence of cholesterol show similar trends overall (Bennett, MacCallum, & Tieleman, 2009). The study, using DPPC, finds that the energy barrier to flip increases at the bilayer center, from ~75 kJ/mol with no cholesterol to ~115 kJmol with 40 mol% cholesterol. The simulations, however, also predict a lowering of the energy barrier to desorb from the bilayer with the addition of cholesterol, which is not detected in the experiments.

In addition to tail variations through saturation state and length, lipids also have different headgroup types, particularly in regard to charge. POPS, POPG (1-palmitoyl-2-oleoyl-*sn*-glycero-3-phosphoglycerol) and POPA (1-palmitoyl-2-oleoyl-sn-glycero-3-phosphate), for example, have the same tail structure as POPC, but the headgroups are negatively charged. POPS, POPG and POPA are important signaling lipids and their distribution in cellular mambranes is specific (van Meer et al., 2008); for example, POPS is found primarily in the inner leaflet of the PM, while POPC, and saturated lipids, are found mostly in the outer leaflet of the PM. In their study of POPA (Nakano, Fukuda, Kudo, Matsuzaki, et al., 2009) and POPG (Nakano, 2019), Nakano and colleagues find that the smaller headgroups in these lipids, though charged, produce a slower inter-vesicluar exchange and in particular faster flip-flop rates, which at 37 °C correspond to halftimes between 420 and 230 min, while for POPC, as shown in Table 1, it takes hundreds of hours. Fig. 9 supports this observation, where POPS, having a similar size headgroup to POPC, has similar exchange and flip-flop rates to those of POPC, varying by less than a factor of 2.

Experimentally, we have also investigated the effect of membrane curvature on the transfer rates (potentially being another source of discrepancy) by comparing the transport of DMPC using 100 nm vesicles and 30 nm vesicles. We found that the rates increase slightly when increasing the curvature of the vesicles, ie, we find faster rates in 30 nm vesicles compared to 100 nm vesicles. The effect, however, is of order ~1. On the other hand, the energetics for both flip-flop and exchange remain essentially the same (Wah et al., 2017). Recent MD simulations find a similar result, where the energetics of lipid flip-flop and lipid desorption are found to be independent of curvature (Jing, Wang, Desai, Ramamurthi, & Das, 2020).

Certainly, the feedback loop between experiments and MD simulations will ultimately reveal an ever more detailed molecular picture underlying various structural and dynamic processes in membranes.

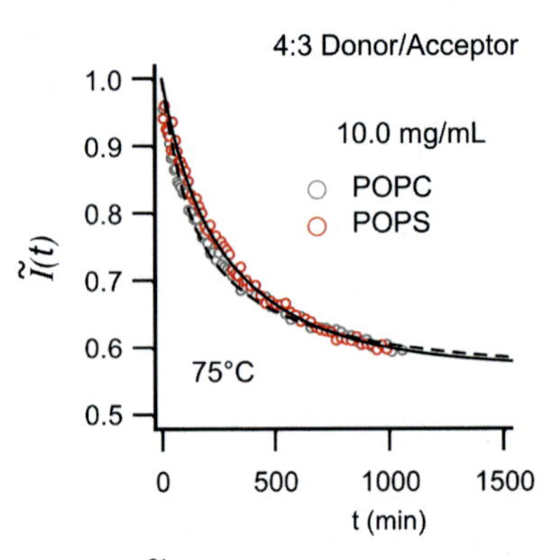

Fig. 9 Normalized total intensity, $\tilde{I}(t)$, for the transfer of hPOPC in CM dPOPC vesicles and hPOPS in CM dPOPS vesicles at $T = 75$ °C. The lines through the data correspond to fits using Eqs. (12a)–(12d). The rates found were for POPC: $k_f = 0.001 \pm 0.0001\ \text{min}^{-1}$ and $k_{desorption}/2 = 0.006 \pm 0.0002\ \text{min}^{-1}$, and for POPS: $k_f = 0.0018 \pm 0.0001\ \text{min}^{-1}$ and $k_{desorption}/2 = 0.0043 \pm 0.0002\ \text{min}^{-1}$.

4.2 Exchange and flip-flop behavior of sterols in model membranes

4.2.1 Cholesterol transfer

Cholesterol is the most abundant lipid of the PM with a 2:1 ratio to the total amount of lipids (Kobayashi & Menon, 2018). Currently, existing evidence on cholesterol's distribution in the PM spans almost the entire range of possible outcomes, from mostly residing in the inner leaflet (Courtney et al., 2018; Mondal, Mesmin, Mukherjee, & Maxfield, 2009; Solanko et al., 2018) to more than 10-fold enrichment in the outer leaflet (Buwaneka, Ralko, Liu, & Cho, 2021; Liu et al., 2017). Therefore, there is significant debate concerning possible biases in understanding these conflicting results. This highlights the challenges of measuring the leaflet occupancy of this molecule (Steck & Lange, 2018). Cholesterol, in contrast to lipids, is seen as a molecule that can traverse the lipid bilayer much faster than lipids (Bruckner et al., 2009; Hamilton, 2003; London, 2019; Steck et al., 2002). Simulations support this result as well (Atkovska, Klingler, Oberwinkler, Keller, & Hub, 2018; Baral et al., 2020; Bennett, MacCallum, Hinner, et al., 2009; Bennett & Tieleman, 2012; Gu et al., 2019). Therefore in order to produce an asymmetric distribution of cholesterol in the PM other mechanisms—yet to be identified—have to play a significant role (Doktorova, Symons, & Levental, 2020).

Our TR-SANS findings suggest that one mechanism facilitating this is, as with lipids, a spassive regulation strategy: if flips slowly (Breidigan et al., 2017; Garg et al., 2011). As shown in Table 1, the flip-flop and exchange rates for cholesterol in POPC membranes are an order of magnitude faster than those for POPC in POPC membranes. In contrast, the corresponding energies of activation and free energy barriers for flip-flop and for exchange between membranes are similar. MD simulations do show that cholesterol's free energy barrier to desorb from the lipid bilayer into the aqueous environment is similar to lipids, such as DPPC and POPC. However, cholesterol's free energy barrier to flip is found to be significantly lower, by factors between 3 and 10 (Bennett, MacCallum, Hinner, et al., 2009; Sapay et al., 2009). One hypothesis we proposed for this inconsistency between TR-SANS and MD simulations was a possible issue with the force fields used to describe cholesterol. For example, using coarse grain MD simulations and the MARTINI force field, we found that the amount of cholesterol incorporated into membranes was overestimated; the simulations would put cholesterol molecules at the bilayer's center rather than expelling these molecules from the membrane altogether(Garg et al., 2014).

In terms of experimental biases, these include the use of analogs (Garg et al., 2011; Nyholm, Jaikishan, Engberg, Hautala, & Slotte, 2019), or the use of extraneous molecules like cyclodextrin (Garg et al., 2011). A bias in the work by Bruckner et al. (Bruckner et al., 2009) using ^{13}C NMR and labeled ^{13}C cholesterol is not obvious. Their work shows very fast flip-flop (ms range) for cholesterol. A possible bias in our TR-SANS measurements were brought up by Kelley et al., who have pointed out that unilamellarity in vesicles between 30 and 100 nm is not always reached and is actually better attained when using at least a small amount of charged lipids (~1–5%) (Scott et al., 2019). Fig. 10 shows a plot for the transfer of cholesterol between CM 50 nm in diameter dPOPC vesicles having 3 mol% of charged lipids at 50 °C where found that the flip-flop and exchange rates are the same as those in Garg et al. (Garg et al., 2011)

Other factors affecting the transfer rates are possible differences in the state of cholesterol within the membrane, such as whether a higher concentration of cholesterol will slow down the process because of synergetic or collaborative motions. Fig. 10 shows that cholesterol concentration effects depend on the lipid environment; we find that while in POPC the change

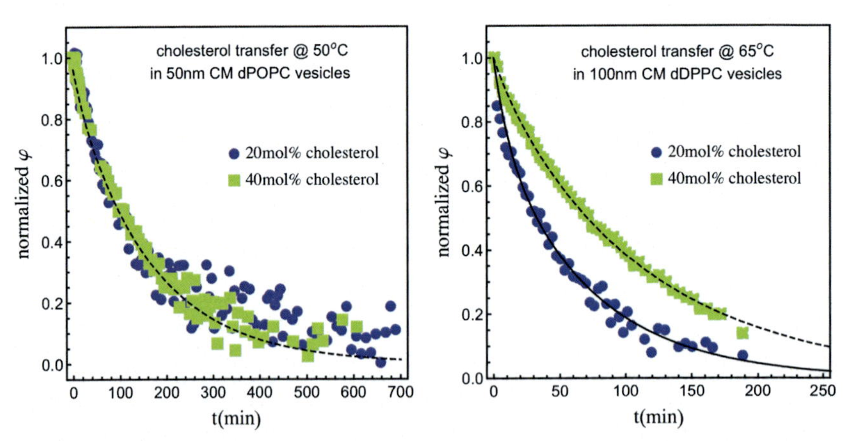

Fig. 10 Normalized cholesterol fraction in donor vesicles as a function of time at two different cholesterol concentrations: 20 and 40 mol%. Left plot, at 50 °C in 50 nm in diameter CM dPOPC vesicles with 3 mol% dPOPG (1-palmitoyl (d31)-2-oleoyl-phosphoglycerol). Right plot, at 65 °C in 100 nm in diameter CM dDPPC vesicles with 2 mol% dDPPG (dipalmitoyl (d62)-phosphoglycerol). Lines through the data correspond to fits with Eqs. (11a)–(11b) and (12a)–(12d). The rates for cholesterol in 50 nm POPC vesicles at 50 °C are, $k_f = 0.01 \pm 0.004\,\text{min}^{-1}$ and $k_{desorption}/2 = 0.008 \pm 0.0005\,\text{min}^{-1}$. The rates for cholesterol in 100 nm DPPC vesicles at 65 °C are $k_f = 0.02 \pm 0.004\,\text{min}^{-1}$ and $k_{desorption}/2 = 0.03 \pm 0.002\,\text{min}^{-1}$ at 20 mol% cholesterol, and $k_f = 0.02 \pm 0.004\,\text{min}^{-1}$ and $k_{desorption}/2 = 0.01 \pm 0.002\,\text{min}^{-1}$ at 40 mol% cholesterol.

in cholesterol concentration from 20 mol% to 40 mol% has no effect on the transport of cholesterol, in DPPC, a saturated lipid, there is. Fits to the normalized intensity (or normalized cholesterol fraction, as shown in the figure) with Eqs. (12a)–(12d) suggest that flip-flop is unaffected by cholesterol concentration but that the exchange rate decreases by a factor of nearly ~3. Comparing the transfer rates of cholesterol between POPC and DPPC in Fig. 10 as well as when varying the ratio of saturated to monounsaturated tails, as shown in Fig. 11A (taken from Breidigan et al. (Breidigan et al., 2017)), we find that the rates for cholesterol increase gradually, by at most a factor of 6, as the fraction of unsaturated tails increase from all DPPC membranes to all POPC membranes. One prediction found by MD simulations that we did not observe in our measurements was a significant slowdown of cholesterol flip-flop rates due to a "raft" effect (a raft being a mixture of sphingomyelin, POPC and cholesterol). In the simulations of a raft mixture, done at 50 °C, cholesterol's flip-flop halftimes increased several orders of magnitude to ~30 min in contrast to milliseconds found in non-raft mixtures. Our measurement of a raft-like system did not slowdown cholesterol's flip flop dramatically; instead the lifetimes (rates) were found to be similar to those in non-raft mixtures as shown in Fig. 11A (Breidigan et al., 2017).

In deciphering cholesterol's location in the cell, we start by noting that the lipid environment across organelle membranes is very different (van Meer et al., 2008). This difference could certainly have an impact on cholesterol transfer and can hold clues as to what may drive cholesterol to one particular environment over the other. Shown in Fig. 11B, are cholesterol's exchange kinetics in POPC and POPS. We find that the rates are not only an order of magnitude slower in POPS than in POPC at near physiological temperatures, but that they exhibit a surprising discontinuous Arrhenius behavior around 48 °C, where cholesterol appears nearly frozen at physiological temperatures. In this case, flip-flop kinetics were not rate limiting and therefore not captured in these measurements. The thermodynamic analysis showed that at biologically relevant temperatures, below the discontinuity, the exchange of cholesterol is entropically dominated while it is enthalpically driven, as is the case in POPC vesicles, above that discontinuity. In this case, the use of Laurdan (see chapter "Evaluating membrane structure by Laurdan imaging: Disruption of lipid packing by oxidized lipids" by Levitan in this volume) provided additional information, pointing to a quasi order-disorder transition in the headgroup region responsible for this effect, even while the lipid tail environment was in the fluid phase.

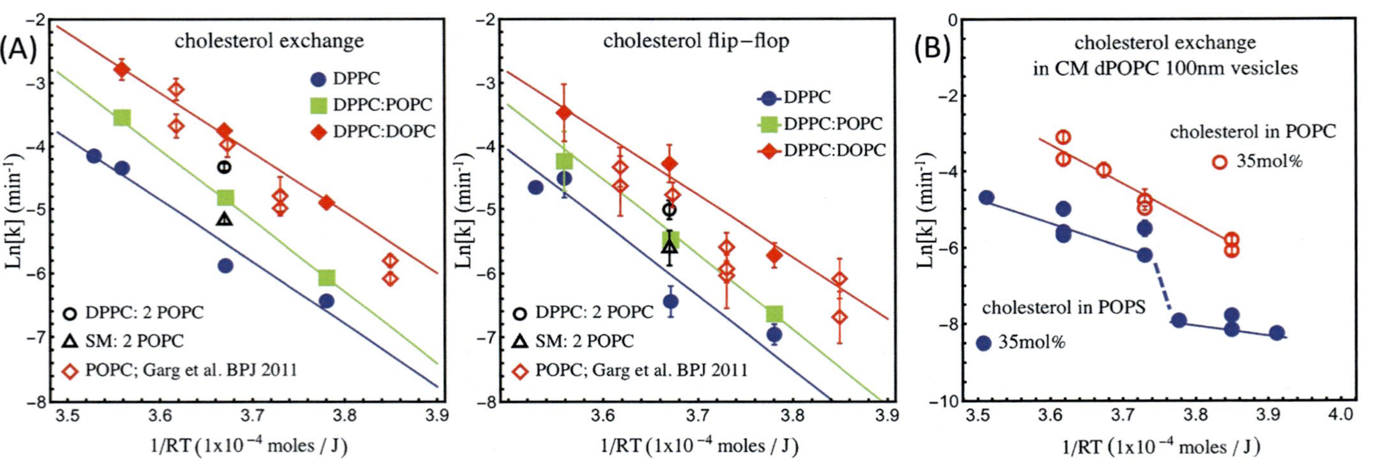

Fig. 11 (A) Arrhenius plots comparing the exchange and flip-flop rates in different membrane environments, where the fraction of saturated lipids is varied relative to the unsaturated lipid fraction. We observe a transfer slowdown with an increase in chain saturation in the membrane. Surprisingly though, the energetics for exchange and flip-flop remains the same as can be assessed from the near parallel Arrhenius behavior. (B) Arrhenius plot comparing the exchange of cholesterol in POPC vesicles to the exchange of cholesterol in POPS vesicles. We find an unusual (anomalous) transition, where the exchange becomes nearly frozen for temperatures below 48 °C. *Panel (A) is reproduced from Breidigan, J. M., Krzyzanowski, N., Liu, Y., Porcar, L., & Perez-Salas, U. (2017). Influence of the membrane environment on cholesterol transfer. Journal of Lipid Research, 58(12), 2255–2263. doi:10.1194/jlr.M077909. Panel (B) is reproduced from S. Garg, S., Liu, Y., Perez-Salas, U., Porcar, L., & Butler, P. D. (2019). Anomalous inter-membrane cholesterol transport in fluid phase phosphoserine vesicles driven by headgroup ordered to disordered entropic transition. Chemistry and Physics of Lipids, 223, 104779. doi:10.1016/j.chemphyslip.2019.05.004.*

This surprising result also triggered research on cholesterol's solubility in POPS membranes. The strength of SANS to study how much cholesterol can be incorporated into membranes comes from our ability to match-out the lipids, and therefore only highlighting cholesterol. Using this approach, we verified that cholesterol can incorporate in POPC membranes up to 61 mol% as previously shown (Huang, Buboltz, & Feigenson, 1999; Stevens, Honerkamp-Smith, & Keller, 2010), while in POPS it was found to be unexpectedly high, 73 mol% (Garg et al., 2014). The consequences of these findings are still being investigated. However, as discussed in Garg et al. (Garg et al., 2014), this finding suggests that a higher than physiologically relevant cholesterol concentration in the PM is not driven to form toxic cholesterol crystals due to the presence of POPS.

4.2.2 Sterol structure effects on transfer

In addition to the effects that the lipid environment has on the transfer rates of cholesterol, we have also investigated the effect of structural changes to the cholesterol molecule to identify correlations between structure and transport properties. Fig. 12A shows the effect of adding double bonds on the steroid ring and its tail—dehydroergosterol (DHE)—while Fig. 12B shows the effect of the replacement of the hydroxyl group by a sulfate group. These sterols have physiological and even beneficial functions; for example, DHE has been found to be help treat cognitive function loss (Ano & Nakayama, 2018) while cholesterol sulfate, which is a component of cell membranes, aids in protecting erythrocytes from osmotic lysis as well as in regulating sperm capacitation (Strott & Higashi, 2003). As shown in the figure, the increase in double bonds in the ring structure as well as the tail in DHE, increases the transport properties dramatically; DHE exchanges 8 times faster and flips 10 times faster than cholesterol (Garg et al., 2011). This is an interesting results since DHE has been found to be asymmetrically distributed in the PM of CHO (Mondal et al., 2009), yeast (Solanko et al., 2018) and synaptic (Wood, Igbavboa, Muller, & Eckert, 2011) cells, which suggests that even though we found that this sterol moves very fast through the lipid bilayer, the lipid and protein compositional asymmetry in combination with sterol transport proteins in the PM, are likely keeping DHE in the cytoplasmic leaflet. The presence of the sulfate moiety in cholesterol sulfate produces a fast transfer through the solvent, shown in the precipitous drop in the normalized total intensity. The value of 0.625 for the normalized intensity, as given by Eq. (9), corresponds exactly to the outer leaflets of donor vesicles and acceptor vesicles attaining the

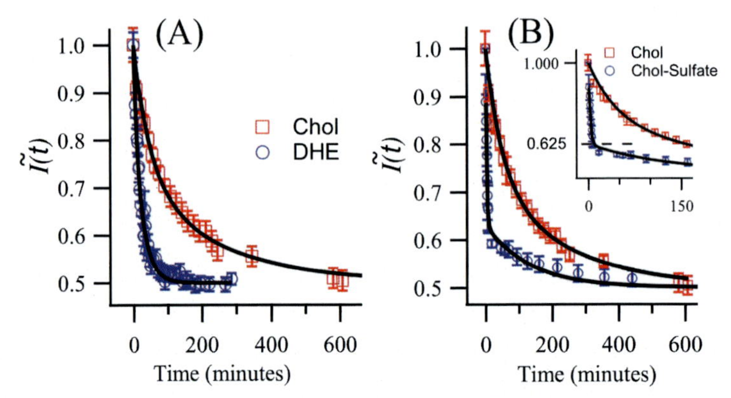

Fig. 12 (A) Comparison of normalized intensity decay curves for DHE and normal cholesterol in POPC vesicles at 50 °C. Both the flip-flop rates and exchange rates are 8–10 times faster than for cholesterol. (B) Comparison of normalized intensity decay curves for normal cholesterol and cholesterol-sulfate in POPC vesicles at 50 °C. The inset highlights the point where the outer leaflets of the donor and acceptor vesicles have the same composition, but the inner leaflets remain unchanged from their initial state, producing an intensity of 0.625. The long tail that follows is due to slow flip-flop of the sterol in the bilayer. *Figure is reproduced from Garg, S., Porcar, L., Woodka, A. C., Butler, P. D., & Perez-Salas, U. (2011). Noninvasive neutron scattering measurements reveal slower cholesterol transport in model lipid membranes. Biophysical Journal, 101(2), 370–377. doi:10.1016/j.bpj.2011.06.014.*

same sterol composition but where the inner leaflets' composition has not changed; i.e., the inner leaflet of donor and acceptor vesicles is the same as it was at $t=0$, and therefore 75% of the sterol is still in the donor vesicles while 25% (residing in the outer leaflet) has moved to the acceptor vesicles. The normalized intensity change from 0.625 to 0.5 is slow, indicative of the slow flip-flop process. Interestingly, we found that the flip-flop rate for cholesterol sulfate was similar to that of cholesterol (Garg et al., 2011).

Sterols are critical for diverse functions in cells: structurally, for signaling as well as biochemically, as these are precursors of, for example, hormones and steroids (Ikonen & Jansen, 2008; Menon, 2018). From the results presented above, we find that the transport properties of sterols can be significantly different. Therefore, how the cell transports and distributes these sterols is possibly not via a "one-size fits all" mechanism. For example, DHE and cholesterol sulfate transfer between membranes very fast while cholesterol is slow, so their transfer mechanisms are surely different. In terms of their distribution across membranes, they may be facilitated by the sterol's structure, like in the case cholesterol sulfate, but other mechanisms—still unknown—are surely provided by other characteristics of the membrane, as is likely the case for DHE. A detail study of sterol structure characteristics and how these impact transport behavior could help elucidate sterol traffic pathways.

MD simulations studying the effects of different chemical modifications on sterols have found also great variability in both flip-flop rates (Atkovska et al., 2018; Dickey & Faller, 2007; Parisio, Sperotto, & Ferrarini, 2012) and their desorption rate from the membrane (Atkovska et al., 2018). The work of Atkovska et al., who studied 26 steroids, highlights a kinetics spectrum that is broad and that varies by orders of magnitude (Atkovska et al., 2018). To compare to the data presented here, they find that changing the hydroxyl group for a sulfate group in the steroid pregnenolone produces it to desorb fast from the membrane, a difference of 8–9 orders of magnitude compared to cholesterol, while the flip-flop rate remains similar to cholesterol (in their study, cholesterol flip-flop is fast). These trends are qualitatively similar to our measurements of cholesterol sulfate and cholesterol. In the case of DHE, their study shows that both the desorption rate from the membrane and the flip-flop rate increase by a factor ~10 compared to cholesterol. Interesitngly, even if the rates do not agree with our measurements, we also obtained that DHE flips and exchanges ~10 faster than cholesterol. Hence, it is expected that the rational for the kinetics they observe, which follow cyclohexane/water and membrane/water partition coefficients—except for long-tailed steroids, which have an increased membrane affinity and therefore a greatly decreased membrane exiting rate—will hold in concomitant, yet to be reported, experiments.

4.3 Decoupling exchange from flip-flop

From TR-SANS measurements, we have shown that it is possible to obtain transport characteristics of the exchange of lipids and sterols between membranes in situ, and, if rate limiting, the flip-flop between leaflets without the need of perturbative tags. Even though the experimental results are robust, shown to have statistical confidence (using tools like the Akaike information criterion (Burnham & Anderson, 2002)), the results do occur while two processes, of possibly similar time scales, are happening simultaneously. Further, despite the evidence that the transport kinetic model based on exchange and flipping from donor-to-acceptor vesicles agrees extremely well with the experimental data, we have not yet unambiguously demonstrated an asymmetric composition in the bilayer while the exchange is occuring. Hence analysis of the compositional changes over time relies more on mathematical fitting having statistical significance than in our ability to distinguish these two processes "by eye." Indeed, in cases where the two kinetic processes are naturally separated by an order of magnitude, flipping being slow and exchange being fast, the two are clearly visually separated, as are the case shown in Figs. 7 and 12B, and consequently prone to agree with our hypothesis. If not, decoupling the two processes becomes an important goal in these types of measurements to confirm the numbers obtained.

4.3.1 *Studying lipid flip-flop in asymmetric vesicles*

Studying composition asymmetry across the lipid bilayer has been an important goal to understand the consequences of two lipid leaflets having different physicochemical properties. Examples are, the mechanisms of interleaflet coupling in the context of overall mechanical properties and well as in the formation of lipid domains (London, 2019; Weiner & Feigenson, 2019) or in membrane dynamics, such as bending fluctuations, in particular as they compare to those of symmetric membranes (Blumer et al., 2020; Rickeard et al., 2020). Several approaches have been developed to create asymmetric membranes to be studied by optical, spectroscopic and scattering techniques and some of these strategies have been recently reviewed by Scott et al. (Scott et al., 2021). Neutron scattering stands out as a particularly well suited technique to studying leaflet compositional asymmetry and its changes through flip-flop, in situ.

SANS/TR-SANS is sensitive at discerning lipid composition differences across the lipid bilayer because the scattering signal from vesicles allows for the retrieval of structure information of individual leaflets in the high Q region of the spectra as shown in Fig. 1B. In order to extract this information, contrast between leaflets, reflecting their compositional asymmetry, is key and this is provided by the use of hydrogenated and deuterated lipids. The high Q of the scattering plot shown in Fig. 13 displays a highly asymmetric dDPPC/hDPPC distribution across the leaflets of the vesicles with the characteristic up-lift in the scattering. In this case, the outer leaflet is enriched with hDPPC while the inner leaflet is enriched with dDPPC as obtained from the fit and shown in the SLD profile plot. Indeed, upon allowing the flip-flop process to proceed, the membrane becomes symmetric in composition and the intensity in this Q region also drops producing the typical signature of symmetric vesicles. Hence SANS can unambiguously determine the lipid composition in each leaflet as shown in Fig. 13.

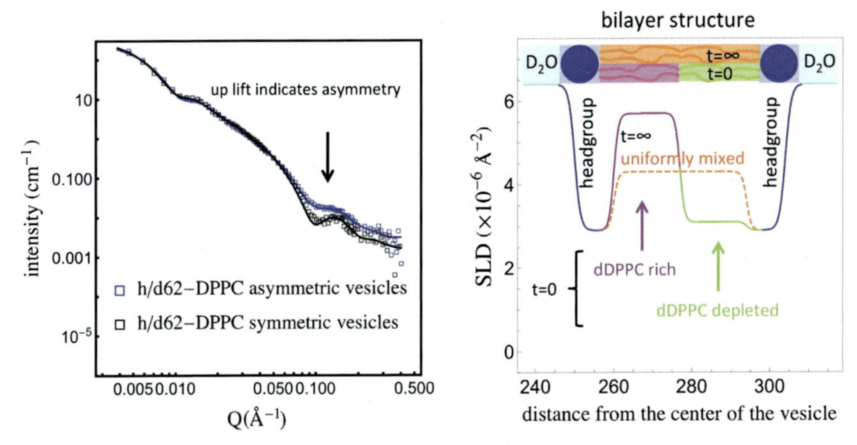

Fig. 13 Left, SANS scattering curves with tight collimation to resolve bilayer features. The vesicles are in D_2O to have the lowest possible background. The scattering from asymmetric vesicles have the characteristic up-lift in the scattering (blue). Upon mixing, due to lipid flip flop, the scattering develops a well-defined minimum characteristic of a symmetric bilayer (black). The lines through the data correspond to fits showing the change in the composition in each leaflet between the initial and final bilayer configurations, as shown on the SLD profile on the right. *Figure on the left is reproduced from Liu, Y., Kelley, E. G., Batchu, K. C., Porcar, L., & Perez-Salas, U. (2020). Creating asymmetric phospholipid vesicles via exchange with lipid-coated silica nanoparticles. Langmuir, 36(30), 8865–8873. doi:10.1021/acs.langmuir.0c01188.*

While the scattering analysis and the corresponding asymmetric composition can be resolved by SANS straightforwardly, the experimental methods to produce asymmetric bilayers are not as straight forward. Several approaches have emerged as pathways to creating a single monodisperse population of unilamellar vesicles having the same asymmetric lipid distribution. One technique that has proven to be robust for SANS/TR-SANS measurements consists of utilizing cyclodextrin (CD) molecules. CDs directly interact with membranes. These compounds are used to deplete or add lipids and sterols to model (StClair, Wang, Li, & London, 2017) and cell membranes (Zidovetzki & Levitan, 2007). CDs are used in the treatment of lipid diseases (Coisne et al., 2016) like Niemann pick disease type C (Matencio, Navarro-Orcajada, Gonzalez-Ramon, Garcia-Carmona, & Lopez-Nicolas, 2020), as well as cancer (Qiu, Li, & Liu, 2017) and SARS-COV-2 (Fatmi, Taouzinet, Skiba, & Iguer-Ouada, 2021). CDs, in fact, have a broad applicability (Fourmentin, 2018) because they are small molecules that are soluble in an aqueous environment but can host hydrophobic molecules in their inner barrel-like core (Scott et al., 2021).

For the purpose of creating asymmetric unilamellar vesicles, the method consists of using a donor lipid population that exchanges lipids with an acceptor vesicle population where the exchange happens through the CD carriers. CDs only change the composition of the outer leaflet of the acceptor vesicles by enriching it with donor lipids (Doktorova et al., 2018; Markones et al., 2018; Scott et al., 2021). It has also been shown that the method can be applied to generate asymmetric vesicles with peptides (Nguyen, DiPasquale, Rickeard, Doktorova, et al., 2019) and fully functional integrated proteins (Markones et al., 2020) pre-embedded in the vesicles.

Notwithstanding, we found that in the case of cholesterol, as shown in Fig. 14, the presence of CD not only produced a highly accelerated exchange rate between donor and acceptor vesicles, it also produced a highly accelerated flip-flop rate. As a result, we devised a completely different strategy to producing asymmetric membranes, one in which we take advantage of the fact that lipids do diffuse freely through an aqueous environment and follow well-defined thermodynamic properties. The strategy consists of combining lipid coated nanoparticles (we used silica nanoparticles) and vesicles at a ratio dominated by the donor population (lipid coated nanoparticles). By using a very unequal ratio of donors to acceptors, as described by Eqs. (13a) and (13c), it is possible to build high compositional asymmetry in the acceptor vesicle population while not affecting the flip-flop process. The lipid-coated nanoparticles, having a higher density

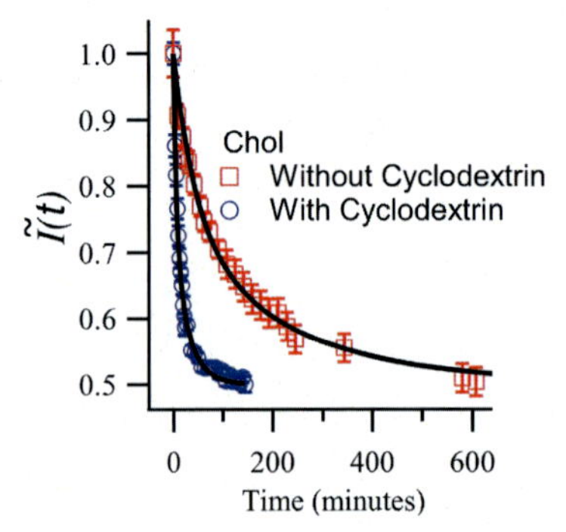

Fig. 14 Comparison of normalized intensity decay curves for cholesterol in POPC vesicles with and without the presence of 2 mM cyclodextrin (CD) at 50 °C. *Figure is reproduced from Garg, S., Porcar, L., Woodka, A. C., Butler, P. D., & Perez-Salas, U. (2011). Noninvasive neutron scattering measurements reveal slower cholesterol transport in model lipid membranes. Biophysical Journal, 101(2), 370–377. doi:10.1016/j.bpj.2011.06.014.*

than the vesicles, are removed by centrifugation, leaving behind the asymmetric vesicle population. Our proof-of-principle measurements are shown in Fig. 13 (Liu et al., 2020). In this study, which was complemented with ^{1}H NMR measurements (which is also sensitive to isotopic labeling and can probe the distribution of hydrogenated and deuterated molecules in a lipid bilayer (Marquardt et al., 2017)), we found that the rate of flip-flop of DPPC, was consistent to the flip-flop rates found in asymmetric vesicles that used the CD approach (Liu et al., 2020). Hence, at least for DPPC and other long chain phospholipids, CDs may not produce additional or significant interleaflet scrambling at low CD concentration (Nakano et al., 2009). However, the case shown in Fig. 14 for cholesterol highlights awareness of possible biases induced by CDs (Zidovetzki & Levitan, 2007) and the need to have alternative approaches.

4.3.2 Studying lipid exchange and flip-flop in single membranes

Techniques that can directly probe the structure of a single lipid bilayer deposited on a surface like neutron reflectometry (NR) or sum frequency generation vibrational spectroscopy (SFGVS) were seen as ideal to follow flip-flop and exchange on a single membrane. In addition, both techniques are sensitive to isotopic labeling and can probe the replacement and displacement of protiated and deuterated molecules in the membrane. With high spatial resolution (<1 nm) and a temporal resolution of minutes, with SFGVS and NR it is possible to follow the spontaneous loss of asymmetry via lipid flip-flop in asymmetric bilayers (Allhusen & Conboy, 2017; Gerelli, Porcar, & Fragneto, 2012). With NR it is also possible to monitor the presence of asymmetric intermediates during experiments involving lipid exchange between vesicles in solution and a single bilayer on a surface (Gerelli et al., 2013).

Using this latter approach, we deposited a symmetric SLB on a solid substrate (usually a large (~40 cm^2) and highly polished (rms roughness ~0.3 nm) silicon crystals with a thin (1 nm) silicon oxide layer) via vesicle fusion which was then exposed to a solution of vesicles composed by the same phospholipid but having the complementary deuteration form. This type of experiment was developed to mimic the measurement of the transfer of lipids between membranes (in vesicles) with TR-SANS with the additional advantage of directly monitoring composition changes in a single membrane. In the case of DMPC, reported to flip slow in DMPC membranes using TR-SANS by Nakano et al. (Nakano et al., 2007), we found that flip-flop was inaccessible. However, we were able to recapitulate the results obtained by Nakano et al. for lipid exchange, including the energetics.

Asymmetric SLB can also be formed, one leaflet at a time, on a solid substrate by Langmuir-Blodgett and Langmuir-Schaefer deposition techniques. Using this approach, we studied asymmetric SLB that either had the same phospholipids (DPPC/dDPPC) or different ones (DMPC/dDSPC). These layers were deposited in the gel phase where they remained asymmetric until lipid flip-flop was activated by increasing the sample's temperature, i.e., by crossing the gel-to-fluid phase transition temperature of the system. The phase transition from gel to the fluid phase behavior, which is normally sharp in vesicles, it is broad in SLBs (Gerelli, 2019), and this influences the activation and progression of lipid flip-flop, which adds complexity to the experimental analysis (Porcar & Gerelli, 2020).

Unfortunately, for both approaches, the "surface" was found to accelerate lipid flipping (Gerelli et al., 2012; Porcar & Gerelli, 2020). While in defect-free supported membranes on silica nanoparticles we find that fast flip-flop is driven by the surface's induced disorder of chain packing (Wah et al., 2017), as evidenced from the surface induced broad melting temperature behavior on flat surfaces (Gerelli, 2019) and as needed in NR experiments (or sum vibrational spectroscopy), SLB have additional unavoidable membrane defects—membrane discontinuities (Marquardt et al., 2017)—that accelerate flipping by several orders of magnitude and producing a lower activation energy (Porcar & Gerelli, 2020). Notwithstanding, in lipids that flip-flop very slowly, like DPPC, it is still possible to capture membrane asymmetry as shown in Fig. 15, and therefore the technique could still provide a platform to study asymmetry in membranes (Porcar & Gerelli, 2020).

4.4 Lipid exchange and flip-flop behavior in the presence of biological agents

Although cells use proteins and other molecular transporters to regulate the distribution of lipids across membranes (Kobayashi & Menon, 2018), membranes are also targets of biological agents which may alter this distribution (Doktorova et al., 2020; Lohner, 2017). An example of important/critical biological agents that interact with membranes are antimicrobial peptides (AMPs). Antimicrobial peptides are, as its name suggests, small molecules, ubiquitous in nature, that are an important part in the immune system of different organisms and which have inhibitory effects against bacteria, fungi,

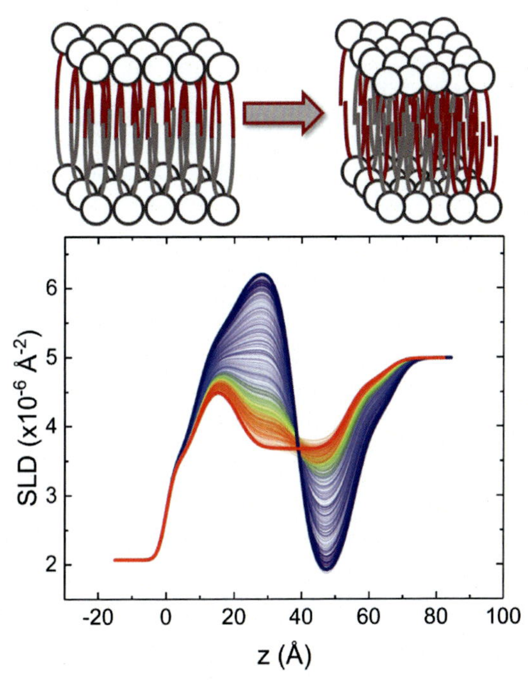

Fig. 15 Top: Schematic representation of initial asymmetric SLB in gel phase and final fully symmetric SLB in fluid phase. Different colors indicate deuterated and protiated phospholipid molecules. Bottom: Collection of scattering length density profiles along the normal direction to the bilayer surface ($z \approx 40$ Å indicate the SLB mid-plane). The initial asymmetric structure is represented with the blue thick line while the final symmetric system is represented by the thick red line. Intermediate curves collected at different times and temperatures clearly indicate the gradual loss of symmetry. *Readapted from Porcar, L., & Gerelli, Y. (2020). On the lipid flip-flop and phase transition coupling. Soft Matter, 16(33), 7696–7703. doi:10.1039/d0sm01161d.*

parasites and viruses. In addition, the advent of antibiotic-resistant microorganisms and the increasing concern in the use of antibiotics, has resulted in the development of de novo AMPs (Vishnepolsky et al., 2019) which have potential protective use in humans, animals and plants (Huan, Kong, Mou, & Yi, 2020). Thus, there is great impetuous to understand their mechanisms of action especially because these do not follow a direct lock-and-key mechanism that makes antibiotics susceptible to the development of resistance (Wimley & Hristova, 2011). As recently shown by Marx et al. (Marx, Frewein, et al., 2021; Marx, Semeraro, et al., 2021), connecting peptide activity in bacteria with its activity on model membrane mimics is intricate. Notwithstanding, as the authors emphasize, membrane characteristics do play an important role in the translocation of the peptide into cells, and hence studies in understanding peptide-lipid-membrane structure interactions are important. TR-SANS studies that have examined how these molecules affect the movement of lipids between and within membranes have found that, independent of AMP structure, their presence accelerates lipid flip-flop (Marx, Frewein, et al., 2021; Marx, Semeraro, et al., 2021; Nguyen, DiPasquale, Rickeard, Doktorova, et al., 2019) as well as the exchange of lipids between membranes (Nakao et al., 2021; Nguyen et al., 2021; Nielsen et al., 2021, 2020). In these studies, the authors looked at pre-inserted peptides as well as free peptides that interact with membranes—which being cationic, interact electrostatically with membranes—and found that in the latter case the effect is near instantaneous, while in the former it proceeds much slower, but faster than when no peptides are present (Nguyen, DiPasquale, Rickeard, Doktorova, et al., 2019). The combined results from the work of these groups of researchers shows that these molecules share certain features that allow them to disrupt membranes by promoting the transport of lipids between and within membranes despite differences in structure and in how they interact with membranes.

As it is always the case, possible sources of bias have to be identified and parsed out from peptide action, such as the—separate—effect of co-solvents commonly used in these studies (Nguyen, DiPasquale, Rickeard, Doktorova, et al., 2019). Indeed solvents, such as short-chain alkanes, are known to modify membrane properties in vitro (Ly & Longo, 2004) and in vivo (Goldstein, 1986) as well as accelerate lipid exchange and flip-flop, as recently reported by Nguyen et al. for methanol (Nguyen, DiPasquale, Rickeard, Stanley, et al., 2019). Indeed solvents having low solubility in water with preferential partitioning into membranes, enhance these effects (Dickey & Faller, 2007). Fig. 16 shows the dramatic increase in the transport characteristics of hDMPC between CM dDMPC membranes in the presence of butanol, a low solubility solvent in water.

Another example of physiological interest is the interaction of plasma lipoprotein particles—low and high-density lipoprotein particles (LDL and HDL)—with plasma membranes. HDL and LDL are currently used as clinical markers for atherosclerosis, a disease in which plaques of lipids and fibrous elements accumulate in the blood vessels (Carmena, Duriez, & Fruchart, 2004; Lusis, 2000). Therefore, the mechanism of lipid exchange between HDL and LDL particles with cellular membranes needs to be carefully examined at a molecular level in order to understand how they

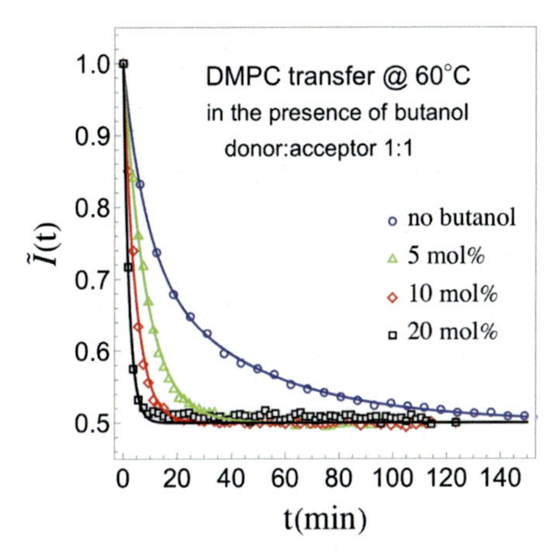

Fig. 16 Normalized total intensity decay curves for hDMPC in CM dDMPC vesicles with and without butanol. Butanol's molar concentration represents the total number of Butanol molecules/(total number of DMPC molecules + Alcohol molecules). DMPC concentration in the solution was 40 mg/mL.

participate in the buildup of arterial plaques. A study by Maric et al., using TR-SANS, followed the lipid exchange between human HDL and LDL particles and cellular membrane mimics (vesicles), which, through deuteration and contrast matching, were "invisible" to neutrons (Maric et al., 2019). The data they obtained shows that, in addition to lipid exchange through monomer diffusion, the exchange also occurs through collisions and tethering, which also depends on the apolipoprotein type. They find that the exchange of lipids between cell membrane mimics and HDL particles is more efficient than with LDL particles. The authors associate tethering efficiency to their envelope protein density. Indeed, HDL have a larger concentration of envelope ApoA proteins while LDL particles have a lower ApoB protein content.

Lipid transport is also an important consideration when using lipid-based scaffolds for membrane protein studies. Lipid nanodiscs consist of a protein or a polymer or a detergent "belt" surrounding a nanometer (~3 to ~30 nm) lipid bilayer patch. Lipid nanodiscs have become the method of choice as a stabilizing scaffold for membrane proteins used in protein structural studies (Denisov & Sligar, 2017). In addition, lipid nanodiscs are used in many other applications, such as high-throughput screening and diagnostics or carriers of hydrophobic therapeutics, to mention some (Ryan, 2010). Differences exist between nanodiscs, particularly as it relates to the belt and its interaction with lipids. Specific lipid associations with the transmembrane domain of membrane proteins, like those commonly found with phosphatidylinositol, may be altered by the belt. TR-SANS has thus been used to assess the ease of movement of lipids between nanodics to understand lipid stability. Overall, it is found that between nanodiscs lipids move faster than between vesicles (Nakano, Fukuda, Kudo, Miyazaki, et al., 2009; Xia et al., 2015). Further, different belt strategies for the nanodiscs can also produce large differences in the exchange kinetics of lipids, as recently highlighted by Cuevas et al. (Cuevas Arenas et al., 2017). They find that polymer-based belt nanodiscs produce the most disorder of the lipid bilayer and therefore the fastest exchange of lipids between nanodiscs, which, they suggest, could be used advantageously to be able to only keep those lipids that strongly associate with the protein while excluding those with weaker contacts.

5. Current and future perspectives

Eukaryotic cells generate thousands of chemically distinct lipids (Sud et al., 2007) from which the membranes of organelles, including the PM, can be built. These lipids confer these membranes with not only different physical properties but also host distinct functions. Interestingly, the mapping of the distribution of lipids across organelle membranes reflects the secretory pathway established by evolution (van Meer, 1989; Voelker, 1991). Key in preserving the homeostatic balance of cell membranes is the machinery that distributes lipids—with high sensitivity—from their place of synthesis, mostly in the endoplasmic reticulum, to their target membrane and eventual disposal (Blom, Somerharju, & Ikonen, 2011; Holthuis & Menon, 2014; Lev, 2010; Voelker, 1990). This machinery, however, is constantly battling equilibrium, where entropy of mixing drives homogenization (Callan-Jones, Sorre, & Bassereau, 2011). Lipid architecture, lipid–lipid and lipid–protein interactions, as we are finding out, have built-in passive regulatory roles in maintaining a relatively stable membrane organization over hours or days, therefore aiding in the cost of its maintenance. It has been recognized for many decades that understanding these interactions including how they change—rapidly—as a result of signaling (Doktorova et al., 2020) or fail with the onset of disease (Goldberg & Riordan, 1986; Maxfield & Tabas, 2005) will provide a molecular-based tool to detect and possibly address health in the membrane.

TR-SANS clearly stands out as a powerful technique to follow the movement of lipids between and within membranes and extract the time-scales and energetics for maintaining compositionally distinct membranes. This is critical information to map out the built-in mechanisms of passive regulation of membranes. In addition, it can also detect and follow in situ how these gradients are affected by the presence of peptides, proteins or other molecules. The work in this regard is new because of breakthroughs in identifying and resolving experimental biases, on the one hand, but also in producing systems that had not been accessible before, like asymmetric vesicles. Deuteration capabilities will expand the complexity of the systems studied, which together with advances in protein expression and purification of membrane proteins, will open the road for detailed work on the lipid transport machinery itself, such as the action of flippases and scamblases, which is ultimately directly responsible of maintaining the homeostatic state of cell membranes.

As already alluded to, scattering techniques have to be pursued in combination with other techniques (e.g., NMR, calorimetry, gas chromatography to name a few (Liu et al., 2020)) that provide additional information for data analysis or checks. Most importantly, as recently shown by Marx et al. (Marx, Frewein, et al., 2021; Marx, Semeraro, et al., 2021), the goal is to seek congruency between experiments at different scales (molecular models to cells). It is certainly now common or even expected that molecular biology studies be combined with a powerful theoretical tool like MD simulations and scattering studies of membranes are no exception (Ashkar et al., 2018; Gupta & Ashkar, 2021). In hand with experiements, advances in MD simulations are allowing current efforts to simulate physiologically relevant membranes (Khakbaz & Klauda, 2015; Marrink et al., 2019).

The scattering techniques described herein are provided by large-scale government sponsored facilities, and appear hard to access, but they are not. Access is free once a peer-reviewed proposal has been allotted beamtime. The scientific staff will help new users build scattering expertize in support of their research program. Admittedly, the largest barrier to using scattering more broadly is the need of extensive data modeling. There is a significant effort to utilize complex algorithms (Treece et al., 2019) to streamline the analysis and to conceivably incorporate multiple data sets from different experimental techniques. This work is being provided by the scattering community at large as well as directly from scientific staff at facilities to make the technique accessible to none experts (Doucet et al., 2021; Lewis-Laurent, Doktorova, Heberle, & Marquardt, 2021). Hence, the future is bright for scattering techniques as they become utilized to their full potential to understand cell membranes and beyond.

Acknowledgments

This work utilized the facilities at the National Institute for Standards and Technology, supported in part by the National Science Foundation under agreement no. DMR-0454672. Commercial materials or equipment if identified in this paper do not imply recommendation or endorsement by the National Institute of Standards and Technology nor should be identified as the best available for the purpose. The authors thank the Institute Laue Langevin for providing neutron beam time under DOI:10.5291/ILL-DATA.9-13-823. U.P.-S. additionally acknowledges travel support from the ILL to promote scientific collaboration with L. Porcar, Y. Gerelli, and G. Fragneto. U.P.-S. gratefully acknowledges the support from the NSF CAREER award DMR-1753238.

References

Allhusen, J. S., & Conboy, J. C. (2017). The ins and outs of lipid Flip-flop. *Accounts of Chemical Research*, *50*(1), 58–65. https://doi.org/10.1021/acs.accounts.6b00435.

Anglin, T. C., & Conboy, J. C. (2009). Kinetics and thermodynamics of flip-flop in binary phospholipid membranes measured by sum-frequency vibrational spectroscopy. *Biochemistry*, *48*(43), 10220–10234. https://doi.org/10.1021/bi901096j.

Ano, Y., & Nakayama, H. (2018). Preventive effects of dairy products on dementia and the underlying mechanisms. *International Journal of Molecular Sciences*, *19*(7). https://doi.org/10.3390/ijms19071927.

Ashkar, R., Bilheux, H. Z., Bordallo, H., Briber, R., Callaway, D. J. E., Cheng, X., et al. (2018). Neutron scattering in the biological sciences: Progress and prospects. *Acta Crystallographica Section D: Structural Biology*, *74*(Pt 12), 1129–1168. https://doi.org/10.1107/S2059798318017503.

Atkovska, K., Klingler, J., Oberwinkler, J., Keller, S., & Hub, J. S. (2018). Rationalizing steroid interactions with lipid membranes: Conformations, partitioning, and kinetics. *ACS Central Science*, *4*(9), 1155–1165. https://doi.org/10.1021/acscentsci.8b00332.

Backer, J. M., & Dawidowicz, E. A. (1981). Transmembrane movement of cholesterol in small unilamellar vesicles detected by cholesterol oxidase. *The Journal of Biological Chemistry*, *256*(2), 586–588 (Retrieved from http://www.ncbi.nlm.nih.gov/pubmed/6935193).

Baral, S., Levental, I., & Lyman, E. (2020). Composition dependence of cholesterol flip-flop rates in physiological mixtures. *Chemistry and Physics of Lipids*, *232*, 104967. https://doi.org/10.1016/j.chemphyslip.2020.104967.

Bennett, W. F., MacCallum, J. L., Hinner, M. J., Marrink, S. J., & Tieleman, D. P. (2009). Molecular view of cholesterol flip-flop and chemical potential in different membrane environments. *Journal of the American Chemical Society*, *131*(35), 12714–12720. https://doi.org/10.1021/ja903529f.

Bennett, W. F., MacCallum, J. L., & Tieleman, D. P. (2009). Thermodynamic analysis of the effect of cholesterol on dipalmitoylphosphatidylcholine lipid membranes. *Journal of the American Chemical Society*, *131*(5), 1972–1978. https://doi.org/10.1021/ja808541r.

Bennett, W. F., & Tieleman, D. P. (2012). Molecular simulation of rapid translocation of cholesterol, diacylglycerol, and ceramide in model raft and nonraft membranes. *Journal of Lipid Research*, *53*(3), 421–429. https://doi.org/10.1194/jlr.M022491.

Blom, T., Somerharju, P., & Ikonen, E. (2011). Synthesis and biosynthetic trafficking of membrane lipids. *Cold Spring Harbor Perspectives in Biology*, *3*(8), a004713. https://doi.org/10.1101/cshperspect.a004713.

Blumer, M., Harris, S., Li, M., Martinez, L., Untereiner, M., Saeta, P. N., et al. (2020). Simulations of asymmetric membranes illustrate cooperative leaflet coupling and lipid adaptability. *Frontiers in Cell and Development Biology*, *8*, 575. https://doi.org/10.3389/fcell.2020.00575.

Brasaemle, D. L., Robertson, A. D., & Attie, A. D. (1988). Transbilayer movement of cholesterol in the human-erythrocyte membrane. *Journal of Lipid Research*, *29*(4), 481–489.

Breidigan, J. M., Krzyzanowski, N., Liu, Y., Porcar, L., & Perez-Salas, U. (2017). Influence of the membrane environment on cholesterol transfer. *Journal of Lipid Research*, *58*(12), 2255–2263. https://doi.org/10.1194/jlr.M077909.

Bretscher, M. S. (1972). Asymmetrical lipid bilayer structure for biological membranes. *Nature: New Biology*, *236*(61), 11–12 (Retrieved from http://www.ncbi.nlm.nih.gov/pubmed/4502419).

Bruckner, R. J., Mansy, S. S., Ricardo, A., Mahadevan, L., & Szostak, J. W. (2009). Flip-flop-induced relaxation of bending energy: Implications for membrane remodeling. *Biophysical Journal*, *97*(12), 3113–3122. https://doi.org/10.1016/j.bpj.2009.09.025.

Bryant, G., Taylor, M. B., Darwish, T. A., Krause-Heuer, A. M., Kent, B., & Garvey, C. J. (2019). Effect of deuteration on the phase behaviour and structure of lamellar phases of phosphatidylcholines - deuterated lipids as proxies for the physical properties of native bilayers. *Colloids and Surfaces. B, Biointerfaces*, *177*, 196–203. https://doi.org/10.1016/j.colsurfb.2019.01.040.

Burnham, K. P., & Anderson, D. R. (2002). *Model selection and inference: A practical information-theoretic approach* (2nd ed.). New York: Springer-Verlag.

Buwaneka, P., Ralko, A., Liu, S.-L., & Cho, W. (2021). Evaluation of the available cholesterol concentration in the inner leaflet of the plasma membrane of mammalian cells. *Journal of Lipid Research*, *62*, 100084. https://doi.org/10.1016/j.jlr.2021.100084.

Callan-Jones, A., Sorre, B., & Bassereau, P. (2011). Curvature-driven lipid sorting in biomembranes. *Cold Spring Harbor Perspectives in Biology*, *3*(2). https://doi.org/10.1101/cshperspect.a004648.

Carbone, C. B., Kern, N., Fernandes, R. A., Hui, E., Su, X., Garcia, K. C., et al. (2017). In vitro reconstitution of T cell receptor-mediated segregation of the CD45 phosphatase. *Proceedings of the National Academy of Sciences of the United States of America*, *114*(44), E9338–E9345. https://doi.org/10.1073/pnas.1710358114.

Carmena, R., Duriez, P., & Fruchart, J. C. (2004). Atherogenic lipoprotein particles in atherosclerosis. *Circulation*, *109*(23 Suppl 1), III2–7. https://doi.org/10.1161/01.CIR.0000131511.50734.44.

Chakraborty, S., Doktorova, M., Molugu, T. R., Heberle, F. A., Scott, H. L., Dzikovski, B., et al. (2020). How cholesterol stiffens unsaturated lipid membranes. *Proceedings of the National Academy of Sciences*, *117*(36), 21896–21905. https://doi.org/10.1073/pnas.2004807117.

Choi, S. H., Bates, F. S., & Lodge, T. P. (2011). Molecular exchange in ordered Diblock copolymer micelles. *Macromolecules*, *44*(9), 3594–3604. https://doi.org/10.1021/ma102788v.

Choi, S. H., Lodge, T. P., & Bates, F. S. (2010). Mechanism of molecular exchange in diblock copolymer micelles: Hypersensitivity to core chain length. *Physical Review Letters*, *104*(4), 047802. https://doi.org/10.1103/PhysRevLett.104.047802.

Coisne, C., Tilloy, S., Monflier, E., Wils, D., Fenart, L., & Gosselet, F. (2016). Cyclodextrins as emerging therapeutic tools in the treatment of cholesterol-associated vascular and neurodegenerative diseases. *Molecules*, *21*(12). https://doi.org/10.3390/molecules21121748.

Courtney, K. C., Pezeshkian, W., Raghupathy, R., Zhang, C., Darbyson, A., Ipsen, J. H., et al. (2018). C24 sphingolipids govern the Transbilayer asymmetry of cholesterol and lateral organization of model and live-cell plasma membranes. *Cell Reports*, *24*(4), 1037–1049. https://doi.org/10.1016/j.celrep.2018.06.104.

Cuevas Arenas, R., Danielczak, B., Martel, A., Porcar, L., Breyton, C., Ebel, C., et al. (2017). Fast collisional lipid transfer among polymer-bounded Nanodiscs. *Scientific Reports*, 7, 45875. https://doi.org/10.1038/srep45875.

Darwish, T. A., Luks, E., Moraes, G., Yepuri, N. R., Holden, P. J., & James, M. (2013). Synthesis of deuterated [D32]oleic acid and its phospholipid derivative [D64]dioleoyl-sn-glycero-3-phosphocholine. *Journal of Labelled Compounds and Radiopharmaceuticals*, *56*(9–10), 520–529. https://doi.org/10.1002/jlcr.3088.

Dawoud, M., & Abdou, R. (2021). Ion exchange column technique as a novel method for evaluating the release of docetaxel from different lipid nanoparticles. *Drug Delivery and Translational Research*. https://doi.org/10.1007/s13346-021-00937-2.

Denisov, I. G., & Sligar, S. G. (2017). Nanodiscs in membrane biochemistry and biophysics. *Chemical Reviews*, *117*(6), 4669–4713. https://doi.org/10.1021/acs.chemrev.6b00690.

Devaux, P. F., & Morris, R. (2004). Transmembrane asymmetry and lateral domains in biological membranes. *Traffic*, *5*(4), 241–246. https://doi.org/10.1111/j.1600-0854.2004.0170.x.

Dickey, A. N., & Faller, R. (2007). How alcohol chain-length and concentration modulate hydrogen bond formation in a lipid bilayer. *Biophysical Journal*, *92*(7), 2366–2376. https://doi.org/10.1529/biophysj.106.097022.

Doktorova, M., Heberle, F. A., Eicher, B., Standaert, R. F., Katsaras, J., London, E., et al. (2018). Preparation of asymmetric phospholipid vesicles for use as cell membrane models. *Nature Protocols*, *13*(9), 2086–2101. https://doi.org/10.1038/s41596-018-0033-6.

Doktorova, M., Symons, J. L., & Levental, I. (2020). Structural and functional consequences of reversible lipid asymmetry in living membranes. *Nature Chemical Biology*, *16*(12), 1321–1330. https://doi.org/10.1038/s41589-020-00688-0.

Doucet, M., Cho, J. H., Alina, G., Attala, Z., Bakker, J., Bouwman, W., et al. (2021). SASview. *Zenodo*.

Efimova, Y. M., Haemers, S., Wierczinski, B., Norde, W., & van Well, A. A. (2007). Stability of globular proteins in H2O and D2O. *Biopolymers*, *85*(3), 264–273. https://doi.org/10.1002/bip.20645.

Eicher, B., Heberle, F. A., Marquardt, D., Rechberger, G. N., Katsaras, J., & Pabst, G. (2017). Joint small-angle X-ray and neutron scattering data analysis of asymmetric lipid vesicles. *Journal of Applied Crystallography*, *50*(Pt 2), 419–429. https://doi.org/10.1107/S1600576717000656.

Eyring, H. (1935). The activated complex in chemical reactions. *Journal of Chemical Physics*, *3*(2). https://doi.org/10.1063/1.1749604.

Fadok, V. A., Voelker, D. R., Campbell, P. A., Cohen, J. J., Bratton, D. L., & Henson, P. M. (1992). Exposure of phosphatidylserine on the surface of apoptotic lymphocytes triggers specific recognition and removal by macrophages. *Journal of Immunology*, *148*(7), 2207–2216 (Retrieved from https://www.ncbi.nlm.nih.gov/pubmed/1545126).

Fatmi, S., Taouzinet, L., Skiba, M., & Iguer-Ouada, M. (2021). The use of Cyclodextrin or its complexes as a potential treatment against the 2019 novel coronavirus: A mini-review. *Current Drug Delivery*, *18*(4), 382–386. https://doi.org/10.2174/1567201817666200917124241.

Fourmentin, S. (2018). *Cyclodextrin applications in medicine, food, environment and liquid crystals*. New York, NY: Springer Berlin Heidelberg.

Garg, S., Castro-Roman, F., Porcar, L., Butler, P., Bautista, P. J., Krzyzanowski, N., et al. (2014). Cholesterol solubility limit in lipid membranes probed by small angle neutron scattering and MD simulations. *Soft Matter*, *10*(46), 9313–9317. https://doi.org/10.1039/c4sm01219d.

Garg, S., Porcar, L., Woodka, A. C., Butler, P. D., & Perez-Salas, U. (2011). Noninvasive neutron scattering measurements reveal slower cholesterol transport in model lipid membranes. *Biophysical Journal, 101*(2), 370–377. https://doi.org/10.1016/j.bpj.2011.06.014.

Garg, S., Porcar, L., Woodka, A. C., Butler, P. D., & Perez-Salas, U. (2012). Response to "how slow is the Transbilayer diffusion (Flip-flop) of cholesterol?". *Biophysical Journal, 102*(4), 947–949.

Gerelli, Y. (2019). Phase transitions in a single supported phospholipid bilayer: Real-time determination by neutron reflectometry. *Physical Review Letters, 122*(24), 248101. https://doi.org/10.1103/PhysRevLett.122.248101.

Gerelli, Y., Porcar, L., & Fragneto, G. (2012). Lipid rearrangement in DSPC/DMPC bilayers: A neutron reflectometry study. *Langmuir, 28*(45), 15922–15928. https://doi.org/10.1021/la303662e.

Gerelli, Y., Porcar, L., Lombardi, L., & Fragneto, G. (2013). Lipid exchange and Flip-flop in solid supported bilayers. *Langmuir, 29*(41), 12762–12769. https://doi.org/10.1021/La402708u.

Goldberg, D. M., & Riordan, J. R. (1986). Role of membranes in disease. *Clinical Physiology and Biochemistry, 4*(5), 305–336 (Retrieved from https://www.ncbi.nlm.nih.gov/pubmed/3022980).

Goldstein, D. B. (1986). Effect of alcohol on cellular membranes. *Annals of Emergency Medicine, 15*(9), 1013–1018. https://doi.org/10.1016/s0196-0644(86)80120-2.

Gu, R. X., Baoukina, S., & Tieleman, D. P. (2019). Cholesterol Flip-flop in heterogeneous membranes. *Journal of Chemical Theory and Computation, 15*(3), 2064–2070. https://doi.org/10.1021/acs.jctc.8b00933.

Gupta, S., & Ashkar, R. (2021). The dynamic face of lipid membranes. *Soft Matter, 17*(29), 6910–6928. https://doi.org/10.1039/d1sm00646k.

Gupta, A., Korte, T., Herrmann, A., & Wohland, T. (2020). Plasma membrane asymmetry of lipid organization: Fluorescence lifetime microscopy and correlation spectroscopy analysis. *Journal of Lipid Research, 61*(2), 252–266. https://doi.org/10.1194/jlr.D119000364.

Hamilton, J. A. (2003). Fast flip-flop of cholesterol and fatty acids in membranes: Implications for membrane transport proteins. *Current Opinion in Lipidology, 14*(3), 263–271. https://doi.org/10.1097/01.mol.0000073507.41685.9b.

Hamley, I. W. (2021). *Small-angle scattering: Theory, instrumentation, data and applications* (1st ed.). Hoboken, NJ: Wiley.

Heberle, F. A., Marquardt, D., Doktorova, M., Geier, B., Standaert, R. F., Heftberger, P., et al. (2016). Subnanometer structure of an asymmetric model membrane: Interleaflet coupling influences domain properties. *Langmuir, 32*(20), 5195–5200. https://doi.org/10.1021/acs.langmuir.5b04562.

Heberle, F. A., & Pabst, G. (2017). Complex biomembrane mimetics on the subnanometer scale. *Biophysical Reviews, 9*(4), 353–373. https://doi.org/10.1007/s12551-017-0275-5.

Heberle, F. A., Petruzielo, R. S., Pan, J., Drazba, P., Kucerka, N., Standaert, R. F., et al. (2013). Bilayer thickness mismatch controls domain size in model membranes. *Journal of the American Chemical Society, 135*(18), 6853–6859. https://doi.org/10.1021/ja3113615.

Heinrich, F., Kienzle, P. A., Hoogerheide, D. P., & Losche, M. (2020). Information gain from isotopic contrast variation in neutron reflectometry on protein-membrane complex structures. *Journal of Applied Crystallography, 53*(Pt 3), 800–810. https://doi.org/10.1107/S1600576720005634.

Holthuis, J. C., & Levine, T. P. (2005). Lipid traffic: Floppy drives and a superhighway. *Nature Reviews. Molecular Cell Biology*, *6*(3), 209–220. https://doi.org/10.1038/nrm1591.

Holthuis, J. C., & Menon, A. K. (2014). Lipid landscapes and pipelines in membrane homeostasis. *Nature*, *510*(7503), 48–57. https://doi.org/10.1038/nature13474.

Homan, R., & Pownall, H. J. (1988). Transbilayer diffusion of phospholipids: Dependence on headgroup structure and acyl chain length. *Biochimica et Biophysica Acta*, *938*(2), 155–166 (Retrieved from http://www.ncbi.nlm.nih.gov/pubmed/3342229).

Huan, Y., Kong, Q., Mou, H., & Yi, H. (2020). Antimicrobial peptides: Classification, design, application and research Progress in multiple fields. *Frontiers in Microbiology*, *11*, 582779. https://doi.org/10.3389/fmicb.2020.582779.

Huang, J., Buboltz, J. T., & Feigenson, G. W. (1999). Maximum solubility of cholesterol in phosphatidylcholine and phosphatidylethanolamine bilayers. *Biochimica et Biophysica Acta*, *1417*(1), 89–100. https://doi.org/10.1016/s0005-2736(98)00260-0.

Ikonen, E., & Jansen, M. (2008). Cellular sterol trafficking and metabolism: Spotlight on structure. *Current Opinion in Cell Biology*, *20*(4), 371–377. https://doi.org/10.1016/j.ceb.2008.03.017.

Jing, H., Wang, Y., Desai, P. R., Ramamurthi, K. S., & Das, S. (2020). Lipid flip-flop and desorption from supported lipid bilayers is independent of curvature. *PLoS One*, *15*(12), e0244460. https://doi.org/10.1371/journal.pone.0244460.

Johansen, N. T., Pedersen, M. C., Porcar, L., Martel, A., & Arleth, L. (2018). Introducing SEC-SANS for studies of complex self-organized biological systems. *Acta Crystallographica Section D: Structural Biology*, *74*(Pt 12), 1178–1191. https://doi.org/10.1107/S2059798318007180.

John, K., Kubelt, J., Muller, P., Wustner, D., & Herrmann, A. (2002). Rapid transbilayer movement of the fluorescent sterol dehydroergosterol in lipid membranes. *Biophysical Journal*, *83*(3), 1525–1534. https://doi.org/10.1016/S0006-3495(02)73922-2.

Jones, J. D., & Thompson, T. E. (1989). Spontaneous phosphatidylcholine transfer by collision between vesicles at high lipid concentration. *Biochemistry*, *28*(1), 129–134. https://doi.org/10.1021/bi00427a019.

Kaplan, M. R., & Simoni, R. D. (1985). Intracellular transport of phosphatidylcholine to the plasma membrane. *The Journal of Cell Biology*, *101*(2), 441–445 (Retrieved from http://www.ncbi.nlm.nih.gov/pubmed/4040519).

Khakbaz, P., & Klauda, J. B. (2015). Probing the importance of lipid diversity in cell membranes via molecular simulation. *Chemistry and Physics of Lipids*, *192*, 12–22. https://doi.org/10.1016/j.chemphyslip.2015.08.003.

Kobayashi, T., & Menon, A. K. (2018). Transbilayer lipid asymmetry. *Current Biology*, *28*(8), R386–R391. https://doi.org/10.1016/j.cub.2018.01.007.

Kucerka, N., Nieh, M. P., & Katsaras, J. (2011). Fluid phase lipid areas and bilayer thicknesses of commonly used phosphatidylcholines as a function of temperature. *Biochimica et Biophysica Acta*, *1808*(11), 2761–2771. https://doi.org/10.1016/j.bbamem.2011.07.022.

Laidler, K. J., & King, M. C. (1983). The development of transition-state theory. *Journal of Physical Chemistry*, *87*, 2657–2664.

Lev, S. (2006). Lipid homoeostasis and Golgi secretory function. *Biochemical Society Transactions*, *34*(Pt 3), 363–366. https://doi.org/10.1042/BST0340363.

Lev, S. (2010). Non-vesicular lipid transport by lipid-transfer proteins and beyond. *Nature Reviews. Molecular Cell Biology*, *11*(10), 739–750. https://doi.org/10.1038/nrm2971.

Leventis, R., & Silvius, J. R. (2001). Use of cyclodextrins to monitor transbilayer movement and differential lipid affinities of cholesterol. *Biophysical Journal*, *81*(4), 2257–2267. https://doi.org/10.1016/S0006-3495(01)75873-0.

Lewis-Laurent, A., Doktorova, M., Heberle, F. A., & Marquardt, D. (2021). Vesicle viewer: Online analysis of small angle scattering from lipid vesicles. *Biophysical Journal*, (ChemRxiv. Cambridge: Cambridge Open Engage; 2021; This content is a preprint and has not been peer-reviewed).

Liu, J., & Conboy, J. C. (2005). 1,2-diacyl-phosphatidylcholine flip-flop measured directly by sum-frequency vibrational spectroscopy. *Biophysical Journal, 89*(4), 2522–2532. https://doi.org/10.1529/biophysj.105.065672.

Liu, Y., Kelley, E. G., Batchu, K. C., Porcar, L., & Perez-Salas, U. (2020). Creating asymmetric phospholipid vesicles via exchange with lipid-coated silica nanoparticles. *Langmuir, 36*(30), 8865–8873. https://doi.org/10.1021/acs.langmuir.0c01188.

Liu, S. L., Sheng, R., Jung, J. H., Wang, L., Stec, E., O'Connor, M. J., et al. (2017). Orthogonal lipid sensors identify transbilayer asymmetry of plasma membrane cholesterol. *Nature Chemical Biology, 13*(3), 268–274. https://doi.org/10.1038/nchembio.2268.

Lohner, K. (2017). Membrane-active antimicrobial peptides as template structures for novel antibiotic agents. *Current Topics in Medicinal Chemistry, 17*(5), 508–519.

London, E. (2019). Membrane structure-function insights from asymmetric lipid vesicles. *Accounts of Chemical Research, 52*(8), 2382–2391. https://doi.org/10.1021/acs.accounts.9b00300.

Luchini, A., Delhom, R., Deme, B., Laux, V., Moulin, M., Haertlein, M., et al. (2018). The impact of deuteration on natural and synthetic lipids: A neutron diffraction study. *Colloids and Surfaces. B, Biointerfaces, 168*, 126–133. https://doi.org/10.1016/j.colsurfb.2018.02.009.

Lund, R., Willner, L., Stellbrink, J., Lindner, P., & Richter, D. (2006). Logarithmic chain-exchange kinetics of diblock copolymer micelles. *Physical Review Letters, 96*(6) (068302. Retrieved from http://www.ncbi.nlm.nih.gov/pubmed/16606054).

Lusis, A. J. (2000). Atherosclerosis. *Nature, 407*(6801), 233–241. https://doi.org/10.1038/35025203.

Ly, H. V., & Longo, M. L. (2004). The influence of short-chain alcohols on interfacial tension, mechanical properties, area/molecule, and permeability of fluid lipid bilayers. *Biophysical Journal, 87*(2), 1013–1033. https://doi.org/10.1529/biophysj.103.034280.

Mahieu, E., & Gabel, F. (2018). Biological small-angle neutron scattering: Recent results and development. *Acta Crystallographica Section D: Structural Biology, 74*(Pt 8), 715–726. https://doi.org/10.1107/S2059798318005016.

Maric, S., Lind, T. K., Raida, M. R., Bengtsson, E., Fredrikson, G. N., Rogers, S., et al. (2019). Time-resolved small-angle neutron scattering as a probe for the dynamics of lipid exchange between human lipoproteins and naturally derived membranes. *Scientific Reports, 9*(1), 7591. https://doi.org/10.1038/s41598-019-43713-6.

Markones, M., Drechsler, C., Kaiser, M., Kalie, L., Heerklotz, H., & Fiedler, S. (2018). Engineering asymmetric lipid vesicles: Accurate and convenient control of the outer leaflet lipid composition. *Langmuir, 34*(5), 1999–2005. https://doi.org/10.1021/acs.langmuir.7b03189.

Markones, M., Fippel, A., Kaiser, M., Drechsler, C., Hunte, C., & Heerklotz, H. (2020). Stairway to asymmetry: Five steps to lipid-asymmetric Proteoliposomes. *Biophysical Journal, 118*(2), 294–302. https://doi.org/10.1016/j.bpj.2019.10.043.

Marquardt, D., Heberle, F. A., Miti, T., Eicher, B., London, E., Katsaras, J., et al. (2017). 1H NMR shows slow phospholipid Flip-flop in gel and fluid bilayers. *Langmuir*. https://doi.org/10.1021/acs.langmuir.6b04485.

Marrink, S. J., Corradi, V., Souza, P. C. T., Ingolfsson, H. I., Tieleman, D. P., & Sansom, M. S. P. (2019). Computational modeling of realistic cell membranes. *Chemical Reviews, 119*(9), 6184–6226. https://doi.org/10.1021/acs.chemrev.8b00460.

Marx, L., Frewein, M. P. K., Semeraro, E. F., Rechberger, G. N., Lohner, K., Porcar, L., et al. (2021). Antimicrobial peptide activity in asymmetric bacterial membrane mimics. *Faraday Discussions*. https://doi.org/10.1039/D1FD00039J.

Marx, L., Semeraro, E. F., Mandl, J., Kremser, J., Frewein, M. P., Malanovic, N., et al. (2021). Bridging the antimicrobial activity of two Lactoferricin derivatives in E. coli and lipid-only membranes. *Frontiers in Medical Technology, 3*(2). https://doi.org/10.3389/fmedt.2021.625975.

Matencio, A., Navarro-Orcajada, S., Gonzalez-Ramon, A., Garcia-Carmona, F., & Lopez-Nicolas, J. M. (2020). Recent advances in the treatment of Niemann pick disease type C: A mini-review. *International Journal of Pharmaceutics*, *584*, 119440. https://doi.org/10.1016/j.ijpharm.2020.119440.

Maxfield, F. R., & Tabas, I. (2005). Role of cholesterol and lipid organization in disease. *Nature*, *438*(7068), 612–621. https://doi.org/10.1038/nature04399.

McLean, L. R., & Phillips, M. C. (1981). Mechanism of cholesterol and phosphatidylcholine exchange or transfer between unilamellar vesicles. *Biochemistry*, *20*(10), 2893–2900. https://doi.org/10.1021/bi00513a028.

Mellman, I., & Emr, S. D. (2013). A Nobel prize for membrane traffic: Vesicles find their journey's end. *The Journal of Cell Biology*, *203*(4), 559–561. https://doi.org/10.1083/jcb.201310134.

Menon, A. K. (2018). Sterol gradients in cells. *Current Opinion in Cell Biology*, *53*, 37–43. https://doi.org/10.1016/j.ceb.2018.04.012.

Mondal, M., Mesmin, B., Mukherjee, S., & Maxfield, F. R. (2009). Sterols are mainly in the cytoplasmic leaflet of the plasma membrane and the endocytic recycling compartment in CHO cells. *Molecular Biology of the Cell*, *20*(2), 581–588. https://doi.org/10.1091/mbc.E08-07-0785.

Moulin, M., Strohmeier, G. A., Hirz, M., Thompson, K. C., Rennie, A. R., Campbell, R. A., et al. (2018). Perdeuteration of cholesterol for neutron scattering applications using recombinant Pichia pastoris. *Chemistry and Physics of Lipids*, *212*, 80–87. https://doi.org/10.1016/j.chemphyslip.2018.01.006.

Nakano, M. (2019). Evaluation of Interbilayer and Transbilayer transfer dynamics of phospholipids using time-resolved small-angle neutron scattering. *Chemical & Pharmaceutical Bulletin (Tokyo)*, *67*(4), 316–320. https://doi.org/10.1248/cpb.c18-00942.

Nakano, M., Fukuda, M., Kudo, T., Endo, H., & Handa, T. (2007). Determination of interbilayer and transbilayer lipid transfers by time-resolved small-angle neutron scattering. *Physical Review Letters*, *98*(23). https://doi.org/10.1103/Physrevlett.98.238101.

Nakano, M., Fukuda, M., Kudo, T., Matsuzaki, N., Azuma, T., Sekine, K., et al. (2009). Flip-flop of phospholipids in vesicles: Kinetic analysis with time-resolved small-angle neutron scattering. *The Journal of Physical Chemistry. B*, *113*(19), 6745–6748. https://doi.org/10.1021/jp900913w.

Nakano, M., Fukuda, M., Kudo, T., Miyazaki, M., Wada, Y., Matsuzaki, N., et al. (2009). Static and dynamic properties of phospholipid bilayer nanodiscs. *Journal of the American Chemical Society*, *131*(23), 8308–8312. https://doi.org/10.1021/ja9017013.

Nakao, H., Kimura, Y., Sakai, A., Ikeda, K., & Nakano, M. (2021). Development of membrane-insertable lipid scrambling peptides: A time-resolved small-angle neutron scattering study. *Structural Dynamics*, *8*(2), 024301. https://doi.org/10.1063/4.0000045.

Nguyen, M. H. L., DiPasquale, M., Rickeard, B. W., Doktorova, M., Heberle, F. A., Scott, H. L., et al. (2019). Peptide-induced lipid Flip-flop in asymmetric liposomes measured by small angle neutron scattering. *Langmuir*, *35*(36), 11735–11744. https://doi.org/10.1021/acs.langmuir.9b01625.

Nguyen, M. H. L., DiPasquale, M., Rickeard, B. W., Stanley, C. B., Kelley, E. G., & Marquardt, D. (2019). Methanol accelerates DMPC Flip-flop and transfer: A SANS study on lipid dynamics. *Biophysical Journal*, *116*(5), 755–759. https://doi.org/10.1016/j.bpj.2019.01.021.

Nguyen, M. H. L., DiPasquale, M., Rickeard, B. W., Yip, C. G., Greco, K. N., Kelley, E. G., et al. (2021). Time-resolved SANS reveals pore-forming peptides cause rapid lipid reorganization. *New Journal of Chemistry*, *45*(1), 447–456. https://doi.org/10.1039/D0NJ04717A.

Nickels, J. D., Chatterjee, S., Stanley, C. B., Qian, S., Cheng, X., Myles, D. A. A., et al. (2017). The in vivo structure of biological membranes and evidence for lipid domains. *PLoS Biology*, *15*(5), e2002214. https://doi.org/10.1371/journal.pbio.2002214.

Nickels, J. D., Cheng, X., Mostofian, B., Stanley, C., Lindner, B., Heberle, F. A., et al. (2015). Mechanical properties of Nanoscopic lipid domains. *Journal of the American Chemical Society*, *137*(50), 15772–15780. https://doi.org/10.1021/jacs.5b08894.

Nicolson, G. L. (2014). The fluid-mosaic model of membrane structure: Still relevant to understanding the structure, function and dynamics of biological membranes after more than 40 years. *Biochimica et Biophysica Acta*, *1838*(6), 1451–1466. https://doi.org/10.1016/j.bbamem.2013.10.019.

Nielsen, J. E., Bjørnestad, V. A., Pipich, V., Jenssen, H., & Lund, R. (2021). Beyond structural models for the mode of action: How natural antimicrobial peptides affect lipid transport. *Journal of Colloid and Interface Science*, *582*, 793–802. https://doi.org/10.1016/j.jcis.2020.08.094.

Nielsen, J. E., Prévost, S. F., Jenssen, H., & Lund, R. (2020). Impact of antimicrobial peptides on E. coli-mimicking lipid model membranes: Correlating structural and dynamic effects using scattering methods. *Faraday Discussions*. https://doi.org/10.1039/D0FD00046A.

Nyholm, T. K. M., Jaikishan, S., Engberg, O., Hautala, V., & Slotte, J. P. (2019). The affinity of sterols for different phospholipid classes and its impact on lateral segregation. *Biophysical Journal*, *116*(2), 296–307. https://doi.org/10.1016/j.bpj.2018.11.3135.

Opdenkamp, J. A. F. (1979). Lipid asymmetry in membranes. *Annual Review of Biochemistry*, *48*, 47–71. https://doi.org/10.1146/Annurev.Bi.48.070179.000403.

Parisio, G., Sperotto, M. M., & Ferrarini, A. (2012). Flip-flop of steroids in phospholipid bilayers: Effects of the chemical structure on transbilayer diffusion. *Journal of the American Chemical Society*, *134*(29), 12198–12208. https://doi.org/10.1021/ja304007t.

Porcar, L., & Gerelli, Y. (2020). On the lipid flip-flop and phase transition coupling. *Soft Matter*, *16*(33), 7696–7703. https://doi.org/10.1039/d0sm01161d.

Poznansky, M. J., & Lange, Y. (1978). Transbilayer movement of cholesterol in phospholipid vesicles under equilibrium and non-equilibrium conditions. *Biochimica et Biophysica Acta*, *506*(2), 256–264 (Retrieved from http://www.ncbi.nlm.nih.gov/pubmed/620032).

Qian, S., Sharma, V. K., & Clifton, L. A. (2020). Understanding the structure and dynamics of complex biomembrane interactions by neutron scattering techniques. *Langmuir*, *36*(50), 15189–15211. https://doi.org/10.1021/acs.langmuir.0c02516.

Qiu, N., Li, X., & Liu, J. (2017). Application of cyclodextrins in cancer treatment. *Journal of Inclusion Phenomena and Macrocyclic Chemistry*, *89*(3), 229–246. https://doi.org/10.1007/s10847-017-0752-2.

Rickeard, B. W., Nguyen, M. H. L., DiPasquale, M., Yip, C. G., Baker, H., Heberle, F. A., et al. (2020). Transverse lipid organization dictates bending fluctuations in model plasma membranes. *Nanoscale*, *12*(3), 1438–1447. https://doi.org/10.1039/c9nr07977g.

Rodrigueza, W. V., Wheeler, J. J., Klimuk, S. K., Kitson, C. N., & Hope, M. J. (1995). Transbilayer movement and net flux of cholesterol and cholesterol sulfate between liposomal membranes. *Biochemistry*, *34*(18), 6208–6217. https://doi.org/10.1021/Bi00018a025.

Ryan, R. O. (2010). Nanobiotechnology applications of reconstituted high density lipoprotein. *Journal of Nanobiotechnology*, *8*(1), 28. https://doi.org/10.1186/1477-3155-8-28.

Sahoo, S., Adamiak, M., Mathiyalagan, P., Kenneweg, F., Kafert-Kasting, S., & Thum, T. (2021). Therapeutic and diagnostic translation of extracellular vesicles in cardiovascular diseases: Roadmap to the clinic. *Circulation*, *143*(14), 1426–1449. https://doi.org/10.1161/CIRCULATIONAHA.120.049254.

Sapay, N., Bennett, W. F. D., & Tieleman, D. P. (2009). Thermodynamics of flip-flop and desorption for a systematic series of phosphatidylcholine lipids. *Soft Matter*, *5*(17), 3295–3302. https://doi.org/10.1039/b902376c.

Schroeder, F., Frolov, A. A., Murphy, E. J., Atshaves, B. P., Jefferson, J. R., Pu, L. X., et al. (1996). Recent advances in membrane cholesterol domain dynamics and intracellular cholesterol trafficking. *Proceedings of the Society for Experimental Biology and Medicine*, *213*(2), 150–177.

Scott, H. L., Kennison, K. B., Enoki, T. A., Doktorova, M., Kinnun, J. J., Heberle, F. A., et al. (2021). Model membrane systems used to study plasma membrane lipid asymmetry. *Symmetry, 13*(8), 1356. https://doi.org/10.3390/sym13081356.

Scott, H. L., Skinkle, A., Kelley, E. G., Waxham, M. N., Levental, I., & Heberle, F. A. (2019). On the mechanism of bilayer separation by extrusion, or why your LUVs are not really Unilamellar. *Biophysical Journal, 117*(8), 1381–1386. https://doi.org/10.1016/j.bpj.2019.09.006.

Seelig, A., & Seelig, J. (1977). Effect of a single cis double bond on the structures of a phospholipid bilayer. *Biochemistry, 16*(1), 45–50. https://doi.org/10.1021/bi00620a008.

Semeraro, E. F., Devos, J. M., Porcar, L., Forsyth, V. T., & Narayanan, T. (2017). In vivo analysis of the Escherichia coli ultrastructure by small-angle scattering. *IUCrJ, 4*(Pt 6), 751–757. https://doi.org/10.1107/S2052252517013008.

Semeraro, E. F., Marx, L., Frewein, M. P. K., & Pabst, G. (2021). Increasing complexity in small-angle X-ray and neutron scattering experiments: From biological membrane mimics to live cells. *Soft Matter, 17*(2), 222–232. https://doi.org/10.1039/c9sm02352f.

Semeraro, E. F., Marx, L., Mandl, J., Frewein, M. P. K., Scott, H. L., Prevost, S., et al. (2021). Evolution of the analytical scattering model of live Escherichia coli. *Journal of Applied Crystallography, 54*(Pt 2), 473–485. https://doi.org/10.1107/S1600576721000169.

Simons, K., & Ikonen, E. (1997). Functional rafts in cell membranes. *Nature, 387*(6633), 569–572. https://doi.org/10.1038/42408.

Sivia, D. S. (2011). Elementary scattering theory: For X-ray and neutron users. In *Oxford*. New York: Oxford University Press.

Solanko, L. M., Sullivan, D. P., Sere, Y. Y., Szomek, M., Lunding, A., Solanko, K. A., et al. (2018). Ergosterol is mainly located in the cytoplasmic leaflet of the yeast plasma membrane. *Traffic, 19*(3), 198–214. https://doi.org/10.1111/tra.12545.

Sprong, H., van der Sluijs, P., & van Meer, G. (2001). How proteins move lipids and lipids move proteins. *Nature Reviews. Molecular Cell Biology, 2*(7), 504–513. https://doi.org/10.1038/35080071.

StClair, J. R., Wang, Q., Li, G., & London, E. (2017). Preparation and physical properties of asymmetric model membrane vesicles. In R. Epand, & J. M. Ruysschaert (Eds.), *Springer series in biophysics: Vol. 19. The biophysics of cell membranes* Springer.

Steck, T. L., & Lange, Y. (2018). Transverse distribution of plasma membrane bilayer cholesterol: Picking sides. *Traffic, 19*(10), 750–760. https://doi.org/10.1111/tra.12586.

Steck, T. L., Ye, J., & Lange, Y. (2002). Probing red cell membrane cholesterol movement with cyclodextrin. *Biophysical Journal, 83*(4), 2118–2125. https://doi.org/10.1016/S0006-3495(02)73972-6.

Stevens, M. M., Honerkamp-Smith, A. R., & Keller, S. L. (2010). Solubility limits of cholesterol, Lanosterol, Ergosterol, Stigmasterol, and beta-Sitosterol in electroformed lipid vesicles. *Soft Matter, 6*(23), 5882–5890. https://doi.org/10.1039/c0sm00373e.

Stone, M. B., Shelby, S. A., Nunez, M. F., Wisser, K., & Veatch, S. L. (2017). Protein sorting by lipid phase-like domains supports emergent signaling function in B lymphocyte plasma membranes. *eLife, 6*. https://doi.org/10.7554/eLife.19891.

Strott, C. A., & Higashi, Y. (2003). Cholesterol sulfate in human physiology: what's it all about? *Journal of Lipid Research, 44*(7), 1268–1278. https://doi.org/10.1194/jlr.R300005-JLR200.

Sud, M., Fahy, E., Cotter, D., Brown, A., Dennis, E. A., Glass, C. K., et al. (2007). LMSD: LIPID MAPS structure database. *Nucleic Acids Research, 35*(Database issue), D527–D532. https://doi.org/10.1093/nar/gkl838.

Sugiyama, M., Yagi, H., Yamaguchi, T., Kumoi, K., Hirai, M., Oba, Y., et al. (2014). Conformational characterization of a protein complex involving intrinsically disordered protein by small-angle neutron scattering using the inverse contrast matching method:

A case study of interaction between [alpha]-synuclein and PbaB tetramer as a model chaperone. *Journal of Applied Crystallography*, *47*(1), 430–435. https://doi.org/10.1107/S1600576713033475.

Svergun, D. I. (2010). Small-angle X-ray and neutron scattering as a tool for structural systems biology. *Biological Chemistry*, *391*(7), 737–743. https://doi.org/10.1515/BC.2010.093.

Svergun, D. I., Feĭgin, L. A., & Taylor, G. W. (1987). *Structure analysis by small-angle X-ray and neutron scattering*. New York: Plenum Press.

Thomson, J. F. (1960). Physiological effects of D20 in mammals. *Annals of the New York Academy of Sciences*, *84*, 736–744. https://doi.org/10.1111/j.1749-6632.1960.tb39105.x.

Treece, B. W., Kienzle, P. A., Hoogerheide, D. P., Majkrzak, C. F., Losche, M., & Heinrich, F. (2019). Optimization of reflectometry experiments using information theory. *Journal of Applied Crystallography*, *52*(Pt 1), 47–59. https://doi.org/10.1107/S1600576718017016.

van Meer, G. (1989). Lipid traffic in animal cells. *Annual Review of Cell Biology*, *5*, 247–275. https://doi.org/10.1146/annurev.cb.05.110189.001335.

van Meer, G., Voelker, D. R., & Feigenson, G. W. (2008). Membrane lipids: Where they are and how they behave. *Nature Reviews. Molecular Cell Biology*, *9*(2), 112–124. https://doi.org/10.1038/nrm2330.

Vance, J. E., Aasman, E. J., & Szarka, R. (1991). Brefeldin a does not inhibit the movement of phosphatidylethanolamine from its sites for synthesis to the cell surface. *The Journal of Biological Chemistry*, *266*(13), 8241–8247 (Retrieved from http://www.ncbi.nlm.nih.gov/pubmed/2022641).

Vishnepolsky, B., Zaalishvili, G., Karapetian, M., Nasrashvili, T., Kuljanishvili, N., Gabrielian, A., et al. (2019). De novo design and in vitro testing of antimicrobial peptides against gram-negative bacteria. *Pharmaceuticals (Basel)*, *12*(2). https://doi.org/10.3390/ph12020082.

Voelker, D. R. (1990). Lipid transport pathways in mammalian cells. *Experientia*, *46*(6), 569–579 (Retrieved from http://www.ncbi.nlm.nih.gov/pubmed/2193820).

Voelker, D. R. (1991). Organelle biogenesis and intracellular lipid transport in eukaryotes. *Microbiological Reviews*, *55*(4), 543–560. https://doi.org/10.1128/mr.55.4.543-560.1991.

Wah, B., Breidigan, J. M., Adams, J., Horbal, P., Garg, S., Porcar, L., et al. (2017). Reconciling differences between lipid transfer in free-standing and solid supported membranes: A time resolved small angle neutron scattering study. *Langmuir*, *33*, 3384–3394. https://doi.org/10.1021/acs.langmuir.6b04013.

Waldie, S., Moulin, M., Porcar, L., Pichler, H., Strohmeier, G. A., Skoda, M., et al. (2019). The production of Matchout-deuterated cholesterol and the study of bilayer-cholesterol interactions. *Scientific Reports*, *9*(1), 5118. https://doi.org/10.1038/s41598-019-41439-z.

Weiner, M. D., & Feigenson, G. W. (2019). Molecular dynamics simulations reveal leaflet coupling in compositionally asymmetric phase-separated lipid membranes. *The Journal of Physical Chemistry. B*, *123*(18), 3968–3975. https://doi.org/10.1021/acs.jpcb.9b03488.

Wimley, W. C., & Hristova, K. (2011). Antimicrobial peptides: Successes, challenges and unanswered questions. *The Journal of Membrane Biology*, *239*(1–2), 27–34. https://doi.org/10.1007/s00232-011-9343-0.

Wimley, W. C., & Thompson, T. E. (1991). Transbilayer and interbilayer phospholipid exchange in dimyristoylphosphatidylcholine/dimyristoylphosphatidylethanolamine large unilamellar vesicles. *Biochemistry*, *30*(6), 1702–1709. https://doi.org/10.1021/bi00220a036.

Wood, W. G., Igbavboa, U., Muller, W. E., & Eckert, G. P. (2011). Cholesterol asymmetry in synaptic plasma membranes. *Journal of Neurochemistry*, *116*(5), 684–689. https://doi.org/10.1111/j.1471-4159.2010.07017.x.

Xia, Y., Li, M., Charubin, K., Liu, Y., Heberle, F. A., Katsaras, J., et al. (2015). Effects of nanoparticle morphology and acyl chain length on spontaneous lipid transfer rates. *Langmuir, 31*(47), 12920–12928. https://doi.org/10.1021/acs.langmuir.5b03291.

Xiao, Q., McAtee, C. K., & Su, X. (2021). Phase separation in immune signalling. *Nature Reviews. Immunology*. https://doi.org/10.1038/s41577-021-00572-5.

Yancey, P. G., Rodrigueza, W. V., Kilsdonk, E. P., Stoudt, G. W., Johnson, W. J., Phillips, M. C., et al. (1996). Cellular cholesterol efflux mediated by cyclodextrins. Demonstration of kinetic pools and mechanism of efflux. *The Journal of Biological Chemistry, 271*(27), 16026–16034. https://doi.org/10.1074/jbc.271.27.16026.

Yeagle, P. L. (1993). *The membrane of cells*. New York: Academic Press, Inc.

Zaccai, N. R., Sandlin, C. W., Hoopes, J. T., Curtis, J. E., Fleming, P. J., Fleming, K. G., et al. (2016). Deuterium labeling together with contrast variation small-angle neutron scattering suggests how Skp captures and releases unfolded outer membrane proteins. *Methods in Enzymology, 566*, 159–210. https://doi.org/10.1016/bs.mie.2015.06.041.

Zidovetzki, R., & Levitan, I. (2007). Use of cyclodextrins to manipulate plasma membrane cholesterol content: Evidence, misconceptions and control strategies. *Biochimica et Biophysica Acta, 1768*(6), 1311–1324. https://doi.org/10.1016/j.bbamem.2007.03.026.

Printed in the United States
by Baker & Taylor Publisher Services